ENTOMOPATHOGENIC BACTERIA:
FROM LABORATORY TO FIELD APPLICATION

Entomopathogenic Bacteria: from Laboratory to Field Application

Edited by

Jean-François Charles
Armelle Delécluse

and

Christina Nielsen-Le Roux

Institut Pasteur,
Paris, France

KLUWER ACADEMIC PUBLISHERS
DORDRECHT / BOSTON / LONDON

A C.I.P. Catalogue record for this book is available from the Library of Congress.

ISBN 0-7923-6523-2

Published by Kluwer Academic Publishers,
P.O. Box 17, 3300 AA Dordrecht, The Netherlands.

Sold and distributed in North, Central and South America
by Kluwer Academic Publishers,
101 Philip Drive, Norwell, MA 02061, U.S.A.

In all other countries, sold and distributed
by Kluwer Academic Publishers,
P.O. Box 322, 3300 AH Dordrecht, The Netherlands.

QR
82
.B3
E55
2000

Printed on acid-free paper

Printed in the Netherlands.

Contents

Contributors   xiii

Preface   xix

SECTION 1 :   THE ENTOMOPATHOGENIC BACTERIA   1

**1.1 Biodiversity of the entomopathogenic, endospore-
      forming bacteria** *(FG Priest)*   1
1. Introduction   2
2. The genus *Bacillus* - a brief history   3
3. The genus *Bacillus* becomes several genera   4
4. *Bacillus sensu stricto* (rRNA group 1 sensu Ash *et al.*, 1991) 7
5. Round-spore-forming bacilli (rRNA group 2 *sensu* Ash *et al.*,
   1991)   10
6. *Paenibacillus* (rRNA group 3 *sensu* Ash *et al.*, 1991)   12
7. *Brevibacillus* (rRNA group 4 *sensu* Ash *et al.*, 1991)   16
8. Concluding remarks   16

**1.2 Natural occurrence and dispersal of *Bacillus thuringiensis*
      in the environment** *(PH Damgaard)*   23
1. Occurrence in soil   23
2. Occurrence on foliage   24
3. Occurrence in specific insects habitats   26
4. Occurrence in foods   28
5. Clinical infections   29
6. Epizootiology of *B. thuringiensis*   30
7. Factors governing *B. thuringiensis*-caused epizootics   32
8. Conclusions   34

1.3 Virulence of *Bacillus thuringiensis*
(BM Hansen & S Salamitou)   41

1. Introduction   42
2. Taxonomy and relations to *B. cereus*   42
3. Non-insect pathogenesis of *B. thuringiensis* and *B. cereus*   43
4. Virulence of *B. thuringiensis* and *B. cereus* spores   44
5. Virulence factors of *B. thuringiensis* and *B. cereus*   46
6. Expression of virulence factors   52
7. *B. thuringiensis* in the environment   53
8. Consequences for application of *B. thuringiensis*   54

SECTION 2 :   TOXINS AND GENES   65

2.1 The diversity of *Bacillus thuringiensis* δ-endotoxins
(N Crickmore)   65

1. Toxin nomenclature   65
2. Sequence comparisons   70
3. Related sequences   77
4. Summary   78

**2.2 Insecticidal proteins produced by bacteria pathogenic to
agricultural pests**   *(T Yamamoto & DH Dean)*   81

1. Description of *B. thuringiensis*   81
2. Extra-cellular insecticidal proteins   83
3. Characterisation of the crystal protein   83
4. Three domain structure of the crystal protein   85
5. Domain I function   86
6. Receptor binding sites in domain II   90
7. Domain III function   93
8. Cry2a family proteins   94
9. Future for the bacterial insecticidal proteins   96

2.3 Vector-active toxins: structure and diversity
(A Delécluse, V Juárez-Pérez & C Berry)   101

1. Introduction   101
2. Dipteran active bacteria   102
3. Mosquitocidal toxins   105
4. Summary and conclusions   114

**2.4 Toxin and virulence gene expression in *Bacillus
thuringiensis*** *(D Lereclus & H Agaisse)*   127

1. Expression of the insecticidal toxin genes   128
2. Virulence gene expression   137
3. Conclusion   139

2.5 Genetic and genomic contexts of toxin genes
(M-L Rosso, J Mahillon & A Delécluse) 143

1. Introduction 143
2. The entomopathogenic bacteria genome 144
3. Genomic location of toxins 149
4. Virulence gene mobility and transfer 151
5. Concluding remarks 156

SECTION 3 :   MODE OF ACTION AND RESISTANCE 167

3.1 Pathogenesis of *Bacillus thuringiensis* toxins
(P Lüthy & MG Wolfersberger) 167

1. Introduction 167
2. From the bacterial inclusion to the active polypeptide 169
3. Reaction of insect larvae to the δ-endotoxin 171
4. The gut epithelium as the target tissue 172
5. *In vitro* studies 175
6. On the origin of the δ-endotoxin: a hypothesis 177

3.2 Investigations of *Bacillus thuringiensis* Cry1 toxin receptor structure and function
(SF Garczynski & MJ Adang) 181

1. Cry1 toxin binding to brush border membrane vesicles 181
2. Cry1 receptor detection using toxin overlays 184
3. Identification of Cry1 receptor proteins 185
4. Characteristics of Cry1a binding molecules in *M. sexta* BBMV 186
5. Cry1a toxin-induced pores 189
6. Functional Cry1 receptors 190
7. Concluding remarks 191

3.3 Membrane permeabilisation by *Bacillus thuringiensis* toxins: protein insertion and pore formation
(J-L Schwartz & R Laprade) 199

1. Introduction 199
2. Toxicity at the molecular level 200
3. Approaches and techniques 203
4. Microenvironment of target cells 206
5. Role of the receptor 206
6. Conformational changes and pore structure 207
7. Postbinding events, resistance 211

**3.4 Insect resistance to *Bacillus thuringiensis* insecticidal
    crystal proteins**   *(J Van Rie & J Ferré)*   219

1. Introduction   219
2. Biochemical basis of resistance   220
3. Genetics of resistance   227
4. Conclusions   232

**3.5 Mode of action of *Bacillus sphaericus* on mosquito larvae:
    incidence on resistance**
    (J-F Charles, MH Silva-Filha & C Nielsen-LeRoux)   237

1. Introduction   237
2. Cytological and physiological effects   238
3. Binding of the binary toxin to a specific receptor   240
4. The toxin receptor in *Culex pipiens*   243
5. Toxin receptor interaction in *B. sphaericus*-resistant colonies
   of *C. pipiens*   246
6. Toxin structure and *in vivo/in vitro* activity   248
7. Conclusions/perspectives   249

SECTION 4 :   SAFETY AND ECOTOXICOLOGY OF
              ENTOMOPATHOGENIC BACTERIA   253

    (LA Lacey & JP Siegel)   253

1. Introduction   253
2. Direct effects of *Bacillus* entomopathogens on invertebrate
   non target organisms   254
3. Indirect effects of *Bt* on nontarget invertebrates   257
4. Effects of *Bacillus* entomopathogens on vertebrates   258
5. Indirect effects of *Bacillus* entomopathogens on vertebrates   263
6. Long term impact of *Bacillus* pathogens used as microbial pest
   control agents   264
7. Conclusion   266

SECTION 5 :   STANDARDISATION, PRODUCTION AND
              REGISTRATION   275

5.1 Is *Bacillus thuringiensis* standardisation still possible?
    (O Skovmand, I Thiéry & G Benzon)   275

1. Introduction   275
2. History of *Bt* standardisation   277
3. Standard procedures   278
4. *In vitro* assays   285
5. Future of bioassays   286
6. Suggestion for new type of standard   287

**5.2 Industrial fermentation and formulation of
entomopathogenic bacteria** *(TL Couch)* 297

1. Introduction 297
2. Culture selection 298
3. Laboratory techniques for culture maintenance 299
4. Fermentation inoculum preparation 299
5. Fermentation medium selection 301
6. Fermentation process 302
7. Recovery of entomopathogenic bacteria 303
8. Formulation 304
9. Quality control requirements 313
10. Conclusion 314

**5.3 Rural production of *Bacillus thuringiensis* by solid state
fermentation**
(E Aranda, A Lorence & M del Refugio Trejo) 317

1. Introduction 317
2. Strategies for insect control 318
3. Solid state fermentation (SSF) 320
4. Concluding remarks 329

5.4 Registration of biopesticides
(GN Libman & SC MacIntosh) 333

1. Introduction 333
2. What are biopesticides? 334
3. Registration of products containing *Bacillus thuringiensis*
toxins as the active ingredient 336

**SECTION 6 :   FIELD APPLICATION AND RESISTANCE
MANAGEMENT** 355

6.1 *Bacillus thuringiensis* application in agriculture *(A Navon)* 355

1. Introduction 355
2. Considerations of *Bt* uses in the field 356
3. Combinations of *Bt* with other means of pest management 362
4. Future prospects 365

6.2 Application of *Bacillus thuringiensis* in forestry
(K van Frankenhuyzen) 371

1. Introduction 371
2. Field development 372
3. The biological interface: reducing the efficacy bottleneck 377

6.3 Bacterial control of vector-mosquitoes and black flies
   (N Becker)   383

   1. Introduction   384
   2. Mosquitoes   387
   3. Blackflies   394
   4. Future prospects   396

6.4 Resistance management for agricultural pests
   (RT Roush)   399

   1. Introduction   399
   2. Factors that influence selection   401
   3. Myths about management of resistance   405
   4. Promising tactics for resistance management for bacteria and
      sprays   408
   5. Resistance monitoring   412
   6. Implementation   413
   7. Conclusions   414

6.5 Management of resistance to bacterial vector control
   (L Regis & C Nielsen-LeRoux)   419

   1. Introduction   419
   2. Case histories of *B. sphaericus* resistance in mosquito
      populations   420
   3. Mechanisms and genetics of resistance in terms of stability and
      reversibility   422
   4. Factors influencing the rate of development of resistance in
      the field   425
   5. Cross-resistance and toxin receptor interaction   427
   6. Strategy for the management of resistance to *B. sphaericus*   429
   7. Conclusions and perspectives   433

SECTION 7 :   BIOTECHNOLOGY AND RISK ASSESSMENT   441

**7.1 Biotechnological improvement of *Bacillus thuringiensis* for
   agricultural control of insect pests: benefits and
   ecological implications** *(V Sanchis)*   441

   1. Introduction   441
   2. Improvement of *Bt* strains   443
   3. Expression of *cry* genes in plants   449
   4. Ecological risks associated with the use of transgenic *Bt* crops 450
   5. Future challenges and prospects   452
   6. Summary and conclusions   455

7.2 Genetic engineering of bacterial insecticides for improved efficacy against medically important Diptera *(B Federici, H-W Park, DK Bideshi & B Ge)* 461

1. Introduction 461
2. Properties of mosquitocidal bacteria 463
3. Factors for enhancing endotoxin synthesis 466
4. Improvement of mosquitocidal bacteria 474
5. Summary and conclusions 479

7.3 *Bacillus thuringiensis* : risk assessment *(A Klier)* 485

1. Introduction 485
2. Taxonomy of *Bacillus thuringiensis* and its occurrence in the environment 485
3. Risk assessment 492
4. Conclusions 499

Index 505

Contributors

Michael J. Adang
Affiliation Departments of Entomology, and Biochemistry and Molecular Biology, University of Georgia, Athens, Georgia 3060, USA
  e-mail: adang@arches.uga.edu

Hervé Agaisse
Biochimie Microbienne, Institut Pasteur, 25 rue du Dr Roux, 75724 Paris Cedex 15, France, and Station de Lutte Biologique, INRA, La Minière, 78285 Guyancourt Cedex, France
  e-mail:agaisse@pasteur.fr

Eduardo Aranda
Centro de Investigación en Biotecnología (CEIB), Universidad Autónoma del Estado de Morelos (UAEM). Av. Universidad 1001, Col. Chamilpa. C.P. 62210, Cuernavaca, Morelos, México
  e-mail: aranda@cib.uaem.mx

Norbert Becker
German Mosquito Control Association, Ludwigstrasse 99, 67165 Waldsee, Germany
  e-mail: kabs-gfs@t-online.de

Gary Benzon
Benzon Research, 208 Burt House Road, Carlisle, PA 17013, USA
  e-mail: gbenzon@aol.com

Colin Berry
Cardiff School of Biosciences, cardiff University, Museum Avenue P.O. Box 911, Cardiff CF10 3US, Wales, UK
  e-mail: berry@cf.ac.uk

Dennis K. Bideshi
Department of Entomology and Interdepartemental Graduate Program in Genetics,
University of California-Riverside, Riverside, California 92521, USA
    e-mail: dbideshi@citrus.ucr.edu

Jean-François Charles
Bactéries et Champignons Entomopathogènes, Institut Pasteur, 25 rue du Dr Roux,
75724 Paris Cedex 15, France
    e-mail: jcharles@pasteur.fr

Terry L. Couch
Beker Microbial Products, 9464 NW 11th st., Plantation, FL 33322, USA
    e-mail: tcouch@gate.net

Neil Crickmore
School of Biological Sciences, University of Sussex, Falmer, Brighton BN1 9QG E
Sussex, UK
    e-mail: n.crickmore@sussex.ac.uk

Per H. Damgaard
The Royal Veterinary and Agricultural University, Department of Ecology,
Thorvaldsensvej 40, 1870 Frederiksberg C., Denmark
    e-mail: pdg@novo.dk

Donald H. Dean
Department of Biochemistry, The Ohio State University, Columbus, OH 43210, USA
    e-mail: dean.10@osu.edu

Armelle Delécluse
Bactéries et Champignons Entomopathogènes, Institut Pasteur, 25 rue du Dr Roux,
75724 Paris Cedex 15, France
    e-mail: armdel@pasteur.fr

Brian A. Federici
Department of Entomology and Interdepartemental Graduate Program in Genetics,
University of California-Riverside, Riverside, California 92521, USA
    e-mail: federici@ucrac1.ucr.edu

Juan Ferré
Department of Genetics, Faculty of Biology, Universitat de Valencia, 46100 Burjassot,
Spain
    e-mail: juan.ferre@uv.es

Stephen F. Garczynski
Affiliation Departments of Entomology, and Biochemistry and Molecular Biology,
University of Georgia, Athens, Georgia 3060, USA
    e-mail: stevegar@arches.uga.edu

Baoxue Ge
Department of Entomology and Interdepartemental Graduate Program in Genetics,
University of California-Riverside, Riverside, California 92521, USA
    e-mail: ge@scripps.edu

Bjarne M. Hansen
Department of Marine Ecology and Microbiology, National Environmental Research
Institute, Frederiksborgvej 399, DK-4000 Roskilde, Denmark
    e-mail: bmh@dmu.dk

Victor Juárez-Pérez
Bactéries et Champignons Entomopathogènes, Institut Pasteur, 25 rue du Dr Roux,
75724 Paris Cedex 15, France
    e-mail: vicjua@pasteur.fr

André Klier
Biochimie Microbienne, URA 1300 CNRS, Institut Pasteur, 25 rue du Dr Roux, 75724
Paris Cedex 15, France
    e-mail: aklier@pasteur.fr

Lawrence A. Lacey
Fruit and Vegetable Research Unit, Yakima Agricultural Research Laboratory,
USDA-ARS, 5230 Konnowac Pass Road, Wapato, WA 98951, USA
    e-mail: llacey@yarl.ars.usda.gov

Raynald Laprade
Groupe de recherche en transport membranaire, Université de Montréal, Montréal,
Quebec, Canada
    e-mail: raynald.laprade@umontreal.ca

Didier Lereclus
Biochimie Microbienne, Institut Pasteur, 25 rue du Dr Roux, 75724 Paris Cedex 15,
France, and Station de Lutte Biologique, INRA, La Minière, 78285 Guyancourt Cedex,
France
    e-mail: lereclus@pasteur.fr

Gary N. Libman
Ecogen Inc. Company, 39 Sage Hill Drive, Placitas, NM 87043, USA
    e-mail: glibman@email.msn.com

Argelia Lorence
Centro de Investigación en Biotecnología (CEIB), Universidad Autónoma del Estado
de Morelos (UAEM). Av. Universidad 1001, Col. Chamilpa. C.P. 62210, Cuernavaca,
Morelos, México
    e-mail: argelia1@cib.uaem.mx

Peter Lüthy
Institute of Microbiology, Swiss Federal Institute of Technology, 8092 Zurich,
Switzerland
    e-mail: luethy@micro.biol.ethz.ch

Susan C. MacIntosh
AgrEvo USA Company, 7200 Hickman Road, Suite 202, Des Moines, IA 50322, USA
    e-mail: susan.macintosh@aventis.com

Jacques Mahillon
Unité de Génétique, Université Catholique de Louvain, Place Croix du Sud, G 1348
Louvain-la-Neuve, Belgium
    e-mail: mahillon@mbla.ucl.ac.be

Amos Navon
Department of Entomology, Agricultural Research Organization, The Volcani Center, Bet Dagan 50250, Israel
    e-mail: navona@netvision.net.il

Christina Nielsen-LeRoux
Bactéries et Champignons Entomopathogènes, Institut Pasteur, 25 rue du Dr Roux, 75724 Paris Cedex 15, France
    e-mail: cnielsen@pasteur.fr

Hyun-Woo Park
Department of Entomology and Interdepartemental Graduate Program in Genetics, University of California-Riverside, Riverside, California 92521, USA
    e-mail: hwpark@citrus.ucr.edu

Fergus G. Priest
Department of Biological Sciences, Heriot-Watt University, Edinburgh EH14 4AS, Scotland, UK
    e-mail: f.g.priest@hw.ac.uk

Ma. del Refugio Trejo
Centro de Investigación en Biotecnología (CEIB), Universidad Autónoma del Estado de Morelos (UAEM). Av. Universidad 1001, Col. Chamilpa. C.P. 62210, Cuernavaca, Morelos, México
    e-mail: mtrejo@cib.uaem.mx

Lêda Regis
Centro de Pesquisas Aggeu Magalhães-FIOCRUZ, Av. Moraes Rêgo s/n 50670-420 Recife PE, Brazil
    e-mail: leda@cpqam.fiocruz.br

Richard T. Roush
Centre for Weed Management Systems, Waite Institute, University of Adelaide, South Australia, Australia
    e-mail: rroush@schooner.waite.adelaide.edu.au

Marie-Laure Rosso
Bactéries et Champignons Entomopathogènes, Institut Pasteur, 25 rue du Dr Roux, 75724 Paris Cedex 15, France

Sylvie Salamitou
Biochimie Microbienne, Institut Pasteur, 25 Rue du Dr Roux, 75724 Paris Cedex 15, France
    e-mail: mitou@pasteur.fr

Vincent Sanchis
Biochimie Microbienne, Institut Pasteur, 25 rue du Dr Roux, 75724 Paris Cedex 15, France and Station de Recherches de Lutte Biologique, INRA, La Minière, 78285 Guyancourt Cedex, France
    e-mail: vsanchis@pasteur.fr

Jean-Louis Schwartz
Biotechnology Research Institute, National Research Council of Canada, Montreal, and Groupe de recherche en transport membranaire, Université de Montréal, Montréal, Quebec, Canada
    e-mail: jean-louis.schwartz@bri.nrc.ca

Joel P. Siegel
Horticultural Crops Research Laboratory, USDA-ARS, 2021 S. Peach Ave., Fresno, CA 93727, USA
    e-mail: siegel@qnis.net

Ole Skovmand
Intelligent Insect Control, 80 rue Paul Ramart, Montpellier 34070, France
    e-mail: ole.skovmand@wanadoo.fr

Maria Helena Silva-Filha
Centro de Pesquisas Aggeu Magalhães-FIOCRUZ, Av. Moraes Rêgo s/n 50670-420 Recife PE, Brazil
    e-mail: mhneves@cpqam.fiocruz.br

Isabelle Thiéry
Bactéries et Champignons Entomopathogènes, Institut Pasteur, 25 rue du Dr Roux, 75724 Paris Cedex 15, France
    e-mail: ithiery@pasteur.fr

Kees van Frankenhuyzen
Canadian Forest Service, Great Lakes Forestry Centre, P.O. Box 490, Sault Ste. Marie, Ontario P6A 5M, Canada
    e-mail: kvanfran@NRCan.gc.ca

Jeroen Van Rie
Aventis CropScience N.V., J. Plateaustraat 22, 9000 Gent, Belgium
    e-mail: jeroen.vanrie@aventis.com

Michael G. Wolfersberger
Biology Department, Temple University, Philadelphia, PA 19122, USA
    e-mail: v2222a@vm.temple.edu

Takashi Yamamoto
Maxygen, Inc., 515 Galveston Drive, Redwood City, CA 94063, USA
    e-mail: takashi_yamamoto@maxygen.com

Preface

The discovery of a bacterium with specific insecticidal activity was made with *Bacillus thuringiensis* (*Bt*) in 1911 and the first attempts to use *Bt* in insect control were reported as early as the 1930's. Today *Bt* is the most successful commercial microbial insecticide, comprising about 90% of the biopesticide market. Its application for the protection of crops and forests and for the prevention of human diseases is occurring world-wide and is replacing chemical insecticides in some areas. This is due to both its specific activity with less damage to the environment and also to insect resistance to chemicals. The development of *Bt* has more recently led to the industrial development of another bacterium, *Bacillus sphaericus* (*Bsp*) as a biopesticide, with restricted activity against some Dipteran larvae. Apart from *Bt* and *Bsp*, other insecticidal bacteria have been identified, but none has yet been industrially developed.

The main interest in biopesticides as opposed to chemical pesticides is their high specificity and thus less damage to non-target fauna and flora. The insecticidal activity of *Bt* and *Bsp* is due to the presence of parasporal protein inclusion bodies, also called crystals, produced during sporulation. These inclusions are composed of one or several specific crystal protoxins (Cry, Cyt and Bin toxins) which act like a stomach poison. Upon ingestion by insect larvae, the inclusions are solubilised in the insect midgut, and protoxins proteolysed into active toxins which interact with specific midgut receptors. Pores and/or channels are formed, creating osmotic imbalance, cell lysis, and larval death.

In this book, we have attempted to bring together all recent studies regarding both fundamental and more applied research aspects related to entomopathogenic bacteria, in order to facilitate their development and further success. The last book covering such various topics was published in 1993 by Entwistle *et al.* (*Bacillus thuringiensis*, An Environmental Biopesticide: Theory and Practice). Since that time, there have been substantial research advances that are included here. The 26 different chapters written by the leading researchers in the field comprehensively update the entire subject.

- Section 1: The entomopathogenic bacteria

The first three chapters provide a wide overview of the entomopathogenic bacteria, including their classification using molecular tools, their occurrence in different environments, and, for *Bt* strains, the nature of virulence factors they produce. This latter point has recently gained importance, as *Bt* seems to produce similar virulence factors as those found in *B. cereus*. That could have implications for human pathogenesis.

- Section 2: Toxins and genes

This section includes five chapters discussing the large diversity of *cry* toxins genes and the presence of important insecticidal polypeptides which do not enter the Cry nomenclature (Vip, Bin, Mtx, Cyt...). For some of the Cry toxins, the tridimensional structure has been determined. It seems to fit a common Cry toxin model of three domains, where domain II is variable and responsible for insect target specificity and receptor interactions. Considerable efforts have been made to elucidate the mechanisms of regulation of the *cry* and virulence genes. Both transcriptional and post-transcriptional controls were found to regulate their expression. Their location (plasmid *versus* chromosome) and association with mobile elements (insertion sequences and transposons) is also presented in relation to their dispersion among the bacteria.

- Section 3: Mode of action and resistance

The chapters of this section deal with the mode of action of the insecticidal toxins and implications for resistance development. In several insects, an aminopeptidase serves as the membrane receptor for the *Bt* Cry toxins, but a cadherine-like protein is also found in one insect species, and membrane lipid interactions are important. On the other hand, the receptor for *B. sphaericus* crystal toxin has been identified as an α-glycosidase. The resistance mechanism, in most cases, is due to the loss of a functional receptor, but proteinase modifications might also be involved.

• Section 4: Safety and ecotoxicology

This unique chapter is specifically devoted to the safety and environmental impact of entomopathogenic bacteria, an emerging concern.

• Section 5: Standardisation, production and registration

This part deals in four chapters with the production of entomopathogenic bacteria, either at large industrial sites or on a rural scale, the latter considering the use of native strains and local wastes to reduce the cost. The registration procedures for use of all organisms expressing *Bt* toxins (either wild-type and recombinant *Bt* strains or transgenic plants) are also clearly defined, and the use of standard procedures for insecticidal potency evaluation is discussed.

• Section 6: Field application and resistance management

These chapters are dedicated to the field application of entomopathogenic bacteria, to control agricultural and forestry pests (Lepidopteran, Coleopteran larvae) as well as vectors of tropical diseases (Dipteran). The impact of large and/or frequent application of *Bt* on the development of insecticidal resistance as well as resistance/susceptibility management are also well documented for both agriculture and health. The advantages and performances of these bacteria in IPM programs are also debated.

• Section 7: Biotechnology and risk assessment

Finally, this last part considers all the advantages and drawbacks of biopesticide development and suggests methods to increase the use of entomopathogenic bacteria through improvement of these bacteria by way of genetic engineering or construction of transgenic plants. Already much has been done and interesting recombinants have been obtained. The last chapter of this section evaluates the benefits for agriculture and the environment of both native and recombinant bacteria and addresses the question of possible adverse environmental impact to both fauna and flora.

Research in all areas of entomopathogenic bacteria has been investigated in recent years. However, lots of work still has to be done to improve the successful use of bacteria in insect control. In fact, the bio-pesticide market is expected to reach at least 10% of the whole pesticide market in the next few years. However, it is important to state that the appropriate measurement of this increase should be based on the amount of treated areas rather than on an economical base. There is a need for users to become aware of the benefits obtained from using biopesticides, especially in IPM programs. It is in forestry where *Bt* has best demonstrated its efficiency and competitivity, mainly because *1)* chemicals were not allowed in these sensitive ecosystems and *2)* the methods and equipments were better developed. The lessons learn in forestry should be extended to agriculture and human health and wherever biopesticides can replace chemicals.

The challenge is great and there is still a need for expanded research and cooperation among a variety of disciplines; entomologists, microbiologists, biochemists, ecologists, pest control executors and authorities need to interact. Our hope is that this book will provide the basic information to stimulate more research in the area of biopesticides and that it will be a useful tool for all, when considering bacteria for both fundamental and applied needs of insect control. We would like to take the opportunity to thank all authors, and hope that their contributions will enhance the interest of the larger public in entomopathogenic bacteria.

Chapter 1.1

Biodiversity of the entomopathogenic, endospore-forming bacteria

Fergus G. Priest
Department of Biological Sciences, Heriot-Watt University, Edinburgh EH14 4AS, Scotland, UK

Key words: *Bacillus cereus, Bacillus sphaericus, Bacillus thuringiensis, Paenibacillus, Brevibacillus*, Phylogeny, Clone

Abstract: Small subunit rRNA gene sequencing has revealed that the aerobic endospore forming bacteria previously accommodated within the genus *Bacillus* are diverse, and more appropriately classified in several genera namely, *Alicyclobacillus, Amphibacillus, Aneurinibacillus, Bacillus sensu stricto, Brevibacillus, Halobacillus, Paenibacillus* and *Virgibacillus*. Organisms pathogenic to insects and other invertebrates have been found within several of these genera, in particular *Bacillus* (rRNA groups 1 and 2), *Brevibacillus* and *Paenibacillus*. It is reassuring that the phylogenetic classification of these bacteria correlates with their pathogenic properties, for example the fastidious obligate pathogens are now reclassified as *Paenibacillus larvae*, *P. lentimorbus* and *P. popilliae*. While the generic classifications have been explored fairly thoroughly, there is still confusion at the species level, particularly with *Bacillus cereus* and *Bacillus thuringiensis*. By most taxonomic criteria these two species are synonymous, but preliminary studies suggest that they may be resolved in a population genetic context as separate clones within the one taxon. Insect pathogenic strains of *Bacillus sphaericus*, on the other hand, are clearly divisible into genetically dissimilar clones with H-serotype 5a5b representing the predominant clonal type. The application of molecular taxonomic methods at both the generic and species level is providing insight into the evolution of insect toxicity and the prevalence of lateral gene transfer both within and between species and even genera.

J.-F. Charles et al. (eds.),
Entomopathogenic Bacteria: From Laboratory to Field Application, 1–22.
© 2000 *Kluwer Academic Publishers. Printed in the Netherlands.*

1.   INTRODUCTION

The bacterial microflora of insects is largely confined to the gut. This flora is rich and diverse and comprises both Gram-positive and Gram-negative bacteria. Many of these gut residents benefit the insect host by assisting with the digestion of food. In contrast, relatively few bacteria are pathogenic to their insect hosts, but they have received a disproportionate amount of attention because of their potential for control of agricultural pests and vectors of disease. Pathogenicity is largely associated with entry to the hemocoel either through a wound in the exoskeleton or more generally through the peritrophic membrane of the gut. Of the Gram-negative bacteria, some members of the family *Enterobacteriaceae* are recognised as insect pathogens. *Serratia* species, in particular, have often been associated with insect disease and a commercial product containing *S. entomophila* is being used to control the grass grub *Costelytra zealandica* in New Zealand. The bacteria turn the larvae a yellow or amber colour, hence the name "amber disease" [32]. The other important class of Gram-negative pathogens, comprises the nematode-borne organisms, *Photorhabus* and *Xenorhabus* which provide a fascinating story of symbiosis and pathogenicity. These closely-related members of the family *Enterobacteriaceae* are carried as symbionts in the intestines of the juvenile stages of certain nematodes. The nematodes infect insect larvae and upon entering the hemocoel release the bacteria which, together with the nematode kill the insect host. The bacteria both release toxins which affect the larva and provide nutrients for the nematodes. During the later stages of the infection, the bacteria and nematodes reassociate to move on to pastures new [22].

Although these Gram-negative organisms have their applications in insect control, the Gram-positive bacteria, have proven to be the most useful pathogens for biological control purposes and form the basis of the microbiological insecticide industry. Among these bacteria, pathogenicity seems to be almost exclusively associated with the endospore-forming bacteria although *Melissococcus pluton* (formerly *Streptococcus pluton)* is the causal agent of European foulbrood in honey bees [30] and some actinobacteria such as *Brevibacterium* and *Corynebacterium* are often isolated from healthy and sick insects [12]. This chapter will focus on the systematics and evolution of the aerobic, endospore-forming bacteria (AEFB) with particular reference to *Bacillus.*

2.   THE GENUS *BACILLUS*-A BRIEF HISTORY

Bacteria which differentiate into endospores have traditionally been assigned to two major genera, *Bacillus* and *Clostridium*. The former is restricted to rod-shaped bacteria which sporulate under aerobic conditions and *Bacillus* species possess aerobic or facultatively anaerobic modes of energy metabolism. *Clostridium* is confined to rod-shaped bacteria with a heterotrophic and obligately anaerobic mode of growth. This simple arrangement made generic identification of most spore-forming isolates a straightforward task, but species identification became increasingly difficult as the number of species grew in each genus. By the 1950s, almost two hundred species of *Bacillus* had been described, often on the basis of scant physiological and morphological criteria and identification of isolates was almost impossible. In order to rationalise the situation, Ruth Gordon and her colleagues examined more than 1000 strains of *Bacillus* in a comparative study and reduced the number of species to just 19 [25]. In achieving this dramatic reorganisation, Gordon *et al.* appreciated that they were "lumping" several taxa into single species or "species complexes" but felt that the techniques available at that time were inadequate to enable classification and identification at a finer level [24]. Descriptions of these species and species complexes were published in a helpful handbook describing procedures for the isolation and identification of AEFB [25].

During the 1980s, molecular techniques, in particular DNA reassociation, were applied to numerous species and species complexes and the number of species began to grow again. Using the generally accepted definition of a species [61] as a group of strains with greater than 70% DNA reassociation under optimal conditions and less than 5°C depression in the melting temperature of the hybrid duplexes (a measure of base sequence mismatching) this time there was a more solid foundation for the taxonomic inflation, and most new species were based on full molecular and phenotypic descriptions. These studies confirmed the earlier suspicions and several of Gordon *et al's* species complexes such as *B. brevis, B. circulans* and *B. megaterium* were redefined as several distinct species. The result is that today about 112 valid species of AEFB have been described (see http://www.dsmz.de/bactnom/nam0379.htm).

A second revolution was taking place during the 1970s and 80s. Measurements of guanine plus cytosine (G+C contents) in chromosomal DNA had indicated that the range displayed by *Bacillus* species was extremely wide with *B. cereus* and *B. thuringiensis* representing the low end (about 33% G+C) and strains of several thermophilic species with more than 60% G+C at the other extreme [49]. This suggests substantial genetic diversity and indicates that these organisms have diverged to such an extent that they should be placed in separate genera; indeed, the maximum variation in G+C of species within a well defined genus is usually around 12% G+C [49]. Further evidence for the diversity of AEFB was provided by two extensive numerical taxonomic studies [41, 52] which revealed that most species could be assigned to six major taxa each with reasonably well conserved morphological (spore shape and position) and physiological attributes. These taxa equated well with genera in other groups of bacteria.

The time was ripe for redefining the ever-increasing number of *Bacillus* species in terms of several genera of AEFB, the problem was how to achieve this on a formal basis. The answer was at least partly provided by developments in molecular systematics.

3.   THE GENUS *BACILLUS* BECOMES SEVERAL GENERA

Sequence comparisons of the small subunit (ssu) ribosomal RNA gene (16S rRNA) have provided bacterial systematics with a sound phylogenetic basis. The ssu rRNA gene is particularly valuable in this context because within its approximate 1600 bases there are pronounced areas of secondary structure which assist sequence alignments, there are areas of high conservation which allow comparisons between distantly related organisms and there are more variable regions that enable comparisons of more closely related organisms. As a result, bacteria within a species defined by high DNA reassociation have essentially identical 16S rRNA genes and strains from closely related species (which would show minimal DNA reassociation) differ in 16S rRNA sequence by around 1 to 3% [61]. Phylogenetic classifications based on ssu rRNA comparisons are particularly valuable at the generic level, since the conserved areas of the gene provide valuable information for revealing relationships at this level while DNA reassociation remains the most appropriate basis for speciation.

The initial reclassification of *Bacillus* was achieved by Carol Ash and her colleagues [6] who compared 16S rRNA sequences of almost 60 representative strains of *Bacillus* and related bacteria. The strains were allocated to five major taxa with "*B. cycloheptanicus*" as a single isolate. The latter has since been assigned to a new genus, *Alicyclobacillus,* which accommodates some acidophilic endospore-formers with unusual membrane lipids [70]. Group 3 species (*sensu* Ash *et al.*) have also been awarded generic status as *Paenibacillus* based on their unique 16S rRNA gene sequences, facultatively anaerobic mode of growth, and ellipsoidal spores which distinctly distend the sporangium or mother cell [7]. The *B. brevis* group (Group 4 *sensu* Ash *et al.*) has been reclassified as *Brevibacillus* with two founder members, *B. brevis* and *B. laterosporous* [59] and the phylogenetic isolation of *B. aneurinolyticus* has been recognised by the allocation of this species to *Aneurinibacillus* [59]. Similarly, *B. pantothenticus* has been reclassified as *Virgibacillus* [27] and halophilic strains as *Halobacillus* (see Table 1). The generic status of *Sporolactobacillus* has been established but although the alkaliphilic and thermophilic bacilli form distinct phylogenetic clusters [45, 55] they have not yet been afforded generic status. Similarly, the round-spore-forming strains, which include *B. sphaericus* (Group 2 *sensu* Ash *et al.*), have not been redefined as a new genus largely because of their polyphyletic origins with some non-spore-forming bacteria such as *Kurthia* and *Planococcus* [20].

Table 1. The range of genera of rod-shaped, aerobic, endospore-forming bacteria and some representative species including all those considered to harbour insect or invertebrate pathogens.

Genus (rRNA group)[1]	Representative species (non-insect pathogens)	Insect pathogenic species	Susceptible insects/ invertebrates	Ref.
Alicyclobacillus	*acidocaldarius*	None		[70]
Aneurinibacillus	*aneurinolyticus*	None		[59]
Bacillus (1)	*firmus*	*circulans*	Diptera	[60]
	licheniformis	*thuringiensis*	Diptera, Coleoptera Lepidoptera	
	subtilis			
Bacillus (2)	*globisporus*	*sphaericus*	Diptera	[16]
Bacillus (5)	*stearothermophilus*	None		
Brevibacillus (4)	*borstelensis*	*brevis*	Molluscs	[60]
	coshinensis	*laterosporus*	Diptera	[21]
Halobacillus	*halophilus*	None		
Paenibacillus (3)	*chibensis*	*alvei*	Diptera, nematodes, molluscs	[60]
	glucanolyticus	*larvae* subsp. *larvae*	honey bees	[28]
	polymyxa	*larvae* subsp. *pulvifaciens*[1]	Honey bees	[23]
		lentimorbus	Coleoptera	[48]
		popilliae	Coleoptera	[48]
Virgibacillus	*pantothenticus*	none		[27]

[1], rRNA groups based on those described by Ash *et al.* [6]

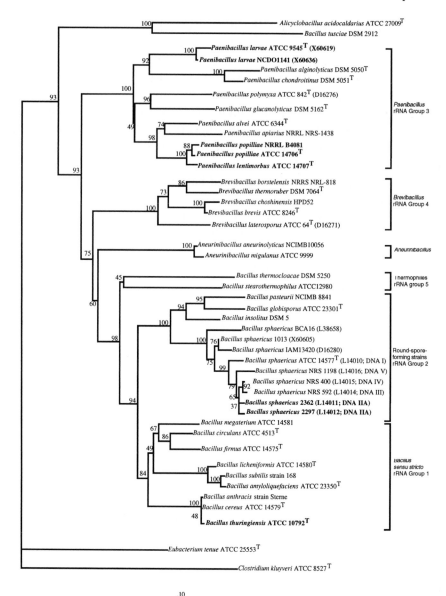

Figure 1. Phylogenetic tree of insect pathogenic, aerobic, endospore-forming bacteria derived from distance data calculated from a non-edited alignment of 16S rRNA gene sequences from insect pathogens and reference strains by using the neighbor-joining method. *Eubacterium tenue* and *Clostridium kluyveri* served as outgroups. The bootstrap values are given as the percentage times out of 500 replicates that a species or a strain to the right of the actual node occurred. The scale bar indicates nucleotide substitutions per site and the distance between two taxa are obtained by adding the horizontal lines connecting the two taxa, vertical lines have no phylogenetic meaning. Adapted from Priest & Dewar [50].

Although the current range of genera accommodating AEFB may be confusing, as we become familiar with the concept of several genera of endospore forming bacteria the situation should be improved so long as the taxonomists follow their task assiduously and do not create anomalous genera with frequent taxonomic revisions. The current range of genera is presented in Table 1 and the relationships shown in Figure 1. These are not exhaustive since the full range of AEFB species can be found on the Web at http://www-sv.cict.fr/bacterio/index.html or at http://www.dsmz.de/bactnom/nam0379.htm, but representative species have been chosen to highlight the new genera and include all known insect pathogens.

The immediate implication is that insect toxicity among AEFB is polyphyletic and has emerged several times among these bacteria. In those cases where the toxins have been studied in detail, such as *B. sphaericus* and *B. thuringiensis*, there is no apparent homology between the toxin genes supporting this view that the origins of insect pathogenicity are several.

4. *BACILLUS SENSU STRICTO* (rRNA GROUP 1 SENSU ASH *et al.*, 1991)

The only well established insect pathogen in the more refined genus of *Bacillus* is the classic biocontrol agent *B. thuringiensis* (*Bt*). *Bt* forms a long independent phylogenetic branch among the Group 1 bacilli in most 16S rRNA-based phylogenetic trees including that shown in Fig. 1 and there could be grounds for separating it from *B. subtilis* and other more typical members of this genus. The only other bacterium in this taxon with anti-invertebrate properties is *B. circulans*, some strains of which are toxic to Dipteran larvae and others to the nematode *Heterodera glycines* and some molluscs [60]. Since these strains of *B. circulans* show low activity and have not been studied to any great extent, they will not be further considered here.

4.1 The *Bacillus thuringiensis* (*Bt*) group

Four species, *Bacillus anthracis,* the causative agent of anthrax in cattle and man, *Bacillus cereus* (*Bc*), responsible for common but mild food poisoning, *Bacillus mycoides,* considered a harmless saprophyte and *Bt* comprise a highly related group of bacteria with almost identical 16S and 23S rRNA sequences [5]. In the past it has been suggested that they should be placed in the same species [25] but the distinctive pathogenicity properties of the organisms have been given priority and the four species have been retained. *B. anthracis* is the most distinctive of these taxa and can usually be distinguished by phenotypic criteria (Table 2). Various nucleic acid typing procedures have characterised *B. anthracis* strains as a homogeneous population distinct from *Bc* and *Bt* strains [26, 37]. Similarly, *B. mycoides* strains have a unique rhizoidal appearance and they can be distinguished from *Bc* by DNA reassociation [44], fatty acid profiles and a molecular probe targeted to a variable region of the 16S rRNA gene [69].

The main cause of confusion is the close similarity between *Bc* and *Bt* strains since the only reliable way to distinguish them is the presence of a parasporal body in the latter. The confusion is unfortunately becoming greater the more we learn about these bacteria. Physical maps of the genomes of several strains of *Bc* and *Bt* indicate that they are so similar that they should be considered the same species [13, 14] and strains of the two species cannot be reliably distinguished by DNA reassociation [43], the "gold standard" of species discrimination [61]. Perhaps the most worrying feature of the close relationship between these two species is that strains of *Bt* have now been associated with the secretion of food poisoning toxins [17, 64], previously the preserve of *Bc*. In the light of these findings, strains applied in the field should be tested for the presence of enterotoxin genes and appropriately-deleted strains prepared.

Table 2. Some diagnostic phenotypic features for the *Bacillus cereus* group of bacteria.

Feature	B. anthracis	B. cereus	B. mycoides	B. thuringiensis
Motility	-	+	-/+	+
Crystal parasporal inclusion	-	-	-	+
Capsule formation on bicarbonate medium	+	-	-	-
Lysis by gamma phage	+	-	-	-
Hemolysis on 5% blood agar (sheep or horse)	-	+	+	(+)
Lecithinase	(+)	+	+	+
Resistance to penicillin (10 units)	-	+	+	+
Rhizoid growth	-	-	+	-
Tyrosine decomposition	-	+	+	+/-

-/+, most strains negative; +/-, most strains positive; (+), weak reaction
Data extracted from references [47, 68]

I believe that the relationships between *Bc* and *Bt* can best be understood in the context of population genetics in which strains are considered as members of clones. Such clones are sub-species groups of bacteria originating from a common ancestor with limited inter-clonal chromosomal gene transfer [32]. Thus all members of a clone are essentially identical when examined using molecular fingerprinting methods. Unfortunately there are no extensive studies of population structures of *Bc* and *Bt,* but in the few strains which have been examined for clonality some agreement does exist. Ribotyping, a form of chromosomal fingerprinting based on the distribution of restriction enzyme sites in and around the rRNA operons of the bacterium, revealed that strains of *Bt* within serotypes generally had consistent ribotype patterns that were distinct from those of other serotypes/ribotypes. In other words, chromosomal structure largely matched serotype [54] and showed that in several instances strains within serotypes originated from a common parent and could be considered genetically similar or identical. However, correlation between serotype and clonal populations based on some other typing procedure may be less obvious and the correlation is weak when assessed by multilocus enzyme electrophoresis (MLEE), the traditional way to reveal clonality in populations. Members of the same serotype were distributed in several clones by this procedure and some MLEE clones contained representatives of several serotypes [13, 74]. Therefore, while flagellum (H) antigens may be consistent with clonality in some serotypes, for example serotype H14 (serovar *israelensis, Bti*) in which all strains are identical in genomic composition, serotype, plasmid composition and toxicity [36], other serotypes may harbour several genomic clones. This has been previously appreciated in a slightly different context with the allocation of different strains within a serotype to separate biotypes such as biotype *subtoxicus* of serotpe H6 (serovar *entomocidus)* [8]. It remains to be seen if these biotypes are consistent with genomic clones.

It seems that *Bc* and *Bt* clones may be intermingled but rarely, if ever, are *Bc* and *Bt* strains included in the same clone. It is therefore tempting to consider *Bc* and *Bt* as a collection of clones which is diverging into species. One possible driving force for this process is the crystal protein. I suggest that strains belonging to separate clones of *Bt* (or *Bc*) could have different abilities to exploit or take advantage of a Cry toxin. In this way, plasmids bearing specific *cry* genes could become established in a clonal lineage just as the plasmids and *cry* genes of *Bti* have become associated with that host. This very successful plasmid/*cry* gene/host combination appears to be in the process of becoming isolated sexually and evolving into a new species [36, 43]. The generation of new *cry* genes could arise through plasmid conjugation resulting in two *cry* genes in the same host, recombination and a new Cry protein [11, 34]. If this protein confers an insect toxicity of which the host can take particular advantage, the combination may become fixed as a new clonal lineage which could subsequently emerge as a new species. In this way, *Bc* clones could accommodate Cry plasmids and evolve as *Bt* while other clones could lose plasmids because they no longer can take advantage of the Cry protein effectively and become *Bc*.

In this context *Bc* and *Bt* are distinct populations within the same species. As discrete clonal populations they do not undergo significant chromosomal recombination. Plasmid transfer, and with it *cry* gene transfer, occurs but is not often successful since the necessity of a suitable host for the plasmid excludes the stable establishment of most plasmid/host combinations. Thus only certain plasmid host combinations survive to become successful clones. This possible scenario requires that certain plasmid types (perhaps classified on the basis of maintenance functions) are associated with specific clones of *Bt* and further research will be needed to establish if this is indeed the case.

5.   ROUND-SPORE-FORMING BACILLI (rRNA GROUP 2 *SENSU* ASH *et al.*, 1991)

Strains of *Bacillus* which differentiate into spherical endospores have been collectively labelled *B. sphaericus* because of the difficulties associated with phenotypic characterisation [25]. These bacteria have a strictly aerobic lifestyle and use various organic acids as carbon and energy sources rather than the typical sugars of most other AEFB [2, 58]. They are consequently negative in the usual taxonomic tests based on polysaccharide and sugar metabolism and all strains are "lumped" into the one species, *B. sphaericus*. However, DNA reassociation [38, 56] 16S rRNA sequence analysis [4] and numerical taxonomy [2] revealed that strains of *B. sphaericus* could be assigned to at least six taxa which could be equated with separate species, but no formal nomenclatural revision has been proposed because the taxa cannot be reliably distinguished phenotypically. Strains of *B. sphaericus* which synthesise mosquitocidal toxins are invariably classified in DNA homology group IIA, a "species" which can be recognised by various molecular typing procedures including ribotyping [3], random amplified polymorphic DNA (RAPD; [71]), M13-based fingerprinting [1] and MLEE [70] but has few distinctive phenotpyic characteristics [2]. Here I will refer to these bacteria as "group IIA".

Table 3. Some representative strains of DNA homology group IIA *Bacillus sphaericus* and their characteristics.

Clone[1]	Strain	Origin	Serotype	Toxin genes[2] bin[3]	mtx1	mtx2	mtx3
1	K	USA	1a	-	+	ND	+
1	Q	USA	1a	-	+	+	+
2	9002	India	1a	A1B1	+	ND	ND
3	SSII-1	India	2a2b	-	+	+	+
3	1883	Israel	2a2b	-	+	ND	ND
5	LP24-4	Singapore	2a2b	-	-	ND	ND
5	LP35-6	Singapore	2a2b	-	-	ND	ND
9	BDG2	France	3	-	-	ND	ND
10	IAB 881	Ghana	3	A1B1	-	ND	ND
11	LP1-G	Singapore	3	A4B4	-	ND	ND
12	1593	India	5a5b	A2B2	+	+	+
12	2362	Nigeria	5a5b	A2B2	+	+	+
12	BSE 18	Scotland	5a5b	A2B2	+	+	ND
13	IAB 59	Ghana	6	A1B1	+	+	+
13	IAB 774	Ghana	6	+	+	ND	ND
15	R-1e	Singapore	6	-	-	ND	ND
17	2297	Sri Lanka	25	A3B3	+	+	+
17	M2-1	Malaysia	25	+	+	ND	ND
19	2377	India	26a26b	-	-	ND	ND
20	2315	Thailand	26a26b	-	-	-	ND
21	IAB 872	Ghana	48	A1B1	+	ND	ND

[1], Clone assignments based on PFGE of *Sma*I-digested chromosomal DNA [73].
[2], Data for toxin gene distribution taken from [40, 51, 66, 73].
[3], *bin* gene designations given where known, otherwise presence/absence indicated by + or -.

Several H-serotypes are recognised within group IIA [8]. These are not numbered sequentially as they are in *Bt* because of early confusion surrounding the classification of mosquitocidal strains of *B. sphaericus*. As with *Bt*, there is some correlation between toxicity and serotype, but group IIA has a pronounced clonal population structure within which toxicity is clearly demarcated. Clones are most readily apparent by pulsed field gel electrophoresis of *Sma*I-digests of chromosomal DNA [73] in which some serotypes comprise single genomic clones (e.g. serotype 5a5b) while others, such as serotypes 3 and 6 each comprise several clonal populations (Table 3).

Mosquitocidal toxicity in group IIA is associated with a crystal or binary toxin (Bin, [9, 10]), and cellular mosquitocidal toxins (Mtx1, [65]; Mtx2, [66] and Mtx3, [40]). Strains which synthesise Bin toxins are generally referred to as high toxicity and those without *bin* genes as low toxicity.

The crystal toxin comprises equimolar amounts of two proteins, BinA and BinB. To date, four toxin types (labelled 1, 2, 3, and 4) have been recognised [31, 51] and their distribution to clonal populations is shown in Table 3. All serotpye 5a5b strains studied to date have identical genomic composition (clonal type 12) and mosquitocidal toxin genes. This undoubtedly represents the most successful lineage of group IIA strains. These bacteria can be isolated from all parts of the world, and the combination of the clonal type 12 host with BinA2B2 toxins and full complement of Mtx toxins results in a very successful bacterium; the *Bti* of *B. sphaericus*. On the other hand, the *binA1B1* combination is the most widely distributed crystal protein combination since the identical genes have been discovered in several clonal lines (2, 10, 13 and 21). The only way in which these identical genes could occur in such genetically divergent populations is by lateral gene transfer and there is now evidence that the *bin* genes may lie within a transposable pathogenicity island (M.J. Humphreys, M. Coleman & C. Berry, personal communication).

Although all mosquitocidal strains of *B. sphaericus* are members of DNA group IIA, the converse is not true and not all members of group IIA are mosquitocidal. Representatives of various clones such as 5 (serotype 2a2b), 9 (serotype 3), 15 (serotype 6) and 19 (serotype 26a26b) lack *bin* and *mtx* genes, although not all strains have been screened for all toxins in every instance. It seems likely that these non-toxic strains of group IIA are far more common in the environment than their pathogenic counterparts [33], and that emphasis on larval toxicity in screening programmes has resulted in mosquitocidal strains being preferentially isolated. Therefore, if homology group IIA strains are to be given species status, pathogenicity is not a good defining feature and we must continue our search for diagnostic phenotypic features.

6.   *PAENIBACILLUS* (rRNA GROUP 3 *SENSU* ASH *et al.*, 1991)

Most species of AEFB which differentiate into ellipsoidal spores which distinctly distend the sporangium form a phylogenetically distinct lineage which has diverged to such an extent from other AEFB that it was the first of the RNA groups to be recognised as a distinct genus, *Paenibacillus* [7]. These bacteria are also physiologically distinctive in being facultative anaerobes able to ferment diverse sugars and carbohydrates. They generally grow slowly, producing small translucent colonies and are particularly associated with plants, composts and insects [53].

P. alvei was originally isolated from honey bee larvae suffering European foulbrood. Although subsequent studies indicated that is was not the causative agent of the disease (now attributed to *Melissococcus pluton*), nevertheless, it is used as an indicator of the disease (M. Gilliam, personal communication). Some strains have been reported to be weakly pathogenic towards dipteran larvae and others to molluscs [60].

6.1 *Paenibacillus* (formerly *Bacillus*) *larvae* and *Paenibacillus* (formerly *Bacillus*) *pulvifaciens*

P. larvae is the causative agent of American foulbrood of honey bee larvae, a globally widespread and economically serious disease of honey bees [62]. This is one of the few AEFB which does not grow readily on nutrient agar and sporulates poorly, if at all, *in vitro* although sporulation is extensive in larvae. The requirement for sporulation *in vivo* to complete the "life cycle" has led to these bacteria being referred to as obligate pathogens. The bacterium is catalase negative, an unusual trait among AEFB which allows reasonably accurate identification. Crystal proteins have not been observed in sporulating cells of *P. larvae*. Specialist procedures for its isolation and identification have been described [15, 29].

Recent studies have revealed a clonal population structure to *P. larvae*. DNA restriction endonuclease profiles have proven useful in this respect and in all, 20 isolates were assigned to five types (clones) by this technique, with isolates from geographically localised regions showing highest similarity [18]. This should lead to opportunities to trace, and hopefully limit, the spread of *P. larvae*.

"*Paenibacillus pulvifaciens*" (originally "*Bacillus pulvifaciens*") can be isolated from "powdery scale" within bee hives. This material is light brown in colour with a powdery texture and comprises the dried remnants of dead larvae. It seems unlikely that the bacterium is responsible for the condition since reintroduction of the bacteria into larvae does not cause the disease [23]. The bacterium shares numerous phenotypic features with *P. larvae* including lack of catalase, but it grows and sporulates on normal media and therefore cannot be considered an obligate pathogen, particularly given its questionable pathogenicity. A recent extensive taxonomic study including genomic and phenotypic comparisons has shown that *P. pulvifaciens* is closely related to *P. larvae* and reclassification of these bacteria as *P. larvae* subspecies *larvae* (Pll) and *P. larvae* subsp. *pulvifaciens* (Plp) was recommended [28]. Full phenotypic descriptions allowing the identification of these bacteria have been published [28] but it is unfortunate that we do not have 16S rRNA sequences for the two subspecies since it would helpful to be able to include Pll in the *Paenibacillus* tree (Fig. 2).

6.2   *Paenibacillus* (formerly *Bacillus*) *lentimorbus* and *Paenibacillus* (formerly *Bacillus*) *popilliae*

P. lentimorbus and *P. popilliae* are obligate pathogens of scarabaeid beetle larvae, notably the Japanese beetle (*Popillia japonica* Newman), an important pest of turf grass in the USA. Like Pll, these facultative anaerobes do not grow in nutrient broth, are catalase negative and sporulate poorly, if at all *in vitro*. Although 16S rRNA sequence analyses initially placed these species in rRNA Group 1 [6], a recent re-examination of their phylogenetic position places them firmly among the paenibacilli (see Fig. 2), hence their transfer to this genus [48].

   P. lentimorbus and *P. popilliae* cause "milky" disease in susceptible larvae, an infection in which spores consumed while feeding germinate in the insect gut. The bacteria invade the hemolymph where they grow vegetatively for up to 5 days. Sporulation ensues, and after about two weeks the grub dies when loaded with up to 10^{10} spores/ml hemolymph. In his original description of the bacteria, Dutky associated *P. popilliae* with type A disease symptoms and *P. lentimorbus* with Type B disease, the latter being characterised by the appearance of brown clots which block the circulation of hemolymph in the larva and lead to gangrenous conditions in the affected parts [19]. It is not clear if the association of Types A and B disease symptoms with species is absolute.

   There has been considerable confusion in the past as to whether *P. lentimorbus* and *P. popilliae* are synonymous but the distinct species status has been resolved recently using modern taxonomic techniques. The small differences in 16S rRNA sequence are consistent with separate species and in a comprehensive DNA reassociation study strains of *P. popilliae* and *P. lentimorbus* were clearly assigned to separate homology groups [57]. Phenotypically, most strains of *P. popilliae* are vancomycin resistant and grow in the presence of 2% NaCl while growth of most strains of *P. lentimorbus* is inhibited by these conditions. Interestingly, crystal protein synthesis, considered to be exclusively associated with *P. popilliae,* occurs in some strains of *P. lentimorbus* (which had been mis-identified on the basis of paraspore synthesis) [57]. It is not known if the paraspore, which has 42% homology with the Cry2Aa endotoxin of *Bt* [75], is involved in pathogencity but it may weaken the gut lining thus facilitating entry of the bacterium to the hemolymph.

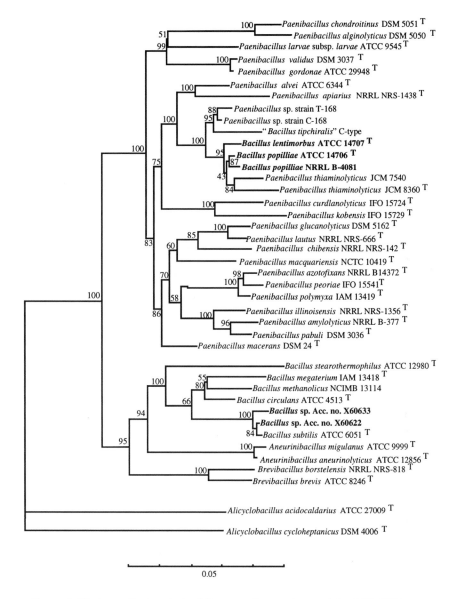

Figure 2. Phylogenetic tree derived from an alignment comprising 16S rRNA gene sequences from *Paenibacillus* species and selected members of closely related genera. *Alicyclobacillus acidocaldarius* ATCC 27009[T] served as outgroup. *P. lentimorbus* ATCC 14707[T], *P. popilliae* ATCC 14706[T] and NRRL B-4081 are shown in bold face as are the previous phylogenetic positions obtained when using the original sequences (accession numbers X60622 and X60633 for *B. lentimorbus* ATCC 14707[T] and *B. popilliae* ATCC 14706[T], respectively) which placed these bacteria in rRNA Group 1 of *Bacillus*. The data set was resampled 500 times by using the bootstrap option and the percentage values are given at the nodes showing the frequency at which the taxa to the right of the actual node cluster together. The scale bar indicates substitutions per nucleotide position (from Pettersson *et al.* [48] with permission).

Bacterial isolates from various insects with milky-type diseases have been afforded separate species or varietal status in particular "*B. popilliae* var. *melolonthae*" and "*B. popilliae* var. *rhopaea*" based on morphology of the spores and crystals and cultural features [42]. The taxonomic positions of these bacteria are largely unknown but DNA reassociation studies and RAPD suggest that "var. *melolonthae*", originally isolated from the common cockchafer (*Melolontha melolontha* L) may represent a valid subspecies [57]. Such molecular techniques will perhaps resolve the status of some of these bacteria when more strains have been isolated and studied. This will be a valuable step forward because it could enable correlation between target pest and variety of bacterium and improvements in biological control could ensue.

7.   *BREVIBACILLUS* (rRNA GROUP 4 *SENSU* ASH et al., 1991)

Brevibacillus (formerly *Bacillus*) *laterosporus* is one of ten species relocated in this genus on the basis of 16S rRNA sequence analysis [59]. All species produce ellipsoidal spores which distend the mother cell, are catalase positive and, with the exception of *B. laterosporus* strains which are facultatively anaerobic, are strictly aerobic. *B. laterosporus* is also distinguished by the synthesis of a canoe-shaped paraspore which cradles the spore and is firmly attached to it. *B. laterosporus* has been isolated from diseased bee larvae but it is not a bee pathogen.

Some strains of *B. laterosporus* are weakly pathogenic for dipteran larvae, (*Aedes aeygpti* and *Culex quinquefasciatus*) but have no activity against *Heliothis virescens* [21]. These early studies suggested that toxicity, which was about 1000 times less than that shown by *Bti*, was not associated with the crystal but rather with stationary phase vegetative cells. However, more recent studies with different crystalliferous strains of *B. laterosporus* showed high toxicity (equivalent to *Bti*) against *Anopheles stephensi* and *A. aegypti* but not *C. quinquefasciatus,* associated with the crystals of these strains [46]. Sequence analysis of these *cry* genes will indicate the relationships of these proteins to those of *B. sphaericus*, *B. thuringiensis* and *P. popilliae.*

8.   CONCLUDING REMARKS

The AEFB comprise the most valuable source of insect toxins for biological control purposes. For many years lepidopteran strains of *Bt*, and the obligate pathogens *P. lentimorbus* and *P. popilliae* were the focus of attention because of their application in insect control. However, in the past 30 years more wide-ranging and effective searches for new biological control agents have resulted in the recognition of the tremendous diversity of *Bt* endotoxins and new, highly toxic species such as *B. sphaericus* and *B. laterosporus* have been characterised. Polyphasic classifications have contributed to the isolation of these organisms in two ways. First, without accurate classifications which clearly demarcate pathogens from non-pathogens, we do not know the target of our isolation processes. For example, the demonstration that only *B. sphaericus* group IIA strains are pathogenic for mosquito larvae narrowed the search for new pathogens and enabled targeted isolation programmes with the successful isolation of new strains [39]. As the detailed classification of *P. popilliae* and *P. lentimorbus* is established it may offer similar opportunities to isolate new varieties of these bacteria with different toxicity ranges.

Second, broad phenotypic descriptions such as those derived from numerical classifications, can be used to formulate selective media such as BATS medium for *B. sphaericus* homology group IIA [72] or vancomycin-based media for *P. popilliae* [63]. Unfortunately, the close relationships between *Bt* and *B. cereus* has precluded effective selective isolation of *Bt* on plates, although methods based on selective germination have been valuable for this bacterium [35, 67].

Classifications based on 16S rRNA sequence analysis have demonstrated the phylogenetic depth and fascinating diversity of the AEFB. This and other molecular techniques have also provided fascinating insight into the evolution of insect toxicity in the AEFB and indicated lateral transfer of toxin genes within species such as *B. sphaericus* group IIA and between genera (the probable transfer of *cry* genes from *Bt* to *P. popilliae*). On a finer scale, it is possible that population genetic studies will allow us to clarify the taxonomic relationships of *Bt* and *B. cereus,* perhaps the single most perplexing taxonomic problem left to resolve among the aerobic, endospore-forming bacteria.

ACKNOWLEDGEMENT

I am grateful to Bertil Pettersson (Royal Institute of Biotechnology, Stockholm) for his invaluable advice and assistance in the preparation of phylogenetic trees.

REFERENCES

[1] Abadjieva A, Miteva V & Grigorova R (1992) Genomic variations in mosquitocidal strains of *Bacillus sphaericus* detected by M13 DNA fingerprinting. J. Invertebr. Pathol. 60, 5-9

[2] Alexander B & Priest FG (1990) Numerical classification and identification of *Bacillus sphaericus* including some strains pathogenic for mosquito larvae. J. Gen. Microbiol. 136, 367-376

[3] Aquino de Muro M, Mitchell WJ & Priest FG (1992) Differentiation of mosquito pathogenic strains of *Bacillus sphaericus* from nontoxic varieties by ribosomal RNA gene restriction patterns. J. Gen. Microbiol. 138, 1159-1166

[4] Aquino de Muro M & Priest FG (1993) Phylogenetic analysis of *Bacillus sphaericus* and development of an oligonucleotide probe specific for mosquito pathogenic strains. FEMS Microbiol. Lett. 112, 205-210

[5] Ash C & Collins MD (1992) Comparative analysis of 23S ribosomal RNA gene sequences of *Bacillus anthracis* and emetic *Bacillus cereus* determined by PCR-direct sequencing. FEMS Microbiol. Lett. 94, 75-80

[6] Ash C, Farrow JA, Wallbanks S & Collins MD (1991) Phylogenetic heterogeneity of the genus *Bacillus* revealed by comparative analysis of small subunit ribosomal RNA sequences. Lett Appl. Microbiol. 13, 202-206

[7] Ash C, Priest FG & Collins MD (1993) Molecular identification of rRNA group 3 bacilli (Ash, Farrow, Wallbanks and Collins) using a PCR probe test. Antonie van Leeuwenhoek 64, 253-260

[8] de Barjac H, Larget-Thiéry I, Cosmao Dumanoir V & Ripouteau H (1985) Serological classification of *Bacillus sphaericus* strains in relation to toxicity to mosquito larvae. Appl. Microbiol. Biotechnol. 21, 85-90

[9] Baumann L, Broadwell AH & Baumann P (1988) Sequence analysis of the mosquitocidal toxin genes encoding the 51.4- and 41.9-kilodalton proteins from *Bacillus sphaericus.* J. Bacteriol. 170, 2045-2050

[10] Berry C, Jackson-Yap J, Oei C & Hindley J (1989) Nucleotide sequence of two toxin genes From *Bacillus sphaericus* IAB59 - sequence comparisons between five highly toxinogenic strains. Nucleic Acids Research 17, 7516-7516

[11] Bravo A (1997) Phylogenetic relationships of *Bacillus thuringiensis* delta-endotoxin family proteins and their functional domains. J. Bacteriol. 179, 2793-2801

[12] Bucher C (1981) Identification of bacteria found in insects, p. 7-33. *In* Burges HD (ed.), Microbial Control of Pests and Plant Diseases 1970-1980, Academic Press

[13] Carlson CR, Caugant DA & Kolstø A-B (1994) Genotypic diversity among *Bacillus cereus* and *Bacillus thuringiensis* strains. Appl. Environ. Microbiol. 60, 1719-1725

[14] Carlson CR, Johansen T & Kolstø A-B (1996) The chromosome map of *Bacillus thuringiensis* subsp. *canadensis* HD224 is highly similar to that of *Bacillus cereus* type strain ATCC 14579. FEMS Microbiol. Lett. 141, 163-167

[15] Carpana E, Marocchi L & Gelmini L (1995) Evaluation of the API-50CHB system for the identification and biochemical characterization of *Bacillus larvae*. Apidologie 46, 270-279

[16] Charles J-F, Nielsen-LeRoux C & Delécluse A (1996) *Bacillus sphaericus* toxins, molecular biology and mode of action. Ann. Rev. Entomol. 41, 451-472

[17] Damgaard PH (1995) Diarrhoeal enterotoxin production by strains of *Bacillus thuringiensis* isolated from commercial *Bacillus thuringiensis*-based insecticides. FEMS Immunol. Med. Microbiol. 12, 245-250

[18] Djordjevic S, Ho-Shon M & Hornitzky M (1994) DNA restriction endonuclease profiles and typing of geographically diverse isolates of *Bacillus larvae*. J. Apicult. Res. 33, 95-103

[19] Dutky SR (1940) Two new spore-forming bacteria causing milky disease of Japanese beetle larvae. J.Agricult. Res. 1940, 57-68

[20] Farrow JAE, Wallbanks S & Collins MD (1994) Phylogenetic interrelationships of round-spore-forming bacilli containing cell walls based on lysine and the non-spore-forming genera *Caryophanon, Exiguobacteirum, Kurthia* and *Planococcus*. Int. J. Syst. Bacteriol. 44, 74-82

[21] Favret ME & Yousten AA (1985) Insecticidal activity of *Bacillus laterosporus*. J. Invertebr. Pathol. 45, 195-203

[22] Forst S & Nealson K (1996) Molecular biology of the symbiotic pathogenic bacteria *Xenorhabdus* spp. and *Photorhabdus* spp. Microbiol. Rev. 60, 21-43

[23] Gilliam M & Dunham DR (1977) Recent isolations of *Bacillus pulvifaciens* from powdery scales of honey bee, *Apis mellifera,* larvae. J. Invertebr. Pathol. 32, 222-223

[24] Gordon RE (1981) One hundred and seven years of the genus *Bacillus*, p. 1-15. *In* Berkeley RCW & Goodfellow M (ed.), The Aerobic Endospore-forming Bacteria: Classification and Identification, Academic Press

[25] Gordon RE, Haynes WC & Pang CH-N (1973) The genus *Bacillus*. Agriculture handbook no. 427. United States Department of Agriculture, Washington, D. C.

[26] Henderson I, Duggleby CJ & Turnbull PCB (1994) Differentiation of *Bacillus anthracis* from other *Bacillus cereus* group bacteria with the PCR. Int. J. Syst. Bacteriol. 44, 99-105

[27] Heyndrickx M, Lebbe L, Kersters K *et al.* (1998) *Virgibacillus* : a new genus to accommodate *Bacillus pantothenticus* (Proom and Knight). Emended description of *Virgibacillus pantothenticus*. Int. J. Syst. Bacteriol. 48, 99-106

[28] Heyndrickx M, Vandemeulebroecke K, Hoste B *et al.* (1996) Reclassification of *Paenibacillus* (formerly *Bacillus*) *pulvifaciens* (Nakamura 1984) Ash et al. 1994, a later subjective synonym of *Paenibacillus* (formerly *Bacillus*) *larvae* White 1906) Ash et al. 1994, as a subspecies of *Paenibacillus larvae,* with emended descriptions of *Paenibacillus larvae* as *Paenibacillus larvae* subsp. *larvae* and *Paenibacillus larvae* subsp. *pulvifaciens*. Int. J. Syst. Bacteriol. 46, 270-279

[29] Hornitzky MAZ & Nicholas PJ (1993) J-medium is superior to sheep blood agar and brain heart infusion agar for the isolation of *Bacillus larvae* from honey samples. J. Apicult. Res. 32, 51-52

[30] Hornitzky MAZ & Smith L (1998) Procedures for the isolation of *Melissococcus pluton* from diseased brood and bulked honey samples. J. Apicult. Res. 37, 267-271

[31] Humphreys MJ & Berry C (1998) Variants of the *Bacillus sphaericus* binary toxins: implications for differential toxicity of strains. J. Invertebr. Pathol. 71, 184-185

[32] Jackson TA, Huger AM & Glare TR (1993) Pathology of amber disease in the grass grub *Costelytra zealandica* (Coleoptera: Scarabaeidae). J. Invertebr. Pathol. 61, 1-8

[33] Jahnz U, Fitch A & Priest FG (1996) Evaluation of an rRNA-targeted oligonucleotide probe for the detection of mosquitocidal strains of *Bacillus sphaericus* in soils - Characterization of novel strains lacking toxin genes. FEMS Microbiol. Ecol. 20, 91-99

[34] Jarrett P & Stephenson M (1990) Plasmid transfer between strains of *Bacillus thuringiensis* infecting *galleria mellonella* and *Spodoptera littoralis*. Appl. Environ. Microbiol. 56, 1608-1614

[35] Johnson C & Bishop AH (1996) A technique for the effective enrichment and isolation of *Bacillus thuringiensis*. FEMS Microbiol. Lett. 142, 173-177

[36] Kaji DA, Rosato YB, Canhos VP & Priest FG (1994) Characterization by polyacrylamide gel electrophoresis of whole cell proteins of some strains of *Bacillus thuringiensis* subsp. *israelensis* isolated in Brazil. Syst. Appl. Microbiol. 17, 104-107

[37] Keim P, Kalif A, Schupp J *et al.* (1997) Molecular evolution and diversity in *Bacillus anthracis* as detected by amplified fragment length polymorphism markers. J. Bacteriol. 179, 818-824

[38] Krych V, Johnson JL & Yousten AA (1980) Deoxyribonucleic acid homologies among strains of *Bacillus sphaericus*. Int. J. Syst. Bacteriol. 30, 476-484

[39] Liu JW, Hindley J, Porter AG & Priest FG (1993) New high-toxicity mosquitocidal strains of *Bacillus sphaericus* lacking a 100-kilodalton-toxin gene. Appl. Environ. Microbiol. 59, 3470-3473

[40] Liu JW, Porter AG, Wee BY & Thanabalu T (1996) New gene from nine *Bacillus sphaericus* strains encoding highly conserved 35.8-kilodalton mosquitocidal toxins. Appl. Environ. Microbiol. 62, 2174-2176

[41] Logan NA & Berkeley RCW (1981) Classification and identification of the genus Bacillus using API tests, p. 106-140. *In* Berkeley RCW & Goodfellow M (ed.), The Aerobic Endospore-forming Bacteria: Classification and Identification, Academic Press

[42] Milner RJ (1981) Identification of the *Bacillus popilliae* group of insect pathogens, p. 45-59. *In* Burges HD (ed.), Microbial Control of Pests and Plant Diseases 1970-1980, Academic Press

[43] Nakamura LK (1994) DNA relatedness among *Bacillus thuringiensis* serovars. Int. J. Syst. Bacteriol. 44, 125-129

[44] Nakamura LK & Jackson MA (1995) Clarification of the taxonomy of *Bacillus mycoides*. Int. J. Syst. Bacteriol. 45, 46-49

[45] Nielsen P, Rainey FA, Outtrup H, Priest FG & Fritze D (1994) Comparative 16S rDNA sequence analysis of some alkaliphilic bacilli and the establishment of a a sixth rRNA group within *Bacillus*. FEMS Microbiol. Lett. 117, 61-66

[46] Orlova MV, Smirnova TA, Ganushkina LA, Yacubovitch VY & Azizbekyan RR (1998) Insecticidal activity of *Bacillus laterosporus*. Appl. Environ. Microbiol. 64, 2723-2725

[47] Parry JM, Turnbull PCB & Gibson JR (1983) A Colour Atlas of *Bacillus* Species. Wolfe Medical Books, London

[48] Pettersson B, Rippere KE, Yousten AA & Priest FG (1999) Transfer of *Bacillus lentimorbus* and *Bacillus popilliae* to the genus *Paenibacillus* with emended descriptions of *Paenibacillus lentimorbus* comb. nov. and *Paenibacillus popilliae* comb. nov. Int. J. Syst. Bacteriol. 49, 531-540

[49] Priest FG (1981) DNA homology in the genus *Bacillus*, p. 33-57. *In* Berkeley RCW & Goodfellow M (ed.), The Aerobic Endospore-forming Bacteria: Classification and Identification, Academic Press

[50] Priest FG & Dewar SJ (1999) Bacteria and insects. *In* (Priest FG & Goodfellow M ed.), Applied Microbial Systematics, Kluyer Academic Publishers, Dordrecht

[51] Priest FG, Ebdrup L, Zahner V & Carter PE (1997) Distribution and characterization of mosquitocidal toxin genes in some strains of *Bacillus sphaericus*. Appl. Environ. Microbiol. 63, 1195-1198

[52] Priest FG, Goodfellow M & Todd C (1988) A numerical classification of the genus *Bacillus*. J. Gen. Microbiol. 134, 1847-1882

[53] Priest FG & Grigorova R (1990) Methods for studying the ecology of endospore-forming bacteria, p. 565-591. *In* Grigorova R & Norris JR (ed.), Methods in Microbiology, Academic press

[54] Priest FG, Kaji DA, Rosato YB & Canhos VP (1994) Characterization of *Bacillus thuringiesis* and related bacteria by ribosomal RNA gene restriction fragment length polymorphisms. Microbiology 140, 1015-1022

[55] Rainey FA, Fritze D & Stackebrandt E (1994) The phylogenetic diversity of thermophilic members of the genus *Bacillus* as revealed by 16S rDNAS analysis. FEMS Microbiol. Lett. 115, 205-212

[56] Rippere KE, Johnson JL & Yousten AA (1997) DNA similarities among mosquito-pathogenic and nonpathogenic strains of *Bacillus sphaericus*. Int. J. Syst. Bacteriol. 47, 214-216

[57] Rippere KE, Tran MT, Yousten AA, Hilu KH & Klein MG (1998) *Bacillus popilliae* and *Bacillus lentimorbus,* bacteria causing milky disease in Japanese beetles and related scarab larvae. Int. J. Syst. Bacteriol. 48, 395-402

[58] Russell BL, Jelley SA & Yousten AA (1989) Carbohydrate metabolism In the mosquito pathogen *Bacillus sphaericus* 2362. Appl. Environ. Microbiol. 55, 294-297

[59] Shida O, Takagi H, Kadowaki K & Komagata K (1996) Proposal for two new genera, *Brevibacillus* gen. nov. and *Aneurinibacillus* gen. nov. Int. J. Syst. Bacteriol. 46, 939-946

[60] Singer S (1996) The utility of strains of morphological group II *Bacillus*. Adv. Appl. Microbiol. 42, 219-261

[61] Stackebrandt E & Goebel BM (1994) Taxonomic note: a place for DNA-DNA reassociation and 16S rRNA sequence analysis in the present species definition in bacteriology. Int. J. Syst. Bacteriol. 44, 846-849

[62] Stahly DP, Andrews RE & Yousten AA (1992) The genus *Bacillus:* insect pathogens, p. 1697-1745. *In* Balows A Trüper HG Dworkin M Harder W & Schleifer K-H (ed.), The Procaryotes, Springer Verlag

[63] Stahly DP, Takefman DM, Livasy CA & Dingman DW (1992) Selective medium for quantitation of *Bacillus popilliae* in soil and in commercial spore powder. Appl. Environ. Microbiol. 58, 740-743

[64] te Giffel MC, Beumer RR, Klijn N, Wagendorp A & Rombouts FM (1997) Discrimination between *Bacillus cereus* and *Bacillus thuringiensis* using DNA probes based on variable regions of 16S rRNA. FEMS Microbiol. Lett. 146, 47-51

[65] Thanabalu T, Berry C & Hindley J (1993) Cytotoxicity and ADP-ribosylating activity of the mosquitocidal toxin from *Bacillus sphaericus* SSII-1, possible roles of the 27-kilodalton and 70-kilodalton peptides. J. Bacteriol. 175, 2314-2320

[66] Thanabalu T & Porter AG (1996) A *Bacillus sphaericus* gene encoding a novel type of mosquitocidal toxin of 31.8 kDa. Gene 170, 85-89

[67] Travers RS, Martin PWA & Reichelderfer CF (1987) Selective process for efficient isolation of soil *Bacillus* species. Appl. Environ. Microbiol. 53, 1263-1266

[68] Turnbull PCB, Kramer J & J. M (1990) *Bacillus*, p. 187-210. *In* Parker MT & Duerdin MI (ed.), Topley & Wilson's Principles of Bacteriology, Virology and Immunity, Vol. 2, Edward Arnold

[69] von Wintzingerode F, Rainey FA, Kroppenstedt RM & Stackebrandt E (1997) Identification of environmental strains of *Bacillus mycoides* by fatty acid analysis and species-specific 16S rRNA oligonucleotide probing. FEMS Microbiol. Ecol. 24, 201-209

[70] Wisotzkey JD, Jurtshuk Jr. P, Fox GE, Deinhard G & Poralla K (1992) Comparative sequence analysis on the 16S rRNA (rDNA) of *Bacillus acidocaldarius*, *Bacillus acidoterristris*, and *Bacillus cycloheptanicus* and proposal for creation of a new genus, *Alicyclobacillus* gen. nov. Int. J. Syst. Bacteriol. 42, 263-269

[71] Woodburn MA, Yousten AA & Hilu KH (1995) Random amplified polymorphic DNA-fingerprinting of mosquito- pathogenic and nonpathogenic strains of *Bacillus sphaericus*. Int. J. Syst. Bacteriol. 45, 212-217

[72] Yousten AA, Fretz SB & Jelley SA (1985) Selective medium for insect pathogenic strains of *Bacillus sphaericus*. Appl. Environ. Microbiol. 49, 1532-1533

[73] Zahner V, Momen H & Priest FG (1998) Serotype H5a5b is a major clone within mosquito-pathogenic strains of *Bacillus sphaericus*. Syst. Appl. Microbiol. 21, 162-170

[74] Zahner V, Momen H, Salles CA & Rabinovitch L (1989) A comparative study of enzyme variation in *Bacillus cereus* and *Bacillus thuringiensis*. J. Appl. Bacteriol. 67, 275-282

[75] Zhang J, Hodgman C, Krieger L, Schnetter W & Schairer HU (1997) Cloning and analysis of the first *cry* gene from *Bacillus popilliae*. J. Bacteriol. 179, 4336-4341.

Chapter 1.2

Natural occurrence and dispersal of *Bacillus thuringiensis* in the environment

Per H. Damgaard
The Royal Veterinary and Agricultural University, Department of Ecology, Thorvaldsensvej 40, 1870 Frederiksberg C., Denmark

Key words: Natural occurrence, Soil, Phylloplane, Insects habitats, Food, Clinical infections, Epizootiology

Abstract: "Although *Bacillus thuringiensis* is widely used to control insect pests, the environmental fate of *B. thuringiensis* is known only in relatively general terms. Until recently, environmental studies looking at the natural distribution and life cycle of this bacterium were conducted only infrequently" [34]. Accordingly, very little is known about the natural transmission and behaviour of *B. thuringiensis*, even though the bacterium has been isolated from a variety of habitats, ranging from soil, phylloplane, and insects to consumables through-out the world. This chapter gives a review of the occurrence and fate of *B. thuringiensis* in different environments, along with a discussion of the epizootiology of *B. thuringiensis*.

1.   OCCURRENCE IN SOIL

The soil habitat is characterised as an environment with extreme diversity in abiotic factors such as water content, ranging from completely arid to swamps, and in amount of nutrients, ranging from "hot spots" with high amounts of nutrients to nutrient-poor habitats. In addition, soil contains a rich fauna of invertebrates, including insects.

In general, *B. thuringiensis* is found all over the globe and in all tested ecosystems, due to its heat- and drought-resistant spore. Chak *et al.* [18] and Martin & Travers [60] indicate that a large fraction of the spore-forming bacteria collected from mountain regions is *B.*

J.-F. Charles et al. (eds.),
Entomopathogenic Bacteria: From Laboratory to Field Application, 23–40.
© 2000 *Kluwer Academic Publishers. Printed in the Netherlands.*

thuringiensis. Both references conclude that this abundance is not connected with any specific insect population, as hardly any insects were found in these regions.

In the past, soil has been the main source for isolation of *B. thuringiensis* strains (e.g. [1, 29, 43, 57, 60, 68, 72, 81]). Nevertheless, information about specific numbers of *B. thuringiensis* in different soils has only been published by a few authors. Martin [59] and Ishii & Ohba [46] found up to 5×10^4 *B. thuringiensis* spores g^{-1} soil. Using colony hybridisation and a probe directed against *cry1A*, Hansen *et al.* [41] found 9×10^2 spores g^{-1} soil containing this gene.

More is known about the numbers of *B. thuringiensis* compared to other spore-forming bacteria from the genus *Bacillus*. The *B. thuringiensis*-index is defined as the number of colonies having *B. thuringiensis* / *B. cereus* colony morphology and containing a crystal inclusion body, divided by the total number of colonies with this colony morphology.

The highest *B. thuringiensis* indices reported from soil samples are 0.28 [60] and 0.27 [19]. The lowest index, 0.005, was found by DeLucca *et al.* [29].

"Non-toxic" strains occur commonly in soil, however, this could be due to the actual definition of "toxicity" or because of the limited number of insect species used in the testing regime and therefore unresolved insect toxicity.

The frequency of serotypes found in soil varies substantially. The most abundant serotype seems to be *kurstaki*, but *galleria* has also been very common in one study. Of all serovars found in soil no specific serovar has been found in frequencies higher than 0.5.

Although the *kurstaki* serotype has been found in soils all over the world, caution should be taken before generalising the causes for this dominance, since 80% of the serovars used for biocontrol are serovar *kurstaki* [64].

2.   OCCURRENCE ON FOLIAGE

The phylloplane is a heterogenous environment in which many species of bacteria can be found, including many *Bacillus* species. Furthermore, the phylloplane is a habitat for a high diversity of leaf-eating insects, including many lepidopteran and coleopteran species.

B. cereus and *B. mycoides* have been isolated from the foliage of different plants, for example broadleaf forest trees and grass foliage, in quantities estimated in the range of 10^3-10^5 bacteria g^{-1} foliage [85], accounting for 15-20% of the bacterial population [16, 83]. As *B. cereus* has been isolated from this habitat it is obvious that *B. thuringiensis* also must be present; however, few studies have been dedicated to the natural occurrence of *B. thuringiensis* in the phylloplane.

In a 3-year study, Smith & Couche [82] isolated *B. thuringiensis* from foliage of different deciduous trees, conifers, and shrubs; the numbers of *B. thuringiensis* ranged from 3 to 100 *B. thuringiensis* spores cm^{-2}. These numbers are much higher than those reported by Ohba [67], although this could be attributed to differences in geographical locations, tree species analysed, as well as technical differences.

Ohba [66] examined 25 mulberry leaves and found a *B. thuringiensis* frequency to vary from 0 to 0.45 on the 25 leaves analysed, with an average index of 0.032. This support the "hot spots" observed by Damgaard *et al.* [23] where the frequency of *B. thuringiensis* on cabbage leaves ranged from 0.02 to 0.67, with an average of 0.11.

Upon serotyping 150 *B. thuringiensis* strains collected from cabbage leaves, Damgaard *et al.* [23] found the majority (64%) to belong to serovar *kurstaki*. Eleven other serotypes were found, but in very low numbers. Ohba [66] found serovar *pakistani* to be the most abundant (46%) whereas *kurstaki* was only found in 7% of the strains. In a subsequent investigation by Damgaard *et al.* [22] of *B. thuringiensis* occurrence on grass foliage, 32 strains of *B. thuringiensis* were isolated, of which 75% belonged to serovar *israelensis*. Another study by Damgaard *et al.* [25] on the natural occurrence of *B. thuringiensis* on pine needles found that of 35 *B. thuringiensis* strains isolated, 31% belonged to serovar *israelensis* and 20% to serovar *aizawai*.

In the study by Smith & Couche [82], 68% of the tested strains reacted in a Western blot with an antibody directed towards whole, purified crystals of *B. thuringiensis* serovar *kurstaki* (HD-1), indicating lepidopteran activity. This is similar to observations by Damgaard *et al.* [23] who found 68% of the strains isolated from cabbage leaves to be lepidopteran. In a survey of grass foliage, 84% of the strains were found to be active against *Aedes aegypti* (Diptera), whereas no lepidopteran active strains found [22]. In contrast to this, Ohba [66] found only 10% of 186 *B. thuringiensis* strains from mulberry leaves to be active against *B. mori* and/or *A. aegypti*.

Studies of *B. thuringiensis* applications in forest ecosystems have shown that spores can be transported by air over long distances. After aerial application during winds of 1.1-1.3 m s^{-1}, dosage and deposition have been detected up to 1,500 m and 3,100 m down wind, respectively [10]. Experiments in cabbage-fields have shown sprayed *B. thuringiensis* spores to be transported to foliage when applied to the soil, and *vice versa* [75]. It is obvious that the naturally-occurring *B. thuringiensis* population isolated from cabbage foliage in the study by Damgaard *et al.* [23] is the same as the population found in soil. In isolations of *B. thuringiensis* from soil, the general trend is that specific serotypes do not occur in frequencies above 50%; single-order insect activity above 50% has seldom been reported. The population found by Damgaard *et al.* [23] on foliage has therefore probably originated in the soil, been transported to the foliage, and then developed into a population different from that found in soil.

In the investigation by Damgaard *et al.* [23], the majority of the isolated *B. thuringiensis* strains had lepidopteran activity and were isolated from cabbage foliage where lepidopteran pest insect was abundant on the sampled plants. In the study of grass foliage [22] the majority of the strains had dipteran activity. Damage from crane fly larvae (*Tipula* spp., Diptera) in the pasture from which the foliage was collected was noted. However a significant contribution of *B. thuringiensis* strains from the soil to the grass foliage is expected; a similar thing has been noted concerning the occurrence of *B. cereus* on grass foliage [85]. Ohba [66], who reports only 10% lepidopteran and/or dipteran activity, gives no information about any special pest insect associated with the analysed phylloplane community. However, in the study by Kaelin *et al.* [51] on the occurrence of *B. thuringiensis* in tobacco residues (leaves), the majority of the strains were found to be Coleoptera active, which they correlate with the occurrence of the tobacco beetle (*L. serricorne*), the most serious and wide-spread tobacco pest.

It therefore seems possible that under certain conditions, the phylloplane can act as a reservoir of *B. thuringiensis* strains active against the insects actively feeding on the plant. A reasonable hypothesis is therefore that insects living on the phylloplane can support an enzootic population of *B. thuringiensis* strains active against the predominant insect species present on the same phylloplane.

3.   OCCURRENCE IN SPECIFIC INSECTS HABITATS

3.1   Insect rearing facilities

Insect rearing facilities are monocultures of an insect species, often with a population density much higher than that found in the natural insect environment.

B. thuringiensis was first described at the beginning of this century from an epizootic of "black larvae" in a silkworm-rearing environment in Japan [48]. Most subsequent reports on *B. thuringiensis* in insect-rearing facilities also refer to silkworm rearing. Besides the records from sericulture, a massive outbreak of "black larvae" in a breeding facility of *P. gossypiella* has been reported, from which the well-known HD-1 strain was isolated [74].

Concentrations of *B. thuringiensis* in silkworm-rearing facilities range from 5×10^2 to 2.9×10^5 spores g^{-1} and with indices varying between 0.05 [72] and 0.36 [70, 71]; these are comparable to the values reported from soil.

A high diversity of serotypes has been found, with the dominating serotypes being *kenyae, alesti, aizawai,* and *sotto* [2, 33, 48, 69-73]. What is striking, in comparison with soil, is that more than 85% of the isolates showed activity against *B. mori.*

3.2   Mosquito breeding habitats

Mosquitoes (Culicidae, Diptera) breed in temporary shallow puddles and ponds in which very high concentrations of fast-developing mosquito larvae are found.

Information about *B. thuringiensis* isolated from mosquito breeding habitats shown that serovar *israelensis* dominates among the investigated material [6, 14, 27, 39]. It is important to note, however, that in several of the investigations, strains with mosquitocidal activity were isolated in the first selection step. Other δ-endotoxin-producing, non-mosquitocidal strains would not be isolated and further characterised (e.g. serotyped).

3.3   Isolation from other insects

In the past, *B. thuringiensis* has been isolated from a number of different insect species, most frequently from Lepidoptera [13, 33, 48, 55], but also from Diptera, Coleoptera, Hemiptera and Hymenoptera [19, 20, 27, 47, 54]. However, this does not imply that lepidopteran insects are more commonly infected by *B. thuringiensis*, rather that these insects are the most analysed. No extensive study on *B. thuringiensis* isolation from a range of insects collected from the same location, followed by a thorough characterisation of the strains has ever been performed.

Most of the important isolates used commercially have been isolated from diseased insects: serovar *kurstaki* was isolated from *E. kühniella* [55], serovar *israelensis* from mosquitoes [27, 39], and "tenebrionis" from *Tenebrio molitor* (Coleoptera) [54].

3.4   The stored product environment

The stored product environment is characterised by a high concentration of a few agricultural products (e.g. grain) in a spatially limited and dry environment. Within this special environment a few pest insect species are found, for example *E. kühniella* in grain storage bins. Isolation of *B. thuringiensis* from this environment has a long history. The type strain of *B. thuringiensis* was isolated from *E. kühniella* [13] and the serovar *kurstaki* was also discovered in grain stores [55].

B. thuringiensis seems to be common in the stored product environment. In a survey of 57 watermills, *B. thuringiensis* was present in all of them [79, 87]. The *B. thuringiensis* index has only been reported by two groups, who found values of 0.31 [28] and 0.67 [62].

Several dominant *B. thuringiensis* serovars have been reported in studies of the stored product environment [17, 28, 30, 51, 62, 65, 79]. Norris [65] examined 25 *B. thuringiensis* strains isolated from stored products or insects associated with these products, mostly from African grain stores; all strains belonged to serovar *kenyae*. DeLucca *et al.* [28] reported that 74 of 79 isolates of *B. thuringiensis* from grain dust belonged to serovar *aizawai*.

4.   OCCURRENCE IN FOODS

From the material presented earlier, it is evident that, in most situations, *B. thuringiensis* occurs together with *B. cereus*, in proportions that depend on the source of isolation material. As *B. cereus* is often associated with foods; due to the close relationship between the two species it is not surprising also to find *B. thuringiensis* in foods. However, limited material has been published on this subject.

Krieg [54] has discussed the natural occurrence of *B. thuringiensis* in grain and grain products. From 22 *B. thuringiensis/B. cereus* strains in flour, he isolated 11 *B. thuringiensis* strains, at an average 9.8×10^2 *B. thuringiensis* spores g^{-1} flour, representing the serovars *thuringiensis*, *galleriae* and *kurstaki*. Significantly higher numbers (10^3-10^5 *B. thuringiensis* cells g^{-1}) were found in pasta products by Damgaard *et al.* [24].

Spices and herbs are known to contain a wide variety of bacteria including *B. cereus* (for a review see [61]). In an outbreak of gastroenteritis at a chronic care institution, *B. thuringiensis* was isolated from stool samples and spices [50]. All isolated strains were found to be enterotoxin positive.

In all the cases of bacterial contamination described above, no data indicate that the organisms originate from pesticide residues. Contamination of milk by *B. cereus* originates mainly from soil and faeces debris from poorly disinfected udders [85]. In the case of both cereals and spices the strains must also come from soil debris or foliage. However, improper pasta production could give conditions favourable for bacterial growth, thereby increasing the contamination load in the final product.

5. CLINICAL INFECTIONS

As *B. cereus* is also involved in clinical infections other than food poisoning, it is therefore reasonable also to believe that *B. thuringiensis* may also be implicated in such infections; however, very limited material investigating this subject has been published. This could be due to lack of information/knowledge on the close association between *B. thuringiensis* and *B. cereus*. For example, Barrie *et al.* [8, 9] report on the occurrence of *B. cereus* in hospital linen and meningitis due to *B. cereus*; however, the techniques used for species identification would not have enabled them to differentiate between *B. cereus* and *B. thuringiensis*.

Clinical infections caused by *B. cereus* include local infections, particularly burns wounds [7, 31, 32]. Opportunistic bacteria in water supplies are potential serious problems in hospital burn units. During an investigation into the cause of *B. cereus* infections in burn wounds at an Italian hospital, *B. cereus* was found in the water supply system used in the treatment of burn wounds [86]. The isolated *B. cereus* strains were later identified as *B. thuringiensis* [21].

Only one well-described case of gastroenteric illness in mammals due to *B. thuringiensis* has been described [50]. However, *B. thuringiensis* may have been involved in many more cases of human diseases, as the identification of the δ-endotoxin for separation of *B. thuringiensis* from *B. cereus* involves specialised techniques not used when identifying cases of food poisoning caused by *B. cereus*. Food poisoning caused by *B. thuringiensis* may therefore have been misdiagnosed as *B. cereus*. The Nordic Committee on Food Analysis writes in its guidelines for examination of foods for *B. cereus*: "Since these species (*B. thuringiensis* and *B. cereus*) are very closely related and both may produce enterotoxins, differentiation is not necessary in foods" [5].

6.   EPIZOOTIOLOGY OF *B. THURINGIENSIS*

Epizootics in entomology are analogous to epidemics in medical science. Fuxa & Tanada [37] define epizootics as "sporadic and limited in duration and characterised by sudden change in prevalence and incidence leading to an unusually large number of cases of diseases", whereas as enzootic diseases are "long in duration, usually low in prevalence, and constantly present in a population". However, there is no clear-cut distinction between enzootic and epizootic; rather, they are opposite ends of a scale. Problems natural arise as to how one should define the onset of an epizootic when observing an unusually large number of cases of a disease. To overcome this problem, one needs data from previous years of sampling for the same insect, at the same location, and infected by the same pathogen. This problem is overcome in human medical science where large and more homogenous material is available and patients can be interviewed about their illness. However several well-documented examples of epizootics in insects caused by fungi [40] and viruses [36] have been published.

When it comes to *B. thuringiensis*, only a few well-described cases are available, because epizootics induced by *B. thuringiensis* are rarely encountered in nature. Data collection in an insect population with the purpose of estimating prevalence and incidence of *B. thuringiensis* would be enormous and would quickly become unwieldy. The data would most probably not even show *B. thuringiensis* to be enzootic, as the prevalence would be under the limit of detection. When situations of high insect mortality due to *B. thuringiensis* have occurred, authors have therefore only used a part of the definition with "sudden change in prevalence", as no data is available on prevalence or incidence.

The different epizootic situations due to *B. thuringiensis* encountered and described in the literature can be divided into three groups based on the habitat in which the scenario is observed: the field, insect rearing, and the stored product environment.

One of the few references in which there is no doubt about the presence of a *B. thuringiensis* epizootic is a paper by Margalit & Dean [58]. They report the isolation of the *B. thuringiensis* serovar *israelensis*-type strain from a mosquito breeding site. Floating on the pond (15 × 60 m) of brackish water was a "thick carpet" of dead and dying larvae of exclusively *C. pipiens*. From this description it is clear that an epizootic was present. As the larvae were decomposing in the pond, the sample taken back to the lab consisted of larvae, mud, silt, and water. Consequently, although serovar *israelensis* was prevalent, it was not possible to tell if all larvae were killed by the same *B. thuringiensis* strain.

Brownbridge & Onyango [15] report the isolation of *B. thuringiensis* from dead larvae of *Heliothis armigera* (Lepidoptera) collected from an apparent epizootic situation in a Kenyan wheat field. Dr. Brownbridge was presented with a large number of dead larvae, which all were in various forms of decay due to septicaemia (personal communication). Isolation of *B. thuringiensis* was only attempted from five individuals, as others were too decayed.

One of the most cited references of epizootics caused by *B. thuringiensis* in the field is the paper by Talalaev [84]. He describes the isolation and characterisation of *B. thuringiensis* from *Dendrolimus sibiricus* (Lepidoptera) larvae collected in the forest of Irkutsk, Russia. It is not clear, however, whether he actually collected the dead larvae during an epizootic in the forest or only observed some mortality in larvae sampled and incubated in the lab. A similar epizootic in *D. sibiricus* has been described from the same forest in Irkutsk by Vasijev in 1898 (Reference unavailable).

Within the realm of insect rearing, several outbreaks of so-called "black larvae" have been reported. The first to describe this phenomenon was Ishiwata, who in 1901 [48] reported "sudden-death *Bacillus*" in silkworm-rearing facilities. A similar mass outbreak of black larvae has been reported in a rearing of *Pectinophora gossypiella* [33]. In both investigations the outbreak was caused by one strain of *B. thuringiensis*, although this does not always have to be the case. From an laboratory colony of the European sunflower moth (*Homoesoma nebulella*, Lepidoptera), Itoua-Apoyolo *et al.* [49], during the span of one week, observed 60-80% of the population to die from a *B. thuringiensis* infection (several hundred larvae). From this outbreak, however, five different serotypes were isolated.

Another protected environment similar to the insectaria is the stored product environment, from which several epizootic situations have been described. As early as 1915, Berliner described *B. thuringiensis* isolated from *E. kühniella* collected from a mill in Thüringen, Germany [13]. It is not clear, however, how many specimens developed this "schlaffsucht" (sluggish disease). In samples collected from a mill in France, Kurstak [55] observed 38-50% mortality caused by *B. thuringiensis* in 686 larvae sampled. Similar prevalence has been observed in *E. kühniella* larvae collected at other mills [79, 87].

Enzootic levels of *B. thuringiensis* have also been found in the stored product environment [62].

7.   FACTORS GOVERNING *B. THURINGIENSIS*-CAUSED EPIZOOTICS

From the data presented it is evident that epizootics caused by *B. thuringiensis* are rare; this rarity can be explained by the *B. thuringiensis* life cycle.

The life cycle of *B. thuringiensis* can be divided into two major phases: vegetative growth and sporulation. During vegetative growth, *B. thuringiensis* multiplies exponentially until nutrient resources becomes limiting. When this occurs, cells start to sporulate and at the same time δ-endotoxin production is initiated. Different factors such as environment and insect species influence the further development of epizootics.

Larvae of *A. aegypti* are known to digest spores of *B. thuringiensis* serovar *israelensis* and incorporate the liberated nutrients (e.g. amino acids) into the larval tissue [53]. This occurs when the larvae ingest spores and at the same time do not ingest enough δ-endotoxin to perforate the midgut epithelium. If enough δ-endotoxin is ingested to cause perforation, the spores germinate and vegetative cells proliferate. When nutrients become limiting in the dead/dying larvae, cells sporulate, producing new spores and δ-endotoxins which fill the carcass [3, 4]. Scavenging larvae that feed on the infected carcasses die [52, 91]; thus producing an accelerating spiral of intoxicated larvae that leads to an epizootic.

In the stored product environment, epizootics caused by *B. thuringiensis* may be favoured by the protected environment. In such dry and light-protected environments, both spores and δ-endotoxin crystals can persist almost inde-finitely, both inside and liberated from insect carcasses. Burges & Hurst [17] showed a limited build up of *B. thuringiensis* spores and crystals appears to be sufficient to provoke an epizootic.

In *B. mori* populations reared for silk production, *B. thuringiensis* active against *B. mori* are apparently kept at an enzootic levels, causing no obvious mortalities in the populations [47, 70, 72, 73].

Following application of *B. thuringiensis* (spores and δ-endotoxin) to foliage, insecticidal activity gradually decreases and is essentially gone after two to four weeks [12, 45, 63]. This is due both to UV-inactivation of the δ-endotoxin [26, 77, 80] and decreasing numbers of spores and δ-endotoxin crystals on the foliage. The halflife (t) of spores on foliage is approximately two days, depending on the plant species and the product formulation applied [75, 76]. This rapid degradation of the δ-endotoxin and the spores effectively inhibits development of natural epizootics by naturally occurring *B. thuringiensis* strains. In addition no evidence has been found to support the idea of spore germination and proliferation of vegetative cells directly on foliage.

As the δ-endotoxin is a protein, West & Burges [89] considered other microorganisms in soil to be the most important factors in the degradation of the δ-endotoxin. Pruett *et al.* [78] showed that the insecticidal activity of inoculated spores and crystals fell quickly to 1% of their original activity, compared to a gradual decline in spore numbers to 25% of inoculum; this indicates that crystals are degraded more quickly than spores. West *et al.* [90] found comparable results, but also showed that crystals inoculated in sterilised soil were not degraded for up to 842 days post inoculation. This shows that microbial degradation of the crystals must take place under field conditions.

As the previous sections shows, only the spores of *B. thuringiensis* last for an extended length of time under field conditions and epizootics can not be initiated only by the ingestion of spores. Something else must assist the spores of *B. thuringiensis* into the host insect, facilitating germination, proliferation, and further sporulation. These factors can be divided into biotic and abiotic factors.

Abiotic factors may include mechanical injury of the peritrophic membrane. In experiments in which abrasive materials were fed to insects, this membrane was injured and an enhanced infection of the bacteria via the midgut was observed [88].

Scavenging has been observed by insects other than mosquito larvae. Several lepidopteran pest species such as *Heliothis zea* (Lepidoptera), *Spodoptera frugiperda* (Lepidoptera), and *Trichoplusia ni* appear to prefer to feed on foliage heavily contaminated with disintegrated cadavers [42]; this could explain the development of the observed epizootic in a population of *H. armigera* [15].

Several reports have shown that physical contaminations of the ovipositor of female parasites could transmit *B. thuringiensis*, but initiation of infection could also be through the ovipositional wound [35, 56].

Different arthropods can also transmit *B. thuringiensis*. Pyemotid and anoetid mites occurring in a reared population of *H. irritans* were found to carry *B. thuringiensis* strains which caused an epizootic in the same population of *H. irritans* [38].

Insect pathogens may also facilitate the transmission of *B. thuringiensis*. Damgaard *et al.* [23] isolated *B. thuringiensis* from a *D. radicum* adult infected with the fungus *Strongwellsea castrans*. This entomopathogenic fungi is characterised by the production of openings on the ventral surface of the host abdomen, from which the fungal spores are discharged. When the fungus grows in the host, it digests the abdominal organs and infects the nervous system [11, 44]. This provides a possible entrance for *B. thuringiensis* spores through the holes, as well as proliferation of spores in the mixed contents of dissolved organs and food eaten by the insect. Even spores ingested together with food could proliferate in the dissolved organs.

8.   CONCLUSIONS

B. thuringiensis has been isolated from soil collected in all parts of the world, from a vide range of ecosystems. The number of *B. thuringiensis* cells varies from 10^2 to 10^4 cfu g^{-1} soil. Many different serovars have been isolated from soil although the most abundant seem to be *kurstaki* and *galleriae*. The abundance of specific serovars in soil has not been found to exceed an index of 0.5. There is no evidence for correlation between either soil ecosystems and quantitative occurrence, serovars, or insecticidal activity.

B. thuringiensis occurs naturally on a range of different phylloplane materials (trees and herbs, deciduous and evergreens). Numbers of *B. thuringiensis* cells on foliage vary from 0 to 100 cfu cm^{-2}; at "hot spots" frequencies of *B. thuringiensis* are much higher than the average. Variation in serovar distribution in terms of both specific serovars and most abundant serovars has also been observed on foliage.

Data from at least two different experiments indicate that the population found on the phylloplane is somewhat different from that found in soil. The strains found in the phylloplane probably originate from the soil, but the phylloplane population may have developed into a population with different serovar composition.

B. thuringiensis is naturally associated with insects. Several well-known serovar type strains, e.g. *kurstaki*, *israelensis*, and "tenebrionis", have all been isolated from insects. There is a tendency that the insect from which the strain is isolated is also susceptible to the isolated strain, although many exceptions have been found.

As *B. thuringiensis* has been found to occur in both soil and on foliage, its presence on/in different consumables is not unexpected. *B. thuringiensis* has been isolated from both processed foods (e.g. pasta) and unprocessed foods (e.g. raw milk, cabbage). It is evident that the food contamination originates from the raw material.

Some of the toxins produced by some strains of *B. thuringiensis* benefit the organism in proliferating not only in insects but also in humans. Involvement of *B. thuringiensis* in clinical infections are therefore evident. There is only one published report of an epidemic caused by *B. thuringiensis*, although it is possible that more have been hidden in cases reported as *B. cereus* infections, due to lack of knowledge in distinguishing *B. thuringiensis* from *B. cereus*.

The insecticidal activity of applied of *B. thuringiensis* products lasts only two to four weeks, even though the preparations contain UV-protectants. The decline is due to a rapid degradation of the δ-endotoxin and, to a lesser extent, the spores. It is therefore clear that δ-endotoxins produced by naturally-occurring strains in an outdoor environment are inactivated even faster than the fortified commercial strains.

The life cycle of *B. thuringiensis* is divided into two phases: sporulation and vegetative growth. Spores can persist under harsh conditions for extended periods of time. When nutrient levels are high, spore dormancy is broken and the spores germinate. Vegetative cells then proliferate exponentially until nutrient levels are exhausted, at which point sporulation and δ-endotoxin production are induced.

Epizootics among insect populations caused by *B. thuringiensis* are rarely reported in the literature. However, under specific conditions epizootics do occur under both field and indoor conditions. Well-described cases of natural epizootics in mosquito-breeding habitats have been reported; experiments have been performed to study the possible mechanisms behind *B. thuringiensis* epizootics.

As the δ-endotoxin is quickly inactivated and only the spores of *B. thuringiensis* survive for any considerable length of time in the environment, initiation of epizootics must be caused by spores alone. The spores have difficulties invading through either the integument or the midgut; therefore, different biotic and abiotic factors must aid the spores in reaching the environment in the insect suitable for germination and completion of its life cycle.

REFERENCES

[1] Abdel-Hameed A & Landén R (1994) Studies on *B. thuringiensis* strains isolated from Swedish soils - Insect toxicity and production of *B. cereus* diarrhoeal-type enterotoxin. World J. Microbiol & Biotech. 10, 406-409

[2] Aizawa K, Takasu T & Kurata K (1961) Isolation of *B. thuringiensis* from the dust of silkworm rearing houses of farmers. J. Sericult. Sci. Jap. 30, 451-455

[3] Aly C (1985) Germination of *B. thuringiensis* var. *israelensis* spores in the gut of *Aedes* larvae (Diptera: Culicidae). J. Invert. Path. 45, 1-8

[4] Aly C, Mulla MS & Federici BA (1985) Sporulation and toxin production by *B. thuringiensis* var. *israelensis* in cadavers of mosquito larvae (Diptera: Culicidae). J. Invert. Path. 46, 251-258

[5] Anonymous (1993) *Bacillus cereus*. Determination in foods. Nordic committee on food analysis 67, pp.4

[6] Asimeng EJ & Mutinga MJ (1992) Isolation of mosquito-toxic bacteria from mosquito-breeding sites in Kenya. J. Am. Mosq. Control Assoc. 8, 86-88

[7] Attwood AI & Evans DM (1983) *Bacillus cereus* infections in burns. Burns 9, 355-357

[8] Barrie D, Hoffman PN, Wilson JA et al (1994) Contamination of hospital linen by *Bacillus cereus*. Epidemiol. & Infections 113, 297-306

[9] Barrie D, Wilson JA, Hoffman PN et al (1992) *Bacillus cereus* meningitis in two neurosurgical patients - an investigation into the source of the organism. J. Infection 25, 291-297

[10] Barry JW, Skyler PJ, Teske ME, et al (1993) Predicting and measuring drift of *B. thuringiensis* sprays. Environ. Tox. & Chem. 12, 1977-1989

[11] Batko A & Weiser J (1965) On the taxonomic position of the fungus discovered by Strong, Wells, and Apple: *Strongwellsea castrans* gen. et sp. nov. (Phycomycetes: Entomophthoraceae). J. Invert. Path. 7, 455-463

[12] Beegle CC, Dulmage HT, Wolfenbarger DA et al (1981) Persistence of *B. thuringiensis* Berliner insecticidal activity on cotton foliage. Environ. Entom. 10, 400-401

[13] Berliner E (1915) Über die Schalffsucht der Mehlmottentaupe und ihren Erreger *B. thuringiensis* n.sp. Zeit. Angew. Entom. 2, 29-56

[14] Brownbridge M & Margalit J (1986) New *B. thuringiensis* strains isolated in Israel are highly toxic to mosquito larvae. J. Invert. Path. 48, 216-222

[15] Brownbridge M & Onyango T (1992) Screening of exotic and locally isolated *B. thuringiensis* (Berliner) strains in Kenya for toxicity to the spotted stem borer, *Chilo partellus* (Swinhoe). Trop. Pest Manag. 38, 77-81

[16] Brunel B, Perissol C, Fernandez M et al (1994) Occurrence of *Bacillus* species on evergreen oak leaves. FEMS Microbiol. Ecol. 14, 331-342

[17] Burges HD & Hurst JA (1977) Ecology of *B. thuringiensis* in storage moths. J. Invert. Path. 30, 131-139

[18] Chak KF, Chao DC, Tseng MY, et al (1994) Determination and distribution of cry-type genes of *B. thuringiensis* isolates from Taiwan. Appl. & Environ. Microbiol. 60, 2415-2420

[19] Chilcott CN & Wigley PJ (1993) Isolation and toxicity of *B. thuringiensis* from soil and insect habitats in New-Zealand. J. Invert. Path. 61, 244-247

[20] Côté JC, Fréchette S & Vincent C (1992) Isolation of *B. thuringiensis* from the tarnished plant bug, *Lygus lineolaris* (Hemiptera: Miridae). 25th Annual Meeting of Society for Invertebrate Pathology, 16-21 August, Heidelberg, Germany Abst # 168

[21] Damgaard PH, Granum, PE, Bresciani J et al (1997) Characterization of *B. thuringiensis* isolated from infections in burn wounds. FEMS Immun. & Medical Microbiol. 18, 47-53

[22] Damgaard PH, Abdel-Hameed A, Eilenberg J et al (1998) Natural occurrence of *B. thuringiensis* on grass foliage. World J. Microbiol. & Biotech 14, 239-242

[23] Damgaard PH, Hansen BM, Pedersen JC et al (1997) Natural occurrence of *B. thuringiensis* on cabbage foliage and in insects associated with cabbage crops. J. Appl. Bact. 82, 253-258

[24] Damgaard PH, Larsen HD, Hansen BW et al (1996) Enterotoxin-producing strains of *B. thuringiensis* isolated from food. Lett. Appl. Microbiol. 23, 146-150

[25] Damgaard PH, Malinowski H, Glowacka B et al (1996) Degradation of *B. thuringiensis* serovar *kurstaki* after aerial application to a Polish pine stand. IOBC / WPRS Bulletin 19, 61-65

[26] Damgaard PH, Skovmand O & Eilenberg J (1992) Protection of the δ-endotoxin of *B. thuringiensis* var. kurstaki HD-1 against sunlight inactivation. 25th Annual Meeting of Society for Invertebrate Pathology, 16-21 August, Heidelberg, Germany Abst # 165

[27] de Barjac H (1978) A new subspecies of *B. thuringiensis* very toxic for mosquitoes *B. thuringiensis* var. *israelensis* sero-type 14. Comptes Rendus de l'Academie des Sciences Paris, ser D 286, 797-800

[28] DeLucca AJ, Palmgren MS & Ciegler A (1982) *B. thuringiensis* in grain elevator dusts. Can. J. Microbiol. 28, 452-456

[29] DeLucca AJI, Simonson JG & Larson AD (1981) *B. thuringiensis* distribution in soils of the United States. Can. J. Microbiol. 27, 865-870

[30] Donovan WP, Gonzalez JM Jr., Gilbert MP et al (1988) Isolation and characterization of EG2158, a new strain of *B. thuringiensis* toxic to coleopteran larvae, and nucleotide sequence of the toxin gene. Mol. & Gen. Gene. 214, 365-372

[31] Dryden MS (1987) Pathogenic role of *Bacillus cereus* in wound infections in the tropics. J. Roy. Soc. Med. 80, 480-481

[32] Dryden MS & Kramer JM (1987) Toxigenic *Bacillus cereus* as a cause of wound infections in the tropics. J. Infec. 15, 207-212

[33] Dulmage HT (1970) Insecticidal activity of HD-1, a new isolate of *B. thuringiensis* var. *alesti*. J. Invert. Path. 15, 232-239

[34] Dulmage HT & Aizawa K (1982) Distribution of *B. thuringiensis* in nature. In: Kurstak E, ed. Microbial and Viral Pesticides. Marcel Dekker (New York), 209-237.

[35] Flanders SE & Hall IM (1965) Manipulated bacterial epizootics in *Anagasta* populations. J. Invert. Path. 7, 368-377

[36] Fuxa JR (1982) Prevalence of viral infections in populations of fall armyworm, *Spodoptera frugiperda*, in southeastern Louisiana. Environ. Entom. 11, 239-242

[37]Fuxa JR & Tanada Y (1987) Epidemiological concepts applied to insect epizootiology. In: Fuxa JR, Tanada Y, eds. Epizootiology of Insect Diseases. John Wiley & Sons (New York), 3-41

[38]Gingrich RE (1984) Control of the horn fly, *Haematobia irritans*, with *B. thuringiensis*. In: Cheng TC, ed. Comparative pathobiology - Pathogens of Invertebrates. Plenum Press (New York), 47-57

[39] Goldberg LJ & Margalit J (1977) A bacterial spore demonstrating rapid larvicidal activity against *Anopheles sergentii, Uranotaenia unguiculata, Culex univitattus, Aedes aegypti* and *Culex pipiens*. Mosquito News 37, 355-358

[40] Hajek AE, Humber RA, Elkinton JS et al (1990) Allozyme and restriction fragment length polymorphism analyses confirm *Entomophaga maimaiga* responsible for 1989 epizootics in North American gypsy moth populations. Proc. Nat. Acad. Sci. USA 87, 6979-6982

[41] Hansen BM, Damgaard PH, Eilenberg J et al (1996) *B. thuringiensis* - Ecology and environmental effects of its use for microbial pest control. Danish EPA Project No. 316, pp. 125

42. Harper JD (1987) Applied epizootiology: Microbial control of insects. In: Fuxa JR, Tanada Y, eds. Epizootiology of Insect Diseases. John Wiley & Sons (New York), 473-496

[43] Hastowo S, Lay BW & Ohba M (1992) Naturally occurring *B. thuringiensis* in Indonesia. J. Appl. Bact. 73, 108-113

[44] Humber RA (1976) The systematics of the genus *Strongwellsea* (Zygomycetes: Entomophthorales). Mycologia 58, 1042-1060

[45] Ignoffo CM, Hostetter DL & Pinnell RE (1974) Stability of *B. thuringiensis* and *Baculovirus heliothis* on soybean foliage. Environ. Entom. 3, 117-119

[46] Ishii T & Ohba M (1993) Characterization of mosquito-specific *B. thuringiensis* strains coisolated from a soil population. Syst. & Appl. Microbiol. 16, 494-499

[47] Ishikawa Y, Hayashida T & Ikawa A (1964) On the isolation of *B. thuringiensis* from silkworm rearing houses in Aichi prefecture. J. of Sericult. Sci. Jap. 33, 480-483

[48] Ishiwata S (1901) On a kind of severe flacherie (sotto diseases) (In Japanese). Dainihon Sanshi Kaiho 114, 1-5

[49] Itoua-Apoyolo C, Drif L, Vassal JM et al (1995) Isolation of multiple subspecies of *B. thuringiensis* from a population of the European sunflower moth, *Homoeosoma nebulella*. Appl. & Environ. Microbiol. 61, 4343-4347

[50] Jackson SG, Goodbrand RB, Ahmed R et al (1995) *Bacillus cereus* and *B. thuringiensis* isolated in a gastroenteritis outbreak investigation. Lett. Appl. Microbiol. 21, 103-105

[51] Kaelin P, Morel P, Gadani F (1994) Isolation of *B. thuringiensis* from stored tobacco and *Lasioderma serricorne*. Appl. & Environ. Microbiol. 60, 19-25

[52] Khawaled K, Ben-Dov E, Zaritsky A et al (1990) The fate of *B. thuringiensis* var. *israelensis* in *B. thuringiensis* var. *israelensis*-killed pupae of *Aedes aegypti*. J. Invert. Path. 56, 312-316

[53] Khawaled K, Cohen T & Zaritsky A (1992) Digestion of *B. thuringiensis* var. *israelensis* spores by larvae of *Aedes aegypti*. J. Invert. Path. 59, 186-189

[54] Krieg A, Huger AM, Langenbruch GA et al (1983) *B. thuringiensis* var. *tenebrionis*: a new pathotype effective against larvae of Coleoptera. Zeitschrift für Angewandte Entomologie 96, 500-508

[55] Kurstak E (1962) Données sur l'epizootie bacterienne naturelle provoguée par un *Bacillus* du type *B. thuringiensis* sur *Ephestia kuhniella*. Entomophaga Memoire Hors Serié 2, 245-247

[56] Kurstak ES (1964) Le processus de l'infection par *B. thuringiensis* Berl. d'*Ephestia kühniella* Zell. déclenché par le parasitisme de *Nemeritis canescens* Grav. (Ichneumonidae). Comptes Rendus Hebdomadaires des Séances de l'Académie des Sciences 259, 211-212

[57] Landén R, Bryne M & Abdel-Hameed A (1994) Distribution of *B. thuringiensis* strains in southern Sweden. World J. Microbiol. & Biotech. 10, 45-50

[58] Margalit J, Dean D (1985) The story of *B. thuringiensis* var. *israelensis*. J Am Mosq Control Assoc 1, 1-7

[59] Martin PAW (1991) Dynamics of *B. thuringiensis* turnover in soil. The General Meeting of The American Society For Microbiol., 1-6 July, Washigton, USA # 315

[60] Martin PAW & Travers RS (1989) Worldwide abundance and distribution of *B. thuringiensis* isolates. Appl. & Environ. Microbiol. 55, 2437-2442

[61] Mckee LH (1995) Microbial contamination of spices and herbs: A review. Food Sci. & Tech. 28, 1-11

[62] Meadows MP, Ellis DJ & Butt J (1992) Distribution, frequency, and diversity of *B. thuringiensis* in an animal feed mill. Appl. & Environ. Microbiol. 58, 1344-1350

[63] Morris ON (1977) Long term study of the effectiveness of aerial application of *B. thuringiensis*-acephate combination against the spruce budworm, *Choristoneura fumiferana* (Lepidoptera: Tortricidae). Can. Entom. 109, 1239-1248

[64] Morris-Coole C (1995) *B. thuringiensis*: Ecology, the significance of natural genetic modification, and regulation. World J. Microbiol. & Biotech. 11, 471-477

[65] Norris JR (1969) The ecology of serotype 4B of *B. thuringiensis*. J. Appl. Bact. 32, 261-267

[66] Ohba M (1996) *B. thuringiensis* populations naturally occurring on mulberry leaves: a possible source of the populations associated with silkworm-rearing insectaries. J. Appl. Bact. 80, 56-64

[67] Ohba M & Aizawa K (1978) Serological identification of *B. thuringiensis* and related bacteria isolated in Japan. J. Invert. Path. 32, 303-309

[68] Ohba M & Aizawa K (1986) Insect toxicity of *B. thuringiensis* isolated from soils of Japan. J. Invert. Path. 47, 12-20

[69] Ohba M & Aizawa K (1989) Distribution of the four flagellar (H) antigenic subserotypes of *B. thuringiensis* H serotype 3 in Japan. J. of Appl. Bact. 67, 505-509

[70] Ohba M, Aizawa K & Furusawa T (1979) Distribution of *B. thuringiensis* serotypes in Ehime prefecture, Japan. Appl. Entom. & Zool. 14, 340-345

[71] Ohba M, Aizawa K & Sudo SI (1984) Distribution of *B. thuringiensis* in sericultural farms of Fukuoka Prefecture, Japan. Proc. Assoc. Plant Protection of Kyushu 30, 152-155

[72] Ohba M & Aratake Y (1994) Comparative study of the frequency and flagellar serotype flora of *B. thuringiensis* in soils and silkworm-breeding environments. J. Appl. Bact. 76, 203-209

[73] Ono K & Watanabe H (1983) Distribution and serological identification of *B. thuringiensis* isolated in Japan. J. Sericult. Sci. Japan 52, 47-50

[74] Padua LE, Gabriel BP, Aizawa K et al (1982) *B. thuringiensis* isolated from the Philippins. The Philippine Entom. 5, 185-194

[75] Pedersen JC, Damgaard PH, Eilenberg J et al (1995) Dispersal of *B. thuringiensis* var *kurstaki* in an experimental cabbage field. Can. J. Microbiol. 41, 118-125

[76] Pinnock DE, Brand RJ & Milstead JE (1971) The field persistence of *B. thuringiensis* spores. J. Invert. Path. 18, 405-411

[77] Pozsgay M, Fast P & Kaplan H, (1987) The effect of sunlight on the protein crystals from *B. thuringiensis* var. *kurstaki* HD1 and NRD12: a Raman spectroscopic study. J. Invert. Path. 50, 246-253

[78] Pruett CJH, Burges HD & Wyborn CH (1980) Effect of exposure to soil on potency and spore viability of *B. thuringiensis*. J. Invert. Path. 35, 168-174

[79] Purrini K (1977) Über die Verbreitung von *B. thuringiensis* Berl. und einiger Sporozoen-Krankheiten bei vorratsschädlichen Lepidopteren im Gebiet von Kosova, Jugoslawien. Anzeiger für Schädlingskunde, Pflanzenschutz, Umweltschutz 50, 169-173

[80] Pusztai M, Fast P, Gringorten L et al (1991) The mechanism of sunlight-mediated inactivation of *B. thuringiensis* crystals. Biochem. J. 273, 43-47

[81] Rongsen L, Shunying D, Xiaogang L et al (1990) Survey of *B. thuringiensis* and *Bacillus sphaericus* from soils of four provinces of China and their principal biological properties (Abstract, Tables and Figures in English). Acta Microbiologica Sinica 30, 380-388

[82] Smith RA & Couche GA (1991) The phylloplane as a source of *B. thuringiensis* variants. Appl. & Environ. Microbiol. 57, 311-315

[83] Stout JD (1961) A bacterial survey of some New Zealand forest lands, grasslands, and peats. New Zealand J. Agric. Res. 4, 1-30

[84] Talalaev EV (1956) Septicemia of the caterpillars of the Siberian silkworm (In Russian). Mikrobiologiya 25, 99-102

[85] te Giffel MC, Beumer RR, Slaghuis BA et al (1995) Occurrence and characterization of (psychrotrophic) *Bacillus cereus* on farms in the Netherlands. Netherlands Milk & Dairy J. 49, 125-138

[86] Valentino L & Torregrossa MV (1995) Risk of *Bacillus cereus* and *Pseudomonas aeruginosa* nosocomial infections in a burns centre: The microbiological monitoring of water supplies for a preventive strategy. Water Sci. & Tech. 31, 37-40

[87] Vankova J & Purrini K (1979) Natural epizooties caused by bacilli of the species *B. thuringiensis* and *Bacillus cereus*. Zeits. Angew. Entom. 88, 216-221

[88] Watanabe H (1987) The host population. In: Fuxa JR, Tanada Y, eds. Epizootiology of Insect Diseases. John Wiley & Sons (New York), 71-112.

[89] West AW & Burges HD (1982) Ecology of *B. thuringiensis* in soil. 15th Annual Meeting of the Society for Invertebrate Pathology, 6-10 September, Brigton, UK Abst # 319

[90] West AW, Burges HD, White RJ et al (1984) Persistence of *B. thuringiensis* parasporal crystal insecticidal activity in soil. J. Invert. Path. 44, 128-133

[91] Zaritsky A & Khawaled K (1986) Toxicity in carcasses of *B. thuringiensis* var. *israelensis*-killed *Aedes aegypti* larvae against scavenging larvae: implications to bioassay. J. Am. Mosq. Control. Assoc. 2, 555-559

Chapter 1.3

Virulence of *Bacillus thuringiensis*

Bjarne Munk Hansen[1] & Sylvie Salamitou[2]
*Department of Microbial Ecology and Biotechnology, National Environmental
Research Institute, Frederiksborgvej 399, DK-4000 Roskilde, Denmark, and [2]Unité de
Biochimie Microbienne, Institut Pasteur, 25 Rue du Dr Roux, 75724 Paris Cedex 15,
France*

Key Words: *Bacillus thuringiensis*, *Bacillus cereus*, spore germination, vegetative
growth, stationary phase, virulence, enterotoxin, emetic toxin, hydrolytic
enzymes, antibiotics, bacteriocins, appendages, risk

Abstract: *Bacillus thuringiensis* and *B. cereus* are genetically and phenotypically
indistinguishable,   except   for   the   plasmid   encoded   ability   of   *B.
thuringiensis* to produce insecticidal parasporal inclusion bodies (Cry
toxins). Some *B. cereus* are known to cause diarrhoeal and emetic
symptoms in humans, but also somatic infections have been reported. The
pathogenesis of *B. cereus* is based on a number of virulence factors
produced during vegetative- and stationary phases. An increasing number
of reports demonstrate the similarity of virulence factors of *B. thuringiensis*
and *B. cereus*, but there are only few reports on *B. thuringiensis* involved
in human pathogenesis. This lack of reports might either be caused by an
real lack of cases, or might be due to diagnostic procedures which do not
distinguish between *B. cereus* and *B. thuringiensis*. Despite the overall
homogeneity of strains within *B. thuringiensis* and *B. cereus*, high
variations in pathogenic potential are found. Clinical *B. cereus* isolates
tend to show higher toxicity than environmental strains generally. Besides
being active in human pathogenesis, these non specific virulence factors
also seem to be of important for effective insect pathogenesis. To avoid
undesired effects from use of *B. thuringiensis*, conditions promoting
vegetative   growth   in   food   products   should   be   avoided,   e.g.   by
introduction of a pre-harvest spray free period. Further, strains used for
commercial   purposes   should   be   tested   for   vegetative   toxicity   in
comparison with pathogenic and apathogenic *B. cereus*. And asprorogenic
commercialised Bt could be promoted. However, evaluation of overall risks
from using *B. thuringiensis* should be a holistic process done by
comparison of risks and inconvenience from other pest control methods,
related to environmental, social and economic parameters.

J.-F. Charles et al. (eds.),
Entomopathogenic Bacteria: From Laboratory to Field Application, 41-64.
© 2000 *Kluwer Academic Publishers. Printed in the Netherlands.*

1. INTRODUCTION

B. thuringiensis is a Gram-positive facultative aerobic spore-forming bacterium which forms a parasporal crystalline protein inclusion body during stationary phase. Apart from this inclusion, which can be bipyramidal, flat rectangular, spherical or without any defined structure [58], *B. thuringiensis* is indistinguishable from *B. cereus*, which generally is considered as being a relatively mild human pathogen. Indeed, using genetic and phenotypic analysis of chromosomal bound characters, *B. cereus* and *B. thuringiensis* appear as identical species [54].

The insecticidal properties of *B. thuringiensis* are largely attributed to the crystalline inclusions, which contain the δ-endotoxins or Cry proteins. This fact led to the development of biopesticides based on *B. thuringiensis* for the control of various insect pests among the orders Lepidoptera, Diptera and Coleoptera. Further, activities against Nematoda, Protozoa, Trematoda, Acari and Hymenoptera have recently been described [36]. *B. thuringiensis* is now the most widely used biologically produced pest control agent. An annual spread of 13,000 tonnes *B. thuringiensis* spores and δ-endotoxins has been estimated [34]. The extensive use of *B. thuringiensis* and the expectation that more new *B. thuringiensis* types [36] will be launched in future, provoke a need to reconsider several aspects of *B. thuringiensis* such as the taxonomic position of these bacteria and the presence of virulence factors in order to evaluate the risk of using *B. thuringiensis* as an insecticide.

Although many studies have been dedicated to various aspects of the biology of *B. thuringiensis*, only little is clearly understood nowadays. In fact, the ecological habitat of *B. thuringiensis*, soil dweller, saprophyte or pathogen, is still controversial although it has been used for more than sixty years as an insecticide [79], and as such, has been spread upon cities, crops, forests and fresh waters.

2. TAXONOMY AND RELATIONS TO *B. CEREUS*

B. thuringiensis is a member of the *B. cereus* group within *Bacillus*, including *B. cereus*, *B. thuringiensis*, *B. mycoides* and *B. anthracis*, and recently it has been proposed to include psychrotolerant species named *B. weihenstephanensis* [72]. Concomitantly with the increasing use of *B. thuringiensis* as a microbial pest control agent, the relationship to *B. cereus* has attracted increasing focus, especially in relation to risk assessment (see Chapter 7.3).

In the fifties it was acknowledged that *B. thuringiensis* was a *B. cereus* with the specific ability to produce parasporal protein inclusions able to kill certain insect larvae. However, more or less for practical reasons it was suggested that the name *B. thuringiensis* was retained, at least until it could be shown that a *B. cereus* could be derived from a *B. thuringiensis* or vice versa [119].

This was actually shown in the beginning of the eighties, where the genes encoding the δ-endotoxins were assigned to specific plasmids [42, 132]. Transfer of the δ-endotoxin encoding plasmids from *B. thuringiensis* to *B. cereus* resulted in production of crystalline inclusions in the recipient *B. cereus* isolates, making them indistinguishable from *B. thuringiensis* [41, 15]. Further, it has been reported that *B. thuringiensis* can loose the ability to produce δ-endotoxins [9, 33, 43, 120], probably either due to loss of plasmid or due to changes in gene expression, making the *B. thuringiensis* indistinguishable from *B. cereus*. Likewise, the pathogenesis of *B. anthracis* has been assigned to two plasmids pXO1 and pXO2 [95].

The close familiarity between *B. thuringiensis* and *B. cereus* group has been established in a number of phenotypic and genotypic studies, reviewed by Hendriksen & Hansen [48]. Basically, it is unimportant whether *B. thuringiensis* and *B. cereus* formally are considered as one or two distinct species, as long as we are aware of their more or less identical functionality in the vegetative stages of life. This is of great importance in the study of *B. thuringiensis* ecology and in the evaluation of effects from use of *B. thuringiensis* as a microbial pest control agent.

3.   NON-INSECT PATHOGENESIS OF *B. THURINGIENSIS* AND *B. CEREUS*

The specific δ-endotoxins that are produced during the stationary phase and the sporulation process have been described in other sections of this book (see Chapters 2.1 and 2.4). There are a few reports available, which describe involvement of *B. thuringiensis* in human pathogenesis. Upon handling a *B. thuringiensis* product, a farm worker developed a corneal ulcer in an eye, which was accidentally splashed with the *B. thuringiensis* subsp. *kurstaki* product. The eye was recovered after medical treatment [103]. In an epidemiological study, *B. thuringiensis* was isolated from patients with pre-existing medical problems. However, it was not inferred that *B. thuringiensis* was involved in the medical problems [46]. Jackson and co-workers isolated *B. thuringiensis* and *B. cereus* in an investigation of a gastro-enteritis outbreak. The isolated *B.*

thuringiensis showed cytotoxic effects known from *B. cereus* [59]. From infections in burn wounds four non-flagellated *B. thuringiensis* strains were isolated. A Vero cell assay for the production of enterotoxins was negative for the four strains [29]. Hernandez and co-workers isolated a *B. thuringiensis* subsp. *konkukian* strain from a war wound. This strain was able to induce myonecrosis in immunosuppressed mice and death in immunocompetent mice after pulmonary infection [55, 56].

Obviously, the handling and spraying operations have a potential for causing allergy after inhalation of *B. thuringiensis*. In a study by Bernstein and co-workers, farm workers exposed to *B. thuringiensis* were positive in skin-prick tests with spore extracts, and IgG and IgE antibodies specific to vegetative *B. thuringiensis* cells were present in all workers. Nevertheless, none of the workers had any symptoms of allergy and medical problems caused by the *B. thuringiensis* products [20].

B. cereus is known for its involvement in relatively mild cases of gastro-intestinal diseases. However, *B. cereus* has also been reported to cause somatic infection and more serious gastro-intestinal diseases. After eating pasta prepared four days earlier a young boy died due to liver failure caused by the *B. cereus* emetic toxin [80], and two patients died due to pneumonia caused by *B. cereus* [88]. Further, two patients died of meningitis after neurological surgery due to heavily *B. cereus* contaminated hospital linen [14]. Likewise, fourteen patients developed *B. cereus* endophthalmitis after cataract surgery [99] and several immunocompromised patients had soft tissue infections caused by *B. cereus* [13, 86]. But not only humans are affected by *B. cereus*. A number of Rottweiler puppies died after consumption of reconstituted dry milk powder [38], and the infectivity of *Fusobacterium necrophorum* to mice was significantly enhanced by the presence of both *B. cereus* and *B. mycoides* [116].

4.   VIRULENCE OF *B. THURINGIENSIS* AND *B. CEREUS* SPORES

4.1   Germination of spores

B. thuringiensis cells produce a complex array of elements the individual role in toxicity of which remains to be elaborated. Most *B. thuringiensis* products consist of crystals and living spores. Germinated spores growing vegetatively have the ability to produce a number of potential virulence products as described below. The *B. cereus* group is generally considered as being a group of saprophytic bacteria, the spores of which are activated to germinate by heat treatment or by nutritionally rich conditions. The alkaline condition in the midgut of *Manduca sexta* was found to be the principal activator of germination of *B. thuringiensis* subsp. *kurstaki* HD-1 spores [133]. Further, there was strong evidence for a possible relation between the ability to germinate at high pH and the ability to produce protoxins that are precipitated in the parasporal crystalline structure. The data indicate that protoxins in the spore coat are responsible for the ability to activate spore germination under alkaline conditions [19]. Further, activated spores have been shown to be triggered to germinate by specific nutritional compounds, as heat activated *B. thuringiensis* subsp. *galleriae* 26 were specifically triggered to germinate by introduction of L-α-alanine, while D-α-alanine inhibited germination, probably due to receptor blocking [6].

4.2   Pathogenesis of spores

B. cereus and *B. thuringiensis* Cry⁻ spores can themselves be pathogenic to insect larvae. A *B. cereus*, isolated from the cigarette beetle, *Lasioderma serricorne*, was found to be insecticidal to *L. serricorne* larvae [125], and *B. cereus* isolates from *Ephestia elutella* and *Spodoptera frugiperda* showed from 0 to 20% mortality when fed to a number of different insect larvae, relative to *B. thuringiensis* subsp. *kurstaki* [130]. Field collected dead and dying larvae and pupae of spruce budworms, *Choristoneura fumiferana* contained *B. thuringiensis* subsp. *kurstaki* and *B. cereus*. Sixth-instar *C. fumiferana* larvae dipped in a culture of one of the *B. cereus* isolates died within 48 hours [120]. But also *B. thuringiensis* defective in production of crystalline inclusion bodies can be toxic to insects. An overnight culture of a δ-endotoxin negative *B. thuringiensis* subsp. *alesti* Bt75 was fed to *Trichoplusia ni* larvae at a density of 2×10^7 bacteria per ml. 48% of the larvae died compared to a 100% mortality when larvae were fed the *B. thuringiensis* subsp. *alesti* wild-type Bt5 [75].

4.3   Synergism between spores and δ-endotoxins

The function of δ-endotoxins in the toxicity of *B. thuringiensis* mixtures of spores/crystals has been reviewed [105], and is described elsewhere in this book (see Chapter 2.4). Heimpel & Angus reported in 1959 [52] that larvae of different Lepidoptera species responded differently to a *B. thuringiensis* subsp. *thuringiensis* spore-crystal mixture. One group of lepidopteran species was paralysed, showed an increase in blood pH and died within a few hours after administration due to the activity of the toxin. Another group of insect species did not show general paralysis and the larvae died first after 2-4 days. For the third group of insect species, spore germination was shown to be an absolute requirement for insecticidal activity. These observations are in correspondence with other reports on the role of the spore in the insecticidal process. Several studies report that *B. thuringiensis* multiply in the cadavers of the insects, sporulate and produce crystals. As an example, *B. thuringiensis* subsp. *israelensis* was in laboratory experiments shown to reach a spore count of more than 10^5 per cadaver of *Aedes aegypti* and *Anopheles albimanus* [7], and even at doses sublethal for larvae, vegetative growth, sporulation and crystal production could occur in dead pupae carcasses, reaching levels of more than 10^6 spores per pupa [63].

Another function of living spores together with δ-endotoxins might be to repress development of insect resistance to δ-endotoxins. In an investigation of the development of *Spodoptera exigua* resistance to pure Cry1Ca toxin, part of the initial sensitivity was recovered when *B. thuringiensis* subsp. *kurstaki* HD-1 spores and Cry1Ca toxins were fed to Cry1Ca resistant *S. exigua* larvae [89]. However, *Plutella xylostella* larvae resistant to Cry1Aa, Cry1Ab and Cry1Ac showed no increased sensitivity to the three Cry1A toxins when *B. thuringiensis* subsp. *kurstaki* HD-1 spores were added in the diet [122].

So the role of the *B. thuringiensis* spore fed to larvae must be concluded to be highly variable, probably depending on type, combination and concentration of Cry toxins, type and density of bacteria, insect species and general condition of the larvae and other micro organisms present in the gut.

5.   VIRULENCE FACTORS OF *B. THURINGIENSIS* AND *B. CEREUS*

The focus in this paragraph will be on non-specific virulence factors produced during vegetative growth and stationary phase of the bacteria. These virulence factors do not show species specificity, as is the case with the δ-endotoxins. Unfortunately, virulence factors produced during the vegetative and stationary phases of *B. thuringiensis* have only been of limited interest in the past. However, the close relationship between *B. cereus* and *B. thuringiensis* and the similar characteristics in the vegetative stages allow us to extrapolate from the experience obtained from studies of the vegetative stages of *B. cereus*. The most prominent characteristics of *B. cereus* are the emetic and diarrhoeal symptoms that can occur in humans after ingestion of food containing at least from 10^4 to 10^8 bacteria per gram. Although *B. thuringiensis* and especially *B. cereus* frequently are found in food, the number of reported cases of food poisoning are limited, although the number of cases tend to increase [121, 134]. The virulence factors comprise a number of toxins, enterotoxins and hydrolytic enzymes, which probably interact in a complex manner. Table 1 lists virulence factors, which are dealt with in this paragraph. The precise distinction between these factors is somehow obscure. Some enterotoxins might be hydrolytic enzymes, and some hydrolytic enzymes might be enterotoxins.

5.1   The emetic toxin

The ability to produce the emetic toxin is limited to a few *B. cereus* serotypes, of which serotype 1 is the most prominent type [67, 90]. The emetic toxin is a cyclic dodecadepsipeptide named cereulide [4] which is highly stable and is resistant to heat, extreme pH values, trypsin and pronase treatments [128]. Recently, it has been shown that its effect is caused by its K^+ ionophore characteristics [87]. The emetic toxin causes vomiting, nausea and malaise, and is primarily produced by vegetative *B. cereus* in cooked rice and pasta, which has been improperly cooked. Toxicity has been observed with bacterial densities from 10^5 to 10^8 *B. cereus* cells per gram food and the symptoms appear within a few hours and last about one day [44, 67].

Table 1. Virulence factors known from *B. cereus* and *B. thuringiensis*

Enterotoxins	*Hydrolytic enzymes*
haemolytic enterotoxin HBL	phoshatidylcholine-preferring phospholipase C
non-haemolytic enterotoxin NHE	phosphatidylinositol-specific phospholipase C
enterotoxin *bceT*	sphingomyelinase
enterotoxin *entFM*	haemolysin II
	haemolysin III
	neutral and alkaline proteinase
Other virulence factors	collagenolytic proteinase
	immune inhibitor
β-exotoxin	vegetative insecticidal protein vip3A
thuricin	chitinase
zwittermicin A	
kanosamine	*Emetic toxin*
antibiotic resistance	
surface structures	cereulide

5.2   **Enterotoxins**

Enterotoxin is a common name for several toxins involved in diarrhoea, but enterotoxins can also be involved in somatic infections. Using a variety of methods, genes for enterotoxins and/or enterotoxin production have been detected in the majority of *B. thuringiensis* and *B. cereus* analysed until now. Six of seven *B. thuringiensis* isolated from food were positive in a Vero cell assay indicating enterotoxin production. All six positive isolates had the *kurstaki* serotype [28]. Most of a collection of Swedish environmental *B. thuringiensis* was positive for enterotoxin production analysed by a commercial immunological kit (BCET-RPLA, Oxoid, UK) [1]. Using another commercial kit (Tecra Diagnostics, Roseville, Australia) several commercial *B. thuringiensis* products growing vegetatively were found to produce enterotoxins [26]. Likewise, due to germination and vegetative growth two commercial *B. thuringiensis* subsp. *kurstaki* products were shown to be cytotoxic to cultured insect cells [124]. The *B. cereus* causing diarrhoea are found in most kinds of food like meat dishes, soups, milk, vegetables and sauces. Infections result in abdominal pain and diarrhoea appearing 8-12 hours after ingestion and lasting 1-2 days. The infective dose varies from 10^4 to 10^7 cells per gram food, indicating high variation in *B. cereus* toxicity. Contrary to the emetic toxin, which is produced by bacteria in the food [66, 91], the enterotoxins are primarily produced by the bacteria during multiplication in the small intestine [44, 67]. The enterotoxins are not as defined as the emetic toxins. Four types of toxins are being considered enterotoxins, but other products like hydrolytic enzymes might add to the total enterotoxic activity.

A haemolytic enterotoxin (HBL) consisting of the three components B, L_1 and L_2 has been purified from a *B. cereus* and characterised. All three components are involved in the enterotoxic activity. The tripartite toxin has besides its haemolytic activity also dermonecrotic and vascular permeability activities. Further, it causes fluid accumulation in ligated rabbit ileal loop assay [16, 17, 18]. The genes for HBL components have been sequenced [53, 100, 138], and PCR analysis showed that all three genes are present in *B. thuringiensis* subsp. *kurstaki* HD-1, while only the L_1 and L_2 genes were detected in *B. thuringiensis* subsp. *israelensis* HD-567 [48].

Recently, a tripartite non-haemolytic enterotoxin (NHE) complex has been described from a *B. cereus*, which was involved in food poisoning. The NHE complex is cytotoxic to Vero cells [78] and the genes encoding the three components (A, B and C) have been sequenced [45].

A single enterotoxin gene, *bceT*, has been cloned and sequenced from *B. cereus* B-4ac. The gene was expressed in *E. coli*, and the gene product exhibited cytotoxicity to Vero cells and was found positive in a vascular permeability assay [3]. PCR with primers deduced from the *bceT* gene gave PCR products of the expected size from *B. thuringiensis* subsp. *kurstaki* HD-1 and *B. thuringiensis* subsp. *israelensis* HD-567 [48].

Another single enterotoxin gene, *entFM*, was cloned and sequenced from *B. cereus* FM1, *B. thuringiensis* subsp. *sotto* and *B. thuringiensis* subsp. *israelensis*. Nucleotide sequences from the two *B. thuringiensis* revealed that the sequence was very similar to that of *B. cereus* FM1. Further PCR analysis of a number of *B. thuringiensis* strains showed the presence of homologous sequences in these strains. Analysis of the toxicity of the product of *entFM* has not been performed [10]

5.3   Hydrolytic enzymes

5.3.1   Phospholipase

B. thuringiensis and *B. cereus* produce a number of hydrolytic enzymes, of which the phospholipase C has been used for the identification of *B. cereus* like isolates [27, 128]. Phospholipases play an important cytotoxic role in the establishment of many infectious bacteria as they show varying ability to hydrolyse membrane phospholipids. The phosphatidylinositol-specific phospholipase C enzymes show a pronounced substrate specificity, while the phosphatidylcholine-preferring phospholipase C enzymes often are able to hydrolyse some other substrates [126, 127]. Phosphatidylinositol-specific phospholipase C has been cloned and sequenced from *B. thuringiensis* ATCC 10792 [71] and from *B. cereus* ATCC 6464 [69]. A phosphatidylcholin-preferring phospholipase C has been cloned from *B. cereus* [39, 40, 60]. With PCR primers deduced from the phosphatidylinositol-specific phospholipase C [71] the gene was shown to be present in all investigated *B. cereus* and *B. thuringiensis* strains [27] and with primers deduced from the phosphatidylcholine-preferring phospholipase C sequence [40] the gene was identified in *B. thuringiensis* subsp *kurstaki* HD-1 and *B. thuringiensis* subsp. *israelensis* HD-567 [48]. Located next to the phosphatidylcholine-preferring phospholipase C of *B. cereus* is a gene encoding a sphingomyelinase [39, 40], which is able to hydrolyse sphingomyelin-rich erythrocytes. The phosphatidylcholine-preferring phospholipase and the sphingomyelinase constitutes the haemolytic   cereolysin   AB   complex.   The   gene   encoding sphingomyelinase has been detected in both *B. thuringiensis* subsp. *kurstaki* HD-1 and *B. thuringiensis* subsp. *israelensis* HD-567 [48].

5.3.2   Hemolysin

Besides the haemolytic BL enterotoxin and the cereolysin AB complex (phosphatidylcholine-preferring phospholipase and sphingomyelinase), two other haemolytic factors haemolysin II and haemolysin III have been characterised and cloned [12, 113]. DNA hybridisation showed that only four of thirteen *B. cereus* had the haemolysin II gene, while the gene was present in thirteen of fourteen *B. thuringiensis* [22]. PCR analysis showed the presence of the haemolysin III gene in both *B. thuringiensis* subsp. *kurstaki* HD-1 and *B. thuringiensis* subsp. *israelensis* HD-567 [48].

5.3.3   Proteinase

Neutral and alkaline proteinases have been detected and described in *B. thuringiensis* and *B. cereus* [11, 25, 57, 111, 138]. Proteinases have not been shown to add to the pathogenicity of *B. cereus*, but as the neutral proteinases are able to hydrolyse haemoglobin and albumin [111] a role in the general pathogenesis of *B. thuringiensis* and *B. cereus* can not be excluded. Likewise, the collagenolytic proteinase described from *B. cereus* is an obvious virulence candidate for somatic infections of all collagen-containing organisms [81, 108]. Another type of virulent proteinase is the neutral metalloproteinase immune inhibitor, which was shown specifically to degrade antibacterial proteins from the silkmoth *Hyalophora cecropia* and to be toxic to several insect species [32, 76]. Among *B. thuringiensis* collected from soil in Sweden, 65% of the isolates secreted immune inhibitor A [70].

5.3.4   Vip3A

A new *B. thuringiensis* vegetative insecticidal protein Vip3A has recently been described. The protein is synthesised during vegetative mid-log phase as well as during stationary phase. The toxin shows activity against a broad spectrum of susceptible lepidopteran insect larvae and data indicate that a specific binding to midgut cells is followed by lysis of midgut epithelial cells [35, 135]. PCR analyses indicate that the gene encoding the Vip3A protein is found in around 30% of *B. thuringiensis* [96].

5.3.5   Chitinase

Chitinases are probably important virulence factors for the insecticidal activity of *B. thuringiensis* [68]. Addition of chitinase to a *B. thuringiensis* formulation enhanced mortality of *C. fumiferana* larvae [114]. Although *B. thuringiensis* subsp. *aizawai* itself produce chitinase, it was possible to show a slightly enhanced mortality against *Spodoptera exigua* larvae when a *Bacillus licheniformis* chitinase was introduced and expressed constitutively in a *B. thuringiensis* subsp. *aizawai* [104, 123].

5.4   Other virulence factors

One of the first recognised broad spectrum vegetative virulence factors produced by *B. thuringiensis* was β-exotoxin, also called thermostable exotoxin, or thuringiensin A and B. The two β-exotoxins are ATP analogues and the B type is converted to the A type by alkaline hydrolyses [61, 117]. Due to the similarity to ATP, β-exotoxins act as general inhibitors of DNA dependent RNA polymerases [106]. When sublethal amounts of β-exotoxin were added to *T. ni* larvae, malformation of adult insects occurred [62]. β-exotoxin is produced in several *B. thuringiensis* isolates, one of which is the *B. thuringiensis* subsp. *thuringiensis* serotype H-1, and the genes encoding the production of β-exotoxin have been assigned to a plasmid [74]. β-exotoxin was produced by 58% of the isolates in a collection of environmental *B. thuringiensis* [92]. *B. thuringiensis* subsp. *thuringiensis* has in Tanzania and Kenya been used as a control agent in compost and latrines, where it grows vegetatively and produces β-exotoxin which effectively controls fly larvae [23].

Bacteriocins, antibiotics and antibiotic resistance genes must also be considered as virulence factors. *B. thuringiensis* subsp. *thuringiensis* HD-2 showed antibacterial activity against most other tested *B. thuringiensis* strains and other Gram-positive species, but not against Gram-negative species. The active bacteriocin, thuricin, was associated with a phospholipase A activity [37]. Vegetative *B. cereus* UW85 suppress several fungal plant diseases due to the production and excretion of two antibiotics, zwittermicin A and kanosamine. About 10% *B. cereus* from soils are able to produce zwittermicin A, and also zwittermicin A producing *B. thuringiensis* isolates have been described, among which is *B. thuringiensis* subsp. *kurstaki* HD-1, which is used commercially for control of lepidopteran pests [109, 112, 118]. Strains that produce zwittermicin A, also have genes encoding resistance to zwittermicin A [94]. Antibiotic resistance genes can also be considered virulence factors, as they are important factors in the interactions/competitions with other microorganisms. The best known antibiotic resistance genes known from both *B. cereus* and *B. thuringiensis* are the β-lactamases which by opening the β-lactam ring confer resistance to penicillins [24, 136]. But not only penicillin resistance has been found in the *B. cereus* group. *B. cereus* strains isolated from pharmaceuticals were tested for resistance to eighteen antibiotics, and resistance to fourteen of these were found among the isolates [30].

It has recently been suggested that hydrophobicity, appendages and the ability of *B. cereus* spores to adhere to epithelial cells should be considered an additional virulence factor, as an extremely virulent *B. cereus* was found to possess appendages. Analysis of the virulent *B. cereus* in comparison to other *B. cereus* isolates showed that the hydrophobicity of the spore coat correlated with the ability of *B. cereus* to bind to cultivated human colon cancer cells [8]. Filamentous appendages have been described in detail from *B. cereus* [65], and spores of several *B. thuringiensis* strains with appendages were able to induce agglutination of red blood cells, which is a general indicator of adherence of bacteria to blood cells [115]. The presence of flagella on vegetative *B. thuringiensis* has also been suggested as a factor of importance for the binding of *B. thuringiensis* to cells, but the data are ambiguous [51, 137]. Further, an S-layer lattice has been reported on the surfaces of some strains of *B. cereus* and *B. thuringiensis* [64, 77]. One function of the S-layer could be to hide immunogenic components in order to elude the immune system of the host, and another function could be adherence to eucaryotic cell matrix proteins. Two clinical *B. cereus* isolates with S-layer showed strong binding to collagen, laminin and fibronectin, which did not bind to two reference strains without an S-layer [64].

6.   EXPRESSION OF VIRULENCE FACTORS

Although strains of *B. cereus* and *B. thuringiensis* are very similar, high variation in pathogenic potency has been observed in both, either due to genetic variation and/or due to differences in regulation of the expression of the involved genes. Several reports agree in that the level of pathogenic potency is variable. Clinical (diarrhoea) and environmental *B. cereus* isolates were analysed for toxicity by feeding mice with bacterial cultures or by intra dermal injection of culture supernatant in adult rabbits (vascular permeability reaction, VPR) followed by intravenous injection of Evans Blue dye after three hours. The clinical *B. cereus* isolates caused a higher incidence of diarrhoea than the environmental isolates. Likewise, the coloured area around the spot of injection of culture supernatant (VPR assay) was significantly higher for the clinical isolates [50]. Recently monoclonal antibodies were produced against the components of the haemolysin BL complex and sphingomyelinase from *B. cereus*. Monoclonal immunological analysis and Vero cell cytotoxicity analysis with a range of *B. cereus* culture supernatants showed first of all variations in toxin level among isolates and second, that isolates with low toxin level in the immunological analysis were also low in cytotoxicity [31]. The same type of observations have been seen with *B. thuringiensis* strains. An immunological analysis of environmental *B. thuringiensis* isolates showed that 43% of the isolates had the same level of enterotoxins as enterotoxic *B. cereus* strains [110]. Besides the natural genetic diversity described above, pleiotropic avirulent *B. thuringiensis* mutants have been described and a pleiotropic transcriptional activator (PlcR) of a phosphatidylinositol-specific phospholipase C was cloned [73]. Besides activation of its own transcription at the onset of the stationary phase, PlcR also activated transcription of phospholipases and enterotoxins as described in chapter 2.5 [2, 73, 138]. A *B. thuringiensis* mutant with a defective PlcR showed 7% mortality when fed to *Galleria mellonella* insect larvae, while the parent *B. thuringiensis* showed 70% mortality under otherwise identical conditions [102].

Further, investigations have shown that some environmental factors can influence the level of virulence factors in *B. cereus* cultures. Using defined media, an emetic *B. cereus* strain was inhibited in its production of emetic toxin by high levels of leucine, isoleucine and glutamic acid, while production was stimulated by the addition of glucose. Further, the level of emetic toxin produced in skimmed milk was 320 times higher than in nutrient broth [5]. Analogous to this, an investigation of one hundred European infant milk products showed that maltodextrin in four of the products induced *B. cereus* enterotoxin production. Again, it was shown that glucose stimulated enterotoxin production [98].

7. *B. THURINGIENSIS* IN THE ENVIRONMENT

Is *B. thuringiensis* an insect pathogenic bacterium, or is it a saprophytic bacterium often found in the digestive tract of insects or other animals, and only occasionally turns into a pathogen? Unfortunately, very little is known about natural function of *B. thuringiensis* and *B. cereus* in the environment. Epizootics of *B. thuringiensis* are almost never reported in the environment, and soil environments are generally not containing sufficient nutrients to support vegetative growth [47]. There exist however, a few recent reports dealing with the ecology of *B. thuringiensis* and *B. cereus,* which elucidate these questions. The gut wall of different arthropods has for 150 years been known to contain segmented filaments with refractile inclusions. When further investigated, these filaments showed up to be composed of strings of *B. cereus* with vegetative cells attached to the intestinal epithelium and sporulated cells at the distal end [83]. To improve the efficiency of *B. thuringiensis* subsp. *israelensis* to mosquito larvae, the bacteria were bioencapsulated in the protozoan *Tetrahymena pyriformis*, where the toxin remains stable. Excreted protozoan food vacuoles containing *B. thuringiensis* were able to support spore germination and vegetative growth and finally re-sporulation and crystal formation. [82]. During a long term field trial with a rifampicin resistant *B. thuringiensis* subsp. *kurstaki*, earthworm gut was investigated for the content of spores and vegetative cells. In the soil, all *B. thuringiensis* remained as spores, while 55% of the bacteria found in the gut system of the earthworm *Aporrectodea caliginosa* were vegetative. It was not possible to calculate whether these vegetative bacteria were multiplying in the gut, but it is an important piece of the puzzle describing the ecology of *B. thuringiensis* [49]. Further, several strains of *B. thuringiensis* and *B. cereus* were isolated from *Ixodes scapularis* (the vector of the Lyme disease) [84] and dead adult cabbage root flies, *Delia radicum*, which were infected by fungi also contained sporulated *B. thuringiensis* subsp. *aizawai* [33]. The last description might be a good illustration of what actually might be going on in the environment: An insect or animal is somehow diseased, and fungi and bacteria multiply in the carcass. One of the bacterial species could be a *B. thuringiensis*, which sporulate and produce crystalline inclusions. If by chance the spores and crystals are ingested by a target organism, a second turn of multiplication can occur. So, maybe *B. thuringiensis* and *B. cereus* actually should be considered enzootic organisms, being ever-present, with limited multiplication and which occasionally become epizootic at a limited scale when δ-endotoxin encoding plasmids are present.

8.   CONSEQUENCES FOR APPLICATION OF *B. THURINGIENSIS*

The short list of cases where *B. thuringiensis* has been involved in human pathogenesis might either be explained by the absence of such cases or it might be due to the procedures used for *B. cereus* diagnosis, which does not involve microscopy for detection of parasporal protein inclusions, and thus does not allow differentiation of *B. cereus* from *B. thuringiensis*.

Looking back on all the toxins and enzymes that are produced by *B. thuringiensis*, it is obvious to ask whether it is safe to spread 13,000 tonnes annually of a bacterium with the potential to cause harm in humans and other mammals? A recent investigation on *B. thuringiensis*, of the same subspecies as used commercially, have tried to enlighten the question. Intraperitoneal injection in mice of spores, crystals and fermentation solids of *B. thuringiensis* subspecies *aizawai*, *israelensis*, *kurstaki* and tenebrionis showed that some strains resulted in mouse mortality, while others did not affect the mice at all [85]. Using cultured insect cells and tetrazolium reduction as a measure for estimation of the fraction of intact mitochondria, it was possible to quantificate the cytotoxic damage of *B. thuringiensis* subsp. *kurstaki* spore-crystal mixtures. Unless blocked by antibiotics, spore germination and vegetative growth was the major cause of cytotoxicity of two commercial *B. thuringiensis* subsp. *kurstaki* products [124]. So, as long as the *B. thuringiensis* remain in the spore stage, the cytotoxicity of the products are limited. Spores and crystals of *B. thuringiensis* subspecies *israelensis*, tenebrionis and *kurstaki*, which are used commercially, were injected subcutaneously in rats (1×10^6 per animal). Two weeks later, the rats were killed, and no *B. thuringiensis* were found in any organs. Further, rats were force fed on a daily basis with the three subspecies for three weeks (total dose was 1×10^{12} spores). No special effect of the *B. thuringiensis* feeding was detected in any of the rats, although all three *B. thuringiensis* were able to produce enterotoxins, determined by two commercial kits for enterotoxin detection. No *B. thuringiensis* were found in any organ of the rats three weeks after feeding had stopped, and only spores were found with colon content. So, at least these *B. thuringiensis* seem to be safe for rats. [21]. In addition, *B. cereus* can have beneficial effect on animals and humans. *B. cereus* isolates have been used as probiotics for domestic animals [97, 101, 107, 129], and a *B. cereus* isolate (IP 5832) has since 1949 been used as a probiotic for human intestinal disease in some European countries [93, 131].

To imitate a real situation of use, spinach plants in glasshouse were sprayed with a commercially recommended dose. After 24 hours the leaves were harvested, washed and the number of remaining spores was estimated to be about 10^6 spores per gram, a dose which is within the range of doses having potential for causing human pathogenesis [21]. However, it needs to be considered whether use of *B. thuringiensis* as pest control agent contributes significantly to the overall toxicity originating from naturally occurring *B. cereus* and *B. thuringiensis* populations.

Anyway, the risk of using *B. thuringiensis* should always be compared to conventional farming practise, where chemical pesticides often are used with negative impacts on the environment in general. Further, by introduction of a sprayfree pre-harvest period with *B. thuringiensis* products, the risk would be substantially lowered, as investigations on persistence in the phylloplane show that *B. thuringiensis* numbers are quickly reduced [47]. Further, before new subspecies are introduced commercially, at least comparative toxicity tests should be performed with several methods for the new subspecies in comparison with known pathogenic and apathogenic *B. cereus* and asporogenous *Bt*-sprays could be promoted.

ACKNOWLEDGEMENTS

We thank Niels Bohse Hendriksen and Didier Lereclus for their critical reading of the manuscript.

REFERENCES

[1] Abdel-Hameed A & Landén R (1994) Studies on *Bacillus thuringiensis* strains isolated from Swedish soils: insect toxicity and production of *B. cereus*-diarrhoeal-type enterotoxin. World J. Microbiol. Biotechnol. 10, 406-409

[2] Agaisse H, Gominet M, Økstad OA, Kolstø AB & Lereclus D (1999) PlcR is a pleiotropic regulator of extracellular virulence factor gene expression in *Bacillus thuringiensis*. Molec. Microbiol. 32, 1043-1053

[3] Agata N, Ohta M, Arakawa Y & Mori M (1995) The *bceT* gene of *Bacillus cereus* encodes an enterotoxic protein. Microbiology 141, 983-988

[4] Agata N, Ohta M, Mori M & Isobe M (1995) A novel dodecadepsipeptide, cereulide, is an emetic toxin of *Bacillus cereus*. FEMS Microbiol. Lett. 129, 17-19

[5] Agata N, Ohta M, Mori M & Shibayama K (1999) Growth condition of and emetic toxin production by *Bacillus cereus* in a defined medium with amino acids. Microbiol. Immunol. 43, 15-18

[6] Alekseev AN, Karabanova LN & Shevtsov VV (1982) Initiators and inhibitors of *Bacillus thuringiensis* spore germination. All-Union Scientific-Research Institute of Applied Microbiology. Translated from Mikrobiologiya 51, 780-783. Plenum Publishing Corporation 1983, pp 628-630

[7] Aly C, Mulla Ms & Federici BA (1985) Sporulation and toxin production by *Bacillus thuringiensis* var. *israelensis* in cadavers of mosquito larvae (Diptera: Culicidae). J. Invertebr. Pathol. 46, 251-258

[8] Andersson A, Granum PE & Rönner U (1998) The adhesion of *Bacillus cereus* spores to epithelial cells might be an additional virulence mechanism. Int. J. Food Microbiol. 39, 93-99

[9] Aronson AI (1993) The two faces of *Bacillus thuringiensis*: insecticidal proteins and post-exponential survival. Molec. Microbiol. 7, 489-496

[10] Asano SI, Nukumizu Y, Bando H, Iizuka T & Yamamoto T (1997) Cloning of novel enterotoxin genes from *Bacillus thuringiensis* and *Bacillus thuringiensis*. Appl. Environ. Microbiol. 63, 1054-1057

[11] Bach HJ, Erampalli D, Leung KT *et al.* (1999) Specific detection of the gene for the extracellular neutral protease of *Bacillus cereus* by PCR and blot hybridisations. Appl. Environ. Microbiol. 65, 3226-3228

[12] Baida GE & Kuzmin NP (1995) Cloning and primary structure of a new hemolysin gene from *Bacillus cereus*. Biochem. Biophys. Acta 1264, 151-154

[13] Banerjee C, Bustamante CI, Wharton R, Talley E & Wade JC (1988) *Bacillus* infections in patients with cancer. Arch. Intern. Med. 148, 769-1774

[14] Barrie D, Wilson JA, Hoffman PN & Kramer JM (1992) *Bacillus cereus* meningitis in two neurosurgical patients: An investigation into the source of the organism. J. Infec. 25, 291-297

[15] Battisti L, Green BD & Thorne CB (1985). Mating system for transfer of plasmids among *Bacillus anthracis*, *Bacillus cereus*, and *Bacillus thuringiensis*. Journal of Bacteriology 162, 543-550

[16] Beecher DJ & Wong ACL (1994) Improved purification and characterisation of Hemolysin BL, a hemolytic dermonecrotic vascular permeability factor from *Bacillus cereus*. Infect. Immun. 62, 980-986

[17] Beecher DJ, Shoeni JL & Wong ACL (1995) Enterotoxin activity of hemolysin BL from *Bacillus cereus*. Infect. Immun. 63, 4423-4428

[18] Beecher DJ & Wong ACL (1997) Tripartite hemolysin BL from *Bacillus cereus*. Hemolytic analysis of component interaction and model for its characteristic paradoxial zone phenomenon. J. Biol. Chem. 272, 233-239

[19] Benoit TG, Newnam KA & Wilson GR (1995) Correlation between alkaline activation of *Bacillus thuringiensis* var. *kurstaki* spores and crystal production. Curr. Microbiol. 31, 301.303

[20] Bernstein IL, Bernstein JA, Miller M, Tierzieva M, Bernstein DI, Lummus Z, Selgrade MJK, Doerfler DL & Seligy VL (1999) Immune responses in farm workers after exposure to *Bacillus thuringiensis* pesticides. Environ. Health Perspect. 107, 575-582

[21] Bishop AH, Johnson C & Perani M (1999) The safety of *Bacillus thuringiensis* to mammals by oral and subcutaneous dosage. W. J. Microbiol. Biotechnol. 15, 375-380

[22] Budarina ZI, Sinev MA, Mayorov SG, Tomashevski AY, Shmelev IV & Kuzmin NP (1994) Hemolysin II is more characteristic of *Bacillus thuringiensis* than *Bacillus cereus*. Arch. Microbiol. 161, 252-257

[23] Carlberg G, Tikkanen L & Abdel-Hameed AH (1995) Safety testing of *Bacillus thuringiensis* preparations, including thuringiensis, using the *Salmonella* assay. J. Invertebr. Pathol. 66, 68-71

[24] Cid H, Carrillo O, Bunster M, Martinez J & Vargas V (1988) The relationship between the structures of four β-lactamases obtained from *Bacillus cereus*. Arch. Biol. Med. Exp. 21, 101-107

[25] Donovan WP, Tan Y & Slaney AC (1997) Cloning of the *nprA* gene for neutral protease A of *Bacillus thuringiensis* and effect of in vivo deletion of *nprA* on insecticidal crystal protein. Appl. Environ. Microbiol. 63, 2311-2317

[26] Damgaard PH (1995) Diarrhoeal enterotoxin production by strains of *Bacillus thuringiensis* isolated from commercial *Bacillus thuringiensis*-based insecticides. FEMS Immunol. Med. Microbiol. 12, 245-250

[27] Damgaard PH, Jacobsen CS & Sørensen J (1996) Development and application of a primerset for specific detection of *Bacillus thuringiensis* and *Bacillus cereus* in soil using magnetic capture-hybridisation and PCR amplification. System. Appl. Microbiol. 19, 436-441

[28] Damgaard PH, Larsen HD, Hansen BM, Bresciani J & Jørgensen K (1996) Enterotoxin-producing strains of *Bacillus thuringiensis* isolated from food. Lett. Appl. Microbiol. 23, 146-150

[29] Damgaard PH, Granum PE, Bresciani J, Torregrossa MV, Eilenberg J & Valentino L (1997) Characterization of *Bacillus thuringiensis* isolated from infections in burn wounds. FEMS Immunol. Med. Microbiol. 18, 47-53

[30] de la Rosa MC, Mosso MA, García ML & Plaza C. (1993) Resistance to the antimicrobial agents of bacteria isolated from non-sterile pharmaceuticals. J. Appl. Bacteriol. 74, 570-577

[31] Dietrich R, Fella C, Strich S & Märtlbauer E. (1999) Production and characterization of monoclonal antibodies against the hemolysin BL enterotoxin complex produced by *Bacillus cereus*. Appl. Environ. Microbiol. 65, 4470-4474

[32] Edlund T, Sidén I & Boman HG (1976) Evidence for two immune inhibitors from *Bacillus thuringiensis* interfering with the humoral defense system of saturniid pupae. Infect. Immun. 14, 934-941

[33] Eilenberg J, Damgaard PH, Hansen BM, Pedersen JC, Bresciani J & Larsson R (*in press*) Natural co-prevalence of *Strongwellsea castrans*, *Cystosporogenes deliaradicae* and *Bacillus thuringiensis* in the host, *Delia radicum*. J. Invertebr. Pathol.

[34] Environmental Health Criteria, *Bacillus thuringiensis* (In press) International Programme on Chemical Safety, WHO, Geneva, Switzerland.

[35] Estruch JJ, Warren GW, Mullins MA *et al.* (1996) Vip3A, a novel *Bacillus thuringiensis* vegetative insecticidal protein with a wide spectrum of activities against lepidopteran insects. Proc. Natl. Acad. Sci. USA 93, 5389-5394

[36] Feitelson JS (1993) The *Bacillus thuringiensis* family tree. *In* (Kim L, ed.) Advanced engineered pesticides. Marcel Dekker, New York, pp 63-71

[37] Favret ME & Yousten AA (1989) Thuricin: The bacteriocin produced by *Bacillus thuringiensis*. J. Invertebr. Pathol. 53, 206-216

[38] Gareis M & Walz A (1994) Vergiftungsfälle bei rottweilerwelpen durch milkpulver, kontaminiert mit *Bacillus cereus*. Tierärzl. Umschau 49, 319-322

[39] Gavrilenko IV, Baida GE, Karpov AP & Kuzmin NP (1993) Nucleotide sequences of the genes of the phospholipase C and the sphingomyelinase of *Bacillus cereus* VKM-B164. Translated from Bioorganicheskaya Khimiya 19 (1) 133-138. Plenum Press Corporation

[40] Gilmore MS, Cruz-Rodz AL, Leimeister-Wächter M, Kreft J & Goebel W (1989) A *Bacillus cereus* cytolytic determinant, cereolysin AB, which comprises the phospholipase C and sphingomyelinase genes: Nucleotide sequence and genetic linkage. J. Bacteriol. 171, 744-753

[41] González JMJ, Brown BJ & Carlton BC (1982) Transfer of *Bacillus thuringiensis* plasmids coding for delta-endotoxin among strains of *Bacillus thuringiensis* and *Bacillus cereus* insecticidal toxin. Proceedings of the National Academy of Sciences of the United States of America 69, 6951-6955

[42] González JMJ, Dulmage HT & Carlton BC (1981) Correlation between specific plasmids and δ-endotoxin production in *Bacillus thuringiensis*. Plasmid 5, 351-365

[43] Gordon, RE (1977) Some taxonomic observations on the genus *Bacillus*. *In*: (Briggs JB ed.) Biological regulations of vectors: The saprophytic and aerobic bacteria and fungi. US Dept. of Health, Education and Welfare, Publ. NIH 77-1180, Washington, pp. 67-82.

[44] Granum PE (1997) *Bacillus cereus*. *In* (Doyle MP, Beuchat LR & Montville TJ, eds.) Food microbiology, fundamentals and frontiers. ASM Press, Washington DC, pp 327-336

[45] Granum PE, O'Sullivan K & Lund T (1999) The sequence of the non-hemolytic enterotoxin operon from *Bacillus cereus* FEMS Microbiol. Lett. 177, 225-229

[46] Green M, Heumann M, Sokolow R, Foster LR, Bryant R & Skeels M (1990) Public health implications of the microbial pesticide *Bacillus thuringiensis*: An epidemiological study, Oregon 1985-86. Am. J. Publ. Health 80, 848-852

[47] Hansen BM, Damgaard PH, Eilenberg J & Pedersen JC (1996) *Bacillus thuringiensis*. Ecology and environmental effects of its use for microbial pest control. Danish Environmental Protection Agency, Environmental Project No. 316, 126 pages

[48] Hansen BM & Hendriksen NB (1998) *Bacillus thuringiensis* and *B. cereus* toxins. IOBC Bulletin 21, 221-224

[49] Hansen BM & Hendriksen NB (1999) Ecological aspects of the survival in soil of spray released *Bacillus thuringiensis* subsp. *kurstaki*. *In* Evaluating indirect ecological effects of biological control, Global IOBC International Symposium, Montpellier, France, October 1999. IOBC wprs Bulletin 22, 21

[50] Hassan G & Nabbut N (1996) Prevalence and characterization of *Bacillus cereus* isolates from clinical and natural sources. J. Food Protec. 59, 193-196

[51] Heierson A, Sidén I, Kivaisi A & Boman GH (1986) Bacteriophage-resistant mutants of *Bacillus thuringiensis* with decreased virulence in pupae of *Hyalophora cecropia*. J. Bacteriol. 167, 18-24

[52] Heimpel am & Angus TA (1959) The site of action of crystalliferous bacteria in Lepidoptera larvae. J. Insect Pathol. 1, 152-170

[53] Heinrichs JH, Beecher DJ, Macmillan JD & Zilinskas BA (1993) Molecular cloning and characterisation of the *hblA* gene encoding the B component of hemolysin BL from *Bacillus cereus*. J. Bacteriol. 175, 6760-6766

[54] Hendriksen NB & Hansen BM (1998) Phylogenetic relations of *Bacillus thuringiensis*: Implications for risks associated to its use as a microbial pest control agent. IOBC Bulletin 21, 5-8

[55] Hernandez E, Ramisse F, Cruel T, Ducoureau JP, Alonso JM & Cavallo JD (1998) *Bacillus thuringiensis* serovar H34-*konkukian* superinfection: report of one case and experimental evidence of pathogenicity in immunosuppressed mice. J. Clin. Microbiol. 36, 2138-2139

[56] Hernandez E, Ramisse F, Cruel T, le Vagueresse R & Cavallo JD (1999) *Bacillus thuringiensis* serotype H34 isolated from human and insecticidal strains serotype 3a3b and H14 can lead to death of immunocompetent mice after pulmonary infection. FEMS immunol. Med. Microbiol. 24, 3-47

[57] Hotha SH & Banik RM (1997) Production of alkaline protease by *Bacillus thuringiensis* H 14 in aqueous two-phase systems. J. Chem. Tech. Biotechnol. 69, 5-10

[58] Höfte H & Whiteley HR (1989) Insecticidal crystal proteins of *Bacillus thuringiensis*. Microbiol. Rev. 53, 242-255

[59] Jackson SG, Goodbrand RB, Ahmed R & Kasatiya S (1995) *Bacillus cereus* and *Bacillus thuringiensis* isolated in a gastro-enteritis outbreak investigation. Lett. Appl. Microbiol. 21, 103-105

[60] Johansen T, Holm T, Guddal PH, Sletten K, Haugli FB & Little C (1988) Cloning and sequencing of the gene encoding the phosphatidylcholine-preferring phospholipase C of *Bacillus cereus*. Gene 65, 293-304

[61] Kim YT & Huang HT (1970) The β-exotoxins of *Bacillus thuringiensis*. I. Isolation and characterization. J. Invertebr. Pathol. 15, 00-108

[62] Kim YT, Gregory BG & Ignoffo CM (1972). The β-exotoxins of *Bacillus thuringiensis*. III. Effects on *in vivo* synthesis of macromolecules in an insect system. J. Invertebr. Pathol. 20, 46-50

[63] Khawaled K, Ben-Dov E, Zaritsky A & Barak Z (1990) The fate of *Bacillus thuringiensis* var. *israelensis* in *B. thuringiensis* var. *israelensis*-killed pupae of *Aedes aegyptii*. J. Invertebr. Pathol. 56, 312-316

[64] Kotiranta A, Haapasalo M, Kari K et al. (1998) Surface structure, hydrophobicity, phagocytosis, and adherence to matrix proteins of *Bacillus cereus* cells with and without the crystalline surface protein layer. Infect. Immun. 66, 4895-4902

[65] Kozuka S & Tochikubo K (1985) Properties and origin of filamentous appendages on spores of *Bacillus cereus*. Microbiol. Immunol. 29, 1-37

[66] Kramer JM & Gilbert RJ (1989) *Bacillus cereus* and other Bacillus species. *In* (Doyle MP, ed.) Foodborne bacterial pathogens. Marcel Dekker, New York, pp 21-70

[67] Kramer JM & Gilbert RJ (1992) *Bacillus cereus* gastro-enteritis. *In* (Tu AT ed.) Handbook of natural toxins volume 7, Marcel Dekker, New York, pp 119-153

[68] Kramer KJ & Muthukrishnan S (1997) Insect chitinases: Molecular biology and potential use as biopesticides. Insect Biochem. Molec. Biol. 27, 887-900

[69] Kuppe A, Evans LM, McMillen DA & Griffith OH (1989) Phosphatidylinositol-specific phospholipase C of *Bacillus cereus*: Cloning, sequencing, and relationship to other phospholipases. J. Bacteriol. 171, 6077-6083

[70] Landén R, Bryne M & Abdel-Hameed A (1994) Distribution of *Bacillus thuringiensis* strains in southern Sweden. World J. Microbiol. Biotechnol. 10, 45-50

[71] Lechner M, Kupke T, Stefanovic S & Götz F (1989) Molecular characterization of phosphatidylinositol-specific phospholipase C of *Bacillus thuringiensis*. Molec. Microbiol. 3, 621-626

[72] Lechner S, Mayr R, Francis KP, et al. (1998) *Bacillus weihenstephanensis* sp. nov. is a new psychrotolerant species of the *Bacillus cereus* group. Int. J. Syst. Bacteriol. 48, 1373-1382

[73] Lereclus D. Agaisse H, Gominet M, Salamitou S & Sanchis V (1996) Identification of a *Bacillus thuringiensis* gene that positively regulates transcription of the phosphatidylinositol-specific phospholipase C gene at the onset of the stationary phase. J. Bacteriol. 178, 2749-2756

[74] Levinson BL, Kasyan KJ, Chiu SS, Currier TC & González Jr. JM (1990) Identification of β-exotoxin production, plasmids encoding β-exotoxin, and a new exotoxin in *Bacillus thuringiensis* by using high-performance liquid chromatography. J. Bacteriol. 172, 3172-3179

[75] Lövgren A, Zhang MY, Engström Å & Landén R (1993) Identification of two expressed flagellin genes in the insect pathogen *Bacillus thuringiensis* subsp. *alesti* J. Gen. Microbiol. 139, 21-30

[76] Lövgren A, Carlson CR, Eskils K & Kolstø AB (1998) Localization of putative virulence genes on a physical map of the *Bacillus thuringiensis* subsp. *gelechiae* chromosome. Curr. Microbiol. 37, 245-250

[77] Luckevich MD and Beveridge TJ (1989) Characterization of a dynamic S layer on *Bacillus thuringiensis*. J. Bacteriol. 171, 6656-6667

[78] Lund T & Granum PE (1996) Characterisation of a non-hemolytic enterotoxin complex from *Bacillus cereus* isolated after a foodborne outbreak. FEMS Microbiol. Lett. 141, 151-156

[79] Lüthy P, Cordier J, & Fischer H (1982) *Bacillus thuringiensis* as a bacterial insecticide: Basic considerations and application, pp. 35-74. *In* (Kurstak E ed.) Microbial and viral pesticides. Marcel Dekker, New York

[80] Mahler H, Pasi A, Kramer JM, Schulte P, Scoging AC, Bär W & Krähenbühl S (1997) Fulminant liver failure in association with the emetic toxin of *Bacillus cereus*. New Engl. J. Med. 336, 1142-1148

[81] Makinen KK & Makinen PL (1987) Purification and properties of an extracellular collagenolytic protease produced by the human oral bacterium *Bacillus cereus* (strain Soc 67). J. Biol. Chem. 262, 12488-12495

[82] Manasherob R, Ben-Dov E, Zaritsky A & Barak Z (1998) Germination, growth, and sporulation of *Bacillus thuringiensis* subsp. *israelensis* in excreted food vacuoles of the protozoan *Tetrahymena pyriformis*. Appl. Environ. Microbiol. 64, 1750-1758

[83] Margulis L, Jorgensen JZ, Kolchinsky R, Rainey FA & Lo SC (1998) The *Arthromitus* stage of *Bacillus cereus*: Intestinal symbionts of animals. Proc. Natl. Acad. Sci. USA 95, 1236-1241

[84] Martin PA & Schmidtmann ET (1998) Isolation of aerobic microbes from *Ixodes scapularis* (Acari: Ixodidae), the vector of Lyme disease in the eastern United States. J. Econ. Entomol. 91, 864-868

[85] McClintock JT, Schaffer CR & Sjoblad RD (1995) A comparative review of the mammalian toxicity of *Bacillus thuringiensis*-based pesticides

[86] Meredith FT, Fowler VG, Gautier M, Corey GR & Reller LB (1997) *Bacillus cereus* necrotizing cellulitis mimicking clostridial myonecrosis: Case report and review of the literature. Scand. J. Infect. Dis. 29, 528-529

[87] Mikkola R, Saris NEL, Grigoriev PA, Andersson MA & Salkinoja-Salonen MS (1999) Ionophoretic properties and mitochondrial effects of cereulide. The emetic toxin of *B. cereus*. Eur. J. Biochem. 263, 112-117

[88] Miller JM, Hair JG, Hebert M, Hebert L & Roberts Jr. FJ (1997) Fulminating bacteremia and pneumonia due to *Bacillus cereus*. J. Clin. Microbiol. 35, 504-507

[89] Moar WJ, Pusztai-Carey M, Van Faasen H *et al.* (1995) Development of *Bacillus thuringiensis* CryIC resistance by *Spodoptera exigua* (Hübner) (Lepidoptera: Noctuidae). Appl. Environ. Microbiol. 61, 2086-2092

[90] Nishikawa Y, Kramer JM, Hanaoka M & Yasukawa A (1996) Evaluation of serotyping, biotyping, plasmid banding pattern analysis, and Hep-2 vacuolation factor assay in the epidemiological investigation of *Bacillus cereus* emetic-syndrome food poisoning. Int. J. Food Microbiol. 31, 149-159

[91] Notermans S & Batt CA (1998) A risk assessment approach for food-borne *Bacillus cereus* and its toxins. J. Appl. Microbiol. Symp. Suppl. 84, 51S-61S

[92] Perani M, Bishop AH & Vaid A (1998) Prevalence of β-exotoxin, diarrhoeal toxin and specific δ-endotoxin in natural isolates of *Bacillus thuringiensis*. FEMS Microbiol. Lett. 160, 55-60

[93] Pillen D (1971) Behandlung von darmerkrankungen mit *Bacillus* stamm 5832. Med. Welt 22, 266-268

[94] Raffel SJ, Stabb EV, Milner JL & Handelsman J (1996) Genotypic and phenotypic analysis of zwittermicin A-producing strains of *Bacillus cereus*. Microbiology 142, 3425-3436

[95] Ramisse V, Patra G, Garrigue H, Guesdon JL & Mock M (1996) Identification and characterization of *Bacillus anthracis* by multiplex PCR analysis of sequences on plasmids pXO1 and pXO2 and chromosomal DNA. FEMS Microbiol. Lett. 145, 9-16.

[96] Rice WC (1999) Specific primers for the detection of vip3A insecticidal gene within a *Bacillus thuringiensis* collection. Lett. Appl. Microbiol. 28, 378-382

[97] Roth FX, Kirchgessner M, Eidelsburger U & Gedek B (1992) Zur nutritiven wirksamkeit von *Bacillus cereus* als probiotikum in der kälbermast. ! Mitteilung: Einfluss auf wachstumsparameter, schlachtleistung und mikrobielle metabolitten im dünndarm. Agribiol. Res. 45, 294-302

[98] Rowan NJ & Anderson JG (1997) Maltodextrin stimulates growth of *Bacillus cereus* and synthesis of diarrheal enterotoxin in infant milk formulae. Appl. Environ. Microbiol. 63, 1182-1184

[99] Roy M, Chen JC, Miller M, Boyaner D, Kasner O & Edelstein E (1997) Epidemic *Bacillus* endophthalmitis after cataract surgery. Ophthalmology 104, 1768-1772

[100] Ryan PA, Macmillan JD & Zilinskas BA (1997) Molecular cloning and characterisation of the genes encoding the L and L components of hemolysin BL from *Bacillus cereus*. J. Bacteriol. 179, 2551-2556

[101] Rychen G & Nunes CS (1995) Effects of three microbial probiotics on postprandial portoarterial concentration differences of glucose, galactose and amino-nitrogen in the young pig. Brit. J. Nutr. 74, 19-26

[102] Salamitou S, Ramisse F, Brehélin M *et al.* (*Unpublished data*)

[103] Samples JR & Buettner H (1983) Corneal ulcer caused by a biological insecticide (*Bacillus thuringiensis*). Am. J. Ophthalmol. 95, 258-260.

[104] Sampson MN & Gooday GW (1998) Involvement of chitinases of *Bacillus thuringiensis* during pathogenesis in insects. Microbiology 144, 2189-2194

[105] Schnepf E, Crickmore N, Van Rie J *et al.* (1998) *Bacillus thuringiensis* and its pesticidal crystal proteins. Microbiol. Mol. Biol. Rev. 62, 775-806

[106] Sebesta K & Horská K (1970) Mechanism of inhibition of DNA-dependent RNA polymerase by exotoxin of *Bacillus thuringiensis*. Biochem. Biophys. Acta 209, 357-376

[107] Seifert HSH & Geissler F (1996) Orale dauerapplikation des probiotischen *B. cereus* − eine alternative zur verhütung der enterotoxämie ? Dtsch. TierÄrztl. Wschr. 103, 386-389

[108] Sela S, Schickler H, Chet I & Spiegel Y (1998) Purification and characterization of a *Bacillus cereus* collagenolytic/proteolytic enzyme and its effect on *Meloidogyne javanica* cuticular proteins. Eur. J. Plant Pathol. 104, 59-67

[109] Shang H, Chen, Handelsman J & Goodman RM (1999) Behaviour of *Pythium torulosum* zoospores during their interaction with tobacco roots and *Bacillus cereus*. Curr. Microbiol. 38, 199-204

[110] Shinagawa K (1990) Purification and characterization of *Bacillus cereus* enterotoxin and its application to diagnosis. *In* (Pohland AE, Dowell jr. VR & Richard JL, eds.) Microbial toxins in foods and feeds - cellular and molecular methods of action. Plenum Press, New York, pp 181-193

[111] Sierecka JK (1998) Purification and partial characterization of a neutral protease from a virulent strain of *Bacillus cereus*. Int. J. Biochem. Cell Biol. 30, 579-595

[112] Silo-Suh LA, Stabb EV, Raffel SJ & Handelsman J (1998) Target range of zwittermicin A, an aminopolyol antibiotic from *Bacillus cereus*. Curr. Microbiol. 37, 6-11

[113] Sinev MA, Budarina ZI, Gavrilenko IV, Tomashevskii AY & Kuzmin NV (1993) Evidence for existence of *Bacillus cereus* hemolysin II: Cloning of hemolysin II genetic determinant. Mol. Biol. 27, 753-760

[114] Smirnoff WA, Randall AP, Martineau R, Haliburton W & Juneau A (1973) Field test of the effectiveness of chitinase additive to *Bacillus thuringiensis* Berliner against *Choristoneura fumiferana*. Can. J. For. Res. 3, 226-236

[115] Smirnova TA, Kulinich LI, Galperin MY & Azizbekyan RR (1991) Subspecies specific haemagglutination patterns of fimbriated *Bacillus thuringiensis* spores. FEMS Microbiol. Lett. 90, 1-4

[116] Smith GR, Barton SA & Wallace LM (1991) Further observations on enhancement of the infectivity of *Fusobacterium necrophorum* by other bacteria. Epidemiol. Infect. 106, 305-310

[117] Solomin AA, Burov GP, Onatskii NM, Firagina OV & Isangaiin FS (1989) Two forms of *Bacillus thuringiensis* β-exotoxin. Biotekhnologiya 6, 792-795

[118] Stabb EV, Jacobson & Handelsman J (1994) ZwittermicinA-producing strains of *Bacillus cereus* from diverse soils. Appl. Environ. Microbiol. 60, 4404-4412

[119] Steinhaus EA (1951) Possible use of *Bacillus thuringiensis* Berliner as an aid in the biological control of the alfalfa caterpillar. Hilgardia 20, 359-381

[120] Strongman DB, Eveleigh ES, van Frankenhuyzen KK & Royama T (1997) The occurrence of two types of entomopathogenic bacilli in natural populations of the spruce budworm, *Choristoneura fumiferana*. Can. J. For. Res. 27, 1922-1927

[121] Tan TM, Wang TK, Lee CL, Chien SW & Horng CB (1997) Foodborne outbreaks due to bacteria in Taiwan, 1986 to 1995. J. Clin. Microbiol. 35, 1260-1262

[122] Tang JD, Shelton AM, Van Rie J *et al*. (1996) Toxicity of *Bacillus thuringiensis* spore and crystal protein to resistant diamondback moth (*Plutella xylostella*). Appl. Environ. Microbiol. 62, 564-569

[123] Tantimavanich S, Pantuwatana S, Bhumiratana A & Panbangred W (1997) Cloning of a chitinase gene into *Bacillus thuringiensis* subsp. *aizawai* for enhanced insecticidal activity. J. Gen. Appl. Microbiol. 43, 341-347

[124] Tayabali AZ & Seligy VL (1995) Semiautomated quantification of cytotoxic damage induced in cultured insect cells exposed to commercial *Bacillus thuringiensis* biopesticides. J. Appl. Toxicol. 15, 365-373

[125] Thompson JV & Fletcher LW (1972) A pathogenic strain of *Bacillus cereus* isolated from the cigarette beetle, *Lasioderma serricorne*. J. Invertebr. Pathol. 20, 341-350

[126] Titball RW (1993) Bacterial phospholipases C. Microbiol. Rev. 57, 347-366

[127] Titball RW (1998) Bacterial phospholipases. J. Appl. Microbiol. Symp. Suppl. 84, 127S-137S

[128] Turnbull PCB (1981) *Bacillus cereus* toxins. Pharmac. Ther. 13, 453-505

[129] Tortuero F & Fernández E (1995) Effects of inclusion of microbial cultures in barley-based diets feed to laying hens. Anim. Feed Sci. Tech. 53, 255-265

[130] Vanková J & Purrini K (1979) Natural epizootics caused by bacilli of the species *Bacillus thuringiensis* and *Bacillus cereus*. Z. Ang. Ent. 88, 216-221

[131] Vidal (1997) «Bactisubtil» in a catalogue of pharmaceuticals, Editions du Vidal, Avenue de Wagram, 75854 Paris Cedex 17

[132] Ward ES & Ellar DJ (1983) Assignment of the δ-endotoxin gene of *Bacillus thuringiensis* var. *israelensis* to a plasmid by curing analysis. FEBS Letters 158, 45-49

[133] Wilson GR & Benoit TG (1993) Alkaline pH activates *Bacillus thuringiensis* spores. J. Invertebr. Pathol. 62, 87-89

[134] Wilson IG, Wilson TS & Kramer JM (1993) Increase in *Bacillus* food poisoning in Northern Ireland. The Lancet 342, 928

[135] Yu CG, Mullins MA, Warren GW, Koziel MG & Estruch JJ (1997) The *Bacillus thuringiensis* vegetative insecticidal protein vip3A lyses midgut epithelium cells of susceptible insects. Appl. Environ. Microbiol. 63, 532-536

[136] Zhang MY & Lövgren A (1995) Cloning and sequencing of a β-lactamase-encoding gene from the insect pathogen *Bacillus thuringiensis*. Gene 158, 83-86

[137] Zhang MY, Lövgren A & Landén R (1995) Adhesion and cytotoxicity of *Bacillus thuringiensis* to cultured *Spodoptera* and *Drosophila* cells. J. Invertebr. Pathol. 66, 46-51

[138] Økstad OA, Gominet M, Purnelle B, Rose M, Lereclus D & Kolstø AB (1999) Sequence analysis of three *Bacillus cereus* loci carrying PlcR-regulated genes encoding degradative enzymes and enterotoxin. Microbiology 145, 3129-3138

Chapter 2.1

The diversity of *Bacillus thuringiensis* δ-endotoxins

Neil Crickmore
School of Biological Sciences, University of Sussex, Falmer, Brighton BN1 9QG E Sussex, UK

Key Words: Cry Toxin, Cyt Toxin, Nomenclature, Phylogeny, Motif

Abstract: Around two hundred pesticidal toxin genes have been cloned from a wide range of *Bacillus thuringiensis* strains. The encoded toxins currently fall into eighty distinct classes, although both the total number of toxins, and the number of classes continue to increase as new toxins are discovered. Despite the large number of classes almost ninety percent of the toxins contain conserved sequence motifs, and are believed to have similar structures. Comparing the toxins' amino acid sequences has resulted in the development of a rigorous nomenclature and allowed correlation's to be made between the sequence, the activity of the toxin, and its evolutionary origin.

1.   TOXIN NOMENCLATURE

In 1989 Höfte and Whiteley [12] proposed a classification scheme for *Bacillus thuringiensis* crystal proteins based on a combination of their deduced amino acid sequence and host range. In that classification thirty eight toxins were placed into fourteen distinct classes. The four primary classes were based on the toxin showing activity against lepidopteran (I), lepidopteran and dipteran (II), coleopteran (III) and dipteran (IV) insects. Inconsistencies did exist in this initial scheme due to attempts to indicate relationships between toxins with similar deduced amino acid sequences but with differing insecticidal activities. Thus the CryIIB toxin was classed with the CryIIA toxin despite showing no dipteran activity. Within the same system CryIVA and CryIVC were classed together due to

J.-F. Charles et al. (eds.),
Entomopathogenic Bacteria: From Laboratory to Field Application, 65–79.
© 2000 *Kluwer Academic Publishers. Printed in the Netherlands.*

the toxins sharing dipteran activity, but despite limited sequence similarity.

During the late 1990's a new nomenclature system was introduced [7] based solely on amino acid sequence relationships between the toxins. This change allowed closely related toxins to be ranked together and removed the necessity to bioassay new toxins against a wide range of possible target organisms before assigning them names. In the revised system Roman numerals were replaced by Arabic ones such that CryIIB became Cry2B. The mnemonic Cyt has been retained to describe toxins which display a general cytolytic activity *in vitro*. The nomenclature contains four ranks allowing each independently isolated toxin gene to be assigned a unique name, and indicating the degree of relatedness between the toxins. Toxins sharing less than approximately 45% sequence similarity are assigned different primary ranks, the boundaries for secondary and tertiary ranks lie at approximately 78% and 95%, respectively. The boundaries were chosen in an attempt to reflect significant evolutionary relationships whilst at the same time minimise the number of name changes from the original system. The relationships are established using the software applications ClustalW [17] and Phylip [10] which are used to first create a multiple sequence alignment and then a distance matrix quantifying the sequence similarities between the set of toxins. A phylogenetic tree is then constructed from this distance matrix using the UPGMA algorithm. Figure 1 shows the outcome of such an analysis on the set of eighty toxins current at the time of writing. This set of toxins is also listed in Table 1 which in addition indicates an accession number for the toxin sequence, the known host range of each toxin, and the strain from which the toxin gene was cloned. An up-to-date list of toxins genes is maintained by the *Bacillus thuringiensis* nomenclature committee and displayed on the World Wide Web [6].

To be included in the nomenclature a Cry toxin must satisfy the following criteria: a parasporal inclusion (crystal) protein from *B. thuringiensis* that exhibits some experimentally verifiable toxic effect to a target organism, or any protein that has obvious sequence similarity to a known Cry protein. Cyt proteins must exhibit haemolytic activity. Thus a novel Cry or Cyt toxin must be found within a *B. thuringiensis* crystalline inclusion and must demonstrate some degree of toxicity. Newly discovered genes which encode proteins with significant sequence similarity to Cry or Cyt proteins already in the nomenclature can also be included without having to demonstrate either a crystal location or toxicity. Furthermore, toxins isolated from sources other than *B. thuringiensis* can also be included if they show sequence similarity. The

Cry16 and Cry17 mosquitocidal toxins from *Clostridium bifermentans* are such examples.

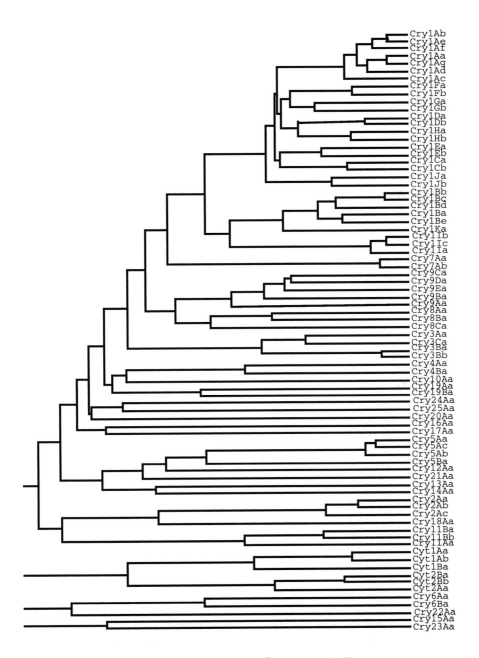

Figure 3. Phylogram of the δ-endotoxin family

Table 1. The δ-endotoxins

Toxin	Holotype acc. no.	Specificity	Parent Strain (serovar where known)
Cry1Aa	M11250	L	*B. thuringiensis* (3a3b3c, 4a4b, 6, 7)
Cry1Ab	M13898	L	*B. thuringiensis* (3a3b3c, 7)
Cry1Ac	M11068	L	*B. thuringiensis* (3a3b3c, 4a4c)
Cry1Ad	M73250	L	*B. thuringiensis* (7)
Cry1Ae	M65252	L	*B. thuringiensis* (3a3c)
Cry1Af	U82003	L	*B. thuringiensis*
Cry1Ag	AF081284	L	*B. thuringiensis*
Cry1Ba	X06711	L,C,D	*B. thuringiensis* (1, 7)
Cry1Bb	L32020	L	*B. thuringiensis*
Cry1Bc	Z46442	L	*B. thuringiensis* (8a8b)
Cry1Bd	U70726	L	*B. thuringiensis* (*wuhanensis*)
Cry1Be	AF077326	L	*B. thuringiensis*
Cry1Ca	X07518	L,D	*B. thuringiensis* (6, 7)
Cry1Cb	M97880	L	*B. thuringiensis* (5a5b)
Cry1Da	X54160	L	*B. thuringiensis* (7)
Cry1Db	Z22511	L	*B. thuringiensis*
Cry1Ea	X53985	L	*B. thuringiensis* (4a4c)
Cry1Eb	M73253	L	*B. thuringiensis*
Cry1Fa	M63897	L	*B. thuringiensis* (7)
Cry1Fb	Z22512	L	*B. thuringiensis* (8a8b)
Cry1Ga	Z22510	L	*B. thuringiensis* (*wuhanensis*)
Cry1Gb	U70725	L	*B. thuringiensis* (*wuhanensis*)
Cry1Ha	Z22513	L	*B. thuringiensis*
Cry1Hb	U35780	L	*B. thuringiensis* (8a8b)
Cry1Ia	X62821	L,C	*B. thuringiensis* (3a3b3c, 6)
Cry1Ib	U07642	L,C	*B. thuringiensis* (3a3b3c, 6)
Cry1Ic	AF056933	L,C	*B. thuringiensis*
Cry1Ja	L32019	L	*B. thuringiensis*
Cry1Jb	U31527	L	*B. thuringiensis*
Cry1Ka	U28801	L	*B. thuringiensis* (8a8b)
Cry2Aa	M31738	L,D	*B. thuringiensis* (3a3b3c, 4a4b, 4a4c)
Cry2Ab	M23724	L	*B. thuringiensis* (3a3b3c)
Cry2Ac	X57252	L	*B. thuringiensis* (5a5b)
Cry3Aa	M22472	C	*B. thuringiensis* (8a8b)
Cry3Ba	X17123	C	*B. thuringiensis* (9)
Cry3Bb	M89794	C	*B. thuringiensis*
Cry3Ca	X59797	C	*B. thuringiensis* (3a3b3c)
Cry4Aa	Y00423	D	*B. thuringiensis* (14)
Cry4Ba	X07423	D	*B. thuringiensis* (14)
Cry5Aa	L07025	N	*B. thuringiensis* (10a10b)
Cry5Ab	L07026	N	*B. thuringiensis*
Cry5Ac	I34543	H	*B. thuringiensis*
Cry5Ba	U19725	H	*B. thuringiensis*
Cry6Aa	L07022	N	*B. thuringiensis*
Cry6Ba	L07024	N	*B. thuringiensis*
Cry7Aa	M64478	C	*B. thuringiensis*
Cry7Ab	U04367	C	*B. thuringiensis* (15, 18a18b)

Table 1 cont.

Toxin	Holotype acc. no.	Specificity	Parent Strain (serovar where known)
Cry8Aa	U04364	C	*B. thuringiensis* (18a18b)
Cry8Ba	U04365	C	*B. thuringiensis* (18a18b)
Cry8Ca	U04366	C	*B. thuringiensis* (23)
Cry9Aa	X58120	L	*B. thuringiensis* (5a5b)
Cry9Ba	X75019	L	*B. thuringiensis* (5a5b)
Cry9Ca	Z37527	L	*B. thuringiensis* (9)
Cry9Da	D85560	L	*B. thuringiensis* (23)
Cry9Ea	AB011496	L	*B. thuringiensis* (7)
Cry10Aa	M12662	D	*B. thuringiensis* (14)
Cry11Aa	M31737	D	*B. thuringiensis* (14)
Cry11Ba	X86902	D	*B. thuringiensis* (28a28c)
Cry11Bb	AF017416	D	*B. thuringiensis* (30)
Cry12Aa	L07027	N	*B. thuringiensis*
Cry13Aa	L07023	N	*B. thuringiensis*
Cry14Aa	U13955	C	*B. thuringiensis* (4a4b)
Cry15Aa	M76442	L	*B. thuringiensis* (12)
Cry16Aa	X94146	D	*C. bifermentans* (*malaysia*)
Cry17Aa	X99478	D	*C. bifermentans* (*malaysia*)
Cry18Aa	X99049	C	*B. popilliae*
Cry19Aa	Y07603	D	*B. thuringiensis* (28a28c)
Cry19Ba	D88381	D	*B. thuringiensis* (44)
Cry20Aa	U82518	D	*B. thuringiensis* (3a3d3e)
Cry21Aa	I32932	N	*B. thuringiensis*
Cry22Aa	I34547	H	*B. thuringiensis*
Cry23Aa	AF03048	C	*B. thuringiensis*
Cry24Aa	U88188	D	*B. thuringiensis* (28a28c)
Cry25Aa	U88189	D	*B. thuringiensis* (28a28c)
Cyt1Aa	X03182	D	*B. thuringiensis* (8a8b, 14)
Cyt1Ab	X98793	D	*B. thuringiensis* (30)
Cyt1Ba	U37196	D	*B. thuringiensis* (24a24b)
Cyt2Aa	Z14147	D	*B. thuringiensis* (11a11c)
Cyt2Ba	U52043	D	*B. thuringiensis* (3a3d3e, 8a8b, 14)
Cyt2Bb	U82519	D	*B. thuringiensis* (28a28c)

The accession number given refers to that of the holotype toxin, usually the first reported sequence for that toxin. The broad specificity of each toxin towards Lepidoptera (L), Diptera (D), Coleoptera (C), Hymenoptera (H) and Nematodes (N) is indicated. The parent strain (and serovar) refers to the strain from which sequenced genes have been isolated.

The strain serovars (subspecies) listed in Table 1 are based on flagella antigens [8]. A poor correlation between subspecies and toxin composition is observed. Analysis of toxin gene complement by PCR confirms that a given toxin gene can often be found in a wide range of subspecies [2, 11, 15]. A PCR screen for the two mosquitocidal toxin genes from *Clostridium bifermentans* also found them in a variety of Clostridia species as well as *B. thuringiensis* [1]. The fact that toxin genes are mostly located on transmissible plasmids, and are often flanked by transposable elements, may account for their widespread distribution (see Chapter 2.5).

2.   SEQUENCE COMPARISONS

In their review Höfte and Whiteley [12] identified five conserved amino acid blocks present in most of the toxin sequences known at that time. The same five blocks have indeed proved to be well conserved and most are found within the sequences of the sixty nine toxins that comprise the main Cry lineage as seen in Figure 1. The determination of the three dimensional structure of several of the Cry toxins has allowed the location of these blocks to be mapped to specific regions of the structure. Table 2 lists the consensus sequence of these five blocks and indicates their structural location.

Figure 2 shows the location of each of these blocks, where present, within each of the known Cry and Cyt toxins. The shaded blocks represent truncated or variant forms of the consensus block. This figure also gives an indication of the relative length of each toxin. The shaded boxes at the C-terminal ends of Cry1Af and Cry1Bd represent the presumed ends of the toxins since only a partial sequence was available. A more detailed analysis of the conserved blocks, including the identification of additional blocks within the C-terminal ends of the longer toxins is given in reference [14].

As can be seen from Figure 2 the lengths of the individual toxins varies quite considerably. Within the sixty nine toxins of the main lineage there are two main classes; those that terminate shortly after conserved block five, and those that contain an extended C-terminus. This extended C-terminus is removed during proteolytic activation within the target organism and does not appear to be directly involved in the toxic mechanism.

Table 2. Conserved amino acid blocks within the δ-endotoxins

Block	Consensus	Location
1	LPVYAQAANLHLXLLRD	Domain I, alpha helix
2	WVRYNQFRREMTLXVLDLVALFPXYDXRXYP	Domain I/II boundary
3	WTHRSADXXNTIXXXXITQIPLVKAXXLXXGXXVVXGPGFTGGDIL	Domain II/III boundary
4	QRYRVRIRYAS	Domain III, beta strand
5	VYIDRIEFVP	End of Domain III

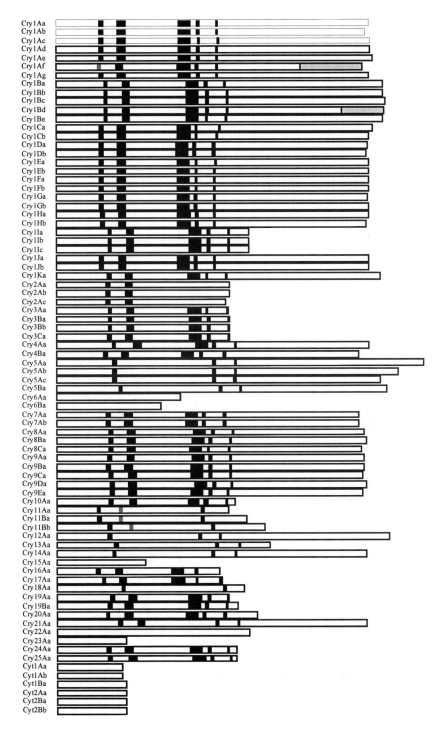

Figure 4. Location of the conserved blocks

   The role of the C-terminal extension is believed to be to direct crystallisation of the toxin within the bacterium. The process used to derive the toxin nomenclature described above uses the full length toxin, i.e. including any C-terminal extension. Since however this extension is not directly involved in toxicity attempts to correlate sequence similarity with toxicity may be more reliable if the extension is not considered. Figure 3 shows a phylogenetic tree based on toxin sequences from which the C-terminal extensions have been removed. The outer ring indicates the main pesticidal activity associated with each group of toxins. A similar analysis was performed by Bravo [3] whose observations are similar to those described below.

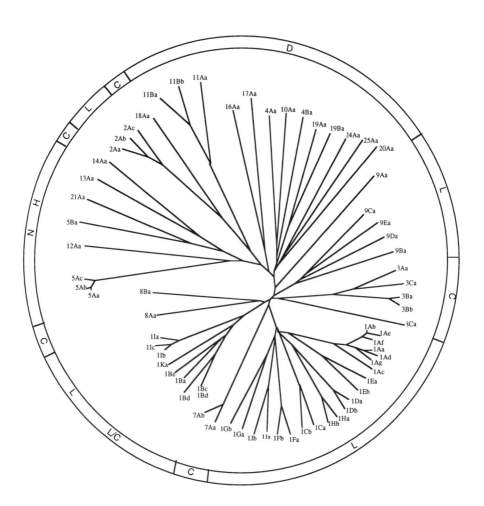

Figure 5. Phylogenetic analysis of the active toxins

Comparing Figure 3 with the dendrogram used to derive the nomenclature (Figure 1) a number of obvious differences are apparent:

a) Cry9Aa no longer clusters with the other Cry9's.
b) Cry5Ba no longer clusters with the other Cry5's.
c) Cry8Ca no longer clusters with the other Cry8's.
d) Cry10Aa clusters more closely with the Cry4's.

The differences in a-c represent the fact that similar C-terminal extensions link toxins in the nomenclature that are less related at the level of 'active toxin'. Example d) is due the presence of a highly similar C-terminal extension in Cry4Aa and Cry4Ba which is absent in Cry10Aa.

When comparing the clustering pattern with insecticidal activity there is a reasonable correlation with toxins sharing a common activity also clustering together. Two notable exceptions can be observed though:

a) The coleopteran active toxin Cry18Aa clusters with the lepidopteran active Cry2's.
b) The coleopteran active toxin Cry14Aa clusters with the nematocidal Cry13Aa.

The structure of most, if not all, of the sixty nine toxins shown in Figure 3 is thought to be very similar to that of Cry3Aa, the first toxin to have its structure solved [13]. Basically it consists of three domains, the N-terminal domain (domain I) is involved in forming a pore in the epithelial membrane of the cells lining the gut of the target pest. Domains II and III have both been implicated in determining specificity, specific loops in domain II for example, are believed to be involved in receptor binding. The next three figures show relationships between the individual domains of the set of sixty nine toxins. Figure 4 shows a phylogenetic tree derived from domain I. The clustering patterns observed are very similar to those found with the active toxin (Figure 3).

Once again there is a good correlation between the clustering pattern and insecticidal activity. This might not have been expected since domain I is not generally thought to be a major determinant of specificity. This correlation might reflect other factors which affect specificity such as membrane composition or the involvement of pest-specific proteinases. Alternatively structural constraints may prevent domains I and II evolving completely independently thus linking domain I to the specificity-determining domain II. One major difference in the clustering patterns between domain I and the active toxin is that domain I from Cry9Ca no longer clusters with the other Cry9's but joins Cry8Aa and Cry8Ba.

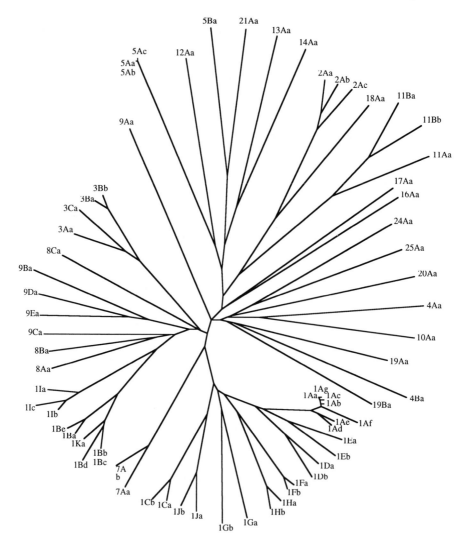

Figure 6. Domain I phylogenetic analysis

Figure 5 shows the clustering pattern of the domain II sections of the toxin genes. Surprising the clustering patterns show less correlation with pesticidal activity than domain I, given that domain II has always been implicated in specificity determination. As is discussed in detail by Bravo [3] there is a greater diversity amongst the domain II's suggesting the possibility of different evolutionary origins. In particular the cluster containing the Cry2's, Cry11's and Cry18Aa and the cluster containing the Cry5's, Cry12Aa, Cry13Aa, Cry14Aa and Cry21Aa are only distantly related to the rest of the set. This is consistent with the observation that all the toxins in these two clusters lack conserved block three which is largely located in domain II.

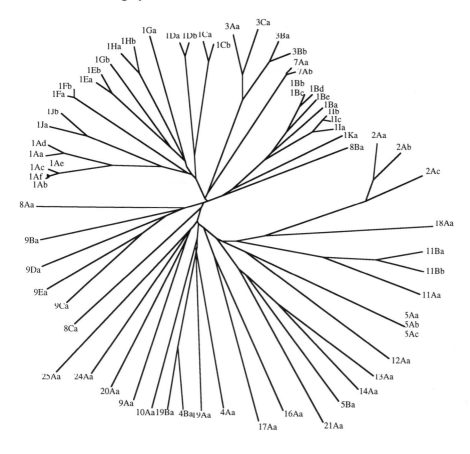

Figure 7. Domain II phylogenetic analysis

Comparing Figure 5 with the previous clustering patterns reveals the following:
a) Cry1Ga has split from Cry1Gb.
b) Cry8Aa has split from Cry8Ba.
c) Cry9Aa clusters more tightly with the dipteran toxins Cry20Aa, Cry24Aa and Cry25Aa.

Clustering of the domain III sections is shown in Figure 6. Of the individual domains, the clustering pattern for domain II is the most different from that observed for the active toxin.

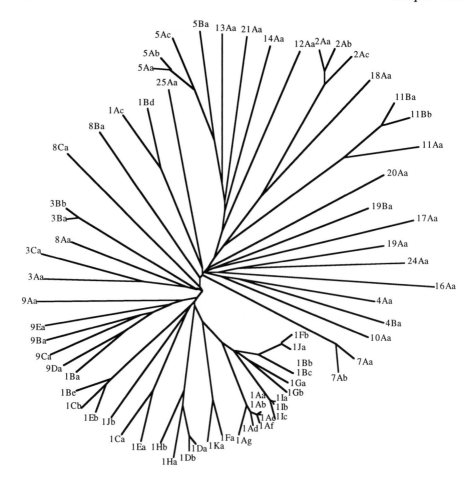

Figure 8. Domain III phylogenetic analysis

The Cry2's, which contain no obvious conserved blocks four or five, cluster with the Cry11's and Cry18Aa which contain block four but not five. The domains from these toxins may have different evolutionary origins [3]. Although the Cry5's, Cry12Aa, Cry13Aa, Cry14Aa and Cry21Aa contained no conserved block three, they all contain conserved blocks four and five and seem related, albeit distantly, from the others in this set. Notable features of the domain III clustering patterns are:

a) Cry1Ea clusters with Cry1Ca
b) Cry1Eb clusters with Cry1Cb
c) Cry1Be clusters with Cry1Cb and Cry1Eb
d) Cry1Fb clusters with Cry1Ja
e) Cry1Ba clusters with the Cry9's
f) Cry1Ac and Cry1Bd cluster together and away from other Cry1's.

These observations suggest swapping of domain III between toxins, in particular there may have been a reciprocal swap between Cry1C and Cry1E.

3. RELATED SEQUENCES

A number of genes have been isolated from *Bacillus thuringiensis*, as potential δ-endotoxin genes, whose encoded proteins do not fit the definition of a δ-endotoxin described earlier. These are listed below:

3.1 Orf1 and Orf2 from the *cry2A* operon

The Cry2A δ-endotoxins are encoded by genes found as the third gene in a three gene operon and preceded by two open reading frames known as *orf1* and *orf2*. The gene product of *orf2* is located within the Cry2A crystal and is required for the efficient crystallisation of the toxin protein [16]. Unlike Orf2, Orf1 does not appear to be necessary for Cry2A crystal formation [5]. Orf1-like proteins have been found in a number of different *B. thuringiensis* strains and their genes are usually found in close association with a toxin gene [16]. However no function has yet been established for this family of proteins, in some strains they are found as very minor components of the crystal whereas in at least one strain the Orf1-like protein is present within the crystal in significant amounts [9].

3.2 The 40 kDa protein from *Bt thompsoni*

The crystal from *B. thuringiensis* subsp. *thompsoni* (HD-542) contains two major polypeptides of 34 kDa and 40 kDa whose deduced amino acid sequence is unrelated to other δ-endotoxins [4]. Because purified 34 kDa protein is toxic towards *Manduca sexta* it was designated Cry15A. However since the purified 40 kDa protein was not toxic towards any insects tested it did not meet the criteria to be named as a Cry toxin. The possibility remains that the 40 kDa is toxic against some species against which it has not yet been tested, or that it might have some other function within the crystal.

3.3 The Vip toxins

Like the Cry toxins the midgut epithelial cells appear to be the primary target for the Vip toxins originally isolated from *B. thuringiensis* strain AB88 [18]. Unlike the Cry toxins though, the Vip toxins are secreted rather than being laid down in a crystalline inclusion. Given that they also share no significant sequence similarity to the Cry toxins they are not considered as part of the δ-endotoxin family.

3.4   Others

A number of other genes have been cloned from *B. thuringiensis* that encode novel proteins that may be δ-endotoxins. Until both a pesticidal activity and crystal location are confirmed these proteins can not be entered into the nomenclature. The genes are listed on the nomenclature committee web site [6].

Finally the sequencing of microbial genomes has identified a number of δ-endotoxin related sequences in other bacterial species. The *ynzF*, *yokG* and *yobE* genes from *Bacillus subtilis* are reportedly similar to known *B. thuringiensis* δ-endotoxins. Also the Y1107 gene found on the *Yersinia pestis* pMT1 plasmid bears some similarities to the *B. thuringiensis* cytolytic toxin genes.

4.   SUMMARY

In the ten years since Höfte and Whiteley first proposed a nomenclature for *B. thuringiensis* crystal toxins the number of distinct toxin classes has increased almost six fold. Despite the large diversity in amino acid sequence the five conserved blocks identified by the above authors are still present in the vast majority of toxins. Phylogenetic analysis has provided evidence of independently evolved domains as well as domain swapping. Apart from the cytolytic toxins at least five toxin classes exist which do not contain any of these conserved blocks. Toxins within these classes are still found within the cytoplasmic crystal and possess a pesticidal activity. Recently toxins highly similar to the *Bacillus thuringiensis* δ-endotoxins have been found in other bacterial species.

REFERENCES

[1] Barloy F, Lecadet M-M & Delécluse A (1998) Distribution of clostridial *cry*-like genes among *Bacillus thuringiensis* and *Clostridium* strains. Curr. Microbiol. 36, 232-237

[2] Ben Dov E, Zaritsky A, Dahan E *et al.* (1997) Extended screening by PCR for seven cry-group genes from field- collected strains of *Bacillus thuringiensis*. Appl. Environ. Microbiol. 63, 4883-4890

[3] Bravo A (1997) Phylogenetic relationships of *Bacillus thuringiensis* delta-endotoxin family proteins and their functional domains. J. Bacteriol. 179, 2793-2801

[4] Brown KL & Whiteley HR (1992) Molecular characterization of 2 novel crystal protein genes from *Bacillus thuringiensis* subsp *thompsoni*. J. Bacteriol. 174, 549-557

[5] Crickmore N & Ellar DJ (1992) Involvement of a possible chaperonin in the efficient expression of a cloned CryIIA delta-endotoxin gene in *Bacillus thuringiensis*. Mol. Microbiol. 6, 1533-1537

[6] Crickmore N, Zeigler DR, Feitelson J *et al. B. thuringiensis* toxin nomenclature. http://www.biols.susx.ac.uk/Home/Neil_Crickmore/Bt

[7] Crickmore N, Zeigler DR, Feitelson J *et al.* (1998) Revision of the nomenclature for the *Bacillus thuringiensis* pesticidal crystal proteins. Microbiol. Mol. Biol. Rev. 62, 807-813

[8] de Barjac H & Frachon E (1990) Classification of *Bacillus thuringiensis* strains. Entomophaga 35, 233-240

[9] Dunn M Personal communication

[10] Felsenstein J (1989) PHYLIP-phylogeny inference package (version 2). Cladistics 5, 164-166

[11] Guerchicoff A, Ugalde RA & Rubinstein CP (1997) Identification and characterization of a previously undescribed *cyt* gene in *Bacillus thuringiensis* subsp. *israelensis*. Appl. Environ. Microbiol. 63, 2716-2721

[12] Hofte H & Whiteley HR (1989) Insecticidal crystal proteins of *Bacillus thuringiensis*. Microbiol. Rev. 53, 242-255

[13] Li J, Carroll J & Ellar DJ (1991) Crystal structure of insecticidal delta-endotoxin from *Bacillus thuringiensis* at 2.5Å resolution. Nature 353, 815-821

[14] Schnepf E, Crickmore N, Van Rie J *et al.* (1998) *Bacillus thuringiensis* and its pesticidal crystal proteins. Microbiol. Mol. Biol. Reviews 62, 775-806

[15] Shin BS, Park SH, Choi SK *et al.* (1995) Distribution of CryV-type insecticidal protein genes in *Bacillus thuringiensis* and cloning of CryV-type genes from *Bacillus thuringiensis* subsp *kurstaki* and *Bacillus thuringiensis* subsp *entomocidus*. Appl. Environ. Microbiol. 61, 2402-2407

[16] Staples NJ (1996) Investigation of *Bacillus thuringiensis* "helper" proteins. PhD dissertation, Cambridge University

[17] Thompson JD, Higgins DG & Gibson TJ (1994) CLUSTAL W: improving the sensitivity of progressive multiple sequence alignment through sequence weighting, position-specific gap penalties and weight matrix choice. Nucleic Acids Res. 22, 4673-4680

[18] Yu CG, Mullins MA, Warren GW, Koziel MG & Estruch JJ (1997) The *Bacillus thuringiensis* vegetative insecticidal protein Vip3A lyses midgut epithelium cells of susceptible insects. Appl. Environ. Microbiol. 63, 532-536

Chapter 2.2

Insecticidal proteins produced by bacteria pathogenic to agricultural pests

Takashi Yamamoto[1] & Donald H. Dean[2]
Maxygen, Inc., Redwood City, CA 94063, USA[1] and Department of Biochemistry, The Ohio State University, Columbus, OH 43210, USA[2]

Key words: insecticidal protein, *Bacillus thuringiensis*, *Bacillus popilliae*, exotoxin, protein structure and function

Abstract: Numerous bacteria pathogenic to insects have been characterised. Some of those are known for the production of insecticidal proteins. The insecticidal protein produced by *Bacillus thuringiensis* has been studied extensively partly due to its commercial value. A large number of *B. thuringiensis* strains have been isolated. Each *B. thuringiensis* isolate has a narrow insect specificity, but the specificity is diversified amongst different isolates. Insects susceptible to *B. thuringiensis* include those in Lepidoptera, Diptera, Hymenoptera and Coleoptera as well as some nematode species. From these isolates, over 100 insecticidal protein genes have been cloned and sequenced. There are several other *Bacillus* species known to synthesise insecticidal proteins. *Bacillus popilliae* produces a proteinaceous inclusion in sporulated cells. The inclusion protein is involved in the pathogenicity of *B. popilliae*. A gene encoding for this inclusion has been cloned. In this chapter, we shall make a brief introduction to insecticidal proteins found in *B. thuringiensis* and *B. popilliae* followed by analysis on the structure and function of *B. thuringiensis* insecticidal proteins.

1.   DESCRIPTION OF *B. THURINGIENSIS*

The first description of *B. thuringiensis* appeared in a Japanese article in 1901 [30]. It was isolated from silkworm (*Bombyx mori*) larvae, which were rapidly dying from a bacterial infection. The cause of death was

81

J.-F. Charles et al. (eds.),
Entomopathogenic Bacteria: From Laboratory to Field Application, 81–100.

attributed to a toxin or toxins associated with the spores of the bacterium isolated from the infected larvae. The bacterium was named as *Bacillus sotto* ("sotto" means sudden collapse in Japanese), because the toxin kills the larvae quickly. A decade later, a similar bacterium was isolated in Germany from *Anagasta kühniella* (Mediterranean flour moth) brought from a town called Thuringia [6]. This isolate was officially named as *Bacillus thuringiensis* Berliner. Since then, many *B. thuringiensis* strains have been isolated. These isolates have been classified into subspecies primarily based on the flagellar antigenicity. One of these subspecies, *B. thuringiensis kurstaki*, has been extensively used in commercial insecticide formulations due to its wide spectrum against important agricultural and forestry pests (see ref. [5] for the history of *B. thuringiensis* commercialisation).

B. *thuringiensis* grows under the aerobic condition in an ordinary artificial culture medium. When a certain condition is met, such as lack of key nutrients or accumulation of undesirable metabolites, it undergoes the sporulation process during the stationary phase. At the onset of sporulation, *B. thuringiensis* synthesises a massive amount of insecticidal proteins. The accumulated proteins form a crystalline inclusion body or bodies in the sporulated cells, which eventually lyse to release the free spores and crystals into the culture medium. Because of this crystal formation, the protein is called crystal (Cry) protein. The gene coding for the crystal protein was first isolated from a *kurstaki* strain [46]. This gene is now called *cry1Aa* according to the latest *B. thuringiensis cry* gene nomenclature [13]. Sequencing *cry1Aa* has revealed a promoter region that is recognised by sporulation-specific sigma factors explaining the expression timing of this gene. The cloned *cry1Aa* gene contains the coding region for a protein of 133 kDa and the terminator. When cloned in *Escherichia coli* with appropriate segments of the promoter and terminator, the gene was highly expressed [46]. At the time when the first *cry* gene was cloned, researchers found that the *cry* genes are mostly on plasmids [24]. A few *cry* genes are reported to be on the chromosome [9]. Some *cry* gene-encoding plasmids were found to be transmissible between *Bacillus* strains by conjugation [23]. These last three points are presented in Chapter 2.5. This discovery explained, at least in part, why some *cry* genes, such as *cry1Ab*, are widely distributed amongst different *B. thuringiensis* strains. Since then, numerous *cry* genes have been cloned. They can be roughly divided into two groups. One encodes a protein of about 135 kDa and the other about one half in size.

2. EXTRA-CELLULAR INSECTICIDAL PROTEINS

Beside endotoxins, which are normally expressed in the late stationary phase and crystallise in the cell along with the spore, many *B. thuringiensis* strains also harbour extra-cellular insecticidal toxin or β-exotoxin genes. Normally, the exotoxin genes are expressed during the vegetative growth stage. One class of the exotoxins is called Vip (vegetative insecticide protein). Vip1 and Vip2 are binary toxins and are specifically active against *Diabrotica* [55]. Interestingly, *B. thuringiensis* tenebrionis, which is active against Coleoptera, contains the *vip1* and *vip2* genes. Vip3 is an 81 kDa protein produced by a variety of *B. thuringiensis* strains including those of the *kurstaki* subspecies [18]. Vip3 was reported to be active against Lepidoptera especially *Spodoptera* but not active against Coleoptera. The mode of action of these proteins has not been reported.

B. thuringiensis is often considered a variant of *B. cereus*. A report indicates that *B. thuringiensis* contains an enterotoxin gene similar to that of *B. cereus* and expresses the gene [10]. The enterotoxin gene has been cloned from several *B. thuringiensis* strains [4]. Due to these reports, a concern of possible infection with *B. thuringiensis* in higher animals has been raised. *B. thuringiensis*, however, is considered to be a safe insecticide, because no significant health problems with the human or domestic animal have been reported even though it is extensively used as commercial insecticides.

3. CHARACTERISATION OF THE CRYSTAL PROTEIN

B. thuringiensis crystals can be dissolved in an alkaline solution. A disulphide bond-reducing agent, such as dithiothreitol, reduces the minimum alkalinity required for the solubilisation. The solubilised crystal proteins can be analysed by an appropriate method such as SDS-PAGE or HPLC. Both methods have been used to quantify the amount of the crystal proteins [26]. For SDS-PAGE, the crystal can be solubilised in a normal SDS-PAGE sample buffer containing SDS and a disulphide bond-reducing agent. Since most Cry1-type proteins are about 135 kDa, SDS-PAGE does not resolve these proteins. Cry1Ca, however, tends to appear at a position slightly above the Cry1A-type proteins. The crystal proteins similar in size can be quantitatively analysed by SDS-PAGE when the proteins are pre-digested with CNBr [37]. CNBr digests the protein at methionine residues. Unlike most proteinases, CNBr produces fragments large enough for SDS-PAGE, because methionine residues occur in the crystal proteins infrequently.

A DEAE ion-exchange column or a reverse phase column can be used to isolate the crystal proteins. Since many *B. thuringiensis kurstaki* strains contain several *cry1A*-type genes, determination of the expression level of each *cry1A* gene is of interest. A reliable analytical method is particularly desired when these strains are used to manufacture commercial insecticides. The separation of individual full-length Cry1A-type proteins is very difficult, if it is indeed possible, because the C-terminal halves of most Cry1-type proteins are highly homologous. When the C-terminus is removed by trypsin digestion, the activated toxins expose the variable regions and can be separated by HPLC. However, the trypsin-activated toxins have the lowest solubility at pH around 7 (pI: isoelectric point, see below). Therefore, an alkaline buffer [43] and urea in a neutral pH buffer [11] have been used to overcome this problem. The toxins are reasonably soluble at a high or low pH or a neutral-pH with a moderate concentration of urea (e.g. 2 M). Separation of Cry1Aa, Cry1Ab, Cry1Ac, Cry1Ca and Cry1Da has been successfully carried out by HPLC using a DEAE column under one of these conditions [36]. Figure 1 shows an example of HPLC separation of the crystal proteins produced by a recombinant *B. thuringiensis kurstaki* strain, which contains the hybrid G27 gene made from *cry1Ea* and *cry1Ca* in addition to the indigenous *cry1A* genes [61].

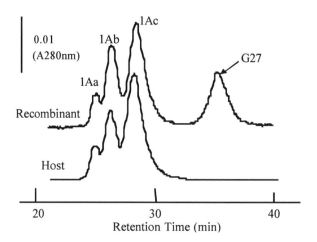

Figure 9. HPLC separation of trypsin-activated crystal proteins. Trypsin-activated crystal proteins were prepared from proteinase-free *B. thuringiensis kurstaki* crystal samples. These proteins, which had been solubilised in 2 M urea-50 mM Tris-HCl, pH 8.8, were separated with a Mono-Q column. An NaCl-gradient from 150 to 200 mM was used to elute the proteins. G27 is a hybrid protein made from Cry1Ea and Cry1Ca.

In the early stage of studying the activation of the crystal protein, the Cry1A-type proteins were often used. The full-length Cry1A protoxins are converted to the active form by insect gut proteinase(s). Since the digestive tract of lepidopteran insects contains an alkaline proteinase, which recognises specific peptide sequences the same as those of trypsin from higher animals, trypsin was often used in the study to find the activation sites. In the case of Cry1Ac, trypsin digests the C-terminal half of the protein molecule in several steps up to K623 [7, 12]. This C-terminus portion of Cry1Ac has several cysteine residues, which contribute to the stability of the crystal matrix by forming intermolecular disulphide bonds. Trypsin also removes the N-terminus of the Cry1A-type proteins up to R28 [39]. The Cry1Ac protein delineated between I29 and K623 is the fully matured toxin and is substantially resistant to any further digestion by trypsin. In the lepidopteran insect gut juice, there are additional proteinase activities like chymotrypsin and terminal peptidases. Therefore, the crystal proteins can be activated and, more importantly, can be inactivated by these proteinases. Several studies indicated that any further digestion of the Cry1A-type protein beyond K623 (i.e. the region of P607-K623) could destroy the activity [28, 47, 54]. The Cry1A-type protoxins are acidic proteins. Calculated pI of the Cry1Ac protoxin is pH 4.95, whereas pI of the trypsin-activated form is pH 6.45. Actually, it was reported that pI of the Cry1A protoxins is about pH 4.4 [62].

4.   THREE DOMAIN STRUCTURE OF THE CRYSTAL PROTEIN

The tertiary structures of three crystal proteins have been determined by X-ray crystallography. These proteins are Lepidoptera-specific Cry1Aa [25], Coleoptera-specific Cry3Aa [34] and Diptera, Lepidoptera-specific Cry2Aa [38]. The activated Cry1Aa and Cry3Aa toxins, not the full-length protoxins, were used to determine the structure by X-ray crystallography, whilst the structure of the undigested, full-length Cry2Aa protein was determined. These proteins share the similar structural characteristics. As shown in Figure 2, the crystal proteins have three structurally distinguishable domains each consisting of about 200 amino acid residues. Domain I is a bundle of seven α-helices. Domains II and III are made of repeating β-strands. In this chapter, we follow the numbering rule of α-helices and β-strands as appeared in the Cry3Aa paper [34].

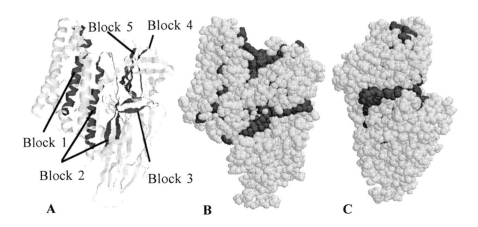

Figure 10. Cry1Aa structure shown in a ribbon model (A) and a space-fill model (B, C). B was rotated by 90° left to produce C. The sequence-conserved blocks 1 to 5 are identified with dark grey.

In the first paper proposing the nomenclature rule for the *B. thuringiensis cry* genes, highly homologous regions amongst the crystal proteins were described [29]. These homologous regions were called "blocks", and five blocks from 1 to 5 were identified. Since then, numerous arguments have been made in reference to these blocks. The locations of these blocks are indicated on the Cry1Aa structure in Figure 2. Block 1 is exclusively α5, the internal, hydrophobic α-helix in the domain I. Block 2 spans from the almost entire α7 to β1 in domain II. Block 3 bridges domain II to domain III stretching from β11 to β15. Blocks 4 and 5 are antiparallel β17 and β23, which are held together with a number of hydrogen bonds in the core of domain III. As shown in Figure 2, all these conserved blocks are embedded in the toxin molecule suggesting that *B. thuringiensis* must conserve these blocks in order to form a common structure motif amongst those which contain these conserved blocks. The conserved blocks may be important for the toxin function in one way or the other. But, it is unlikely that the blocks are involved in the receptor binding, because the crystal proteins having diversified insect specificity share these homologous blocks.

5.   DOMAIN I FUNCTION

B. thuringiensis toxins have two major functions related to the insecticidal activity, receptor binding and membrane insertion. Domain I is considered to be involved in the membrane insertion. Domain I alone, when expressed from a truncated gene, has shown to be able to insert into the phospholipid vesicles, although its ion channel activity is quite different from the whole toxin [53]. Structural resemblance between domain I of the crystal proteins and membrane-spanning domains of other bacterial toxins such as colicin A supports this theory. Membrane insertion mechanisms of bacterial toxins other than *B. thuringiensis* helped to propose several membrane insertion models for the crystal proteins. Two models called "umbrella" and "penknife" models are based on an idea that only a part of domain I goes into the membrane. According to the umbrella model, the hydrophobic hairpin between α4 and α5 initiates the membrane insertion [34]. The penknife model is adapted from colicin A [27]. In this model, the hydrophobic helices, α5 and α6, protrude from the toxin molecule like a penknife and insert into the membrane. In these cases, the structural integrity of the α-helix cluster must be broken requiring a high level of energy. An alternative model that virtually the entire domain I portion or the entire toxin molecule penetrates into the membrane has been proposed [16].

In support of the umbrella model is a series of papers on the partitioning of synthetic peptides into liposomes [19, 20]. Another supporting paper utilises disulphide bridges between the α-helices of domain I of Cry1Aa [48]. When the toxin molecule was restrained by introducing mutations with cysteine residues that form intra- or inter-domain disulphide bonds, no membrane insertion was observed. Unfortunately, biochemical studies indicating structural stability of the disulphide bridge constructs were not reported nor were insecticidal activities of the constructs. Besides restraining the α-helices by disulphide linkages, such mutations may cause a substantial structural modification, which can make the whole molecule unable to insert into the membrane.

There are a few published papers and several unpublished reports that contradict the umbrella and penknife models. One published note [58] reported that Cry1Ab inserted into insect BBMV (brush border membrane vesicles) was not susceptible to proteinase digestion or detectable by monoclonal antibodies. This experiment was repeated in the laboratory of one of the authors (DHD) using proteinase K [2]. Proteinase K can completely digest Cry1Ab in solution and can remove all of the surface proteins from *Manduca sexta* (tobacco hornworm) BBMV. However, when bound to *M. sexta* BBMV, Cry1Ab was no longer susceptible to proteinase K, and solubilisation of the BBMV sample revealed a 60 kDa protein. This 60 kDa protein was identified as a major part (less α1) of the Cry1Ab toxin by sequencing and probing the protein with various antisera including those made against Cry1Ab, *M. sexta* aminopeptidase N and 210 kDa cadherin-like receptors. The conclusion is that the whole toxin, not the whole or a part of domain I, is inserted in the BBMV

membrane without the receptor molecule. Other data that support the
whole-insertion model have been presented in doctoral dissertations.
Mutations in α7 alter the ion channel activity [1]. Mutations in α2
effect the membrane insertion, inhibition of short-circuit current and
insecticidal activity [2]. Disulphide bridges between α2 and α5 and
between α5 and α7 have been introduced into the Cry3Aa toxin [60].
These mutations result in stable toxins that have increased Tm (75°C
compared to 55°C for the wild-type toxin), but no reduction in
insecticidal activity. A higher Tm figure means a more rigid protein.
These data contradict the partial insertion models, which predict that
only α4 and α5 (umbrella model) or α5 and α6 (penknife model)
partition into the membrane and form the ion channel.

Mutations in domain I have shown to decrease or increase the
insecticidal activity. Site-directed mutagenesis made on Cry1Ab revealed
residues around R93 are important for the insecticidal activity against
M. sexta. Any amino acid substitutions on R93 except for lysine caused a
loss of the activity [59]. Arginine and lysine have positively charged,
bulky side groups. The orientation of the R93 side group is important to
understand the result of this mutagenesis study. Since the peptide
sequence of domain I of Cry1Ab is almost identical to that of Cry1Aa, we
can comfortably predict the orientation of the Cry1Ab R93 side group.
According to the Cry1Aa structure as shown in Figure 3, the R93 side
group appeared to fit snugly in a space between α2 and α3. R93 seems
important for the configuration of α2 and α3.

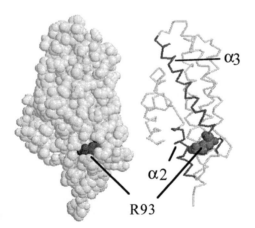

Figure 11. Orientation of the R93 side group showing in both space-fill and backbone
models of Cry1Aa domain I, which is presumably almost identical to Cry1Ab domain I.
Two figures are shown in an identical orientation.

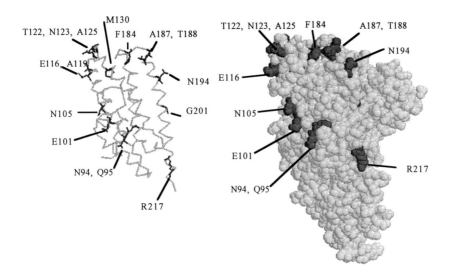

Figure 12. Cry1Aa model showing the location of mutated residues, which were found in the Cry1Ab up-mutants. Mutated residues are shown in grey. The left figure is a backbone model of domain I. Note that most side groups are extruding to the solvent. The right figure is a space-fill model of the entire Cry1Aa toxin. Two figures are shown in an identical orientation.

The same domain of Cry1Ab has been studied by chemically induced, random mutations [31]. Several thousand mutants were made and screened for increased activity against *H. virescens*. A small number of up-mutants, some of which showed activity increases as much as five times of that of the wild-type protein, were found, and their peptide sequences were determined. Mutations in up-mutants include N94K, Q95K, E101K, N105Y, E116K, A119T, T122I, N123Y, A125V, M130I, F184I, A187T, T188S, N194K, G201D and R217H. No up-mutants were found with mutations in domain II. At the time when this study was done about 10 years ago, no structure of crystal proteins was available. Now, we are able to examine the results on the Cry1Aa structure. Interestingly, most mutations were found in α3 (E90-A119), α4 (P124-L148) and the loop connecting these two α-helices. There is a cluster of mutations found in α6 (A186-V218) and the loop connecting α5 to α6 (F184I). All these mutated residues except for M130 and G201 are exposed to the solvent (Figure 4). Mutations occurred in α3 are particularly interesting.

Four of the 6 residues mutated in this helix were changed to the positively charged lysine from either neutral or negatively charged amino acid residues. Mutations in two loops (α3/α4 and α5/α6 loops) appear to make the loops more hydrophobic. Although we do not know exactly what this finding means, there are several important points to note. Presumably, the exposed residues are more likely to be involved in the insecticidal activity. In this case, the protein functions that modify the activity include the membrane insertion and ion-channel activity. On the other hand, the mutations made on the embedded amino acid residues tend to modify the backbone structure of the protein causing the protein to lose the activity. Together with the R93 mutation study, this chemically induced mutagenesis study seems to support the model that the whole toxin molecule, not a part of domain I, inserts into the membrane.

6.   RECEPTOR BINDING SITES IN DOMAIN II

Domain II appears to contain a receptor binding region or regions. Domain exchange between two crystal proteins having different insect specificity has demonstrated that insect specificity resides in domain II [22]. A subsequent report showed that the receptor binding determines the insect specificity [52]. Further studies, in particular those using site-directed mutagenesis as described below, clearly indicate that this domain is involved in the receptor binding. Domain II is consisting of repeating β-strands, which form a triangular column made of three sheets. There is a striking similarity between domain II and plant lectin jacalin [45]. Jacalin binds carbohydrate with loops at the end of the triangular column. One of the *B. thuringiensis* receptors, aminopeptidase N, is reported to contain carbohydrate, and the *B. thuringiensis* crystal toxins are believed to recognise the carbohydrate moiety of the receptor [32]. Details on *B. thuringiensis* receptors are reviewed elsewhere in this book.

There are numerous reports empirically indicating that specific amino acid residues within domain II are involved in the receptor binding. Some of these reports will be reviewed below. When the structural determination of Cry3Aa made it possible to predict the structure of other proteins, three loops at the bottom end of the domain II triangular column were subjected to extensive studies. Unfortunately, only limited studies have been done to date on Cry3Aa and Cry1Aa of which structures have been determined. When proteins other than Cry3Aa and Cry1Aa are used, lack of information concerning the location and orientation of specific amino acid residues limits the understanding of the results of these studies. Most studies made mutations on a block of several amino acid residues. The block mutation has higher chance to modify the backbone structure of the protein than a single amino acid residue mutation does. This approach, therefore, causes additional difficulties in interpreting the result.

Single residue substitutions were done on CryIAb domain II loops [44]. Using *M. sexta*, F371 in putative loop 2 was found to be significant for "irreversible" binding. In *M. sexta*, CryIAb recognises the 210 kDa cadherin-like receptor. The receptor binding study was done with BBMV prepared from the midgut. The crystal toxin acts on BBMV by a two-step process. First, the toxin binds to the receptor. This binding is reversible, as an excess amount of competitive toxin molecules replaces the bound toxin. Second, the receptor-toxin binding becomes irreversible. This irreversible binding can be explained by a model that the toxin inserts into the membrane.

As described above, the tertiary structure of CryIAb can be predicted from that of CryIAa. The structure of CryIAb is considered to be very similar to that of CryIAa, because two sequences are quite homologous. The structure, particularly the backbone structure, must be highly conserved. As shown in Figure 5, the sequence heterogeneity is more obvious in loop 2 connecting β6 and β7 but not within these β-strands. The loop 3 sequence also substantially differs between these two proteins. CryIAa loop 2 has a stretch of hydrophobic residues of – I369-I370-L371– whilst the corresponding residues in CryIAb are relatively hydrophilic except for F371. It is very important to note that a mutation at F371 causes a reduction in "irreversible" binding suggesting that this residue may be involved in the membrane insertion not in the receptor binding. This speculation is further supported by the location of F371. Loop 2 is likely to be protruding into the solvent more than loops 1 and 3. Particularly, the hydrophobic F371 residue appears to be exposed.

Several mutagenesis studies were done on CryICa. For several reasons, CryICa is an interesting protein, but its structure has not been determined. Two putative loops, loop 1 and loop 2 of the CryICa domain II as shown in Figure 5, were examined by block mutagenesis [49]. Using degenerate oligonucleotides, mutations were made on a block of amino acid residues in these putative loops. Although no receptor binding was examined, the mutants were evaluated on Sf9 (*Spodoptera frugiperda:* fall armyworm) cells and *Aedes aegypti* (yellow fever mosquito) larvae. The report clearly showed that a single amino acid substitution (i.e. R318) changes the insect specificity. The R318I substitution killed the dipteran activity whilst maintaining the Sf9 cytotoxicity. Importance of loop 1 in the dipteran activity will be discussed later with Cry2Aa.

```
              β2          Loop 1      β3
CrylAa    DILNSITIYTDV  HRG      FNYWSGHQITASP
CrylAb    DILNSITIYTDA  HRG      EYYWSGHQIMASP
CrylCa    DILNNLTIFTDW  FSVG R   NFYWGGHRVISSL


              β6              Loop 2          β7
CrylAa    IFRTLSSPLYR  RIILGSGPNN Q  ELFVLDGTEFSF
CrylAb    VYRTLSSTLYR  RP  FNIGINNQ  QLSVLDGTEFAY
CrylCa    VFRTLSNPTLR  ILQ QPWPAPPF  NLRGVEGVEFST


              β10         Loop 3     β11
CrylAa    HRLSHVTMLS  QA   AGAVY  TLRAPTFSWQH
CrylAb    HRLSHVSMFR  SGFSNSSVS  IIRAPMFSWIH
CrylCa    HRLCHATFVQ  RS  G TPF   LTTGVVFSWTD
```

Figure 5. Comparison of the loop sequences amongst CrylAa, CrylAb and CrylCa.
I369-I370-L371 of CrylAa, F371 of CrylAb and R318, Q374, T440 of CrylCa are
underlined. Boxed residues in CrylCa were studied by alanine scanning.

Three domain II loops of CrylCa were extensively studied in the
laboratory of one of the authors (TY) using a technique called alanine
scanning [42]. This technique substitutes amino acid residues in an
interesting part of the protein with alanine one at a time. Each mutant is
then tested for the protein function to see which residue substitution
would modify the function. Alanine substitution effectively removes the
bulky amino acid side group, which may be involved in the protein
function. Since it is a single residue mutation, and alanine maintains the
flexibility of the peptide chain, alanine scanning minimises the
modification of the protein backbone. Amongst the mutants made on the
residues (boxed in Figure 5), which are predicted to be in and near the
loops, Q374A in loop 2 and T440A in loop 3 were found to be
substantially less active against *S. exigua* (beet armyworm). A
competitive binding study using *S. exigua* BBMV indicated that Q374A
and T440A still retain the binding capability, albeit at a reduced level.
But, the binding was completely abolished when these two residues were
mutated to alanine at the same time (Figure 6). A structural model of
CrylCa was made using Insight II computer software package. According
to the model, Q374 and T440 are at the tips of loop 2 and loop 3 and are
fully exposed. This conclusion and the molecular model presented in the
figure, have some limitations. Loop regions are relatively flexible.
Therefore, structural prediction is very difficult, if possible at all. Also,

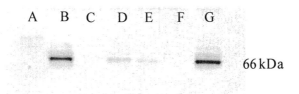

Figure 6. Competitive binding of trypsin-activated Cry1Ca and its mutants on *S. exigua* BBMV. Biotin-labelled, wild-type Cry1Ca toxin was allowed to bind to BBMV and challenged with 500-fold excess, unlabelled mutant toxins. BBMV was then analysed by SDS-PAGE followed by blotting to detect the biotin-labelled toxin remaining on BBMV. A: BBMV only, B: no competition, C: wild-type toxin, D: Q374A, E: T440A, F: Q373A (a fully active mutant), G: Q374A-T440A double mutant. Note the faint band at 66 kDa on D and F.

we found Q374 and T440 are next to prolines. Substitutions of these residues in juxtaposition to proline may cause undesired structural changes to the protein structural backbone more than expected from other residues. Further study is needed to overcome these limitations. It is particularly sensible to conduct more extensive mutagenesis study on a structurally known protein such as Cry1Aa.

7.   DOMAIN III FUNCTION

The function of domain III was originally suggested to protect the toxin molecule from proteinase digestion [34]. However, a domain exchange study showed that domain III of Cry1Ac is required, in addition to domain II, for *H. virescens* specificity [21]. Further study suggested that this domain contains a receptor binding site or sites [33]. When domain III was exchanged between Cry1Aa and Cry1Ac, the receptor specificity of these proteins was transferred with domain III. Cry1Aa binds to a 210 kDa cadherin-like protein, and Cry1Ac recognises the 120 kDa aminopeptidase N in *M. sexta*. Another study using site-directed mutagenesis demonstrated that S503 and S504 in Cry1Ac domain III are involved in the receptor binding [3]. When superimposed on the Cry1Aa structure, these residues are presumably situated in the loop between β15 and β16 and are fully exposed.

Domain III of Cry1Ca seems to contain a receptor binding site or sites. A hybrid protein consisting of domains I and II of Cry1Ea and domain III of Cry1Ca showed substantial increase in the activity against *S. exigua* over Cry1Ca [8]. A similar hybrid between Cry1Ab and Cry1Ca demonstrated six times higher activity to *S. exigua* than Cry1Ca [14]. The activity improvements observed in these hybrid proteins can be accomplished by increasing the number of receptor binding sites on one molecule. Domain II of Cry1Ea and Cry1Ab binds *S. exigua* BBMV. Combining the receptor-binding site in domain II with the site in domain III, the hybrid proteins may bind to the *S. exigua* cells at multiple sites. Interestingly, no activity improvement was seen with the similar Cry1Ac-Cry1Ca hybrid. This multiple site theory needs further scrutiny [15].

8.   CRY2A FAMILY PROTEINS

The first dual-specificity crystal protein, Cry2A, was found in *B. thuringiensis kurstaki* strains and isolated as a protein responsible for the Diptera activity of these strains [62]. There are several Cry2A-type proteins. Cry2Ab, which is highly homologous to Cry2Aa, is not toxic to Diptera [57]. Efforts were made by domain exchange to locate regions responsible for the dipteran and lepidopteran activity [35, 56]. The results indicate that the Cry2Aa region from I307 to V340 is responsible for the dipteran activity. When the region of Cry2Aa from N341 to R412 was transferred to Cry2Ab, Cry2Ab showed the high lepidopteran activity of Cry2Aa. Interestingly, only a limited number of residues differ within the region from I307 to R412 between these two proteins. There are 9 unique residues in the dipteran-specific region and 14 unique residues in the lepidopteran-specific region (Figure 7).

```
Dipteran-specific region
2Aa:   ILSGISGTRLS ITFPNI GG  LPG    STTTHSLNSARV
2Ab:   V.N.F..A...N.....V.  ...    .....A.LA...
18Aa:  V...V..IGARF.YSTVL.RY.HDDLKNII.TYVGGTQG


Lepidopteran -specific region
2Aa:   NYSGG VSSG LIG   AT NL  NH NFNCSTVLP PLSTPFVRSWLDSGTDREGVAT STNWQTESFQT
2Ab:   ..... I...D..  .SPF  .Q ......F.. ..L...........S.......V........E.
18Aa:  PNI.VQL.TTELD **LQQVDFQFFTL..  M..N.ITA.YFATS.YES  .YS SIGGYLRKDV.KS
```

Figure 7. Insect specific sequences of Cry2Aa and Cry2Ab in comparison with Cry18Aa. Insertion at **: ELKKQQQATRDS. Dots signify the conserved residue. I318, G324, L350, N355 and S390 of Cry2Aa are underlined. The sequence alignment of Cry18Aa to Cry2A was done by Dr. Richard Morse, University of California at San Francisco and is shown here with his kind permission.

The newly available tertiary structure of this protein appears to be contributing to the further understanding of the functional regions of Cry2A proteins [38]. Although the sequence of Cry2Aa is quite different from those of Cry1Aa and Cry3Aa, Cry2Aa showed the typical three-domain structure. As shown in Figure 8, only two unique residues out of 9 mosquito specific residues appeared to be exposed. One of those, G324 is in loop 1. A similarly situated R318 of Cry1Ca has been identified as a residue involved in the dipteran activity of Cry1Ca. In the Lepidoptera-specific region of Cry2Aa, there are 14 residues, which differ from the corresponding residues of Cry2Ab. Out of these 14, only a few residues seem to cause a substantial difference in the protein function. When these residues were identified on the Cry2Aa structure, only two residues appeared to be exposed. As shown in Figure 8, they locate on the side of domain II opposite from domain I, not in the bottom loops. Further study is needed to understand the significance of these residues, which may be involved in the receptor binding.

An insecticidal toxin gene from *B. popilliae*, which is a pathogen of *Popillia japonica* (Japanese beetle), has been cloned and sequenced [63]. Surprisingly, this gene is substantially homologous to the *cry2* genes. The homology is particularly held up to the beginning of domain II. High heterogeneity between Cry2Aa and Cry18Aa is seen in the Lepidoptera-specific region (Figure 7). Such a sequence uniqueness of this Cry18Aa region may contribute to the specificity of this protein.

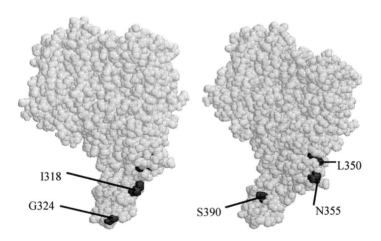

Figure 8. Cry2Aa structure determined by X-ray crystallography showing amino acid residues presumably involved in the insect specificity determination. The unpublished Cry2Aa structure is shown here with a kind permission given by Drs. Richard Morse and Robert Stroud, University of California at San Francisco.

9.   FUTURE FOR THE BACTERIAL INSECTICIDAL PROTEINS

Bacterial insecticidal proteins can be applied for pest control in two major ways, spray-on and transgenic plants. Since the protein is active by ingestion, it must be delivered to the habitat where insects eat. *B. thuringiensis* has been used in commercial insecticide formulations that are sprayed onto the crop plant like synthetic chemical insecticides. Although *B. thuringiensis* insecticidal proteins have very high specific activity (e.g., LC_{50} can be as low as 0.01 ppm) in comparison with typical chemical insecticides, it is expensive to produce and relatively unstable in the field. A typical *B. thuringiensis* formulation contains the insecticidal proteins at a concentration of only a few percents.

It seems that a better future for bacterial insecticides will be accomplished with transgenic crops. Several *B. thuringiensis* *cry* genes have been used in crop plants for commercial applications. For example, *cry1Ab* was introduced to corn [17], *cry1Ac* to cotton [40] and *cry3Aa* to potato [41]. Initially, the *cry* gene was used without nucleotide sequence modification. The low expression of the original *cry* gene was then overcome by making the A/T-rich *Bacillus* sequence to fit better for plants. Producing the crystal proteins in crop plants has several advantages. For example, the protein is continuously produced to protect the crop. Obviously, this approach eliminates or reduces the need to spray chemical insecticides, which may kill beneficial insects. On the other hand, the continuous gene expression causes a concern of insect resistance. Insects resistance to *B. thuringiensis* crystal proteins became an important issue after the first discovery of a resistant *Plutella xylostella* population [50]. Several strategies to prevent or to delay the resistance development have been proposed. One of these strategies is to alternate the gene. The other is to use multiple genes. These ideas are supported by the finding that some crystal proteins such as Cry1Ba and Cry1Ca are still active against the Cry1A-resistant *P. xylostella* population [51]. Under the circumstance like this, possible use of *B. thuringiensis vip* genes is sensible. We believe that a new discovery of more potent bacterial insecticidal proteins and engineering of the existing genes for higher activity and wider insect specificity are dearly needed.

REFERENCES

[1] Alcantara EP (1997) Electrophysiology, receptor binding kinetics and mutational analysis of *Bacillus thuringiensis* delta-endotoxin. PhD Dissertation, The Ohio State University

[2] Alzate O (1998) Structural and functional studies on the *Bacillus thuringiensis* Cry1Ab δ-endotoxin and its membrane bound state to *Manduca sexta* brush border membrane vesicles. PhD Dissertation, The Ohio State University

[3] Aronson AI, Wu D & Zhang C (1995) Mutagenesis of specificity and toxicity regions of a *Bacillus thuringiensis* protoxin gene. J. Bacteriol. 177, 4059-4065

[4] Asano SI, Nukumizu Y, Bando H, Iizuka T & Yamamoto T (1997) Cloning of novel enterotoxin genes from *Bacillus cereus* and *Bacillus thuringiensis*. Appl. Environ. Microbiol. 63, 1054-1057

[5] Beegle CC & Yamamoto T (1992) Invitation paper (C.P. Alexander Fund): History of *Bacillus thuringiensis* Berliner research and development. Can. Ent. 124, 587-616

[6] Berliner E (1911) Die "Schlaffsucht" der Mehlmottenraupe. Z. Gesamte Getreidewes. 3, 63-70

[7] Bietlot HP, Carey PR, Pozsgay M & Kaplan H (1989) Isolation of carboxyl-terminal peptides from proteins by diagonal electrophoresis: application to the entomocidal toxin from *Bacillus thuringiensis*. Anal. Biochem. 181, 212-215

[8] Bosch D, Schipper B, van der Kleij H, de Maagd RA & Stiekema WJ (1994) Recombinant *Bacillus thuringiensis* crystal proteins with new properties: possibilities for resistance management. Biotechnology 12, 915-918

[9] Carlson CR, Johansen T & Kolstø AB (1996) The chromosome map of *Bacillus thuringiensis* subsp. *canadensis* HD224 is highly similar to that of the *Bacillus cereus* type strain ATCC 14579. FEMS Microbiol. Lett. 141, 163-167

[10] Carlson CR & Kolstø AB (1993) A complete physical map of a *Bacillus thuringiensis* chromosome. J. Bacteriol. 175, 1053-1060

[11] Chestukhina GG, Kostina LI, Zalunin IA *et al.* (1994) Production of multiple delta-endotoxins by *Bacillus thuringiensis*: delta-endotoxins produced by strains of the subspecies *galleriae* and *wuhanensis*. Can. J. Microbiol. 40, 1026-1034

[12] Choma CT & Kaplan H (1990) Folding and unfolding of the protoxin from *Bacillus thuringiensis*: evidence that the toxic moiety is present in an active conformation. Biochem. 29, 10971-10977

[13] Crickmore N, Zeigler DR, Feitelson J *et al.* (1998) Revision of the nomenclature for the *Bacillus thuringiensis* pesticidal crystal proteins. Microbiol. Mol. Biol. Rev. 62, 807-813

[14] de Maagd RA, Kwa MS, van der Klei H *et al.* (1996) Domain III substitution in *Bacillus thuringiensis* delta-endotoxin CryIA(b) results in superior toxicity for *Spodoptera exigua* and altered membrane protein recognition. Appl. Environ. Microbiol. 62, 1537-1543

[15] de Maagd RA, van der Klei H, Bakker PL, Stiekema WJ & Bosch D (1996) Different domains of *Bacillus thuringiensis* delta-endotoxins can bind to insect midgut membrane proteins on ligand blots. Appl. Environ. Microbiol. 62, 2753-2757

[16] Dean DH, Rajamohan F, Lee MK *et al.* (1996) Probing the mechanism of action of *Bacillus thuringiensis* insecticidal proteins by site-directed mutagenesis-a minireview. Gene 179, 111-117

[17] Estruch JJ, Carozzi NB, Desai N *et al.* (1994) The expression of a synthetic *cryIAb* gene in transgenic maize confers resistance to European corn borer. International symposium on insect resistant maize: recent advances and utilization, p.172-174, Mexico

[18] Estruch JJ, Warren GW, Mullins MA *et al.* (1996) Vip3A, a novel *Bacillus thuringiensis* vegetative insecticidal protein with a wide spectrum of activities against lepidopteran insects. Proc. Natl. Acad. Sci. USA 93, 5389-5394

[19] Gazit E & Shai Y (1993) Structural and functional characterization of the alpha 5 segment of *Bacillus thuringiensis* delta-endotoxin. Biochem. 32, 3429-3436

[20] Gazit E & Shai Y (1993) Structural characterization, membrane interaction, and specific assembly within phospholipid membranes of hydrophobic segments from *Bacillus thuringiensis* var. *israelensis* cytolytic toxin. Biochem. 32, 12363-12371

[21] Ge AZ, Rivers D, Milne R & Dean DH (1991) Functional domains of *Bacillus thuringiensis* insecticidal crystal proteins. Refinement of *Heliothis virescens* and *Trichoplusia ni* specificity domains on CryIA(c). J. Biol. Chem. 266, 17954-17958

[22] Ge AZ, Shivarova NI & Dean DH (1989) Location of the *Bombyx mori* specificity domain on a *Bacillus thuringiensis* delta-endotoxin protein. Proc. Natl. Acad. Sci. USA 86, 4037-4041

[23] González JM, Jr, Brown BJ & Carlton BC (1982) Transfer of *Bacillus thuringiensis* plasmids coding for delta-endotoxin among strains of *B. thuringiensis* and *B. cereus*. Proc. Natl. Acad. Sci. USA 79, 6951-6955

[24] González JM, Jr, Dulmage HT & Carlton BC (1981) Correlation between specific plasmids and delta-endotoxin production in *Bacillus thuringiensis*. Plasmid 5, 352-365

[25] Grochulski P, Masson L, Borisova S *et al.* (1995) *Bacillus thuringiensis* CryIA(a) insecticidal toxin: crystal structure and channel formation. J. Mol. Biol. 254, 447-464

[26] Hickle LA & Fitch WL (1990) Analytical Chemistry of *Bacillus thuringiensis*. American Chemical Society

[27] Hodgman TC & Ellar DJ (1990) Models for the structure and function of the *Bacillus thuringiensis* delta-endotoxins determined by compilational analysis. DNA Seq. 1, 97-106

[28] Höfte H, de Greve H, Seurinck J *et al.* (1986) Structural and functional analysis of a cloned delta endotoxin of *Bacillus thuringiensis* berliner 1715. Eur. J. Biochem. 161, 273-280

[29] Höfte H & Whiteley HR (1989) Insecticidal crystal proteins of *Bacillus thuringiensis*. Microbiol. Rev. 53, 242-255

[30] Ishiwata S (1901) On a severe flacherie (sotto disease). Dainihon Sanshi Kaiho 114, 1-5

[31] Jellis C, Bassand D, Beerman N *et al.* (1989) Molecular biology of *Bacillus thuringiensis* and potential benefits to agriculture. Israel. J. Entomol. 23, 189-199

[32] Knight PJ, Crickmore N & Ellar DJ (1994) The receptor for *Bacillus thuringiensis* CryIA(c) delta-endotoxin in the brush border membrane of the lepidopteran *Manduca sexta* is aminopeptidase N. Mol. Microbiol. 11, 429-436

[33] Lee MK, Young BA & Dean DH (1995) Domain III exchanges of *Bacillus thuringiensis* CryIA toxins affect binding to different gypsy moth midgut receptors. Biochem. Biophys. Res. Commun. 216, 306-312

[34] Li JD, Carroll J & Ellar DJ (1991) Crystal structure of insecticidal delta-endotoxin from *Bacillus thuringiensis* at 2.5 A resolution. Nature 353, 815-821

[35] Liang Y & Dean DH (1994) Location of a lepidopteran specificity region in insecticidal crystal protein CryIIA from *Bacillus thuringiensis*. Mol. Microbiol. 13, 569-575

[36] Masson L, Erlandson M, Puzstai-Carey M *et al.* (1998) A holistic approach for determining the entomopathogenic potential of *Bacillus thuringiensis* strains. Appl. Environ. Microbiol. 64, 4782-4788

[37] Masson L, Prefontaine G, Peloquin L, Lau PC & Brousseau R (1990) Comparative analysis of the individual protoxin components in P1 crystals of *Bacillus thuringiensis* subsp. *kurstaki* isolates NRD-12 and HD-1. Biochem. J. 269, 507-512

[38] Morse RJ, Powell G, Ramalingam V, Yamamoto T & Stroud RM (1998) Crystal structure of Cry2Aa from *Bacillus thuringiensis* at 2.2 Angstromes: structural bases of dual specificity. IVth International Conference on *Bacillus thuringiensis*, p.1, Sapporo

[39] Nagamatsu Y, Itai Y, Hatanaka C, Funatsu G & Hayashi K (1984) A toxic fragment from the entomocidal crystal protein of *Bacillus thuringiensis*. Agric. Biol. Chem. 48, 611-619

[40] Perlak FJ, Deaton RW, Armstrong TA *et al.* (1990) Insect resistant cotton plants. Biotechnology 8, 939-943

[41] Perlak FJ, Stone TB, Muskopf YM *et al.* (1993) Genetically improved potatoes: protection from damage by Colorado potato beetles. Plant. Mol. Biol. 22, 313-321

[42] Powell G, Asano S, Libs J *et al.* (1996) Effect of domain II mutations on binding of a *Bacillus thuringiensis* crystal protein to *Spodoptera exiguea* BBMV. IIIrd International Conference on *Bacillus thuringiensis*, p. 65, Cordoba

[43] Pusztai-Carey M (1994) A novel method for quantitation and isolation of individual toxins from multi-gene *B. thuringiensis* strains. IInd International Conference on *Bacillus thuringiensis*, p. 51, Montpellier

[44] Rajamohan F, Cotrill JA, Gould F & Dean DH (1996) Role of domain II, loop 2 residues of *Bacillus thuringiensis* CryIAb delta-endotoxin in reversible and irreversible binding to *Manduca sexta* and *Heliothis virescens*. J. Biol. Chem. 271, 2390-2396

[45] Sankaranarayanan R, Sekar K, Banerjee R *et al.* (1996) A novel mode of carbohydrate recognition in jacalin, a Moraceae plant lectin with a beta-prism fold. Nat. Struct. Biol. 3, 596-603

[46] Schnepf HE & Whiteley HR (1981) Cloning and expression of the *Bacillus thuringiensis* crystal protein gene in *Escherichia coli*. Proc. Natl. Acad. Sci. USA 78, 2893-2897

[47] Schnepf HE & Whiteley HR (1985) Delineation of a toxin-encoding segment of a *Bacillus thuringiensis* crystal protein gene. J. Biol. Chem. 260, 6273-6280

[48] Schwartz JL, Juteau M, Grochulski P *et al.* (1997) Restriction of intramolecular movements within the Cry1Aa toxin molecule of *Bacillus thuringiensis* through disulfide bond engineering. FEBS Lett. 410, 397-402

[49] Smith GP & Ellar DJ (1994) Mutagenesis of two surface-exposed loops of the *Bacillus thuringiensis* CryIC delta-endotoxin affects insecticidal specificity. Biochem. J. 302, 611-616

[50] Tabashnik BE, Cushing NL, Finson N & Johnson MW (1990) Field development of resistance to *Bacillus thuringiensis* in diamondback moth (Lepidoptera: Plutellidae). J. Econ. Entomol. 83, 1671-1676

[51] Tabashnik BE, Malvar T, Liu YB *et al.* (1996) Cross-resistance of the diamondback moth indicates altered interactions with domain II of *Bacillus thuringiensis* toxins. Appl. Environ. Microbiol. 62, 2839-2844

[52] Van Rie J, Jansens S, Hofte H, Degheele D & Van Mellaert H (1989) Specificity of *Bacillus thuringiensis* delta-endotoxins. Importance of specific receptors on the brush border membrane of the mid-gut of target insects. Eur. J. Biochem. 186, 239-247

[53] Von Tersch MA, Slatin SL, Kulesza CA & English LH (1994) Membrane-permeabilizing activities of *Bacillus thuringiensis* coleopteran-active toxin CryIIIB2 and CryIIIB2 domain I peptide. Appl. Environ. Microbiol. 60, 3711-3717

[54] Wabiko H, Held GA & Bulla LA, Jr (1985) Only part of the protoxin gene of *Bacillus thuringiensis* subsp. *berliner* 1715 is necessary for insecticidal activity. Appl. Environ. Microbiol. 49, 706-708

[55] Warren GW, Koziel MG, Mullins MA *et al.* (1998): Auxiliary proteins for enhancing the insecticidal activity of pesticidal proteins. Novartis, US Patent 5770696

[56] Widner WR & Whiteley HR (1990) Location of the dipteran specificity region in a lepidopteran-dipteran crystal protein from *Bacillus thuringiensis*. J. Bacteriol. 172, 2826-2832

[57] Widner WR & Whiteley HR (1989) Two highly related insecticidal crystal proteins of *Bacillus thuringiensis* subsp. *kurstaki* possess different host range specificities. J. Bacteriol. 171, 965-974

[58] Wolfersberger MG, Luthy P, Maurer A *et al.* (1987) Interaction of *Bacillus thuringiensis* delta-endotoxin with membrane vesicles isolated from lepidopteran larval midgut. Comp. Biochem. Physiol. 86A, 301-308

[59] Wu D & Aronson AI (1992) Localized mutagenesis defines regions of the *Bacillus thuringiensis* delta-endotoxin involved in toxicity and specificity. J. Biol. Chem. 267, 2311-2317

[60] Wu S-J (1996) Domain-function studies of *Bacillus thuringiensis* Cry3A δ-endotoxin; a molecular genetic apporach. PhD dissertation, The Ohio State University

[61] Yamamoto T, Kalman S, Powell G, Cooper N & Cerf D (1998) Environmental release of genetically engineered *Bacillus thuringiensis*. Rev. Toxicol. 2, 157-166

[62] Yamamoto T & McLaughlin RE (1981) Isolation of a protein from the parasporal crystal of *Bacillus thuringiensis* var. *kurstaki* toxic to the mosquito larva, *Aedes taeniorhynchus*. Biochem. Biophys. Res. Commun. 103, 414-421

[63] Zhang J, Hodgman TC, Krieger L, Schnetter W & Schairer HU (1997) Cloning and analysis of the first *cry* gene from *Bacillus popilliae*. J. Bacteriol. 179, 4336-4341

Chapter 2.3

Vector-active toxins: structure and diversity

Armelle Delécluse[1], Victor Juárez-Pérez[1] & Colin Berry[2]

[1]*Laboratoire des Bactéries et Champignons Entomopathogènes, Institut Pasteur, 25 rue du Dr Roux, 75724 Paris Cedex 15, France and* [2]*Cardiff School of Biosciences, Cardiff University, Museum Avenue PO Box 911, Cardiff CF10 3US, Wales, UK*

Key words: Mosquitocidal toxin, *Bacillus thuringiensis, Bacillus sphaericus, Brevibacillus laterosporus, Clostridium bifermentans*

Abstract: Bacteria active against Dipteran larvae – mosquitoes and blackflies – include a wide variety of *Bacillus thuringiensis* and *Bacillus sphaericus* strains, as well as isolates of *Brevibacillus laterosporus* and *Clostridium bifermentans*. All display different spectra and levels of activity correlated with the nature of the toxins produced during the sporulation process. This paper presents an overview of all mosquitocidal strains reported to date and describes the numerous toxins – including both Cry and Cyt proteins, and others – in terms of primary structure and activity against mosquito larvae.

1.   INTRODUCTION

Mosquitoes and blackflies (Diptera: Nematocera) are responsible for many important tropical diseases such as malaria, yellow fever, dengue and filarial conditions including onchocerciasis. The widespread use of chemical insecticides to control these insects has led to the development of many resistant populations. Therefore, much attention has been given to more environmentally friendly insecticides, and particularly to the use of microbial control agents that cause disease in the target insects.

Entomopathogenic bacteria, namely *Bacillus thuringiensis*, have been known from the early 1900's (see Chapter 2.2), but the control of Dipteran species has been envisaged only since the discovery of *B. thuringiensis* serovar *israelensis*. The high activity of *Bt israelensis* led

J.-F. Charles et al. (eds.),
Entomopathogenic Bacteria: From Laboratory to Field Application, 101–125.
© 2000 *Kluwer Academic Publishers. Printed in the Netherlands.*

to an increased interest in another bacterium, *B. sphaericus,* for which, initially, only weakly active strains were known. Since that time, several screening programmes, aimed at isolating different mosquitocidal strains, have been developed. These have resulted in the identification of a wide variety of Gram-positive bacteria, including both *B. thuringiensis* and *B. sphaericus* isolates and new bacteria, for which no insecticidal activity had been reported previously: *Brevibacillus laterosporus* and *Clostridium bifermentans.*

All strains isolated have been characterised for their level of mosquitocidal activity and specificity. For most of these bacteria, the factors responsible for the insecticidal activity have then been identified and characterised. As shown below, although not all is known about these mosquitocidal bacteria, many Cry, Cyt, and other toxins – differing in their primary structure, specificity and level of activity – have now been characterised.

2.   DIPTERAN ACTIVE BACTERIA

2.1   *Bacillus thuringiensis* strains

The first mosquitocidal *B. thuringiensis* strain was isolated in 1976 from dead *Culex* larvae [39], serotyped as H14 [6], and designated serovar *israelensis* (*Bti*). Composite crystals, produced from stage II of sporulation, are responsible for toxicity. These inclusions show specificity for larval stages of many Culicidae (mosquito) genera: *Culex, Aedes, Anopheles,* with a higher specificity for the two former [39]; crystals are also toxic against Simuliidae [114], Tipulidae [116] and Chironomidae [91] larvae. In contrast, they have no effect on vertebrates and non-target invertebrates [68, 99]. However, upon solubilisation, the crystals display non-specific cytolytic and haemolytic activities [112]. The *Bti* inclusions are composed of four major proteins with molecular weight of 135, 125, 68, and 28 kDa.

Since the discovery of *Bti*, several other *Bt* strains displaying mosquitocidal activity have been isolated [51, 66, 88]. These strains, differing from *Bti* by either their serotype, mosquitocidal activity or polypeptide composition, can be classed into 3 groups (Table 1, page 123): class 1 includes 8 strains with larvicidal and haemolytic activities as well as crystal polypeptides similar to those of *Bti*; class 2 contains 2 strains which are nearly as toxic as *Bti* but produce different polypeptides; class 3 includes 9 strains which synthesise polypeptides different from those found in *Bti* but are only weakly active. On solubilisation, crystals from some of the class 2 and class 3 strains display some degree of haemolytic activity. The polypeptide composition of crystals produced by strains in all 3 classes and their relative toxicity (compared to that found for *Bti*) are given in Table 1.

2.2 *Bacillus sphaericus* strains

Since the first mosquito-active strain of *B. sphaericus*, was reported in 1965 by Kellen *et al.* [54] our knowledge of this bacterium as an insect pathogen and its usefulness as an agent for biological control of the mosquito vectors of disease has increased greatly. The strain described by Kellen *et al.* showed only weak toxicity and was thus unsuitable for use in the field. As a result, there was little interest in *B. sphaericus* as a mosquito pathogen until the isolation of strains with significantly greater larvicidal activity in 1973 [100]. Although these strains were still insufficiently active for commercial use, their discovery stimulated the search for still more active strains and in 1976 the first high-toxicity *B. sphaericus* (strain 1593) was isolated from Indonesia. Since then many mosquitocidal strains have been reported and over 560 are now held by the International Entomopathogenic Bacillus Centre at the Institut Pasteur in Paris.

B. sphaericus strains show specificity for the larval stages of many mosquito genera (*Culex, Anopheles, Aedes, Mansonia* and *Psorophora*) with larval sensitivity varying between individual mosquito species. For instance, *Aedes aegypti* has low sensitivity whereas *Aedes atropalpus*, *Aedes nigromaculis* and *Aedes intrudens* are highly susceptible [15, 71]. In addition to the direct toxic effects of exposure to *B. sphaericus*, longer-term effects on the survival and fitness of treated larvae have been observed [59]. *Toxorhynchites* mosquitoes are insensitive to *B. sphaericus* [60]. These larvae prey on the larvae of other mosquitoes and thus, can be used in combination with *B. sphaericus* in control programmes. Tests with a range of other organisms have shown *B. sphaericus* to have little or no effect on non-target species and no effect on mammals [67, 70, 97].

The toxicity of *B. sphaericus* is mainly linked to the sporulation, during which crystals are produced, composed of 2 polypeptides with molecular weights of 51 and 42 kDa that together form the so-called binary toxin. However, sporulation independent-toxicity has also been reported, and may be attributable to one or more vegetative toxin: proteins of 100, 31.8, and 35.8 kDa, designated Mtx1, Mtx2, and Mtx3, respectively. Not all *B. sphaericus* produce all toxins (see Table 3, Chapter 1.1) but all highly toxic strains synthesise the binary toxin.

2.3   *Brevibacillus laterosporus*

Brevibacillus laterosporus, formerly *Bacillus laterosporus*, is a spore-forming bacterium characterised by its ability to produce canoe-shaped lamellar bodies, adjacent to the spore [69]. Several *B. laterosporus* strains were already reported to be mosquitocidal [35, 90], but with a toxicity at least 100 fold lower than that found for *Bti*, and independent of the presence of crystals. Recently, 2 strains – 16-92 (isolate 921) and LAT006 (isolate 615) have been identified – that are highly toxic for mosquitoes [83, 101]. These two strains are more toxic for the mosquitoes in the species *Aedes aegypti* and *Anopheles stephensi*, and less active against *C pipiens*. Both produce crystalline inclusions of various shapes and sizes [101] responsible for the toxicity [83]. Purified *B. laterosporus* 615 crystals are as toxic as those from *Bti*, to *A. aegypti* and *A. stephensi*. The nature of the crystal components responsible for this activity have not yet been elucidated.

2.4   *Clostridium bifermentans*

Recent screening programmes, aimed to identify new strains of insecticidal bacteria, led to the description, in 1990, of the first anaerobic isolate, CH18, having a high mosquitocidal activity. This strain was isolated from a mangrove swamp soil from Malaysia [62] and identified as *C. bifermentans* subsp. *malaysia* (*Cbm*) [7]. More recently, a new mosquitocidal *C. bifermentans* strain was reported, *C. bifermentans* subsp. *paraiba* (*Cbp*), isolated from a secondary forest floor in 1997 [94]. The insecticidal spectrum of both *C. bifermentans* strains is similar and their toxicity comparable to *Bti* strains, although the toxic factor(s) are not the same as in *Bti*; for these reasons they resemble class-2 *B. thuringiensis* strains. Both strains are characterised by a high toxicity to *Anopheles* larvae [111, 95] followed, in order of increasing larval susceptibility, by *Aedes* and *Culex*. *Simulium* sp. are also susceptible to *Cbm* whole cultures, but in a much less extent than the other Culicidae species [7]. *Clostridium* is a very diverse group of bacteria that includes some human pathogenic species. For this reason, special considerations were undertaken to determine the safety of the *Cbm* strain as a potential bioinsecticide. All tests conducted with non-target invertebrates and vertebrates confirmed the innocuity of this isolate [110].

Little is known concerning the insecticidal components of either *C. bifermentans* strain and, at present, only the *Cbm* strain has been extensively studied, with some controversial results. First, and in contrast to *Bt*, *B. sphaericus* or *B. laterosporus* strains, the *Cbm* isolate does not produce parasporal bodies or any other morphological phenotype that could be associated with toxicity. Amorphous structures are observed close to the spores and it was suggested that they could be responsible for toxicity [7]. However, similar structures are also present in non-mosquitocidal *C. bifermentans* strains (J.-F. Charles, personal communication). Nevertheless, mosquitocidal activity has been associated with sporulated cultures. Toxicity is greatly reduced or inactivated by physical or chemical treatments [7]. In addition, treatment of crude sporulated cell extracts with proteinases or with antibodies raised against a crude extract of *Cbm* cells leads to a total inactivation of the mosquitocidal activity [74], indicating a proteinaceous origin of the toxic factor(s). In this way, and by comparison of the crude extracts of one non-mosquitocidal *C. bifermentans* strain and the *Cbm* strain, four major proteins were identified. Those putative toxins include a doublet of 66-68 kDa and two other small proteins of 18 and 16 kDa [73]. The genes encoding those four proteins have already been cloned and sequenced [9]. PCR and western blotting analysis, showed that those four genes are also present and functional in the *Cbp* strain but have not yet been cloned [8].

3.   MOSQUITOCIDAL TOXINS

3.1   *B. thuringiensis* toxins

B. thuringiensis mosquitocidal toxins belong to the 2 classes previously mentioned (see Chapter 2.1): the Cry family with specific activity, and the Cyt family, polypeptides of which display cytolytic and haemolytic properties. All characteristics of these proteins (structure, molecular mass, toxicity) are reported in Table 2 (pages 124-125) and their structure is presented in Figure 2 from Chapter 2.1.

3.1.1   *B. thuringiensis israelensis* toxins

As mentioned before, *Bti* crystals are composed of 4 major polypeptides of 135, 125, 68, and 28 kDa, representing 5%, 5%, 35%, and 54% of total crystal proteins, respectively (A. Delécluse, unpublished results). All 4 proteins are protoxins, which are processed, after ingestion by target insects due to the action of midgut proteinases under alkaline pH, into smaller fragments ranging around 40-45 kDa for the 125-135 kDa proteins [4], 30-35 kDa for the 68 kDa [25], and 25 kDa for the 28 kDa [58].

Genes encoding these major polypeptides have been cloned, sequenced, and the expression products analysed. According to the new nomenclature, these polypeptides are designated Cry4Aa (125 kDa), Cry4Ba (135 kDa), Cry11Aa (68 kDa), and Cyt1Aa (28 kDa) ([24] and Chapter 2.1). Sequence analysis allowed the identification of 2 other toxin genes: *cry10Aa*, encoding a minor 58 kDa polypeptide downstream of *cry4Ba* [38, 113], and *cyt2Ba*, encoding a polypeptide of 29.8 kDa upstream of *cry4Ba* [40]. Both Cry4Aa and Cry4Ba display a similar structure with i) an identical carboxy-terminal part, which could be involved in the crystallisation, and ii) a variable amino-terminal part (40% similarity) containing the 5 blocks conserved in most Cry toxins (see Chapter 2.1), and which corresponds to the toxic fragment [28, 85, 127]. Cry10Aa lacks the carboxy-terminal part of the Cry4 toxins, and is composed of variable amino-terminal domains containing the 5 conserved blocks, thus resembling a Cry4-activated toxin. Cry11Aa completely differs from Cry4 and Cry10, and includes only the first block. Cyt1Aa and Cyt2Ba do not contain any of these blocks and totally differ in structure.

All *Bti* proteins have been found to be toxic for mosquito larvae, although their level of activity and specificity might differ (Table 2). No individual component was as active as the native *Bti* crystals (between 30 and 7,000 fold less active, depending on the protein and the mosquito species), and only synergistic effects between these toxins can explain the high toxicity of the native crystals [4, 23, 29, 86, 125]. In contrast, the cytolytic and haemolytic activities are due to the only presence of the Cyt proteins, with Cyt1Aa being the major component responsible: preliminary experiments indicate that the haemolytic activity of Cyt2Ba is about 1,000 fold less important than that of Cyt1Aa (A. Delécluse, unpublished).

3.1.2   Toxins from class 1-strains

Toxins from class 1-strains appear to be the same as those found in *Bti*. Indeed, crystals from these strains include polypeptides immunologically related to those of *Bti*, and with similar molecular weight. Moreover, in strain *morrisoni* PG14, genes identical to *cry4B* [115], *cry11A* [36], and a sequence differing from *cyt1Aa* by only 3 nucleotides (leading to one substitution at the protein level) have been found [34, 37]. This latter gene also encodes a polypeptide with haemolytic activity. Synergistic interactions between the different polypeptides are probably responsible for the high mosquitocidal activity of the class 1-strain crystals.

3.1.3   Toxins from class 2-strains

Crystals from class 2-strains have a polypeptide composition different from that found in *Bti* crystals, but also contain multiple proteins. In *Bt jegathesan* (*Btjeg*), genes encoding the 80, 74-70, 65, and 26 kDa polypeptides have been cloned. The 80 kDa protein is 58% identical to the *Bti* Cry11Aa, and has therefore been designated Cry11Ba; it also contains the first conserved block [30]. The 65 kDa protein has a structure similar to Cry10A (presence of the 5 conserved blocks, and absence of carboxy-terminal part) but is less than 30% similar, and thus belongs to a new class, designated Cry19Aa [92]. The 70 kDa protein, designated ORF2, the gene for which lies 145 bp downstream of the *cry19Aa*, is similar to carboxy terminus of Cry4 proteins [92]. The 26 kDa protein is 66% identical to Cyt2Ba from *Bti* and has been named Cyt2Bb [20]. Two proteins of around 74 kDa, with a structure similar to Cry10Aa and Cry19Aa but displaying no significant primary similarity with these proteins, have also been identified and designated as Cry24Aa and Cry25Aa (S. Gill, unpublished results). The *Btjeg* crystal proteins all seem to be involved in the toxicity, to various extents. Cry11Ba is the most active protein on the 3 mosquito species tested *A. aegypti*, *C. pipiens* and *A. stephensi*: it is more toxic (from 5 to 30 fold, depending on the mosquito species) than Cry11Aa, despite their similarities, and as toxic as the native *Btjeg* crystals [30]. Cry19Aa is much less active than Cry11Ba, and its activity is restricted to *C. pipiens* and *A. stephensi*. The presence of ORF2, which is not toxic by itself, increases the activity of Cry19Aa, without changing its specificity [92]. Cyt2Bb is only weakly active against mosquito larvae (600 fold less than native *Btjeg* crystals), but after activation with proteinases, it displays a high haemolytic activity, comparable to that found for Cyt1Aa [20]. The mosquitocidal activities of Cry24Aa and Cry25Aa have not yet been reported.

The *Bt medellin* (*Btmed*) crystals are also composed of several Cry and Cyt proteins: the 94 kDa protein, named Cry11Bb (as it is 83% identical to Cry11Ba); Cry29Aa and Cry30Aa, new Cry-families with structure similar to Cry10Aa, Cry19Aa, Cry24Aa and Cry25Aa and corresponding to proteins of 74.4 and 77.8 kDa, respectively; polypeptides of 28-30 kDa designated Cyt1Ab (86% identical to Cyt1Aa) and Cyt2Bc (91.5% identical with Cyt2Ba, and 85% identical with Cyt2Bb). Cry11Bb is the most toxic component, whatever the mosquito species is, with a level of activity comparable to that of Cry11Ba [81]. No mosquitocidal activity has been found yet for Cry29A and Cry30A, but it cannot be excluded that they interact with other *Btmed* crystal components to enhance their activity (A. Delécluse, unpublished results).

Cyt1Ab is as haemolytic as Cyt1Aa, but less active against mosquitoes (from 3 to 55 fold, depending on the species) [109]. Cyt2Bc has not yet been tested for its biological activity.

It seems that, as found for *Bti*, *Btjeg* and *Btmed* crystal activity results from the presence of several toxic components, which may act synergistically.

3.1.4   Toxins from class 3-strains

Class 3-strains have been poorly studied, due to their low toxicity, and only few proteins have been characterised through gene cloning and analysis. The 27 kDa protein from *Bt kyushuensis*, designated Cyt2Aa as it is 32% identical to Cyt1Aa, is haemolytic and larvicidal, at a level comparable to that of Cyt1Aa. However, the haemolytic activity is only observed after activation by proteinases [57, 58]. A similar polypeptide could be present in strain *darmstadiensis* 73-E10-2 [55], as it is immunologicaly related to Cyt2Aa [56]. In both strains, this polypeptide could be the only toxic component, and the reason for the weak toxicity [52]. The three-dimensional structure of Cyt2Aa has recently been solved [63]. In contrast to Cry1 and Cry3 toxins, Cyt2Aa has a single domain of α/β architecture, comprising two outer layers of α-helix hairpins wrapped around a mixed β-sheet. In the protoxin form, Cyt2Aa is a dimer linked by the intertwined amino-terminal strands in a continuous, 12-stranded β-sheet. Proteolytic processing cleaves the intertwined β-strands to release the active Cyt2Aa as a monomer, as well as removing the carboxy-terminal tail to uncover the three-layered core. This study, in addition to previous mutagenesis experiments [121] indicate that segments forming the sheet are responsible for membrane binding and pore formation.

The gene encoding the 78 kDa from *Bt higo* has been characterised [49]. The protein, displaying 49% identity with Cry19Aa, has been designated Cry19Ba. Its weak activity, restricted to *C pipiens* larvae, suggests that other polypeptides are involved in the mosquitocidal activity of native *Bt higo* crystals. A gene encoding a 94 kDa protein has also been cloned (H. Saitoh, unpublished results) and designated *cry27Aa*, but the mosquitocidal activity of Cry27Aa has not yet been reported.

The 86 kDa protein gene from strain *fukuokaensis* has been characterised and found to encode a polypeptide belonging to a novel family, designated Cry20Aa, with a structure similar to Cry10A and Cry19A [61]. Cry20Aa is only weakly active against *Aedes aegypti* and *C. quinquefasciatus*, probably due to its rapid degradation into 2 inactive components of 56 and 43 kDa.

The gene encoding the 66 kDa protein from *Bt kurstaki* HD-1 has also been cloned [32, 123]. The corresponding polypeptide has a structure similar to the Cry11 family (presence of the first block), but is only distantly related (15% identity), and therefore belongs to another family, designated Cry2Aa. It has a dual specificity against dipteran larvae (*Aedes* and *Anopheles*) and lepidopteran larvae (*Manduca sexta*), due to a short internal segment (amino acids 307 to 382) [124]. A polypeptide immunologically related to Cry2Aa is also present in strain *galleriae* 916, although its molecular weight differs (61 kDa). The corresponding gene has been cloned [1], but no information is available about its nomenclature. It seems that the Cry2A toxins are solely responsible for the mosquitocidal activity of these weekly active strains.

Two other proteins, ranging around 135 kDa, have also been found to be toxic for both dipteran and lepidopteran larvae: the Cry1Ca, produced by strain *aizawai* HD-229 [102], and Cry1Ab7 from *aizawai* IC1 [41, 44]. Although their size is similar, both proteins completely differ from the Cry4 family, except for the presence of the 5 conserved blocks. Cry1Ab7 is highly similar to other Cry1Ab proteins that are only active against Lepidoptera (displaying only 3 amino acid difference). One of these amino acids is crucial to the determination of dipteran specificity [42].

3.2   *B. sphaericus* toxins

3.2.1   The Binary toxin

The binary toxin (Bin) is the most important of the *B. sphaericus* toxins owing to its predominant role in determining the overall toxicity of strains. All high-toxicity strains produce binary toxin as a part of their mosquitocidal arsenal. This toxin (also known as the crystal toxin) is composed of two separate proteins BinA (42 kDa) and BinB (51 kDa) that are produced at the onset of sporulation [17] and are deposited in the form of a parasporal crystalline inclusion body within the exosporium. This is in contrast to *B. thuringiensis* crystal toxins that are independent of the spore following lysis of the sporangium. However, the location of the toxin crystals in *B. sphaericus* appears to play no role in toxin stability [76].

When the spore and its associated crystal are ingested by a susceptible insect, the alkaline environment of the gut solubilises the crystal and releases the toxin (in a manner similar to the solubilisation of the crystal toxins of *B. thuringiensis*). Each of the toxin components is then acted upon by gut proteinases to produce a small change in size [3, 18, 27] and a coordinate increase in toxic activity by approximately 50 fold [18]. The Bin toxin (like the other *B. sphaericus* toxins) is active against the larval stages of the mosquito, reflecting the natural interaction of the filter feeding larvae with *B. sphaericus* in the aquatic flora. The opportunity for this bacterium to enter the guts of adult mosquitoes is likely to be very limited. It is perhaps not surprising therefore that the toxins have no effect when fed to adults [103] (although toxicity to *C. quinquefasciatus* can be achieved when toxin is administered by enema!).

Several studies have shown that, when produced in *E. coli*, both BinA and BinB are required for toxicity to mosquito larvae and that an equimolar ratio of both proteins yields the greatest toxicity [75]. It is interesting to note however, that the 42 kDa BinA produced alone in recombinant *B. thuringiensis* is toxic to mosquitoes [75]. A similar situation is observed in mosquito cells in culture. BinA alone is able to kill *C. quinquefasciatus* cells in the absence of BinB [11]. This implies that, in great enough concentrations, BinA is able to overcome the need for BinB and to act as the toxic moiety of the binary toxin (see Chapter 3.5).

Investigations by Davidson [26] of the roles of the two Bin proteins, clearly showed binding of a BinA/BinB mix to discrete regions of the gut in the mosquito *C. quinquefasciatus*. Subsequent experiments by Oei *et al.* [78] elucidated the roles of the individual toxin components in this binding. In *C. quinquefasciatus*, BinB alone exhibits the same regional binding as whole binary toxin, whereas BinA shows only weak binding throughout the gut. However, in the presence of BinB, binding of BinA becomes strongly regionalised to the gastric caecum and posterior midgut, indicating that binding of BinA is directed by the binding of BinB in this mosquito.

Further information on binding was derived from systematic series of deletions from the N- and C-termini of both Bin proteins. Deletion of more than 16 residues from the N-terminus or 17 residues from the C-terminus of BinA results in loss of toxicity [17, 77]. Similarly, removal of more than 34 or 54 residues from the N- and C-termini respectively, of BinB eradicates activity [22, 77]. Oei *et al.* [78] studied the binding characteristics of representative non-toxic deletions and demonstrated that loss of the N-terminal portion of BinB prevented the regional binding of this protein. In contrast, non-toxic deletions of the C-terminus of BinB or either end of BinA prevented the association of the two components to cause regionalisation of BinA. Thus, these results may indicate regions of each protein with roles in protein-protein interactions. Shanmugavelu [98] showed by mutational analysis that toxicity could also be abolished by alanine replacement of specific amino acids within the N- and C-terminal domains of each protein. A further finding of this study was that whilst variants mutated at the N- and C-termini of BinB were non-toxic alone, they were able to complement each other functionally to restore activity. A similar result was obtained with BinA mutants suggesting that the Bin subunits form multimeric structures in which such complementation can occur.

The *bin* genes encoding the binary toxin proteins have been sequenced from several strains of *B. sphaericus* and reveal a high degree of sequence conservation. To date, only 4 *bin* sequence variants have been described [48, 87] with variation in only 6 possible amino acids in each of the BinA and BinB proteins. It has been shown, however, that these changes in sequence can have significant effects on levels of toxicity and host range. The BinA4/BinB4 proteins from *B. sphaericus* LP1-G confer on this strain a lower toxicity against *C. quinquefasciatus* than strains producing the other Bin variants [64]. In addition, the Bin1, Bin2 and Bin3 variants also have distinct activities. The individual proteins from these Bin types were expressed separately in *E. coli* and assayed against larvae of *C. quinquefasciatus*, *Aedes atropalpus* and *A. aegypti* [15]. In all assays the results for *C. quinquefasciatus* were essentially the same as those for *A. atropalpus*. Bin2 toxins gave the greatest activity against *A. aegypti* and those larvae not killed in the assay failed to develop normally through the larval instars. Bin3 toxins were weakly active and Bin1 showed no toxicity against this mosquito. When assayed against *C. quinquefasciatus* the overall effectiveness of all three Bin types at 48 hours was very similar. At 24 hours however, Bin2 produced most mortality whereas in contrast, larvae exposed to Bin1 and Bin3 exhibited delayed mortality with many dying between 24 and 48 hours. To determine the underlying cause of these phenomena, a programme of site-directed mutagenesis was undertaken to determine which residues were responsible [15]. These experiments showed that alteration of either residue 99 or 104 in the BinA toxin was sufficient to alter the toxicity to all three mosquito species from the Bin3 to the Bin2 pattern.

3.2.2 The Mtx toxins

In an effort to understand the larvicidal activity of the 'low toxicity' *B. sphaericus* strains that lack the genes encoding the binary toxin, Thanabalu *et al.* [106] produced cosmid clones of *B. sphaericus* strain SSII-1 DNA in *E. coli* and analysed their ability to kill *C. quinquefasciatus* larvae in bioassay. The *mtx1* gene identified in this work is distributed widely amongst *B. sphaericus* strains including most (but not all) 'high toxicity' isolates [64].

The *mtx1* gene encodes a 100 kDa protein [106] that is expressed in vegetative *B. sphaericus* cells [2]. On ingestion by the larvae, gut proteinases cleave Mtx1 into two subunits: a 27 kDa moiety with regional homology to ADP-ribosyl transferase bacterial toxins and a 70 kDa subunit with features of a glycoprotein-binding protein [45, 104].

When expressed in *E. coli* and purified, Mtx1 is a highly (LC$_{50}$: 15 ng/ml) potent mosquitocidal agent [105]. Why then does it only confer low toxicity on the *B. sphaericus* that produce it? The explanation appears to have two parts. First, in *B. sphaericus*, the levels of transcription from the *mtx1* promoter appear to be very low [2], possibly owing to repression at an inverted repeat sequence just upstream of the coding region [106]. Second, the low levels of Mtx1 protein produced are unstable owing to the activity of an endogenous proteinase [122] and expression in strains lacking this enzyme results in significantly improved stability [107]. In spite of the above problems, the high potency of Mtx1, especially against *A. aegypti*, makes it a promising agent for development in the future.

The Mtx2 and Mtx3 toxins were both discovered using a strategy similar to that used to identify Mtx1 above [65, 108]. These toxins are related to each other and to the ε-toxin of *Clostridium perfringens* and the cytotoxin of *Pseudomonas aeruginosa*. The latter toxins are both believed to act by forming pores in target cell membranes, thus, it is likely that Mtx2 and Mtx3 also cause toxicity through a similar pore forming mechanism. Mtx2 and Mtx3 are distributed throughout several strains of *B. sphaericus*. Mtx2, which is produced during the vegetative stage of the life cycle, has an LC$_{50}$ for *C. quinquefasciatus* of 320 ng/ml when derived from *B. sphaericus* strain SSII-1 [108]. However, sequence variations in the Mtx2 toxin between strains were found to have significant effects on mosquito target range. Substitutions of Lys at position 224 with Thr abolishes *C. quinquefasciatus* toxicity and leads to a coordinate 100-fold increase in toxicity to *A. aegypti* [19]. In contrast to Mtx2, Mtx3 shows little inter-strain variability [65].

3.2.3   Other toxins

Whilst the four toxins described above appear to account for the toxic properties of the majority of mosquito-pathogenic *B. sphaericus* strains, a few low toxicity strains (e.g. in serotype H26a, 26b) have been isolated that lack the genes encoding Bin, Mtx1, Mtx2 and Mtx3. It would appear, therefore, that at least one other *B. sphaericus* toxin remains to be identified in the future.

3.3 *C. bifermentans malaysia* toxins

As mentioned above, four proteins, with sizes of 68, 66 18 and 16 kDa, were identified as the putative toxic factors of *Cbm*. The genes encoding these proteins are physically close on the chromosome and were cloned on the same DNA fragment [9]. Comparison of the deduced protein sequences of these putative toxins produced some interesting results. First, in both 66-68 kDa proteins, some regions were identified that share similarities with the conserved blocks of the Cry family of *Bt* toxins. For the 66 kDa protein, the similarities are mostly restricted to conserved blocks I to IV of the Cry proteins. It was therefore called Cry16A and was the first report of a Cry protein with a non-*Bt* origin [9]. Similar regions between the 68 kDa peptide and the Cry proteins are more limited, only blocks I, IV and V can be found in this protein. However, the homology level, 20% with respect to other Cry proteins, is enough to include this polypeptide in the Cry family and it was therefore designated Cry17A. Second, the 18 and 16 kDa proteins, called hereafter Cbm17.1 and Cmb17.2, respectively, are encoded by the same gene which is duplicated in tandem. From sequence analysis it appears that the 16 kDa protein results from the cleavage of the first 18 amino acids of the 18 kDa polypeptide. No similarity with any insecticidal protein was found. However, 44.6% similarity restricted to the first 131 amino acids of the haemolysin of *Aspergillus fumigatus* was obtained after comparison with sequences present in the data banks [10].

All this information: presence of the two Cry proteins as well as the putative haemolytic action the Cbm17 proteins supported the idea that they, alone or in combination, could be responsible for the overall toxicity displayed by *Cbm*. In fact, in *Bti*, both Cry proteins and haemolysins coexist contributing to the high toxicity of this isolate; a similar mode of action was therefore suspected for *Cbm*. In order to elucidate the role of each protein in toxicity, expression experiments of *cry16A*, *cry17A*, *cbm17.1* and *cmb17.2*, alone or in combination, in a crystal minus *Bt* strain, were conducted. Results were not successful: 1) expression of *cry16A* in *Bt* resulted in a poor production of Cry16A which was found secreted and to display a low biological activity [9]; 2) no protein expression was achieved with either *cry17A* or both *cbm17* genes under the same experimental conditions [10].

Therefore, expression of these genes was improved either in *E. coli* (as His$_6$-tagged proteins) for the four genes, or in *Bt* by placing the *cry16A* and *cry17A* genes under the control of the *cry1C* gene promoter. All these experiments revealed results contradictory to those described above. First, and in spite of the high level of expression of the *cry16A* gene, the corresponding protein remained in a soluble form in the cytosol of *Bt* sporulated cultures; no trace of Cry16A was found in these culture supernatants using immunodetection. Second, neither *E. coli* nor *Bt* recombinant strains expressing this gene showed any level of toxicity even when high culture concentrations (corresponding approximately to 10-15 µg/ml of Cry16A) were assayed (V. Juárez-Pérez, unpublished results). Small amounts of the Cry17A protein was obtained only when the *cry16A* gene was also expressed in the same *Bt* strain; no production was achieved in the *E. coli* strain. The recombinant *Bt* strain expressing both Cry16A and Cry17A proteins was totally inactive against larvae of Culicidae (V. Juárez-Pérez, unpublished results). This result demonstrates both the lack of toxicity of the Cry17A protein and also indicates that Cry17A does not synergise with Cry16A as previously suggested [10].

An efficient expression of both the His_6-*cbm17.1* and His_6-*cbm17.2* was obtained in *E. coli*. No toxicity against mosquito larvae was scored when these proteins were tested, alone, together or in combination with either the *E. coli* strain expressing the Cry16A or the recombinant *Bt* strains producing the Cry16A or the Cry16A-Cry17A proteins. The similarities found between the Cbm17 proteins and the haemolysin of *A. fumigatus* suggested a putative haemolytic effect of those two proteins. However, no haemolytic action was recorded when as much as 100 µg of purified (activated or not) Cbm17.1 or Cbm17.2 proteins were tested, either alone or in combination, with sheep red blood cells (V. Juárez-Pérez, unpublished results).

All this information suggests the lack of biological activity of the Cry16A, Cry17A, Cbm17.1 and Cbm17.2 proteins, at least when produced in *E. coli* or *Bt*. However, we cannot exclude the possibility that modifications of these proteins, occurring only in the *Clostridium* strain, are necessary for activity. *In vivo* deletion of each gene from the *Cbm* strain should clarify the role of each polypeptide. On the other hand, the search for other factor(s) that may play a more direct role in toxicity needs to be initiated.

4.   SUMMARY AND CONCLUSIONS

Since the discovery of *Bti*, many strains, including *Bacilli* and members of other genera, have been shown to possess some degree of mosquitocidal activity. Whereas the mosquitocidal activity of

B. thuringiensis, *B. sphaericus*, *B. laterosporus* and *C. bifermentans* has been clearly evaluated against different mosquito species, the identification of their toxic components has been, in some cases, a subject of controversy. Presently, most of the *Bt* and *B. sphaericus* toxins have been characterised, and only minor toxins remain to be identified. In contrast, those from *B. laterosporus* are still completely unknown, and although Cry toxins have been found in *C. bifermentans*, their participation in the larvicidal activity of this strain is yet unclear.

The cloning and characterisation of mosquitocidal toxin genes indicate that the same *cry* genes, or variants, may be present in different *B. thuringiensis* strains; the same is also true for the *B. sphaericus* toxin genes. Since mobile elements have been found in the close vicinity of the toxin genes (see Chapter 2.5), it can be suggested that intraspecies genetic transfer might occur between the different *Bt* or *B. sphaericus* strains. The fact that *cry* genes are not only found in *Bt* strains but also in *Clostridium* indicates that inter as well as intraspecies genetic transfer might occur.

A great diversity of mosquitocidal toxins has been found amongst the *Bt* and *B. sphaericus* strains, with proteins having different structure, and/or displaying various levels of activity and specificity. This diversity could be used to improve the 2 bacteria presently used in the field, *Bti* and *B. sphaericus*. Indeed, although both strains can efficiently control vectors (see Chapter 6.2), there are some major drawbacks such as the low persistence of *Bti*, the narrow spectrum of activity of *B. sphaericus*, and the development of insect resistance to the toxins from the latter. Introducing appropriate combinations of the different genes in these bacteria would certainly improve their properties (see Chapter 7.2).

REFERENCES

[1] Ahmad W, Nicholls C & Ellar DJ (1989) Cloning and expression of an entomocidal protein gene from *Bacillus thuringiensis galleriae* toxic to both lepidoptera and diptera. FEMS Microbiol. Lett. 59, 197-202

[2] Ahmed HK, Mitchell WJ & Priest FG (1995) Regulation of mosquitocidal toxin synthesis in *Bacillus sphaericus*. Appl. Microbiol. Biotechnol. 43, 310-314

[3] Aly C, Mulla MS & Federici BA (1989) Ingestion, dissolution and proteolysis of the *Bacillus sphaericus* toxin by mosquito larvae. J. Invertebr. Pathol. 53, 12-20

[4] Angsuthanasombat C, Crickmore N & Ellar DJ (1992) Comparison of *Bacillus thuringiensis* subsp. *israelensis* CryIVA and CryIVB cloned toxins reveals synergism in vivo. FEMS Microbiol. Lett. 94, 63-68

[5] Arapinis C, de la Torre F & Szulmajster J (1988) Nucleotide and deduced amino acid sequence of the *Bacillus sphaericus* 1593 M gene encoding a 51.4 kD polypeptide which acts synergistically with the 42 kD protein for expression of the larvicidal toxin. Nucleic Acids Res. 16, 7731

[6] de Barjac H (1978) Une nouvelle variété de *Bacillus thuringiensis* très toxique pour les moustiques : *B. thuringiensis* var. *israelensis* sérotype H14. C. R. Acad. Sci., Paris, (Série D). 286, 797-800

[7] de Barjac H, Sebald M, Charles JF, Cheong WH & Lee HH (1990) *Clostridium bifermentans* serovar *malaysia*, une nouvelle bactérie anaérobie pathogène des larves de moustiques et de simulies. C. R. Acad. Sci., Paris, (Série III). 310, 383-387

[8] Barloy F (1997) Caractérisation des déterminants génétiques impliqués dans la toxicité de *Clostridum bifermentans* subsp. *malaysia*, souche CH18, sur les larves de moustiques. Thèse de Doctorat, Université Paris VII

[9] Barloy F, Delécluse A, Nicolas L & Lecadet M-M (1996) Cloning and expression of the first anaerobic toxin gene from *Clostridium bifermentans* subsp. *malaysia*, encoding a new mosquitocidal protein with homologies to *Bacillus thuringiensis* delta-endotoxins. J. Bacteriol. 178, 3099-3105

[10] Barloy F, Lecadet M-M & Delécluse A (1998) Cloning and sequencing of three new putative toxin genes from *Clostridium bifermentans* CH18. Gene 211, 293-299

[11] Baumann L & Baumann P (1991) Effects of components of the *Bacillus sphaericus* toxin on mosquito larvae and mosquito-derived tissue culture-grown cells. Curr. Microbiol. 23, 51-57

[12] Baumann L, Broadwell AH & Baumann P (1988) Sequence analysis of the mosquitocidal toxin genes encoding 51.4- and 41.9- kilodalton proteins from *Bacillus sphaericus* 2362 and 2297. J. Bacteriol. 170, 2045-2050

[13] Baumann P, Baumann L, Bowditch RD & Broadwell AH (1987) Cloning of the gene for the larvicidal toxin of *Bacillus sphaericus* 2362 : evidence for a family of related sequences. J. Bacteriol. 169, 4061-4067

[14] Berry C & Hindley J (1987) *Bacillus sphaericus* strain 2362 : identification and nucleotide sequence of the 41.9 kDa toxin gene. Nucleic Acids Res. 15, 5891

[15] Berry C, Hindley J, Ehrhardt AF et al (1993) Genetic determinants of host ranges of *Bacillus sphaericus* mosquito larvicidal toxins. J. Bacteriol. 175, 510-518

[16] Berry C, Jackson-Yap J, Oei C & Hindley J (1989) Nucleotide sequence of two toxin genes from *Bacillus sphaericus* IAB59 : sequence comparisons between five highly toxinogenic strains. Nucleic Acids Res. 17, 7516

[17] Broadwell AH & Baumann P (1986) Sporulation-associated activation of *Bacillus sphaericus* larvicide. Appl. Environ. Microbiol. 52, 758-764

[18] Broadwell AH & Baumann P (1987) Proteolysis in the gut of mosquito larvae results in further activation of the *Bacillus sphaericus* toxin. Appl. Environ. Microbiol. 53, 1333-1337

[19] Chan SW, Thanabalu T, Wee BY & Porter AG (1996) Unusual amino acid determinants of host range in the Mtx2 family of mosquitocidal toxins. J. Biol. Chem. 271, 14183-14187

[20] Cheong H & Gill SS (1997) Cloning and characterization of a cytolytic and mosquitocidal δ-endotoxin from *Bacillus thuringiensis* subsp. *jegathesan*. Appl. Environ. Microbiol. 63, 3254-3260

[21] Chungjatupornchai W, Höfte H, Seurinck J, Angsuthanasombat C & Vaeck M (1988) Common features of *Bacillus thuringiensis* toxins specific for Diptera and Lepidoptera. Eur. J. Biochem. 173, 9-16

[22] Clark MC & Baumann P (1990) Deletion analysis of the 51-kilodalton protein of the *Bacillus sphaericus* 2362 binary mosquitocidal toxin: construction of derivatives equivalent to the larva-processed toxin. J. Bacteriol. 172, 6759-6763

[23] Crickmore N, Bone EJ, Williams JA & Ellar DJ (1995) Contribution of the individual components of the δ -endotoxin crystal to the mosquitocidal activity of *Bacillus thuringiensis* subsp. *israelensis*. FEMS Microbiol. Lett. 131, 249-254

[24] Crickmore N, Zeigler DR, Feitelson J *et al.* (1998) Revision of the nomenclature for the *Bacillus thuringiensis* pesticidal crystal proteins. Microbiol. Mol. Biol. Rev. 62, 807-813

[25] Dai S-M & Gill SS (1993) *In vitro* and *in vivo* proteolysis of the *Bacillus thuringiensis* subsp. *israelensis* CryIVD protein by *Culex quinquefasciatus* larval midgut proteases. Insect. Biochem. Molec. Biol. 23, 273-283

[26] Davidson EW (1988) Binding of the *Bacillus sphaericus* (Eubacteriales: Bacillaceae) toxin to midgut cells of mosquito (Diptera: Culicidae) larvae: relationship to host range. J. Med. Entomol. 25, 151-157

[27] Davidson EW, Bieger AL, Meyer M & Shellabarge RC (1987) Enzymatic activation of the *Bacillus sphaericus* mosquito larvicidal toxin. J. Invertebr. Pathol. 50, 40-44

[28] Delécluse A, Bourgouin C, Klier A & Rapoport G (1988) Specificity of action on mosquito larvae of *Bacillus thuringiensis israelensis* toxins encoded by two different genes. Mol. Gen. Genet. 214, 42-47

[29] Delécluse A, Poncet S, Klier A & Rapoport G (1993) Expression of *cryIVA* and *cryIVB* genes, independently or in combination, in a crystal-negative strain of *Bacillus thuringiensis* subsp. *israelensis*. Appl. Environ. Microbiol. 59, 3922-3927

[30] Delécluse A, Rosso M-L & Ragni A (1995) Cloning and expression of a novel toxin gene from *Bacillus thuringiensis* subsp. *jegathesan*, encoding a highly mosquitocidal protein. Appl. Environ. Microbiol. 61, 4230-4235

[31] Donovan WP, Dankocsik CC & Gilbert MP (1988) Molecular characterization of a gene encoding a 72-kilodalton mosquito-toxic crystal protein from *Bacillus thuringiensis* subsp. *israelensis*. J. Bacteriol. 170, 4732-4738

[32] Donovan WP, Dankocsik CC, Gilbert MP *et al.* (1988) Amino acid sequence and entomocidal activity of the P2 crystal protein. An insect toxin from *Bacillus thuringiensis* var. *kurstaki*. J. Biol. Chem. 263, 561-567

[33] Drobniewski FA & Ellar DJ (1989) Purification and properties of a 28-kilodalton hemolytic and mosquitocidal protein toxin of *Bacillus thuringiensis* subsp. *darmstadiensis* 73-E10-2. J. Bacteriol. 171, 3060-3067

[34] Earp DJ & Ellar DJ (1987) *Bacillus thuringiensis* var. *morrisoni* strain PG14: nucleotide sequence of a gene encoding a 27 kDa crystal protein. Nucleic Acids Res. 15, 3619

[35] Favret ME & Yousten AA (1985) Insecticidal activity of *Bacillus laterosporus*. J. Invertebr. Pathol. 45, 195-203

[36] Frutos R, Chang C, Gill SS & Federici BA (1991) Nucleotide sequences of genes encoding a 72,000 molecular weight mosquitocidal protein and an associated 20,000 molecular weight protein are highly conserved in subspecies of *Bacillus thuringiensis* from Israel and the Philippines. Biochem. System. Ecol. 19, 599-609

[37] Galjart NJ, Sivasubramanian N & Federici BA (1987) Plasmid location, cloning, and sequence analysis of the gene encoding a 27.3-kilodalton cytolytic protein from *Bacillus thuringiensis* subsp. *morrisoni* (PG-14). Curr. Microbiol. 16, 171-177

[38] Garduno F, Thorne L, Walfield AM & Pollock TJ (1988) Structural relatedness between mosquitocidal endotoxins of *Bacillus thuringiensis* subsp. *israelensis*. Appl. Environ. Microbiol. 54, 277-279

[39] Goldberg LH & Margalit J (1977) A bacterial spore demonstrating rapid larvicidal activity against *Anopheles sergentii, Uranotaenia unguiculata, Culex univitatus, Aedes aegypti* and *Culex pipiens*. Mosq. News 37, 355-358

[40] Guerchicoff A, Ugalde RA & Rubinstein CP (1997) Identification and characterization of a previously undescribed *cyt* gene in *Bacillus thuringiensis* subsp. *israelensis*. Appl. Environ. Microbiol. 63, 2716-2721

[41] Haider MZ & Ellar DJ (1988) Nucleotide sequence of a *Bacillus thuringiensis aizawai* IC1 entomocidal crystal protein gene. Nucleic Acids Res. 16, 10927

[42] Haider MZ & Ellar DJ (1989) Functional mapping of an entomocidal δ-endotoxin. Single amino acid changes produced by site-directed mutagenesis influence toxicity and specificity of the protein. J. Mol. Biol. 208, 183-194

[43] Haider MZ, Knowles BH & Ellar DJ (1986) Specificity of *Bacillus thuringiensis* var. *colmeri* insecticidal δ-endotoxin is determined by differential proteolytic processing of the protoxin by larval gut proteases. Eur. J. Biochem. 156, 531-540

[44] Haider MZ, Ward ES & Ellar DJ (1987) Cloning and heterologous expression of an insecticidal delta-endotoxin gene from *Bacillus thuringiensis* var. *aizawai* IC1 toxic to both lepidoptera and diptera. Gene 52, 285-290

[45] Hazes B & Read RJ (1995) A mosquitocidal toxin with a ricin-like cell-binding domain. Structural Biol. 2, 358-359

[46] Hindley J & Berry C (1987) Identification, cloning and sequence analysis of the *Bacillus sphaericus* 1593 41.9 kD larvicidal toxin gene. Mol. Microbiol. 1, 187-194

[47] Hindley J & Berry C (1988) *Bacillus sphaericus* strain 2297 : nucleotide sequence of 41.9 kDa toxin gene. Nucleic Acids Res. 16, 4168

[48] Humphreys MJ & Berry C (1998) Variants of the *Bacillus sphaericus* binary toxins: implications for differential toxicity of strains. J. Invertebr. Pathol. 71, 184-185

[49] Hwang SH, Saitoh H, Mizuki E, Higuchi K & Ohba M (1998) A novel class of mosquitocidal delta-endotoxin, Cry19B, encoded by a *Bacillus thuringiensis* serovar *higo* gene. Syst. Appl. Microbiol. 21, 179-184

[50] Ibarra JE & Federici BA (1986) Parasporal bodies of *Bacillus thuringiensis* subsp. *morrisoni* (PG-14) and *Bacillus thuringiensis* subsp. *israelensis* are similar in protein composition and toxicity. FEMS Microbiol. Lett. 34, 79-84

[51] Ishii T & Ohba M (1993) Diversity of *Bacillus thuringiensis* environmental isolates showing larvicidal activity specific for mosquitoes. J. Gen. Microbiol. 139, 2849-2854

[52] Ishii T & Ohba M (1994) The 23-kilodalton protein is solely responsible for mosquito larvicidal activity of *Bacillus thuringiensis* serovar *kyushuensis*. Curr. Microbiol. 29, 91-94

[53] Ishii T & Ohba M (1997) Investigation of mosquito-specific larvicidal activity of a soil isolate of *Bacillus thuringiensis* serovar *canadensis*. Curr. Microbiol. 35, 40-43

[54] Kellen W, Clark T, Lindergren J et al (1965) *Bacillus sphaericus* Neide as a pathogen of mosquitoes. J. Invertebr. Pathol. 7, 442-448

[55] Kim K-H, Ohba M & Kim B-W (1996) Cloning of a hemolytic mosquitocidal delta-endotoxin gene (*cyt*) of *Bacillus thuringiensis* 73E10-2 (serotype 10) into *Bacillus subtilis* and characterization of the *cyt* gene product. J. Microbiol. Biotechnol. 6, 326-330

[56] Knowles BH, White PJ, Nicholls CN & Ellar DJ (1992) A broad-spectrum cytolytic toxin from *Bacillus thuringiensis* var. *kyushuensis*. Proc. R. Soc. London 248, 1-7

[57] Koni PA & Ellar DJ (1993) Cloning and characterization of a novel *Bacillus thuringiensis* cytolytic delta-endotoxin. J. Mol. Biol. 229, 319-327

[58] Koni PA & Ellar DJ (1994) Biochemical characterization of *Bacillus thuringiensis* cytolytic δ-endotoxins. Microbiol. 140, 1869-1880

[59] Lacey LA, Day J & Heitzman CM (1987) Long-term effects of *Bacillus sphaericus* on *Culex quinquefasciatus*. J. Invertebr. Pathol. 49, 116-123

[60] Lacey LA, Lacey CM, Peacock B & Thiéry I (1988) Mosquito host range and field activity of *Bacillus sphaericus* isolate 2297 (serotype H25). J. Am. Mosq. Control Assoc. 4, 51-56

[61] Lee H-K & Gill SS (1997) Molecular cloning and characterization of a novel mosquitocidal protein gene from *Bacillus thuringiensis* subsp. *fukuokaensis*. Appl. Environ. Microbiol. 63, 4664-4670

[62] Lee HL & Seleena P (1990) Isolation of indigenous larvicidal microbial control agents of moquitoes: the Malaysian experience. Southeast Asian J. Trop. Med. Public Health 21, 281-287

[63] Li J, Koni PA & Ellar DJ (1996) Structure of the mosquitocidal δ-endotoxin CytB from *Bacillus thuringiensis* sp. *kyushuensis* and implications for membrane pore formation. J. Mol. Biol. 257, 129-152

[64] Liu J-W, Hindley J, Porter AG & Priest FG (1993) New high-toxicity mosquitocidal strains of *Bacillus sphaericus* lacking a 100-kilodalton-toxin gene. Appl. Environ. Microbiol. 59, 3470-3473

[65] Liu J-W, Porter AG, Wee BY & Thanabalu T (1996) New gene from nine *Bacillus sphaericus* strains encoding highly conserved 35.8 kilodalton mosquitocidal toxins. Appl. Environ. Microbiol. 62, 2174-2176

[66] López-Meza J, Federici BA, Poehner WJ, Martinez-Castillo A & Ibarra JE (1995) Highly mosquitocidal isolates of *Bacillus thuringiensis* subspecies *kenyae* and *entomocidus* from Mexico. Biochem. Syst. Ecol. 23, 461-468

[67] Mathavan S & Velpandi A (1984) Toxicity of *Bacillus sphaericus* strains to selected target and non-target aquatic organisms. Ind. J. Med. Res. 80, 653-657

[68] Miura T, Takahashi RM & Mulligan FS (1980) Effects of the bacterial mosquito larvicide, *Bacillus thuringiensis* serotype H-14 on selected aquatic organisms. Mosq. News 40, 619-622

[69] Montaldi FA & Roth JL (1990) Parasporal bodies of *Bacillus laterosporus* sporangia. J. Bacteriol. 172, 2168-2171

[70] Mulla MS, Darwazeh HA, Davidson EW, Dulmage HT & Singer S (1984) Larvicidal activity and field efficacy of *Bacillus sphaericus* strains against mosquito larvae and their safety to nontarget organisms. Mosq. News 44, 336-342

[71] Mulligan MS, Schaefer CH & Miura T (1978) Laboratory and field evaluation of *Bacillus sphaericus* as a mosquito control agent. J. Econ. Entomol. 71, 774-777

[72] Nicholls CN, Ahmad W & Ellar DJ (1989) Evidence for two different types of insecticidal P2 toxins with dual specificity in *Bacillus thuringiensis* subspecies. J. Bacteriol. 171, 5141-5147

[73] Nicolas L, Charles J-F & de Barjac H (1993) *Clostridium bifermentans* serovar *malaysia*: characterization of putative mosquito larvicidal proteins. FEMS Micobiol. Lett. 113, 23-28

[74] Nicolas L, Hamon S, Frachon E, Sebald M & de Barjac H (1990) Partial inactivation of the mosquitocidal activity of *Clostridium bifermentans* serovar *malaysia* by extracellular proteinases. Appl. Microbiol. Biotechnol. 34, 36-41

[75] Nicolas L, Nielsen-Leroux C, Charles J-F & Delécluse A (1993) Respective role of the 42- and 51-kDa components of the *Bacillus sphaericus* toxin overexpressed in *Bacillus thuringiensis*. FEMS Microbiol. Lett. 106, 275-280

[76] Nicolas L, Regis LN & Rios EM (1994) Role of the exosporium in the stability of the *Bacillus sphaericus* binary toxin. FEMS Microbiol. Lett. 124, 271-276

[77] Oei C, Hindley J & Berry C (1990) An analysis of the genes encoding the 51.4- and 41.9-kDa toxins of *Bacillus sphaericus* 2297 by deletion mutagenesis: the construction of fusion proteins. FEMS Microbiol. Lett. 72, 265-274

[78] Oei C, Hindley J & Berry C (1992) Binding of purified *Bacillus sphaericus* binary toxin and its deletion derivatives to *Culex quinquefasciatus* gut: elucidation of functional binding domains. J. Gen. Microbiol. 138, 1515-1526

[79] Ohba M, Sitoh H, Miyamoto K, Higuchi K & Mizuki E (1995) *Bacillus thuringiensis* serovar *higo* (flagellar serotype 44), a new serogroup with a larvicidal activity preferential for the anopheline mosquito. Lett. Appl. Microbiol. 21, 316-318

[80] Orduz S, Diaz T, Thiéry I, Charles J-F & Rojas W (1994) Crystal protein from *Bacillus thuringiensis* serovar. *medellin*. Appl. Microbiol. Biotechnol. 40, 794-799

[81] Orduz S, Realpe M, Arango R, Murillo LA & Delécluse A (1998) Sequence of the *cry11Bb1* gene from *Bacillus thuringiensis* subsp. *medellin* and toxicity analysis of its encoded protein. Biochim. Biophys. Acta 1388, 267-272

[82] Orduz S, Rojas W, Correa MM, Montoya AE & de Barjac H (1992) A new serotype of *Bacillus thuringiensis* from Colombia toxic to mosquito larvae. J. Invertebr. Pathol. 59, 99-103

[83] Orlova MV, Smirnova TA, Ganushkina LA, Yacubovich VY & Azizbekyan RR (1998) Insecticidal activity of *Bacillus laterosporus*. Appl. Environ. Microbiol. 64, 2723-2725

[84] Padua LE, Ohba M & Aizawa K (1984) Isolation of a *Bacillus thuringiensis* strain (serotype 8a:8b) highly and selectively toxic against mosquito larvae. J. Invertebr. Pathol. 44, 12-17

[85] Pao-Intara M, Angsuthanasombat C & Panyim S (1988) The mosquito larvicidal activity of 130 kDa delta-endotoxin of *Bacillus thuringiensis* var. *israelensis* resides in the 72 kDa amino-terminal fragment. Biochem. Biophys. Res. Commun. 153, 294-300

[86] Poncet S, Delécluse A, Klier A & Rapoport G (1995) Evaluation of synergistic interactions among the CryIVA, CryIVB and CryIVD toxic components of *B. thuringiensis* subsp. *israelensis* crystals. J. Invertebr. Pathol. 66, 131-135

[87] Priest FG, Ebdrup L, Zahner V & Carter PE (1997) Distribution and characterisation of mosquitocidal toxin genes in some strains of *Bacillus sphaericus*. Appl. Environ. Microbiol. 63, 1195-1198

[88] Ragni A, Thiéry I & Delécluse A (1996) Characterization of six highly mosquitocidal *Bacillus thuringiensis* strains that do not belong to H-14 serotype. Curr. Microbiol. 32, 48-54

[89] Restrepo N, Gutierrez D, Patino MM *et al.* (1997) Cloning, expression and toxicity of a mosquitocidal toxin gene of *Bacillus thuringiensis* subsp. *medellin*. Mem. Inst. Oswaldo Cruz 92, 257-262

[90] Rivers DB, Vann CN, Zimmack HL & Dean DH (1991) Mosquitocidal activity of *Bacillus laterosporus*. J. Invertebr. Pathol. 58, 444-447

[91] Rodcharoen J, Mulla MS & Chaney JD (1991) Microbial larvicides for the control of nuisance aquatic midges (Diptera: Chironomidae) inhabiting mesocosms and man-made lakes in California. J. Am. Mosq. Control Assoc. 7, 56-62

[92] Rosso M-L & Delécluse A (1997) Contribution of the 65-kilodalton protein encoded by the cloned gene *cry19A* to the mosquitocidal activity of *Bacillus thuringiensis* subsp. *jegathesan*. Appl. Environ. Microbiol. 63, 4449-4455

[93] Saitoh H, Higuchi K, Mizuki E, Hwang SH & Ohba M (1998) Characterization of mosquito larvicidal parasporal inclusions of a *Bacillus thuringiensis* serovar *higo* strain. J. Appl. Microbiol. 84, 883-888

[94] Seleena P, Lee HL & Lecadet M-M (1995) A new serovar of *Bacillus thuringiensis* possessing 28a28c flagellar antigenic structure: *Bacillus thuringiensis* serovar *jegathesan*, selectively toxic against mosquito larvae. J Am Mosq Control Assoc. 11, 471-473

[95] Seleena P, Lee H L and M-M Lecadet (1997) A novel insecticidal serotype of *Clostridium bifermentans*. J Am Mosq Control Assoc. 13, 395-397

[96] Sen K, Honda G, Koyama N *et al.* (1988) Cloning and nucleotide sequences of the two 130 kDa insecticidal protein genes of *Bacillus thuringiensis* var. *israelensis*. Agric. Biol. Chem. 52, 873-878

[97] Shadduck JA, Singer S & Lause S (1980) Lack of mammalian pathogenicity of entomocidal isolates of *Bacillus sphaericus*. Environ. Entomol. 9, 403-407

[98] Shanmugavelu M, Rajamohan F, Kathirvel M *et al.* (1998) Functional complementation of nontoxic mutant binary toxins of *Bacillus sphaericus* 1593M generated by site-directed mutagenesis. Appl. Environ. Microbiol. 64, 756-759

[99] Siegel JP & Shadduck JA (1990) Mammalian safety of *Bacillus thuringiensis israelensis*, p. 202-217. *In* de Barjac H & Sutherland DJ (ed.), Bacterial control of mosquitoes and blackflies, Rutgers University Press

[100] Singer S (1973) Insecticidal activity of recent bacterial isolates and their toxins against mosquito larvae. Nature 244, 110-111

[101] Smirnova TA, Minenkova IB, Orlova MV, Lecadet M-M & Azizbekyan RR (1996) The crystal-forming strains of *Bacillus laterosporus*. Res. Microbiol. 147, 343-350

[102] Smith GP, Merrick JD, Bone EJ & Ellar DJ (1996) Mosquitocidal activity of the CyIC δ-endotoxin from *Bacillus thuringiensis* susp. *aizawai*. Appl. Environ. Microbiol. 62, 680-684

[103] Stray JE, Klowden MJ & Hurlbert RE (1988) Toxicity of *Bacillus sphaericus* crystal toxin to adult mosquitoes. Appl. Environ. Microbiol. 54, 2320-2321

[104] Thanabalu T, Berry C & Hindley J (1993) Cytotoxicity and ADP-ribosylating activity of the mosquitocidal toxin from *Bacillus sphaericus* SSII-1: possible roles of the 27- and 70-kilodalton peptides. J. Bacteriol. 175, 2314-2320

[105] Thanabalu T, Hindley J & Berry C (1992) Proteolytic processing of the mosquitocidal toxin from *Bacillus sphaericus* SSII-1. J. Bacteriol. 174, 5051-5056

[106] Thanabalu T, Hindley J, Jackson-Yap J & Berry C (1991) Cloning, sequencing, and expression of a gene encoding a 100-kilodalton mosquitocidal toxin from *Bacillus sphaericus* SSII-1. J. Bacteriol. 173, 2776-2785

[107] Thanabalu T & Porter AG (1995) Efficient expression of a 100-kilodalton mosquitocidal toxin in protease-deficient recombinant *Bacillus sphaericus*. Appl. Environ. Microbiol. 61, 4031-4036

[108] Thanabalu T & Porter AG (1996) A *Bacillus sphaericus* gene encoding a novel type of mosquitocidal toxin of 31.8 kDa. Gene 170, 85-89

[109] Thiéry I, Delécluse A, Tamayo MC & Orduz S (1997) Identification of a *cytIA*-like hemolysin gene from *Bacillus thuringiensis* subsp. *medellin* and expression in a crystal-negative *B. thuringiensis* strain. Appl. Environ. Microbiol. 63, 468-473

[110] Thiéry I, Hamon S, Cosmao Dumanoir V & de Barjac H (1992) Vertebrate safety of *Clostridium bifermentans* serovar *malaysia*, a new larvicidal agent for vector control. J. Econ. Entomol. 85, 1618-1623

[111] Thiéry I, Hamon S, Gaven B & de Barjac H (1992) Host range of *Clostridium bifermentans* serovar *malaysia*, a mosquitocidal anaerobic bacterium. J. Am. Mosq. Control Assoc. 8, 272-277

[112] Thomas WE & Ellar DJ (1983) *Bacillus thuringiensis* var. *israelensis* crystal δ-
endotoxin : effects on insect and mammalian cells *in vitro* and *in vivo*. J. Cell. Sci. 60,
181-197

[113] Thorne L, Garduno F, Thompson T *et al.* (1986) Structural similarity between the
Lepidoptera- and Diptera-specific insecticidal endotoxin genes of *Bacillus
thuringiensis* subsp. "*kurstaki*" and"*israelensis*". J. Bacteriol. 166, 801-811

[114] Undeen A & Nagel W (1978) The effect of *Bacillus thuringiensis* ONR60A strain
(Goldberg) on *Simulium* larvae in the laboratory. Mosq. News 38, 524-527

[115] Waalwijk C, Dullemans A & Maat C (1991) Construction of a bioinsecticidal
rhizosphere isolate of *Pseudomonas fluorescens*. FEMS Microbiol. Lett. 77, 257-264

[116] Waalwijk C, Dullemans A, Wiegers G & Smits P (1992) Toxicity of *Bacillus
thuringiensis* variety *israelensis* against tipulid larvae. J. Appl. Ent. 114, 415-420

[117] Waalwijk C, Dullemans AM, vanWorkum MES & Visser B (1985) Molecular
cloning and the nucleotide sequence of the M_r 28000 crystal protein gene of *Bacillus
thuringiensis* subsp. *israelensis*. Nucleic Acids Res. 13, 8207-8217

[118] Ward ES & Ellar DJ (1986) *Bacillus thuringiensis* var. *israelensis* δ-endotoxin.
Nucleotide sequence and characterization of the transcripts in *Bacillus thuringiensis*
and *Escherichia coli*. J. Mol. Biol. 191, 1-11

[119] Ward ES & Ellar DJ (1987) Nucleotide sequence of a *Bacillus thuringiensis* var.
israelensis gene encoding a 130 kDa delta-endotoxin. Nucleic Acids Res. 15, 7195

[120] Ward ES & Ellar DJ (1988) Cloning and expression of two homologous genes of
Bacillus thuringiensis subsp. *israelensis* which encode 130-kilodalton
mosquitocidal proteins. J. Bacteriol. 170, 727-735

[121] Ward ES, Ellar DJ & Nicholls CN (1988) Single amino acid changes in the
Bacillus thuringiensis var. *israelensis* δ-endotoxin affect the toxicity and expression
of the protein. J. Mol. Biol. 202, 527-535

[122] Wati MR, Thanabalu T & Porter AG (1997) Gene from tropical *Bacillus sphaericus*
encoding a protease closely related to subtilisins from Antarctic bacilli. Biochim.
Biophys. Acta 1352, 56-62

[123] Widner WR & Whiteley HR (1989) Two highly related insecticidal crystal
proteins of *Bacillus thuringiensis* subsp. *kurstaki* possess different host range
specificities. J. Bacteriol. 171, 965-974

[124] Widner WR & Whiteley HR (1990) Location of the dipteran specificity region in a
lepidopteran-dipteran crystal protein from *Bacillus thuringiensis*. J. Bacteriol. 172,
2826-2832

[125] Wu D, Johnson JJ & Federici BA (1994) Synergism of mosquitocidal toxicity
between CytA and CryIVD proteins using inclusions produced from cloned genes of
Bacillus thuringiensis. Mol. Microbiol. 13, 965-972

[126] Yamamoto T & McLaughlin RE (1981) Isolation of a protein from the parasporal
crystal of *Bacillus thuringiensis* var. *kurstaki* toxic to the mosquito larva *Aedes
taeniorhynchus*. Biochem. Biophys. Res. Commun. 103, 414-421

[127] Yoshida K, Matsushima Y, Sen K, Sakai H & Komano T (1989) Insecticidal
activity of a peptide containing the 30th to 695th amino acid residues of the 130-kDa
protein of *Bacillus thuringiensis* var. *israelensis*. Agric. Biol. Chem. 53, 2121-2127

[128] Yu Y-M, Ohba M & Gill SS (1991) Characterization of mosquitocidal activity of
Bacillus thuringiensis subsp. *fukuokaensis* crystal proteins. Appl. Environ.
Microbiol. 57, 1075-1081

Table 1. Mosquitocidal *Bacillus thuringiensis* strains

Class	Strains[1]	Serotype	Mosquitocidal activity[2]			Crystal polypeptides (kDa)[3]	Reference[4]
			A. aegypti	*A. stephensi*	*C. pipiens*		
1	*israelensis* 1884	H14	+++	+++	+++	135, 125, 68, 28	[39, 112]
1	*morrisoni* PG14	H8a, 8b	+++	+++	+++	144, 135, 125, 68, 28	[50, 84]
1	*kenyae* LBIT-52	H4a, 4c	+++	ND	+++	135, 125, 68, 28	[66]
1	*canadensis* 11S2-1	H5a, 5c	+++	+++	+++	135, 125, 68, 28	[88]
1	*entomocidus* LBIT-58	H6	++	ND	+++	135, 125, 68, 28	[66]
1	*thompsoni* B175	H12	+++	+++	+++	135, 125, 68, 28	[88]
1	*malaysiensis* IMR81.1	H36	+++	+++	+++	135, 125, 68, 28	[88]
1	AAT028 K6		+++	+++	+++	135, 125, 68, 28	[88]
1	AAT021 B51		+++	+++	+++	135, 125, 68, 28	[88]
2	*jegathesan* 367	H28a, 28c	++	+++	++	80, 74-70, 65, 37, 26, 16	[88, 94]
2	*medellin* 163-131	H30	++	+++	++	94, 70-68, 30, 28	[80, 82]
3	*kurstaki* HD-1	H3a, 3b, 3c	+/-	+	ND	135-130, 66	[72, 126]
3	*fukuokaensis* 84-1-1-13	H3a, 3d, 3e	+/-	ND	+/-	90, 86, 82, 72, 50, 48, 37, 27	[51, 128]
3	*galleriae* 916	H5a, 5b	+/-	+/-	ND	135-130, 61	[72]
3	*canadensis* 89-T-5-9	H5a, 5c	+/-	ND	ND	65, 53, 28	[51, 53]
3	*aizawai* IC1	H7	+/-	ND	ND	135-130	[43]
3	*darmstadiensis* 73-E10-2	H10a, 10b	+/-	+	+	125, 83, 79, 77, 69, 50, 27	[33, 51, 88]
3	*kyushuensis* 74 F6-18	H11a, 11c	+/-	+/-	+/-	140, 85, 80, 70, 66, 50, 27, 15, 14	[51, 56, 88]
3	*shandongiensis* 89-ST-1-25	H22	+/-	ND	ND	150, 70-60, 25	[51]
3	*higo* 92-KU-137-4	H44	ND	+	+/-	98, 91, 71, 63, 59, 50, 44, 27	[79, 93]

[1] AAT: autoagglutinated strains that cannot be serotyped by flagellar agglutination.

[2] +++: LC_{50} values similar to those found for *Bti* crystals; ++, +, and +/-: LC_{50} values higher than those found for *Bti* crystals with ratios comprised between 2-10, 10-50, and 50-1,500 fold, respectively; -: non-toxic crystals for the species tested; ND: not determined.

[3] Underlined numbers indicate molecular weight of polypeptides immunologically related to those from *Bti*.

[4] Reference for activity and polypeptide composition of the strain

Table 2. Mosquitocidal toxins

Name	Acc. no.	Molecular Weight (kDa)		Residues	Mosquitocidal activity[1]			Found in	Reference
		apparent	estimated		A. aegypti	A. stephensi	C. pipiens		
Cry1Ab7	X13233	130	130.7	1155	+	ND	ND	*Bt aizawai* IC1	[41, 42]
Cry1Ca	X07518	130	134.7	1189	+++	+++[2]	+++[3]	*Bt aizawai* HD-229 and 7.29	[102]
Cry2Aa	M31738	66	70.8	633	++	ND	ND	*Bt kurstaki* HD-1	[123]
Cry4Aa	Y00423	125	134.4	1180	++	++	+++	*Bti* and all class-1 strains	[23, 29, 86, 96, 119, 120]
Cry4Ba	X07423	135	127.8	1136	+++	+++	-	*Bti* and all class-1 strains	[21, 23, 29, 86, 96, 120]
Cry10Aa	M12662	58	77.8	675	+/-	+/-	+/-	*Bti*	[28, 113]
Cry11Aa	M31737	68	72.4	643	+++	+++	+++	*Bti* and all class-1 strains	[23, 31, 36, 86]
Cry11Ba	X86902	80	81.3	724	++++	++++	++++	*Bt jegathesan* 367	[30]
Cry11Bb	AF017416	94	88.2	783	++++	+++[4]	++++[3]	*Bt medellin* 163-131	[81, 89]
Cry16Aa	X94146	60	71.1	614	-	-	-	*C. bifermentans malaysia* CH18	[9], Juárez-Pérez unpublished
Cry17Aa	X99478	63	71.7	619	-	-	-	*C. bifermentans malaysia* CH18	[10], Juárez-Pérez unpublished
Cry19Aa	Y07603	65	74.7	648	-	+	+	*Bt jegathesan* 367	[92]
Cry19Ba	D88381	78	78.5	682	ND	-	+[5]	*Bt higo*	[49]
Cry20Aa	U82518	86	86.1	753	+/-	ND	+/-[3]	*Bt fukuokaensis*	[61]
Cry24Aa	U88188	ND	75.9	674	ND	ND	ND	*Bt jegathesan* 367	Gill unpublished
Cry25Aa	U88189	ND	75.6	675	ND	ND	ND	*Bt jegathesan* 367	Gill unpublished
Cry27Aa	AB023293	94	97.8	854	ND	ND	ND	*Bt higo*	Saitoh unpublished
Cry29Aa	AJ251977	68	74.4	650	-	-	-	*Bt medellin* 163-131	Delécluse unpublished
Cry30Aa	AJ251978	70	77.8	688	-	-	-	*Bt medellin* 163-131	Delécluse unpublished
Cyt1Aa	X03182	28	27.3	249	++	++	++	*Bti* and all class-1 strains	[23, 58 109, 117, 118]
Cyt1Ab	X98793	30	27.5	250	+/-	+/-	+	*Bt medellin* 163-131	[109]
Cyt2Aa	Z14147	25	29.2	259	++	++[2]	++	*Bt kyushuensis*	[57, 58]
Cyt2Ba	U52043	27	29.8	263	++	+	++	*Bti*	[40], Delécluse unpublished
Cyt2Bb	U82519	26	30.1	263	+	ND	ND	*Bt jegathesan* 367	[20]
Cyt2Bc	AJ251979	29	29.7	260	ND	ND	ND	*Bt medellin* 163-131	Delécluse, unpublished

Table 2 cont.

Name	Acc. no.	Molecular Weight (kDa)		Residues	Mosquitocidal activity[1]			Found in	Reference
		apparent	estimated		A. aegypti	A. stephensi	C. pipiens		
Mtx1	M60446	100	100.6	870	ND	ND	++++[3]	*B. sphaericus*: most toxic strains	[105, 106]
Mtx2	U19898, U47299-302, U41822	ND	31.8	292	+	ND	++[3]	*B. sphaericus*: most toxic strains	[19, 108]
Mtx3	U42328-35	ND	35.8	326	+/-	ND	+/-[3]	*B. sphaericus*: most toxic strains	[65]
Cbm17.1	Y10457	19	17.2	153	-	-	-	*C. bifermentans malaysia* CH18	[10], Juárez-Pérez unpublished
Cbm17.2	Y10457	18	17.4	152	-	+/-[2]	-	*C. bifermentans malaysia* CH18	[10], Juárez-Pérez unpublished
P61	ND	61	ND	ND	+/-	+/-[2]	ND	*Bt galleriae* 916	[1]
P51 (BinB)	X07992, Y13311-315	51	51.4	448	ND	ND	-	*B. sphaericus*: 'highly toxic' strains	[5, 12, 13, 75]
P42 (BinA)	Y00528, Y00378, X07025, X14964, Y13316-320	42	41.9	370	ND	ND	+++	*B. sphaericus*: 'highly toxic' strains	[12-14, 16, 46, 47, 75]

[1] LC_{50} values of 10-50 ng/ml: ++++; 50-500 ng/ml: +++; 500-5,000 ng/ml: ++; 5,000-50,000 ng/ml: +; >50,000ng/ml: +/-; -: not toxic; ND: not determined.

2, 3, 4, and 5 larvicidal activity determined on *A. gambiae*, *C. quinquefasciatus*, *A. albimanus*, and *C. molestus*, respectively.

Chapter 2.4

Toxin and virulence gene expression in *Bacillus thuringiensis*

Didier Lereclus & Hervé Agaisse

Unité de Biochimie Microbienne, Centre National de la Recherche Scientifique (URA 1300), Institut Pasteur, 25 rue du Docteur Roux, 75724 Paris cedex 15, France, and Station de Lutte Biologique, INRA, La Minière, 78285 Guyancourt cedex, France

Key words: *Bacillus*, crystal protein, mRNA stability, regulator, sigma factors, sporulation, stationary phase, transcription.

Abstract: At the end of vegetative growth and in response to various nutrient stresses, *Bacillus* species produce specific components (degradative enzymes, antibiotics, toxins) allowing the bacteria to rapidly adapt to the environment, to eliminate competitors and to gain access to novel sources of nutrients. The entomopathogenic properties of *Bacillus thuringiensis* are partly due to the production of larvicidal toxins, called δ-endotoxins (the Cry and Cyt proteins). These proteins are synthesised during the stationary phase and/or during sporulation. They accumulate in the mother cell to form a crystal inclusion which can account for up to 25 % of the dry weight of the cells. This massive production of proteins and its coordination with the stationary phase result from a variety of mechanisms occurring at the transcriptional, post-transcriptional and post-translational levels. Some *B. thuringiensis* strains produce another set of insecticidal toxins, called the Vip proteins. These toxins are synthesised from the vegetative phase to the early stages of the stationary phase. In addition to these insecticidal proteins, *B. thuringiensis* produces a variety of extracellular proteins which might be involved in the virulence of the bacterium and specifically in its ability to provoke a septicaemia in susceptible insects. These potential virulence factors are degradative enzymes (proteinases, phospholipases...), secondary metabolites (antibiotics, β-exotoxin...), cell surface proteins (flagellin, S-layer proteins...) and cytotoxic components (haemolysins, enterotoxins). The expression of the genes encoding these compounds is generally activated at the onset of the stationary phase.

J.-F. Charles et al. (eds.),
Entomopathogenic Bacteria: From Laboratory to Field Application, 127–142.
© 2000 *Kluwer Academic Publishers. Printed in the Netherlands.*

1.   EXPRESSION OF THE INSECTICIDAL TOXIN GENES

1.1   *cry* and *cyt* gene expression

1.1.1   Transcriptional mechanisms

The Cry and Cyt toxins are responsible for the specific insecticidal properties of *Bacillus thuringiensis* [33]. These toxins are generally produced as a crystal inclusion which is the major distinctive trait of this bacterial species (the presence or absence of the crystal distinguishes the two closely related species, *B. thuringiensis* and *B. cereus*). Crystal formation in *B. thuringiensis* cells requires the accumulation of toxin proteins in a non-dividing cell (the mother cell compartment) to avoid protein dilution by cell division. This suggests that *cry* and *cyt* gene expression should specifically occur during the stationary phase and therefore be spatially and temporally regulated.

Under favourable environmental conditions, *Bacillus* spp. grow exponentially and transcription from vegetative gene promoters is driven by RNA polymerase containing the major sigma factor, σ^A. When encountering less favourable conditions (*i.e.* nutrient deprivation), the bacteria enter a transition state which may lead to the sporulation process. Genetic and biochemical studies show that, in response to environmental signals, a phosphorelay signal transduction system is the key determinant governing the initiation of sporulation in *Bacillus subtilis* [19, 29]. The ultimate reaction of the phosphorelay is the phosphorylation of the protein Spo0A. The phosphorylated form of this protein (Spo0A~P) acts as a positive regulator by activating the transcription of various genes including those encoding the sporulation-specific sigma factors σ^H, σ^E and σ^F [35]. Thus, the concentration of Spo0A~P in the cell will determine whether the bacterium will divide or sporulate. The transition from vegetative growth to the stationary phase and the development process leading to the formation of the spore are thus temporally controlled at the transcriptional level and require the activation of new sigma factors (Fig. 1). The sigma factors σ^A and σ^H are present during the vegetative growth; σ^A remains active in the predivisional cell during the early stages of sporulation, and the concentration of σ^H significantly increases at the onset of the stationary phase. During the sporulation process the sigma factors σ^E and σ^K are successively active in the mother cell compartment, whereas σ^F and σ^G are successively active in the forespore [17].

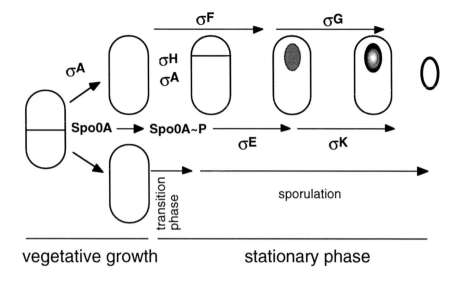

Figure 13. Genetic control of sporulation in *B. subtilis*. During the vegetative growth, the major sigma factor is σ^A. At the onset of the stationary phase, the key event responsible for the initiation of sporulation is the phosphorylation of the Spo0A protein. The increasing concentration of Spo0A~P during the transition phase leads to the synthesis of σ^H, σ^E and σ^F. The sigma factors σ^E and σ^K are successively active in the mother cell compartment, whereas the sigma factors σ^F and σ^G are successively active in the forespore (see reference [17] for review).

Studies on *cry* and *cyt* gene expression show that we can distinguish two different mechanisms responsible for the transcription of these genes during the stationary phase. One is dependent on the sporulation-specific sigma factors, whereas the second is independent of the sporulation process.

The transcription of the *cry1Aa* gene, encoding a toxin active against Lepidoptera, was extensively analysed by biochemical and genetic approaches. Two transcription start sites have been mapped, defining two overlapping promoters (BtI and BtII) which are sequentially functional during the sporulation process [37]. The genes encoding the sigma factors responsible for the transcription from BtI and BtII have been cloned and sequenced; their products are homologous to σ^E and σ^K of *Bacillus subtilis*, respectively [1].

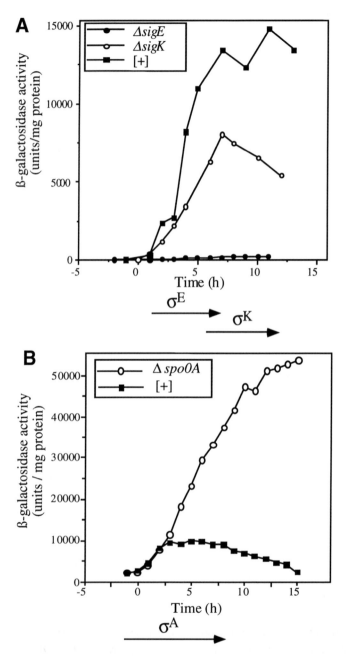

Figure 14. Time course expression of the *cry1Aa* and *cry3A* promoters in *B. thuringiensis*. A) β-galactosidase expression of a *cry1Aa'-lacZ* transcriptional fusion in a wild type *B. thuringiensis* strain [+] and in *sigE* and *sigK* mutants [10]. B) β-galactosidase expression of a *cry3A'-lacZ* transcriptional fusion in a wild type *B. thuringiensis* strain [+] and in a *spo0A* mutant [31].

The role of these two sigma factors was confirmed by analysing *cry1Aa* transcription in σ^E and σ^K mutants of *B. thuringiensis* [10]. The promoter activity of BtI and BtII was monitored using a *cry1Aa'-lacZ* transcriptional fusion (Fig. 2A). In a *B. thuringiensis* wild type strain, *cry1Aa* transcription starts at T_1 (1 hour after the onset of the sporulation) and continues during most of the sporulation process: two successive peaks of expression are observed at T5 and T11, corresponding to the maximum of σ^E and σ^K activity, respectively. The transcription stops earlier in the σ^K mutant and is completely abolished in the σ^E mutant. This indicates that both sigma factors are involved in *cry1Aa* transcription.

Determination and comparison of several *cry* and *cyt* gene promoters show that most of them are similar to the *cry1Aa* promoters (Fig. 3). These promoters are therefore presumably transcribed by RNA polymerase containing σ^E or σ^K. The transcription of the *cry18Aa* gene isolated from *Bacillus popilliae* was analysed in *B. thuringiensis*. This gene is transcribed by σ^E and σ^K forms of RNA polymerase from a single initiation site throughout the sporulation process [42]. The *cry4B* gene isolated from *B. thuringiensis israelensis* is only transcribed by the σ^E form of RNA polymerase. The activity of the *cry4B* promoter is reduced in a *B. subtilis spoIIID* mutant [39]. This suggests that the SpoIIID transcriptional activator positively regulates σ^E-dependent transcription of *cry4B*. A low-level transcription of *cry4A*, *cry4B* and *cry11A* has been detected during the transition state, between the end of vegetative growth and the onset of sporulation [30]. This transcriptional activity may be due to the σ^H factor.

The *cry3A* gene, encoding a coleopteran-active toxin is a typical example of a sporulation independent *cry* gene. *cry3A* expression was originally observed at a low level during the vegetative growth [11, 34]. Analysis with transcriptional fusions to the *lacZ* reporter gene indicated that the *cry3A* promoter is weakly functional during the vegetative phase of growth and is activated at the end of exponential growth (Fig. 2B). Genetic analysis of *cry3A* expression demonstrated that its transcription is not dependent on sporulation-specific sigma factors either in *B. subtilis* [3] or in *B. thuringiensis* [31]. Moreover, *cry3A* expression is increased in *spo0* mutants which are unable to initiate sporulation [3, 22, 25, 26, 31] (Fig. 3B). Expression of the *cry3A* gene in a *spo0A* mutant of *B. thuringiensis* leads to an overproduction of crystal proteins which remain encapsulated in the non sporulating cell [22]. Activation of *cry3A* transcription at the onset of the stationary phase is therefore due to a non-sporulation-dependent mechanism. However, the genetic determinants involved in this activation are not known. The promoters of the *cry3A* and *cry3B* genes were determined [4, 8]: they resemble promoters recognised by the σ^A factor (Fig. 3).

```
cry1A      GTGCATTTTTTCATAAGATGAGTCATATGTTTTAAATTG
cry1B      ATGAATATTTTCATAAGCTGAACCATATGATTTAAACT
cry1C      TTGTATTTTTTCATAAGATGTGTCATATGTATTAAATCG
cry2A      TGGCATATACCGTTTACTCCCCACATAGAATGTGCAGA
cry4A      AGGAATAGATTATTTAAATTACGAATACTTTAAAATG
cry4B      TGGAATAATTATATTGGTACAGAAATATGATTGGGATTA
cry11A     ATGCATCGTTTTTATACAAGTAACATATATTTGTTATG
cry18A     TGGAACGTTATTGCCTTAACCTGAATAGAATGCATTAA
cyt1A      AGGCATCTTTCGAACTATAGC-GCATAGAATACTA

consensus   - 35                      - 10
SigE       GNATNNTT-----N(13)---CATANNNT
```

```
cry1A      TGTTGCACTTTG-TGCATTTTTTCATAAGATGAGTCATA
cry1B      TAGTAGAATTTTATGAATATTTTCATAAGCTGAACCA
cry1C      TGTTACGTTTTT-TGTATTTTTTCATAAGATGTGTCATA
cry4A      CAGAACCTTTGATGTTATTAAGGCGTAGAATATCCAGA
cry11A     AGATACATTTTTGTTTATACATGCATCGTTTTTATA
cry18A     TGGAACGTTATTGCCTTAACCTGAATAGAATGCATTAA
cyt1A      ATGCACCAATGTATACATTAAATAATATTATGTGAATTA

consensus   - 35                      - 10
SigK       ACNNTT-----N(13)---CATANNNT
```

```
cry3A      TTGTTGCAATTGAAGAATTATTAATGTTAAGCTTAATTA
cry3B      TTCTTGCCAATATAGAGCTTAACGTGTTAGGGTGAAATT

consensus   - 35                      - 10
SigA       TTGACA------N(18)-------TATAAT
(B. subtilis)
```

Figure 15. cry gene promoters. Alignment of the σ^E- and σ^K-dependent *cry* gene promoters (see [41] for references) establishing the consensus sequence of the -35 and -10 boxes recognised by the σ^E and σ^K forms of RNA polymerase, respectively. The putative -35 and -10 boxes of the *cry3A* and *cry3B* genes [4, 8] are compared to the σ^A consensus reported for *B. subtilis* [27]. The conserved nucleotides are boldfaced. The transcriptional start sites are underlined.

1.1.2   mRNA stability

The mRNAs encoding the crystal proteins have an average half-life of about 10 minutes [16]. It is likely that this relatively high stability of *cry* gene mRNAs is partly responsible for the high level of Cry toxin production. Two specific regions of the *cry* gene mRNAs are involved in this remarkable stability in the 3' and 5' untranslated regions of the transcripts.

The DNA region corresponding to the 3' untranslated sequence of the *cry1Aa* transcript acts as a positive retroregulator [38]. The fusion of this DNA fragment with the 3' end of heterologous genes increases the half-life of their transcripts and consequently increases their expression level. This untranslated sequence includes the potential transcriptional terminator of the *cry1Aa* gene, an inverted repeat sequence with the potential to form a stable stem-loop structure in the mRNA. It is assumed that this structure may act as a mRNA stabiliser by protecting the molecule from a 3' to 5' exonucleolytic degradation.

An extensive study of *cry3A* gene expression showed that the 5' untranslated region of the *cry3A* mRNA has all the characteristics of a 5' mRNA stabiliser. The fusion of this 5' untranslated sequence to the *lacZ* reporter gene results in a 10-fold increase in β-galactosidase production [4]. This effect is correlated with a 10-fold increase in the stability of the corresponding fusion transcript [6]. Deletion analysis revealed that the stability determinant corresponds to the GAAAGGAGGGATG sequence located four nucleotides downstream of the 5' end of the stable form of *cry3A* mRNA. Because of the striking similarity between this stabilising sequence and the Shine-Dalgarno sequence involved in the initiation of translation in prokaryotes, this mRNA stabiliser was called STAB-SD.

It was hypothesised that binding of a 30S ribosomal subunit may be involved in the stabilisation process. Confirming this assumption, it was shown that purified 30S ribosomal subunits are able to bind *in vitro* to a synthetic STAB-SD sequence (H. Agaisse and G. Rapoport, unpublished data). Furthermore, introduction of mutations designed to abolish the putative interaction between STAB-SD and the 3' end of the 16S rRNA in a *cry3A'-lacZ* fusion results in a drastic decrease in mRNA stability and β-galactosidase production (Fig. 4). Moreover, mutations thought to reduce (but not abolish) specific interaction with the 16S rRNA result in intermediate β-galactosidase production corresponding to an intermediate mRNA stability (Fig. 4). Therefore, it was suggested that binding of a 30S ribosomal subunit to the STAB-SD sequence may stabilise the *cry3A* transcript by protecting its 5' end from a ribonuclease activity (Fig. 5).

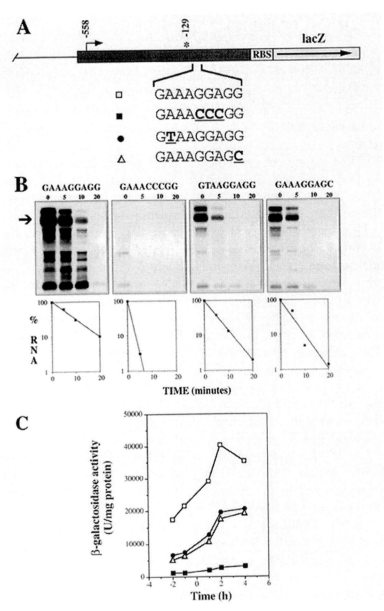

Figure 16. Mutations in the STAB-SD sequence of *cry3A*. A) Mutations in STAB-SD.
The broken arrow indicates the transcriptional start site (558 bp upstream of the start
codon) and the asterisk represents the 5' end of the stable transcript (129 bp upstream of
the start codon). □, original STAB-SD sequence; ●, ■ and Δ, modified STAB-SD
sequences. B) Effect of the mutations on mRNA stability. Northern blot analysis of
cry3A'-lacZ mRNA extracted at various intervals after addition of rifampicin. The arrow
indicates the *cry3A'-lacZ* transcript. C) β-galactosidase production in *Bacillus* strains
carrying *cry3A'-lacZ* fusions with the original or modified STAB-SD sequences [6].

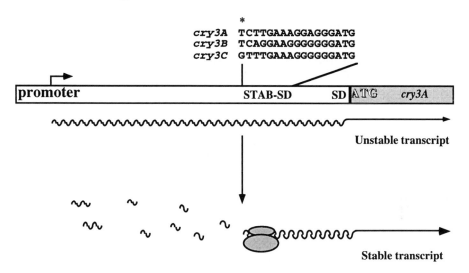

Figure 17. Model for 5' stabilisation of *cry3A* mRNA. The 5' untranslated regions carrying the potential STAB-SD sequences are as previously described for *cry3A* [3, 6], *cry3B* [13] and *cry3C* [21]. The broken arrow indicates the transcriptional start site and the asterisk represents the 5' end of the stable transcript. A ribosome or the 30S ribosomal subunit bound to the STAB-SD sequence may stabilise the *cry3A* transcript by protecting its 5' end from a ribonuclease.

Putative STAB-SD sequences are also present in similar positions upstream of the coding sequence of the *cry3B* and *cry3C* genes (Fig. 5). This suggests that the corresponding transcripts are stabilised by a similar mechanism. It is not known whether 5' mRNA stabilisers are present upstream of the coding sequence of the other *cry* genes. However, due to the high stability of the *cry* mRNAs and to the important role of the 5' end in mRNA stability, we can speculate that mechanisms involved in 5' end protection of the transcripts may exist in other *cry* genes.

1.1.3   Post-translational modifications

The ability to form a crystal undoubtedly contributes to the stability of the protoxins by preventing premature proteolysis of the proteins. This is also a convenient mechanism to concentrate the protoxins and thus to deliver high doses of active compounds. It is generally assumed that most of the 130-140 kDa protoxins (e.g. Cry1, Cry4, Cry5 and Cry7) can spontaneously form crystals. The cysteine-rich C-terminal region of these protoxins may participate in crystal structure, presumably through the formation of disulphide bonds. The 73 kDa Cry3 proteins also seem to crystallise spontaneously, since these proteins form a typical flat rectangular crystal inclusion in *B. subtilis* [3]. However, the Cry3 polypeptides are not linked by disulphide bridges, as they lack the cysteine-rich C-terminal region [9]. Analysis of the three-dimensional structure of the Cry3A protein suggests that four intermolecular salt bridges might be involved in the formation of the crystal inclusion [24].

Several studies indicate that crystallisation of Cry2A (71 kDa), Cry11A (72 kDa) and Cyt1A (27 kDa) requires the presence of accessory proteins which might assist the crystallisation process (see references [5, 8] for review). The genes encoding these helper proteins are organised in an operon and are cotranscribed with the *cry2A* and *cry11A* genes [12, 36].

Two Cry proteins (Cry1Ia and Cry16Aa) are exported proteins which do not form a crystal inclusion. The Cry1Ia toxin is produced early during sporulation and is found in the culture supernatant as a processed polypeptide of 60 kDa [20]. The Cry16Aa toxin from *Clostridium bifermentans* is also synthesised during sporulation and secreted into the culture supernatant [7]. In both cases, the process involved in the secretion of these proteins is unknown.

1.2 *vip* gene expression

Genes coding for a novel family of insecticidal toxins were found in several *B. thuringiensis* strains. These toxins, designated Vip (for Vegetative insecticidal proteins), are active against a broad spectrum of lepidopteran insects [15]. In a manner reminiscent of the Cry toxin mode of action, the ingestion of Vip proteins by the susceptible larvae results in gut paralysis, disruption of the gut epithelium and death [40].

Interestingly, unlike the Cry toxins whose expression is mostly restricted to the stationary phase (with the exception of the Cry3 proteins), the Vip proteins are produced during the vegetative phase of growth as well as during early stages of stationary phase [15]. The promoter of the *vip* genes was not determined. However, it can be assumed from the early production of the Vip proteins that the transcription of the corresponding genes is dependent, at least in part, on the major sigma factor, σ^A.

Unlike the Cry toxins, the Vip proteins do not accumulate in the cytoplasm of non-dividing cells, but are instead secreted into the medium. Surprisingly, despite the presence of a typical N-terminal signal peptide, the Vip toxins are not processed during secretion, suggesting that an unknown mechanism of protein export is responsible for the secretion of these toxins [15].

2. VIRULENCE GENE EXPRESSION

B. thuringiensis and *B. cereus* are known to produce a variety of extracellular compounds that might contribute to virulence by conferring to the bacteria the ability to multiply in their hosts (most generally insect larvae) and to provoke septicaemia [14, 18, 32]. These compounds (see Chapter 1.3), which are not specific to entomopathogenic bacteria, are degradative enzymes (phospholipases C, sphingomyelinase, proteinases, collagenase, chitinase, nucleases...), cytotoxic proteins (haemolysins and enterotoxins), cell surface proteins (flagellin, S-layer proteins) and secondary metabolites (antibiotics, β-exotoxin).

The expression of various extracellular components (flagellin, β-lactamase and two phospholipases PI-PLC and PC-PLC) is reduced in an avirulent and non motile mutant of *B. thuringiensis* which is unable to kill insect larvae and cause septicaemia after injection in the hemocoel [41]. This suggests that these potential virulence factors may be coregulated. However, the mutation involved in the lack of virulence was not characterised.

2.1 The *plcR* regulon

Analysis of the expression of the *plcA* gene, encoding the phosphatidylinositol-specific phospholipase C (PI-PLC), showed that *plcA* transcription starts at the onset of the stationary phase. The positive regulator (PlcR) responsible for this transcriptional activation was identified [23]. PlcR displays polypeptide sequence homology with the transcriptional activators of proteinase production PreL and NprA from *Lactobacillus* sp. and *B. stearothermophilus*, respectively. Similarity between PlcR and these two transcriptional activators is limited to the N-terminal part of the protein which is presumed to contain the DNA-binding domain of these transcriptional activators. The *plcR* gene was also found in the *B. thuringiensis*-related bacteria, *B. anthracis* and *B. cereus* [2]. Comparison of the deduced amino acid sequence of the *B. anthracis*, *B. cereus* and *B. thuringiensis* PlcR polypeptides revealed a high degree of conservation amongst these regulators. However, whilst the *B. cereus plcR* gene encodes a functional transcriptional activator, the *plcR* gene isolated from *B. anthracis* is truncated and its product is not active [2].

PlcR positively regulates its own expression which starts at the onset of the stationary phase. This suggests that additional factors might be required to trigger *plcR* expression during the transition between the vegetative growth and the stationary phase. PlcR also controls the transcription of a two gene operon which encodes the phosphatidylcholine-specific phospholipase C (PlcB or PC-PLC) and the sphingomyelinase CerB (S. Salamitou, H. Agaisse & D. Lereclus, unpublished results). A genetic screen was performed to identify different *B. thuringiensis* genes regulated by PlcR [2]. This study revealed that PlcR is a pleiotropic regulator which controls the transcription of at least 14 genes. These PlcR-regulated genes encode potential virulence factors including a ribonuclease, a S-layer protein and the haemolytic and non-haemolytic enterotoxins, Hbl and Nhe. Except for the *plcR* gene itself, all the identified PlcR-regulated genes code for exported proteins [2]. Alignment of the PlcR-regulated promoters reveals the presence of a highly conserved palindromic sequence which is presumably the specific recognition site for PlcR (Fig. 6).

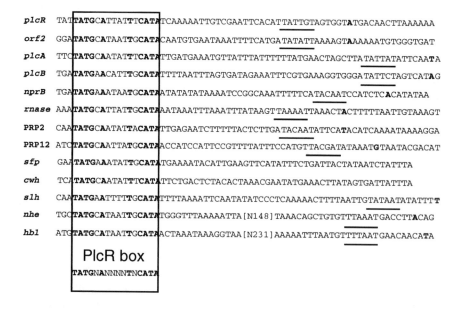

Figure 18. PlcR regulated promoters. *B. thuringiensis* or *B. cereus* genes regulated by PlcR are indicated in italics when their product is characterised; they are designated as PRP when their product is unknown. The palindromic region presumed to be the PlcR recognition site is boxed and the highly conserved nucleotides are boldfaced. The transcription start sites are boldfaced and the -10 boxes are underlined. References: *plcA* and *plcR* [23], *hbl, nhe, npr, rnase, slh*, *orf2*, PRP2 and PRP12 [2], *plcB* (Salamitou, Agaisse & Lereclus, unpublished data), *cwh* and *sfp* [28].

Determination of the transcriptional start of these genes showed that the -10 region of the promoters resembles that of promoters recognised by σ^A. However, the -35 box, characteristic of the σ^A RNA polymerase-dependent promoters, is not found. This suggests that the binding of PlcR upstream of the -10 box may compensate for the absence of a consensus -35 box.

Although some PlcR-regulated genes are grouped two by two, an analysis of their distribution indicates that they are not clustered in a pathogenicity island but scattered throughout the chromosome [2]. In addition to genes detected by a genetic approach, the sequencing of three DNA regions, surrounding *plcA, plcR* and the *hbl* operon of *B. cereus*, led to the detection of three new putative PlcR-regulated genes, encoding two proteinases and a cell wall hydrolase [28].

3.   CONCLUSION

B. thuringiensis produces different types of insecticidal proteins which reflect the capacity of this *Bacillus* species to develop within the large ecological niche constituted by insects. In addition, the production of a large variety of potential virulence factors confers to the bacterial family (including *B. thuringiensis* and *B. cereus*) the properties of opportunistic pathogens. The biological mechanisms used for the toxin and virulence factor synthesis can be considered from a physiological point of view: the characteristics of a given expression system may be related to the function of the corresponding toxins or virulence factors. The genetic and molecular systems devoted to massive production of the Cry proteins during sporulation may serve to deliver a toxin dose that is sufficient to kill or at least weaken an insect larva. Thus, the toxaemic effect might allow the germination of the spores and the multiplication of the bacterial cells in the insect gut environment, as long as nutrients are available. The production of other toxins (*i.e.* the Vip proteins) during the vegetative growth may contribute to reinforce the toxaemic effect. Upon encountering nutrient deprivation, the triggering of stationary phase-specific mechanisms (*i.e.* PlcR-dependent expression) leads to the production of extracellular compounds allowing the bacteria to damage and invade host tissues. Thus, the bacteria gain access to novel sources of nutrients which create favourable growing conditions. These various expression systems constitute original models to study genetic expression of adaptative genes during the stationary phase of Gram-positive sporulating bacteria.

ACKNOWLEDGEMENTS

We thank Tarek Msadek for his critical reading of the manuscript.

REFERENCES

[1] Adams LF, Brown KL & Whiteley HR (1991) Molecular cloning and characterization of two genes encoding sigma factors that direct transcription from a *Bacillus thuringiensis* crystal protein gene promoter. J. Bacteriol. 173, 3846-3854

[2] Agaisse H, Gominet M, Økstad OA, Kolstø AB & Lereclus D (1999) PlcR is a pleiotropic regulator of extracellular virulence factor gene expression in *Bacillus thuringiensis*. Mol. Microbiol. 32, 1043-1053

[3] Agaisse H & Lereclus D (1994) Expression in *Bacillus subtilis* of the *Bacillus thuringiensis cryIIIA* toxin gene is not dependent on a sporulation-specific sigma factor and is increased in a *spoOA* mutant. J. Bacteriol. 176, 4734-4741

[4] Agaisse H & Lereclus D (1994) Structural and functional analysis of the promoter region involved in full expression of the *cryIIIA* toxin gene of *Bacillus thuringiensis*. Mol. Microbiol. 13, 97-107

[5] Agaisse H & Lereclus D (1995) How does *Bacillus thuringiensis* produce so much insecticidal crystal protein ? J. Bacteriol. 177, 6027-6032

[6] Agaisse H & Lereclus D (1996) STAB-SD: a Shine-Dalgarno sequence in the 5' untranslated region is a determinant of mRNA stability. Mol. Microbiol. 20, 633-643

[7] Barloy F, Delécluse A, Nicolas L & Lecadet M-M (1996) Cloning and expression of the first anaerobic toxin gene from *Clostridium bifermentans* subsp. *malaysia*, encoding a new mosquitocidal protein with homologies to *Bacillus thuringiensis* delta-endotoxins. J. Bacteriol. 178, 3099-3105

[8] Baum JA & Malvar T (1995) Regulation of insecticidal crystal protein production in *Bacillus thuringiensis*. Mol. Microbiol. 18, 1-12

[9] Bernhard K (1986) Studies on the delta-endotoxin of *Bacillus thuringiensis* var. *tenebrionis*. FEMS Microbiol. Lett. 33, 261-265

[10] Bravo A, Agaisse H, Salamitou S & Lereclus D (1996) Analysis of *cryIAa* expression in *sigE* and *sigK* mutants of *Bacillus thuringiensis*. Mol. Gen. Genet. 250, 734-741

[11] De Souza MT, Lecadet M-M & Lereclus D (1993) Full expression of the *cryIIIA* toxin gene of *Bacillus thuringiensis* requires a distant upstream DNA sequence affecting transcription. J. Bacteriol. 175, 2952-2960

[12] Dervyn E, Poncet S, Klier A & Rapoport G (1995) Transcriptional regulation of the *cryIVD* gene operon from *Bacillus thuringiensis* subsp. *israelensis*. J. Bacteriol. 177, 2283-2291

[13] Donovan WP, Rupar MJ, Slaney AC *et al.* (1992) Characterization of two genes encoding *Bacillus thuringiensis* insecticidal crystal proteins toxic to *Coleoptera* species. Appl. Environ. Microbiol. 58, 3921-3927

[14] Drobniewski FA (1993) *Bacillus cereus* and related species. Clin. Microbiol. Rev. 6, 324-338

[15] Estruch JJ, Warren GW, Mullins MA *et al.* (1996) Vip3A, a novel *Bacillus thuringiensis* vegetative insecticidal protein with a wide spectrum of activities against lepidopteran insects. Proc. Natl. Acad. Sci. USA 93, 5389-5394

[16] Glatron MF & Rapoport G (1972) Biosynthesis of the parasporal inclusion of *Bacillus thuringiensis* : half-life of its corresponding messenger RNA. Biochimie 54, 1291-1301

[17] Haldenwang WG (1995) The sigma factors of *Bacillus subtilis*. Microbiol. Rev. 59, 1-30

[18] Heierson A, Sidén I, Kivaisi A & Boman HG (1986) Bacteriophage-resistant mutants of *Bacillus thuringiensis* with decreased virulence in pupae of *Hyalophora cecropia*. J. Bacteriol. 167, 18-24

[19] Hoch JA (1993) Regulation of the phosphorelay and the initiation of sporulation in Bacillus subtilis. Annu. Rev. Microbiol. 47, 441-465

[20] Kostichka K, Warren GW, Mullins M *et al.* (1996) Cloning of a *cryV*-type insecticidal protein gene from *Bacillus thuringiensis*: the *cryV*-encoded protein is expressed early in stationary phase. J. Bacteriol. 178, 2141-2144

[21] Lambert B, Theunis W, Aguda R *et al.* (1992) Nucleotide sequence of gene *cryIIID* encoding a novel coleopteran-active crystal protein from strain BTI109P of *Bacillus thringiensis* subsp. *kurstaki*. Gene 110, 131-132

[22] Lereclus D, Agaisse H, Gominet M & Chaufaux J (1995) Overproduction of encapsulated insecticidal crystal proteins in a *Bacillus thuringiensis spo0A* mutant. Bio/Technology 13, 67-71

[23] Lereclus D, Agaisse H, Gominet M, Salamitou S & Sanchis V (1996) Identification of a gene that positively regulates transcription of the phosphatidylinositol-specific phospholipase C gene at the onset of the stationary phase. J. Bacteriol. 178, 2749-2756

[24] Li J, Carroll J & Ellar DJ (1991) Crystal structure of insecticidal δ-endotoxin from *Bacillus thuringiensis* at 2.5 A resolution. Nature 353, 815-821

[25] Malvar T & Baum J (1994) Tn*5401* disruption of the *spo0F* gene, identified by direct chromosomal sequencing, results in *cryIIIA* overproduction in *Bacillus thuringiensis*. J. Bacteriol. 176, 4750-4753

[26] Malvar T, Gawron-Burke C & Baum J (1994) Identification of HknA: a KinA-like protein capable of bypassing early Spo⁻ mutations that result in CryIIIA overproduction in *Bacillus thuringiensis*. J. Bacteriol. 176, 4742-4749

[27] Moran CP (1993) RNA polymerase and transcription factors, p. 653-667. *In* Sonenshein AL, Hoch JA & Losick R (ed.), *Bacillus subtilis* and other gram-positive bacteria, American Society for Microbiology

[28] Økstad OA, Gominet M, Purnelle B *et al.* (1999) Sequence analysis of three *Bacillus cereus* loci carrying PlcR-regulated genes encoding degradative enzymes and enterotoxin. Microbiol. 145, 3129-3138

[29] Perego M (1998) Kinase-phosphatase competition regulates *Bacillus subtilis* development. Trends Microbiol. 6, 366-370

[30] Poncet S, Dervyn E, Klier A & Rapoport G (1997) Spo0A represses transcription of the *cry* toxin genes in *Bacillus thuringiensis*. Microbiol. 143, 2743-2751

[31] Salamitou S, Agaisse H, Bravo A & Lereclus D (1996) Genetic analysis of *cryIIIA* gene expression in *Bacillus thuringiensis*. Microbiology 142, 2049-2055

[32] Salamitou S, Marchal M & Lereclus D (1996) *Bacillus thuringiensis* : un pathogène facultatif. Ann. Inst. Pasteur 7, 285-296

[33] Schnepf E, Crickmore N, Van Rie J *et al.* (1998) *Bacillus thuringiensis* and its pesticidal crystal proteins. Microbiol. Mol. Biol. Rev. 62, 775-806

[34] Sekar V (1988) The insecticidal crystal protein gene is expressed in vegetative cells of *Bacillus thuringiensis* var. *tenebrionis*. Curr. Microbiol. 17, 347-349

[35] Smith I (1993) Regulatory proteins that control late-growth development, p. 785-800. *In* Sonenshein AL, Hoch JA & Losick R (ed.), *Bacillus subtilis* and other Gram-positive bacteria, American Society for Microbiology

[36] Widner WR & Whiteley HR (1989) Two highly related insecticidal crystal proteins of *Bacillus thuringiensis* subsp. *kurstaki* possess different host range specificities. J. Bacteriol. 171, 965-974

[37] Wong HC, Schnepf HE & Whiteley HR (1983) Transcriptional and translational start sites for the *Bacillus thuringiensis* crystal protein gene. J. Biol. Chem. 258, 1960-1967

[38] Wong HC & Chang S (1986) Identification of a positive retroregulator that stabilizes mRNAs in bacteria. Proc. Natl. Acad. Sci. USA. 83, 3233-3237

[39] Yoshisue H, Nishimoto T & Komano T (1993) Idendification of a promoter for the crystal protein-encoding gene *cryIVB* from *Bacillus thuringiensis* subsp. *israelensis*. Gene 137, 247-251

[40] Yu CG, Mullins MA, Warren GW, Koziel MG & Estruch JJ (1997) The *Bacillus thuringiensis* vegetative insecticidal protein Vip3A lyses midgut epithelium cells of susceptible insects. Appl. Environ. Microbiol. 63, 532-536

[41] Zhang M-Y, Lövgren A, Low MG & Landén R (1993) Characterization of an avirulent pleitropic mutant of the insect pathogen *Bacillus thuringiensis*: reduced expression of flagellin and phospholipases. Infect. Immun. 61, 4947-4954

[42] Zhang J, Schairer HU, Schnetter W, Lereclus D & Agaisse H (1998) *Bacillus popilliae cry18Aa* operon is transcribed by σE and σK forms of RNA polymerase from a single initiation site. Nucleic Acids Res. 26, 1288-1293

Chapter 2.5

Genetic and genomic contexts of toxin genes

Marie-Laure Rosso[1], Jacques Mahillon[2] & Armelle Delécluse[1]
Laboratoire des Bactéries et Champignons Entomopathogènes, Institut Pasteur, 25 rue du Dr Roux, 75724 Paris Cedex 15, France[1] and Laboratoire de Génétique Microbienne, Université catholique de Louvain, B-1348 Louvain-la-Neuve, Belgium[2]

Key words: *Bacillus cereus sensu lato, Bacillus popilliae, Bacillus sphaericus, Clostridium bifermentans malaysia*, Conjugation, Insertion Sequences, Mobile Elements, Transduction, Transposition, Transposons

Abstract: This chapter focuses primarily on the genomic distribution (plasmid *vs* chromosome) and the genetic environments of the different toxin genes found in entomopathogenic bacteria, mainly in *Bacillus thuringiensis* and *Bacillus sphaericus*. A special attention is brought to their association with mobile genetic elements (Insertion Sequences, Transposons and conjugative plasmids) and their possible clustering to other, more generic, virulence factors, in the scope of genome variability and plasticity through gene transfer and rearrangements.

1.   INTRODUCTION

How do entomopathogenic bacteria organise their toxin and virulence genes? This issue is particularly relevant to the adaptability of the bacteria to their hosts. Indeed, one can easily foresee the never-ending relationship of mutual/reciprocal adaptation between the pathogenic bacteria and their insect host. It is therefore a necessity for the former to set up genetic and genomic environments that would provide the appropriate flexibility, not only in terms of toxin gene multiplicity and diversity (see Chapters 2.2 and 2.3), but also with regard to the intracellular mobility and rearrangement of these genes and their exchange (spread and acquisition) among other related bacteria sharing the same ecological niches.

J.-F. Charles et al. (eds.),
Entomopathogenic Bacteria: From Laboratory to Field Application, 143–166.
© *2000 Kluwer Academic Publishers. Printed in the Netherlands.*

After a brief description of the apparent genomic flexibility of entomopathogenic bacteria, illustrated by the plasticity of their chromosome and the diversity of their extrachromosomal molecules, the next section will review the various genetic arrangements that have been encountered during the numerous molecular studies on insecticidal toxins and their genes. One of the common features of these contexts was the presence of repeated DNA sequences, varying from 1.6 to 4.5 kb in size. Detailed analyses of several of these elements have since revealed they were transposable elements, either Insertion Sequences (IS) or class II transposons (Tn). Their functional diversity as well as their structural relationship with the toxin genes are reported, together with individual transposition activities obtained for a few elements. The intercellular mobility of insecticidal-active toxin genes is thought to rely mainly on conjugation. Although a limited number of these genetic exchange systems have been thoroughly studied, their potential contribution to the horizontal gene dispersal of the toxin genes is also documented.

2. THE ENTOMOPATHOGENIC BACTERIA GENOME

2.1 *Bacillus thuringiensis*

Bacillus thuringiensis is a Gram-positive spore-forming bacteria that belongs to the *Bacillus cereus* complex composed of 4 species: *Bacillus thuringiensis, Bacillus cereus, Bacillus anthracis* and *Bacillus mycoides*. Most particularly, *B. thuringiensis* is very closely related to *B. cereus* and can only be distinguished from the latter by its capacity to produce parasporal protein during sporulation [40]. Although high genomic variability is observed both within and among each species, no striking genetic differences were found, suggesting that these two bacteria should be considered as a single species [25, 28, 29, 58]. The genome size of *B. thuringiensis* strains varies from 2.4 to 5.7 Mb [25]. As was shown for several *B. cereus* strains, some parts of the chromosomes seem to be constant while other regions are variable, with deletions, additions, and inversions [26].

Pulsed Field Gel Electrophoresis (PFGE) analysis indicated that most *B. thuringiensis* strains have several extrachromosomal elements, some of them circular and others linear [25]. Numerous resident plasmids were found in *B. thuringiensis* strains, ranging from 2 kb to more than 200 kb [18, 37, 72, 111]. The number and the size of plasmids vary considerably between strains, but are independent of the serotypes and pathotypes. Hybridisation experiments indicate that the plasmids fall into two size groups, small plasmids (< 20 kb) and large plasmids (> 20 kb), in which there is partial conservation of DNA sequences [72] ; some of them have been shown to harbour mobile elements (see below and Table 1, pages 165-166). Among the plasmids larger than 20 kb, some harbour insecticidal δ–endotoxin genes (see below), whereas no metabolic or virulence functions for the plasmids less than 20 kb have been shown.

Because of their capacity to encode δ–endotoxin genes, most studies have focused on the large plasmids, including restriction mapping (for one *Bt israelensis* plasmid) and localisation of the replication origin. In addition, some of the large plasmids have been shown to be self transmissible to *Bt* and other species by a conjugation-like mechanism [36, 120]. Large self-transmissible plasmids are also able to mobilise other small or large plasmids [16, 120]. Two of the conjugative plasmids, pHT73 from *B. thuringiensis* HD-73, and pXO12 from *Bt thuringiensis*, carry a gene for crystal production [16]. For others, such as pXO11 from *Bt thuringiensis* [16], pXO13 from *Bt morrisoni*, pXO14 from *Bt toumanoffi*, pXO15 from *Bt alesti* [98], and pXO16 from *Bt israelensis* [49] no function other than those associated with conjugation have yet been unravelled.

Although only a few, small, cryptic *B. thuringiensis* plasmids have been described in detail [3, 22, 30, 71, 83, 89], some have already been sequenced: pTX14-3 [6], pHT1000 and pHT1030 [71] from *Bt israelensis*, pGI2 [86] and pGI3 [46] from *Bt thuringiensis*, pHD2 [90], and a 2 kb plasmid [89] from *Bt kurstaki*. Analysis of these small plasmids suggested that they all replicated by a rolling circle mechanism [46]. Although no function has been clearly determined, it has been suggested that the presence of small plasmids decreases the UV resistance of spores by increasing the amount of dipicolinic acids or by altering the patterns of the small acid-soluble proteins [21]. Moreover, plasmid-transfer experiments between *Bt kurstaki* and *B. cereus* showed that the transfer of a small plasmid from the donor *B. thuringiensis* strain contributed to an increase of protoxin synthesis in the recipient *B. cereus* strain, suggesting its possible involvement in the regulation of protoxin production [92].

In addition to plasmids, many types of virulent and temperate bacteriophages have been reported to be associated with *B. thuringiensis* strains [2, 13, 47, 52-54, 99, 118]. Some of these are associated with plasmids in *B. thuringiensis* strains. For example, the temperate prophage J7W-1, was found to be integrated in the single plasmid pAF101 of strain *B. thuringiensis* AF101 [54]. Integration of the phage J7W-1 into a plasmid was also observed after phage infection of a *Bt israelensis* strain [54], and regions homologous to this phage genome were detected in the largest plasmids of three others *B. thuringiensis* strains (*dendrolimus*, *aizawai* and *indiana*) [52]. A specialised transducing phage, TP21, carried as an autonomously replicating plasmid, was identified in strain *Bt kurstaki* HD-1 [118], likewise, the extrachromosomal prophage SU-11 was found naturally associated with *Bt israelensis* [53].

2.2   **Other entomopathogenic bacteria**

Bacillus sphaericus is a heterogeneous species, with strains assigned to five clearly defined DNA homology groups [62]. The type strain of *B. sphaericus* is equated with DNA homology group I, but group II is of more interest because it contains all the strains that are toxic to mosquitoes. All toxic strains have been classified in group IIA, and differ slightly from group IIB strains in DNA sequence homology (average 60-70% homology), rRNA gene Restriction Fragment Length Polymorphism (RFLP) [9], rRNA gene sequence [10], Random Amplified Polymorphic DNA (RAPD) fingerprinting [123], and rep-PCR [93]. Up to 11 bacteriophages could be isolated from different *B. sphaericus* strains [125] and have been useful for strain typing. For example, most of the toxic strains belong to bacteriophage group 3 [124]. As in *B. thuringiensis*, the presence of plasmids has also been reported in *B. sphaericus*: those range from 3 kb to 51 kb [1]. Although up to five plasmids could be found in a single strain, others especially the highly toxic isolate 1593, were found plasmidless.

Several researchers have also reported the presence of various plasmids in *B. popilliae*, with sizes ranging from 3.6 kb to 15 kb [33, 79, 116]. The coleopteran-active strain *B. popilliae* subsp. *melolonthae* H1 harbours a single 7.5 kb plasmid [127].

The genome of *Brevibacillus laterosporus* has been poorly characterised, and the presence of plasmids in this species has not yet been investigated. Recently, the genomic diversity among various *B. laterosporus* strains has been examined using a combination of PFGE, RAPD, and MultiLocus Enzyme Electrophoresis (MLEE) techniques [126]: only minor variations were found within this species.

Little information is available on the *Clostridium bifermentans malaysia* genome, and the only reports concerning the presence of plasmids in these strains are still controversial. Seleena and Lee [109] found no plasmids in this strain, while Barloy *et al.* [14] showed the presence, in the same isolate, of five plasmids ranging from 4 to 20 kb.

thuringiensis Berliner 1715

    IS232A (IS231B)   cry1Ab IS231C Tn4430 IS231A   IS232B   p65kb

kurstaki HD-1

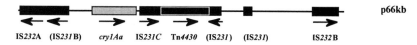

    IS232A (IS231B)   cry1Aa IS231C Tn4430 (IS231) (IS231)   IS232B   p66kb

aizawai 7-29

    IS232A (IS231B)   cry1Ab (IS231) (IS231) Tn4430   IS231A   IS232B   p68kb

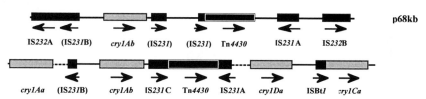

  cry1Aa (IS231B)   cry1Ab IS231C Tn4430 IS231A cry1Da   ISBtI cry1Ca

jegathesan T28A001

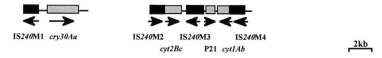

    cry11Ba IS240J1   cry19Aa ORF2 IS240J2 (TnBthI)

medellin T30 001

    IS240M1 cry30Aa   IS240M2   IS240M3   IS240M4

                              cyt2Bc   P21 cyt1Ab   2kb

Figure 1. Genetic context and diversity of toxin genes among representative strains of *Bacillus thuringiensis*. Mobile elements (IS and Tn, see Table 1) are displayed as grey rectangles while the toxin genes are shown as black rectangles. Arrows indicate the relative orientation of the various genes. Names within brackets refer to structurally inactive mobile elements. Dotted lines separate DNA segments whose relative orientation and distance have not been established. Whenever known, the plasmid location (p, followed by its estimated size in kilobases) is also indicated. Description of the p137kb plasmid from *israelensis* has been adapted from [20].

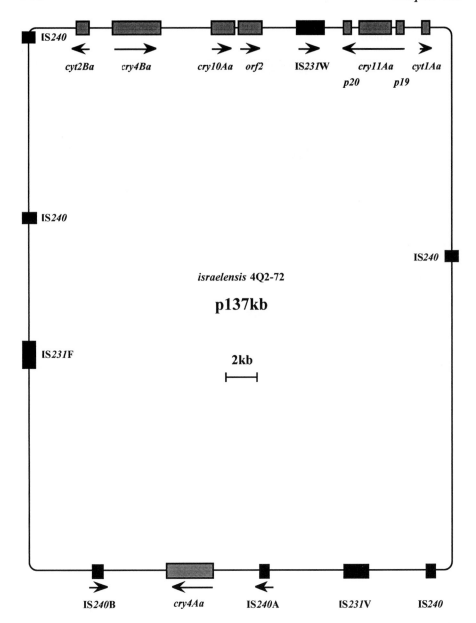

Figure 1 cont.

3. GENOMIC LOCATION OF TOXINS

3.1 *Bacillus thuringiensis*

3.1.1 Toxin strategies: plasmid or chromosome

Stahly *et al.* [111] were the first to correlate the presence of plasmids with crystal production observing that crystal minus mutants caused by heat have lost all six resident plasmids. Later, González *et al.* demonstrated by curing experiments [39] or by transferring plasmid by a conjugation-like process [36], the involvement of specific large plasmids in the production of crystals by several *B. thuringiensis* strains. The cloning of the first crystal protein gene from *Bt kurstaki* HD-1 Dipel® [107], now called *cry1Aa* (see Chapter 2.1 for nomenclature), further confirmed the validity of these observations. Since then, multiple crystal genes cloned from various *B. thuringiensis* strains, have been localised on large resident plasmids (for review, see [70]).

Upon cloning a *cry1* gene from *Bt thuringiensis* Berliner 1715, it was suggested that this type of toxin genes could also be located on the chromosome [56]. Other *cry1* genes from *B. thuringiensis* strains *dendrolimus, entomocidus* and *wuhanensis* have been identified only on the chromosome [56, 57, 59, 106]. However, the possibility cannot be excluded that these genes are in fact carried by large plasmids (>230 kb) which are indistinguishable from the chromosome by the usual DNA preparation techniques.

Gene hybridisation experiments provided evidence for the existence of multiple crystal protein genes within a single bacterium that could be located (see Figure 1) both on plasmids (one or more replicons) and on the chromosome [57, 59, 106]. For instance, as many as five genes could be located at multiple sites in strain *aizawai* HD-7.29 [70, 106].

3.1.2 *Bacillus thuringiensis israelensis*

As shown above, there is a multiplicity of toxin genes in the lepidopteran active *B. thuringiensis* strains. This multiplicity is also found in the dipteran-active *Bt israelensis*, which harbours four *cry* genes and two *cyt* genes (see Chapter 2.3). In contrast to the multiple location of the lepidopteran-active toxin genes previously described, all the dipteran-specific toxin genes of *Bt israelensis* are located on one 137 kb resident plasmid (originally described as a 72 MDa plasmid) [38, 119]. Recently, the restriction map of this plasmid was finalised, and hybridisation experiments allowed the precise location of the different genes [19, 20]. One cluster of crystal toxin genes resembling a "pathogenicity island", that contains all but the *cry4A* gene, was located on the same fragment of about 26 kb. The *cry4A* gene was found opposite to this cluster on the plasmid molecule.

3.1.3   Virulence factors

Many factors (phospholipases, haemolysins...) different from the Cry toxins that could be involved in the virulence of *B. thuringiensis* strains have been identified and characterised (see Chapter 1.3). As is true for the *cry* genes, the virulence genes can also be located on the chromosome or on plasmids. However, both locations for a single gene has never been reported. From cloning experiments, Lechner *et al.* [67] located the phosphatidylinositol-specific phospholipase C gene on the chromosome. Determination of the genetic organisation of the *B. thuringiensis* chromosome through PFGE followed by hybridisation confirmed this observation [78] and also located the phosphatidylcholine phospholipase C [78], and the haemolytic enterotoxin HblA [27, 28] on the chromosome. The genetic mapping of the *B. thuringiensis* chromosome revealed that these virulence genes are all scattered over the chromosome [78]. Another enterotoxin gene has been amplified from various *B. thuringiensis* strains, but its location has not been determined [11]. A similar gene was also found in *B. cereus*, and due to the similarities between the genomes of *B. cereus* and *B. thuringiensis*, it can be expected that a similar location would be found in *B. thuringiensis*. In contrast to all the virulence genes mentioned above, the β-exotoxin determinants were found located on plasmids that also bear the *cry* genes [76].

3.2   Location of toxin genes in other entomopathogenic bacteria

It has been reported that the *B. sphaericus* genes coding for the binary toxin of several strains are located on the bacterial chromosome [77], in agreement with the observation that the "highly" toxic strain 1593 lacks detectable plasmids. Further confirmation of the chromosomal location came from Priest *et al.* [97], who analysed 53 *B. sphaericus* strains including "low", "moderate", and "high" toxicity isolates. However, as for the previous reports on *Bt* gene localisation, the possibility of very large plasmids carrying the *B. sphaericus* toxin genes cannot be excluded. In addition to the binary toxin gene the gene encoding the so-called

Mtx1 toxin (see Chapter 2.3) also appears on the chromosome [77, 97]. The size of the chromosomal fragment bearing the binary toxin gene may vary, depending on the strain tested [77]. However, a greater variation was found for the location of the *mtx1* gene [97].

In strain *B. popilliae melolonthae* H1, the *cry18A* gene was clearly located on the chromosome, although a single plasmid is present in this strain [127].

The genes encoding Cry-like proteins from *C. bifermentans malaysia* have also been shown to be present on the chromosome [14], despite the presence of plasmids in this strain. However, since the nature of toxic factors has not yet been clearly identified (see Chapter 2.3), we cannot exclude the presence of toxin genes on these plasmids.

4.   VIRULENCE GENE MOBILITY AND TRANSFER

4.1   Intracellular gene translocation: the transposons

The entomopathogenic bacteria harbour a large variety of transposable elements. Particularly, several mobile elements belonging to class I (IS*231*, IS*232*, IS*233*, IS*240*, IS*Bt1*, IS*Bt2*) or class II (Tn*4430*, Tn*5401*, Tn*Bth1*) were found often localised in the vicinity of the toxin genes of *B. thuringiensis*, a situation that should greatly facilitate intra- and inter molecular gene mobility. The list of the different mobile elements found in entomopathogenic bacteria is given in Table 1, and their association with toxin genes is illustrated in Figure 1.

4.1.1   The Insertion Sequences (IS)

The first studies of the environment of the δ−endotoxin genes, in lepidopteran toxic *B. thuringiensis* strains showed that *cry1A* genes were flanked by two sets of inverted repeated (IR) sequences (Fig. 1) [60, 61, 75]. These elements were identified as insertion sequences by nucleotide sequence analysis and subsequently designated IS*231* [88] and IS*232* [91]. Ten IS*231*-relatives were identified in four different *B. thuringiensis* strains: IS*231* A, B, and C in *B. thuringiensis* Berliner 1715 [87, 88] ; IS*231*D and E in *Bt finitimus* [100] ; IS*231*G and H in *Bt darmstadiensis* 73-E-10-2 [105]; and IS*231*F, V and W in *Bt israelensis* [100, 101]. These elements belong to the IS*4* family of insertion sequence [82] and can be classified into two groups depending on their length. The first group, of approximately 1650 bp, comprises the elements IS*231*A to G that possess a single open reading frame encoding a putative transposase (with the exception of the IS*231*G element that contains an open reading frame interrupted by several stop codons). The second group, of approximately 1960 bp includes the sequences IS*231*V and W, with transposases thought to be the result of a frameshift between two overlapping open reading frames. The IS*231* elements were frequently found in the vicinity of *cry1* genes (toxic against lepidopteran) and were often associated with the Tn*4430* transposon [85]. One variant, IS*231*W, was found downstream of the mosquitocidal toxin gene *cry11Aa* [101]. IS*231* sequences were detected in more than 50% of the *B. thuringiensis* serotypes tested [68] and also in the other species of the *Bacillus cereus* complex *i.e.*, *B. cereus*, *B. anthracis*, and *B. mycoides* [44, 68]. Transposition of the IS*231* elements was demonstrated in both Gram negative (*E. coli*) and Gram-positive (*B. subtilis*, *B. thuringiensis*, and *B. cereus*) bacteria [42, 69, 85]. These elements transpose exclusively by a conservative transposition mechanism [69] into preferential targets, such as the IR of transposon Tn*4430* [42].

The second insertion element found in the vicinity of the *cry1* genes corresponds to IS*232* [91]. This sequence, that belongs to the IS*21* insertion sequence family, and encodes two polypeptides of 50 kDa (IstA) and 29 kDa (IstB), which are presumably involved in the transposition process [82]. Three iso-elements, IS*232*A, B, and C, have been identified. The transposition of the IS*232*A copy, probably through a conservative mechanism, was demonstrated in *E. coli* [91]. The IS*232* elements are frequently structurally associated with the IS*231* sequences, but unlike IS*231*, IS*232* elements are not well distributed among *B. thuringiensis* and *B. cereus* strains ; moreover they were not found in the 33 strains of *B. mycoides* tested [68].

Other inverted repeated sequences, different from those described above, were characterised in *Bt israelensis* [24]. Two of these sequences were found flanking the *cry4A* toxin gene in a composite transposon-like structure. These sequences, corresponding to two iso-elements (99% similarity) were designated IS*240*A and B and belong to the IS*6* family [31]. The IS*240* elements are widely distributed among *B. thuringiensis* and *B. mycoides* strains [68, 104], but only one iso-element was identified in *B. cereus* [68, 82]. All strains toxic against dipteran larvae possess at least one IS*240*-related sequence [104], often localised in the vicinity of toxin genes. Four IS*240*-like elements were found in *Bt medellin*: IS*240*M1 upstream of the *cry30Aa* gene, and IS*240*M2, M3, and M4, close to *cyt1Ab* and *cyt2Bc* genes (A. Delécluse, unpublished results). In addition, a fifth IS*240*-related element was partially identified 588 bp downstream of the *cry11Bb* of *Bt medellin* [95]. Two related sequences, IS*240*J1 and IS*240*J2 were characterised from *Bt jegathesan*, and were also found localised in the vicinity of toxin genes, *cry11Ba* and *cry19Aa*, respectively [32, 102, 103]. Another variant, IS*240*F, was isolated from a 130 kDa plasmid-encoding toxin gene of *Bt fukuokaensis* 84I113 [35]. No direct repeats (DR) were identified for these IS*240* elements, and although expression of the IS*240* putative transposase has been observed in *E. coli* [31], demonstration of the IS mobility remains to be done. Interestingly, in addition to these IS*240*-elements, IS*240*-like ends were found flanking the *cry29A* gene from *Bt medellin* (Delécluse, unpublished results); these ends were found each located at the internal side of a larger inverted repeat of 116 nucleotides. Two IS*231*-like ends were already found flanking an endopeptidase gene in *B. cereus*. This structure, designed "Mobile Insertion Cassette" turned out to be trans-activated by the transposase of IS*231*A, in a mechanism called "teletransposition" [81]. Such a phenomenon could also occur with the IS*240* ends, thus enabling some toxin genes to move.

Insertion sequences belonging to the IS*3* family were also identified in *B. thuringiensis*. IS*Bt1* is localised between the *cry1C* and *cry1D* genes in the strains *Bt aizawai* HD-229, *B. thuringiensis aizawai* HD-7.29 and *Bt entomocidus* 60.1 toxic against Lepidopteran [110]. IS*Bt2* is found downstream of the cryptic gene *cry2Ab* in *B. thuringiensis* YBT 226, *Bt galleriae* 916, *Bt thuringiensis* HD-201 and *Bt thuringiensis* HD-770 [45]. In addition, two overlapping open reading frames similar to other IS*3* family transposases were described in *Bt fukuokaensis* 83I113, but no IR could be identified [35]. A last insertion sequence, designated IS*233*, from the IS*982* family, was recently found in *Bt galleriae* (C. Leonard, Y. Chen, K. Jendem & J. Mahillon, unpublished results). Other sequences related to mobile elements have also been found in *B. sphaericus*, in the vicinity of the *bin* and *mtx2* toxin genes. However, these elements have not yet been characterised (C. Berry, unpublished results).

4.1.2   Class II transposons

The first transposable element from *B. thuringiensis* was isolated during the course of conjugation experiments with the *Enterococcus faecalis* plasmid pAMβ1 in *Bt kurstaki* Kto [74]. This element, designated Tn*4430*, can be carried by both large and small plasmids, mostly those associated with the *cry1A*-like genes, and on chromosomal DNA of *B. thuringiensis* [17, 29, 41, 74, 75, 83, 86, 106, 120]. Tn*4430* elements were also detected in different strains of *B. cereus* [25]. The transposon encodes a transposase similar to the transposase of the Tn*3* family [84]. In contrast to Tn*3*, the site-specific-recombinase that resolves the cointegrate is not a resolvase but an integrase [84]. It was shown that the Tn*4430* element can transpose by a replicative mechanism not only in its original host [74], but also in *E. coli* [73] and in *B. anthracis* [41]. In *B. anthracis*, Tn*4430* participates in the transfer of the toxin and capsule plasmid pXO1 and pXO2. This transfer takes place *via* Tn*4430*-mediated cointegrate formation between these small plasmids and the large *B. thuringiensis* conjugative plasmid pXO12, followed by subsequent conjugation-like transfers among various strains of *B. anthracis* and *B. cereus* [41]. As was mentioned above, the Tn*4430* inverted repeats are recognised as insertion hot spots by IS*231*A, both in *B. thuringiensis* and *E. coli* [17, 43]. In *Bt thuringiensis* strain Berliner 1715, this has resulted in a peculiar structural association where two IS*231* elements have inserted into the two Tn*4430* inverted repeats, just after the first base. The integrity of Tn*4430* remains, however, intact, because the last nucleotide of IS*231* is identical to the first one of Tn*4430* [87].

A second transposon, Tn*5401*, was isolated from *Bt morrisoni* EG2158, after its spontaneous insertion into a recombinant plasmid [17]. Nucleotide analysis of this element revealed that it shares the same organisation as Tn*4430*, with both a transposase and a site-specifique recombinase belonging to the integrase family. However, transposases and integrases of the two transposons are weakly related with only 28% and 24% identity, respectively [17]. Like Tn*4430*, Tn*5401* can transpose by a replicative mechanism to both chromosomal and plasmid sites in *B. thuringiensis* [17]. Tn*5401* was found in just a few number of *B. thuringiensis* strains and never associated with Tn*4430* in the same strain [17]. In addition, Tn*5401* does not seem to be associated with toxin genes except in *Bt morrisoni* tenebrionis NB176, a strain toxic for coleopteran, where it was found downstream of the *cry3Aa* [85].

Recently, a third transposon, designated TnBth*1*, was partially identified downstream of the *cry19A* gene in *Bt jegathesan* [102]. In contrast to the two other transposons of *B. thuringiensis*, TnBth*1* contains a site-specific recombinase similar to the resolvase of Tn*3*.

The presence of a fourth transposon, TnCbi*1*, was also found in another insecticidal bacteria, *Clostridium bifermentans malaysia*, toxic for Dipteran larvae. In this strain, a sequence homologue to the transposon Tn*1546* of *Enterococcus faecium* was partially identified upstream of the *cry16Aa* toxin gene (V. Juárez-Pérez, unpublished data).

4.2   Intercellular transfer: conjugation and transduction

In 1982, González and Carlton [36] reported the conjugative transfer of *B. thuringiensis* plasmids in broth mating experiments. Since then, several reports on intraspecies plasmid transfer in *B. thuringiensis* and among different species in laboratory conditions have been published [16, 23, 48, 55, 115, 121]. Moreover, several researchers showed that conjugation transfer of *B. thuringiensis* plasmid can also occur in soil or in insect larvae [48, 113, 117], suggesting that the conjugation system could contribute to the dispersion of the *B. thuringiensis* toxins among different bacteria species in nature. Different studies on the conjugation system in *B. thuringiensis* were performed, with particular attention to the *Bt israelensis* strain which contains at least 8 plasmids [38]. Two of the small plasmids, pTX14-1 and pTX14-3 have been cloned and characterised [6, 22, 80]. One large plasmid has been shown to encode the toxin genes [34, 38, 108, 119] while another large plasmid is associated with satellite inclusion [112]. A third large plasmid of about 200 kb, identical to pXO16 reported by Reddy *et al.* [98] was found to be self-transmissible [38] and to mobilise smaller plasmids [4, 8]. It was also shown that mobilisation of small plasmids between strains of *Bt israelensis* is accompanied by non-pheromone-induced and protease sensitive co-aggregation between donor and recipient cells [7]. In contrast to the conjugation systems in Gram-negative bacteria, those discovered in Gram-positive bacteria display no pili. Wiwat *et al.* [122] suggested that the S-layer protein may play a crucial role in this conjugation-like system. Two aggregation phenotypes are characterised by the appearance of macroscopic aggregates when cells of exponentially growing strains of Agr$^+$ and Agr$^-$ bacteria are mixed in broth. The mobilisation of small plasmids was found to be unidirectional from Agr$^+$ to Agr$^-$ cells [7]. This aggregation system is encoded by the large self-transmissible plasmid pXO16 [50]. Jensen *et al.* [49] showed that the aggregation-mediated conjugation system encoded by pXO16 is fully functional in several *B. thuringiensis* and *B. cereus*. Furthermore, the pXO16 plasmid is able to mobilise small "non-mobilisable" plasmids lacking both *mob* gene and *oriT* site, the typical characteristics of mobilisable plasmids [8, 64].

Other studies with pAW63 and pHT73 from *Bt kurstaki* HD-73 described a mediated-mobilisation system different from the aggregation-mediated conjugation system encoded on pXO16 in *Bt israelensis* [120]. In this case, mobilisation of non-conjugative plasmids requires the presence of a *mob* gene on the non-conjugative plasmid. Such a *mob* gene was identified in the mobilisable plasmid pTX14-3 [5].

A different genetic transfer system between *B. thuringiensis* cells, involving transducing phage particles (CP-54, CP-51), was described for the first time by de Barjac in 1970 [12]. This transduction system was initially used for gene mapping [15, 63, 65, 96, 114]. It served then to produce recombinant *Bt* strains expressing novel δ-endotoxin genes: the phage CP-54Ber mediated transduction of recombinant plasmids carrying *cry1A* genes into *Bt morrisoni* tenebrionis and *Bt entomocidus* strains [66]. Transfer of a *cry1C* gene from *Bt aizawai* onto the chromosome of strain *Bt kurstaki* was also achieved through CP-51 mediated transduction [51].

5.   CONCLUDING REMARKS

A major step in the understanding of the actual plasticity of *B. thuringiensis* genome would be achieved with the determination of its chromosomal sequence. Yet, to our knowledge, no such project has been undertaken. Similarly, the full genetic description of large crystal-bearing plasmids should shed light on the local contexts of the *cry* genes and give some hints as to their evolution. Again, such research has not been reported so far. It is important to emphasise, though, that the comparison of these DNA sequence data with those obtained from both *B. anthracis* plasmids [94] and chromosome (project under way, R. Okinaka, personal communication) should also help us in deciphering the similarities and differences between these two specialised pathogens.

Intercellular transfer of toxin genes by either conjugation or mobilisation have been reported both under laboratory condition and in the natural environment of an insect. In the case of intracellular gene mobility, however, converging evidence suggested the role of transposons in the dispersal and possible rearrangement of pesticidal toxin genes, but the demonstration of an actual toxin gene translocation by composite or class II transposable elements is still missing. Also missing is information on the potential triggers of this putative toxin gene mobility and the frequency at which it could occur.

In parallel to this, but from a more fundamental point of view, one can also wonder why these bacteria can harbour several related copies of crystal genes without apparently undergoing the deleterious effects of homologous recombination. Similarly, the same observation could be made for certain transposable elements, such as the IS*231* iso-elements, present in several copies in the same bacteria, on the same DNA molecules.

Finally, it is important to remember that, as in the entomopathogen toxin genes themselves (*cry*, *cyt* or *vip*), most of our current knowledge of their genetic and genomic contexts has been gathered from the crystal-forming *B. thuringiensis* and, to a lesser extent, from the mosquitocidal *B. sphaericus*. Inferences from these observations to other groups of entomopathogenic bacteria so far remain speculative.

AKNOWLEDGEMENTS

We thank Patricia Vary for critical reading of the manuscript and English corrections. We also want to thank Mimi Deprez for her contribution to Figure 1.

REFERENCES

[1] Abe K, Faust RM & Bulla Jr LA (1983) Plasmid deoxyribonucleic acid in strains of *Bacillus sphaericus* and in *Bacillus moritai*. J. Invertebr. Pathol. 41, 328-335

[2] Ackerman H-W, Azizbekyan RR, Berliner RL *et al.* (1995) Phage typing of *Bacillus subtilis* and *Bacillus thuringiensis*. Res. Microbiol. 146, 643-657

[3] Afkami P, Ozcengiz G & Alaeddinoglu NG (1993) Plasmid patterns of *Bacillus thuringiensis*-81 and its crystal-negative mutants. Biotechnol. Lett. 15, 1247-1252

[4] Andrup L, Bendixen HH & Jensen GB (1995) Mobilization of *Bacillus thuringiensis* plasmid pTX14-3. Plasmid 33, 159-167

[5] Andrup L, Bolander G, Boe L *et al.* (1991) Identification of a gene (*mob14-3*) encoding a mobilization protein from the *Bacillus thuringiensis* subsp. *israelensis* plasmid pTX14-3. Nucleic Acids Res. 19, 2780

[6] Andrup L, Damgaard J, Wasserman K *et al.* (1994) Complete nucleotide sequence of the *Bacillus thuringiensis* subsp. *israelensis* plasmid pTX14-3 and its correlation with biological properties. Plasmid 31, 72-88

[7] Andrup L, Damgaard J & Wassermann K (1993) Mobilization of small plasmids in *Bacillus thuringiensis* subsp. *israelensis* is accompanied by specific aggregation. J. Bacteriol. 175,

[8] Andrup L, Jorgensen O, Wilcks A, Smidt L & Jensen GB (1996) Mobilization of "non-mobilizable" plasmids by the aggregation-mediated conjugation system of *Bacillus thuringiensis*. Plasmid 36, 75-85

[9] Aquino de Muro M, Mitchell WJ & Priest FG (1992) Differentiation of mosquito-pathogenic strains of *Bacillus sphaericus* from non-toxic varieties by ribosomal RNA gene restriction patterns. J. Gen. Microbiol. 138, 1159-1166

[10] Aquino de Muro MA & Priest FG (1993) Phylogenetic analysis of *Bacillus sphaericus* and development of an oligonucleotide probe specific for mosquito-pathogenic strains. FEMS Microbiol. Lett. 112, 205-210

[11] Asano S-I, Nukumizu Y, Bando H, Iizuka T & Yamamoto T (1997) Cloning of novel enterotoxin genes from *Bacillus cereus* and *Bacillus thuringiensis*. Appl. Environ. Microbiol. 63, 1054-1057

[12] de Barjac H (1970) Transduction chez *Bacillus thuringiensis*. C. R. Acad. Sc. Paris 270 D, 2227-2229

[13] de Barjac H, Sisman J & Cosmao Dumanoir V (1974) Description de 12 bactériophages isolés à partir de *Bacillus thuringiensis*. C. R. Acad. Sc. Paris 279 D, 1939-1942

[14] Barloy F, Delécluse A, Nicolas L & Lecadet M-M (1996) Cloning and expression of the first anaerobic toxin gene from *Clostridium bifermentans* subsp. *malaysia,* encoding a new mosquitocidal protein with homologies to *Bacillus thuringiensis* delta-endotoxins. J. Bacteriol. 178, 3099-3105

[15] Barsomian GD, Robillard NJ & Thorne CB (1984) Chromosomal mapping of *Bacillus thuringiensis* by transduction. J. Bacteriol. 157, 746-750

[16] Battisti L, Green BD & Thorne CB (1985) Mating system for transfer of plasmids among *Bacillus anthracis, Bacillus cereus,* and *Bacillus thuringiensis.* J. Bacteriol. 162, 543-550

[17] Baum JA (1994) Tn*5401*, a new class II transposable element from *Bacillus thuringiensis.* J. Bacteriol. 176, 2835-2845

[18] Baum JA & Gonzalez Jr JM (1992) Mode of replication, size and distribution of naturally occurring plasmids in *Bacillus thuringiensis.* FEMS Microbiol. Lett. 96, 143-148

[19] Ben-Dov E, Einav M, Peleg N, Boussiba S & Zaritsky A (1996) Restriction map of the 125-kilobase plasmid of *Bacillus thuringiensis* subsp. *israelensis* carrying the genes that encode delta-endotoxins active against mosquito larvae. Appl. Environ. Microbiol. 62, 3140-3145

[20] Ben-Dov E, Nissan G, Pelleg N *et al.* (1999) Refined, circular restriction map of the *Bacillus thuringiensis* subsp. *israelensis* plasmid carrying the mosquito larvicidal genes. Plasmid 42, 186-191

[21] Benoit TG, Wilson GR, Bull DL & Aronson AI (1990) Plasmid associated sensitivity of *Bacillus thuringiensis* to UV light. Appl. Environ. Microbiol. 56, 2282-2286

[22] Boe L, Nielsen TT, Madsen SM, Andrup L & Bolander G (1991) Cloning and characterization of two plasmids from *Bacillus thuringiensis* in *Bacillus subtilis.* Plasmid 25, 190-197

[23] Bora RS, Murty MG, Shenbagarathai R & Sekar V (1994) Introduction of a lepidopteran-specific insecticidal crystal protein gene of *Bacillus thuringiensis* subsp. *kurstaki* by conjugal transfer into a *Bacillus megaterium* strain that persists in the cotton phyllosphere. Appl. Environ. Microbiol. 60, 214-222

[24] Bourgouin C, Delécluse A, Ribier J, Klier A & Rapoport G (1988) A *Bacillus thuringiensis* subsp. *israelensis* gene encoding a 125-kilodalton larvicidal polypeptide is associated with inverted repeat sequences. J. Bacteriol. 170, 3575-3583

[25] Carlson CR, Caugant D & Kolsto AB (1994) Genotypic diversity among *Bacillus cereus* and *Bacillus thuringiensis* strains. Appl. Environ. Microbiol. 60, 1719-1725

[26] Carlson CR, Gronstad A & Kolsto AB (1992) Physical maps of the genomes of three *Bacillus cereus* strains. J. Bacteriol. 174, 3750-3756

[27] Carlson CR, Johansen T & Kolsto AB (1996) The chromosome map of *Bacillus thuringiensis* subsp. *canadensis* HD224 is highly similar to that of the *Bacillus cereus* type strain ATCC 14579. FEMS Microbiol. Lett. 141, 163-167

[28] Carlson CR, Johansen T, Lecadet M-M & Kolsto AB (1996) Genomic organization of the entomopathogenic bacterium *Bacillus thuringiensis* subsp. *berliner* 1715. Microbiol. 142, 1625-1634

[29] Carlson CR & Kolsto AB (1993) A complete physical map of a *Bacillus thuringiensis* chromosome. J. Bacteriol. 175, 1053-1060

[30] Clark BD, Boyle TM, Chu C-Y & Dean DH (1985) Restriction endonuclease mapping of three plasmids from *Bacillus thuringiensis*var. *israelensis.* Gene 36, 169-171

[31] Delécluse A, Bourgouin C, Klier A & Rapoport G (1989) Nucleotide sequence and characterization of a new insertion element, IS240, from *Bacillus thuringiensis israelensis*. Plasmid 21, 71-78

[32] Delécluse A, Rosso M-L & Ragni A (1995) Cloning and expression of a novel toxin gene from *Bacillus thuringiensis* subsp. *jegathesan*, encoding a highly mosquitocidal protein. Appl. Environ. Microbiol. 61, 4230-4235

[33] Dingman DW (1994) Physical properties of three plasmids and the presence of interrelated plasmids in *Bacillus popilliae* and *Bacillus lentimorbus*. J. Invertebr. Pathol. 63, 235-243

[34] Donovan WP, Dankocsik CC & Gilbert MP (1988) Molecular characterization of a gene encoding a 72-kilodalton mosquito-toxic crystal protein from *Bacillus thuringiensis* subsp. *israelensis*. J. Bacteriol. 170, 4732-4738

[35] Dunn MG & Ellar DJ (1997) Identification of two sequence elements associated with the gene encoding the 24-kDa crystalline component in *Bacillus thuringiensis* ssp. *fukuokaensis*: an example of transposable element archeology. Plasmid 37, 205-215

[36] González JM Jr, Brown BJ & Carlton BC (1982) Transfer of *Bacillus thuringiensis* plasmids coding for delta-endotoxin among strains of *B. thuringiensis* and *B. cereus*. Proc. Natl. Acad. Sci. USA 79, 6951-6955

[37] González JM Jr & Carlton BC (1980) Patterns of plasmid DNA in crystalliferous and acrystalliferous strains of *Bacillus thuringiensis*. Plasmid 3, 92-98

[38] González JM Jr & Carlton BC (1984) A large transmissible plasmid is required for crystal toxin production in *Bacillus thuringiensis* variety *israelensis*. Plasmid 11, 28-38

[39] González JM Jr, Dulmage HT & Carlton BC (1981) Correlation between specific plasmids and δ-endotoxin production in *Bacillus thuringiensis*. Plasmid 5, 351-365

[40] Gordon R, Haynes W & Pang C (1973) The genus *Bacillus*. US Dept. of Agriculture Handbook n°427

[41] Green BD, Battisti L & Thorne CB (1989) Involvement of Tn4430 in transfer of *Bacillus anthracis* plasmids mediated by *Bacillus thuringiensis* plasmid pXO12. J. Bacteriol. 171, 104-113

[42] Hallet B, Rezsöhazy R & Delcour J (1991) IS231A from *Bacillus thuringiensis* is functional in *Escherichia coli*: transposition and insertion specificity. J. Bacteriol. 173, 4526-4529

[43] Hallet B, Rezsöhazy R, Mahillon J & Delcour J (1994) IS231A insertion specificity: consensus sequence and DNA bending at the target site. Mol. Microbiol. 14, 131-139

[44] Henderson I, Dongzheng Y & Turnbull PCB (1995) Differenciation of *Bacillus anthracis* and other '*Bacillus cereus* group' bacteria using IS231-derived sequences. FEMS Microbiol. Lett. 128, 113-118

[45] Hodgman TC, Ziniu Y, Shen J & Ellar DJ (1993) Identification of a cryptic gene associated with an insertion sequence not previously identified in *Bacillus thuringiensis*. FEMS Microbiol. Lett. 114, 23-30

[46] Hoflack L, Seurinck J & Mahillon J (1997) Nucleotide sequence and characterization of the cryptic *Bacillus thuringiensis* plasmid pGI3 reveal a new family of rolling circle replicons. J. Bacteriol. 179, 5000-5008

[47] Inal JR, Karunakaran V & Burges DD (1990) Isolation and propagation of phages naturally associated with the *aizawai* variety of *Bacillus thuringiensis*. J. Appl. Bacteriol. 68, 17-21

[48] Jarrett P & Stephenson M (1990) Plasmid transfer between strains of *Bacillus thuringiensis* infecting *Galleria mellonella* and *Spodoptera littoralis*. Appl. Environ. Microbiol. 56, 1608-1614

[49] Jensen GB, Andrup L, Wilcks A, Smidt L & Poulsen OM (1996) The aggregation-mediated conjugation system of *Bacillus thuringiensis* subsp. *israelensis*: host range and kinetics. Curr. Microbiol. 33, 228-236

[50] Jensen GB, Wilcks A, Pertersen SS *et al.* (1995) The genetic basis of the aggregation system in *Bacillus thuringiensis* subsp. *israelensis* is located on the large conjugative plasmid pXO16. J. Bacteriol. 177, 2914-2917

[51] Kalman S, Kiehne KL, Cooper N, Reynoso MS & Yamamoto T (1995) Enhanced production of insecticidal proteins in *Bacillus thuringiensis* strains carrying an additional crystal protein gene in their chromosomes. Appl. Environ. Microbiol. 61, 3063-3068

[52] Kanda K, Kitajima Y, Moriyama Y, Kato F & Murata A (1998) Association of plasmid integrative J7W-1 prophage with *Bacillus thuringiensis* strains. Acta Virol. 42, 315-318

[53] Kanda K, Ohderaotoshi T, Shimojyo A, Kato F & Murata A (1999) An extrachromosomal prophage naturally associated with *Bacillus thuringiensis* serovar *israelensis*. Lett. Appl. Microbiol. 28, 305-308

[54] Kanda K, Tan Y & Aizawa K (1989) A novel phage genome integrated into a plasmid in *Bacillus thuringiensis* strain AF101. J. Gen. Microbiol. 135, 3035-3041

[55] Klier A, Bourgouin C & Rapoport G (1983) Mating between *Bacillus subtilis* and *Bacillus thuringiensis* and transfer of cloned crystal genes. Mol. Gen. Genet. 191, 257-262

[56] Klier A, Fargette F, Ribier J & Rapoport G (1982) Cloning and expression of the crystal protein genes from *Bacillus thuringiensis* strain *berliner* 1715. EMBO J. 1, 791-799

[57] Klier A, Lereclus D, Ribier J *et al.* (1985) Cloning and expression in *Escherichia coli* of the crystal protein gene from *Bacillus thuringiensis* strain *aizawai* 7-29 and comparison of the structural organization of genes from different serotypes, p. 217-224. *In* Hoch JA & Setlow P (ed.), Molecular Biology of Microbial differentiation, American Society for Microbiology

[58] Kolsto AB, Gronstad A & Oppegaard H (1990) Physical map of the *Bacillus cereus* chromosome. J. Bacteriol. 172, 3821-3825

[59] Kronstad JW, Schnepf HE & Whiteley HR (1983) Diversity of locations for *Bacillus thuringiensis* crystal protein genes. J. Bacteriol. 154, 419-428

[60] Kronstad JW & Whiteley HR (1984) Inverted repeat sequences flank a *Bacillus thuringiensis* crystal protein gene. J. Bacteriol. 160, 95-102

[61] Kronstad JW & Whiteley HR (1986) Three classes of homologous *Bacillus thuringiensis* crystal-protein genes. Gene 43, 29-40

[62] Krych VK, Johnson JL & Yousten AA (1980) Deoxyribonucleic acid homologies among strains of *Bacillus sphaericus*. Int. J. Syst. Bact. 30, 476-482

[63] Landen R, Heierson A & Boman HG (1981) A phage for generalized transduction in *Bacillus thuringiensis* and mapping of four genes for antibiotic resistance. J. Gen. Microbiol. 123, 49-59

[64] Lanka E & Wilkins BM (1995) DNA processing reactions in bacterial conjugation. Annu. Rev. Biochem. 64, 141-169

[65] Lecadet M-M, Blondel M-O & Ribier J (1980) Generalized transduction in *Bacillus thuringiensis* var. Berliner 1715, using bacteriophage CP54 Ber. J. Gen. Microbiol. 121, 202-212

[66] Lecadet M-M, Chaufaux L, Ribier J & Lereclus D (1992) Construction of novel *Bacillus thuringiensis* strains with different insecticidal activities by transduction and transformation. Appl. Environ. Microbiol. 58, 840-849

[67] Lechner M, Kupke T, Stefanovic S & Götz F (1989) Molecular characterization and sequence of phosphatidylinositol-specific phospholipase C of *Bacillus thuringiensis*. Molec. Micobiol. 3, 621-626

[68] Léonard C, Chen Y & Mahillon J (1997) Diversity and differential distribution of IS*231*, IS*232* and IS*240* among *Bacillus cereus, Bacillus thuringiensis* and *Bacillus mycoides*. Microbiol. 143, 2537-2547

[69] Léonard C & Mahillon J (1998) IS231A transposition: conservative versus replicative pathway. Res. Microbiol. 149, 549-555

[70] Lereclus D, Bourgouin C, Lecadet M-M, Klier A & Rapoport G (1989) Role, structure, and molecular organization of the genes coding for the parasporal δ-endotoxins of *Bacillus thuringiensis*, p. 255-276. *In* Smith I, Slepecky RA & Setlow P (ed.), Regulation of Procaryotic Development, American Society for Microbiology

[71] Lereclus D, Guo S, Sanchis V & Lecadet M-M (1988) Characterization of two *Bacillus thuringiensis* plasmids whose replication is thermosensitive in *B. subtilis*. FEMS Microbiol. Lett. 49, 417-422

[72] Lereclus D, Lecadet M-M, Ribier J & Dedonder R (1982) Molecular relationships among plasmids of *Bacillus thuringiensis* : conserved sequences through 11 crystalliferous strains. Mol. Gen. Genet. 186, 391-398

[73] Lereclus D, Mahillon J, Menou G & Lecadet M-M (1986) Identification of Tn*4430*, a transposon of *Bacillus thuringiensis* functional in *Escherichia coli*. Mol. Gen. Genet. 204, 52-57

[74] Lereclus D, Menou G & Lecadet M-M (1983) Isolation of a DNA sequence related to several plasmids from *Bacillus thuringiensis* after a mating involving the *Streptococcus faecalis* plasmid pAMβ1. Mol. Gen. Genet. 191, 307-313

[75] Lereclus D, Ribier J, Klier A, Menou G & Lecadet M-M (1984) A transposon-like structure related to the δ−endotoxin gene of *Bacillus thuringiensis*. EMBO J. 3, 2561-2567

[76] Levinson BL, Kasyan KJ, Chiu SS, Currier TS & González JM Jr (1990) Identification of β-exotoxin production, plasmids encoding β-exotoxin, and a new exotoxin in *Bacillus thuringiensis* by using high-performance liquid chromatography. J. Bacteriol. 172, 3172-3179

[77] Liu J-W, Hindley J, Porter AG & Priest FG (1993) New high-toxicity mosquitocidal strains of *Bacillus sphaericus* lacking a 100-kilodalton-toxin gene. Appl. Environ. Microbiol. 59, 3470-3473

[78] Lövgren A, Carlson CR, Eskils K & Kolsto AB (1998) Localization of putative virulence genes on a physical map of the *Bacillus thuringiensis* subsp. *gelechiae* chromosome. Curr. Microbiol. 37, 245-250

[79] Macdonald R & Kalmakoff J (1995) Comparison of pulse-field gel electrophoresis DNA fingerprints of field isolates of the entomopathogen *Bacillus popilliae*. Appl. Environ. Microbiol. 61, 2446-2449

[80] Madsen S, Andrup L & Boe L (1993) Fine mapping and DNA sequence of replication functions of *Bacillus thuringiensis* plasmid pTX14-3. Plasmid 30, 119-130

[81] Mahillon J (1998) Transposons as gene haulers. APMIS 106, 29-36

[82] Mahillon J & Chandler M (1998) Insertion sequences. Microbiol. Mol. Biol. Rev. 62, 725-774

[83] Mahillon J, Hespel F, Pierssens AM & Delcour J (1988) Cloning and partial characterization of three small cryptic plasmids from *Bacillus thuringiensis*. Plasmid 19, 169-173

[84] Mahillon J & Lereclus D (1988) Structural and functional analysis of Tn*4430* : identification of an integrase-like protein involved in the co-integrate-resolution process. EMBO J. 7, 1515-1526

[85] Mahillon J, Rezsöhazy R, Hallet B & Delcour J (1994) IS*231* and other *Bacillus thuringiensis* transposable elements: a review. Genetica 93, 13-26

[86] Mahillon J & Seurinck J (1988) Complete nucleotide sequence of pGI2, a *Bacillus thuringiensis* plasmid containing Tn*4430*. Nucleic Acids Res. 16, 11827-11828

[87] Mahillon J, Seurinck J, Delcour J & Zabeau M (1987) Cloning and nucleotide sequence of different iso-IS*231* elements and their structural association with the Tn*4430* transposon in *Bacillus thuringiensis*. Gene 51, 187-196

[88] Mahillon J, Seurinck J, Rompuy LV, Delcour J & Zabeau M (1985) Nucleotide sequence and structural organization of an insertion sequence element (IS*231*) from *Bacillus thuringiensis* strain berliner 1715. EMBO J. 4, 3895-3899

[89] Marin R, Tanguay RT, Valéro J, Letarte R & Bellemare G (1992) Isolation and sequence of a 2-kbp miniplasmid from *Bacillus thuringiensis* var. *kurstaki* HD-3a3b : relationship with miniplasmids of other *B. thuringiensis* strains. FEMS Microbiol. Lett. 94, 263-270

[90] McDowell DG & Mann NH (1991) Characterization and sequence analysis of a small plasmid from *Bacillus thuringiensis* var. *kurstaki* strain HD1-DIPEL. Plasmid 25, 113-120

[91] Menou G, Mahillon J, Lecadet M-M & Lereclus D (1990) Structural and genetic organization of IS*232*, a new insertion sequence of *Bacillus thuringiensis*. J. Bacteriol. 172, 6689-6696

[92] Minnich SA & Aronson AI (1984) Regulation of protoxin synthesis in *Bacillus thuringiensis*. J. Bacteriol. 158, 447-454

[93] Miteva V, Selenska-Pobell S & Mitev V (1999) Random and repetitive primer amplified polymorphic DNA analysis of Bacillus sphaericus. J. Appl. Microbiol. 86, 928-936

[94] Okinaka RT, Clound K, Hampton O *et al.* (1999) Sequence and organization of pX01, the large *Bacillus anthracis* plasmid harboring the anthrax toxin genes. J. Bacteriol. 181, 6509-6515

[95] Orduz S, Realpe M, Arango R, Murillo LA & Delécluse A (1998) Sequence of the *cry11Bb1* gene from *Bacillus thuringiensis* subsp. *medellin* and toxicity analysis of its encoded protein. Biochim. Biophys. Acta 1388, 267-272

[96] Perlak FJ, Mendelsohn CL & Thorne CB (1979) Converting bacteriophage for sporulation and crystal formation in *Bacillus thuringiensis*. J. Bacteriol. 140, 699-706

[97] Priest FG, Ebdrup L, Zahner V & Carter PE (1997) Distribution and characterisation of mosquitocidal toxin genes in some strains of *Bacillus sphaericus*. Appl. Environ. Microbiol. 63, 1195-1198

[98] Reddy A, Battisti L & Thorne CB (1987) Identification of self-transmissible plasmids in four *Bacillus thuringiensis* subspecies. J. Bacteriol. 169, 5263-5270

[99] Reynolds RB, Reddy A & Thorne CB (1988) Five unique temperate phages from polylysogenic strain of *Bacillus thuringiensis* subsp. *aizawai*. J. Gen. Microbiol. 134, 1577-1585

[100] Rezsöhazy R, Hallet B & Delcour J (1992) IS*231*D, E and F, three new insertion sequences in *Bacillus thuringiensis* : extension of the IS*231* family. Mol. Microbiol. 6, 1959-1967

[101] Rezsöhazy R, Hallet B, Mahillon J & Delcour J (1993) IS*231*V and W from *Bacillus thuringiensis* subsp. *israelensis*, two distant members of the IS*231* family of insertion sequences. Plasmid 30, 141-149

[102] Rosso M-L (1999) Caractérisation des toxines de *Bacillus thuringiensis* serovar *jegathesan*, actif sur les larves de Diptères vecteurs de maladies tropicales. Thèse de Doctorat, Université Paris VI

[103] Rosso M-L & Delécluse A (1997) Contribution of the 65-kilodalton protein encoded by the cloned gene *cry19A* to the mosquitocidal activity of *Bacillus thuringiensis* subsp. *jegathesan*. Appl. Environ. Microbiol. 63, 4449-4455

[104] Rosso M-L & Delécluse A (1997) Distribution of the insertion element IS*240* among *Bacillus thuringiensis* strains. Curr. Microbiol. 34, 348-353

[105] Ryan M, Johnson JD & Bulla LAJ (1993) Insertion sequence elements in *Bacillus thuringiensis* subsp. *darmstadiensis*. Can. J. Microbiol. 39, 649-658

[106] Sanchis V, Lereclus D, Menou G, Chaufaux J & Lecadet M-M (1988) Multiplicity of δ-endotoxin genes with different specificities in *Bacillus thuringiensis aizawai* 7.29. Mol. Microbiol. 2, 393-404

[107] Schnepf HE & Whiteley HR (1981) Cloning and expression of the *Bacillus thuringiensis* crystal protein gene in *Escherichia coli*. Proc. Natl. Acad. Sci. USA 78, 2893-2897

[108] Sekar V & Carlton BC (1985) Molecular cloning of the delta-endotoxin gene of *Bacillus thuringiensis* var. *israelensis*. Gene 33, 151-158

[109] Seleena P & Lee HL (1994) Absence of plasmids in mosquitocidal *Clostridium bifermentans* serovar *malaysia*. Southeast Asian J. Trop. Med. Public Health 25, 394-396

[110] Smith GP, Ellar DJ, Keeler SJ & Seip CE (1994) Nucleotide sequence and analysis of an insertion sequence from *Bacillus thuringiensis* related to IS*150*. Plasmid 32, 10-18

[111] Stahly DP, Dingman DW, Bulla LA Jr & Aronson AI (1978) Possible origin and function of the parasporal crystals in *Bacillus thuringiensis*. Biochem. Biophys. Res. Comm. 84, 581-588

[112] Tam A & Fitz-James P (1986) Plasmids associated with a phagelike particle and with a satellite inclusion in *Bacillus thuringiensis* ssp. *israelensis*. Can. J. Microbiol. 32, 382-388

[113] Thomas DJI, Morgan JAW, Whipps JM & Saunders JR (2000) Plasmid transfer between the *Bacillus thuringiensis* subspecies *kurstaki* and *tenebrionis* in laboratory culture and soil and in lepidopteran and coleopteran larvae. Appl. Environ. Microbiol. 66, 118-124

[114] Thorne CB (1978) Transduction in *Bacillus thuringiensis*. Appl. Environ. Microbiol. 38, 1109-1115

[115] Trieu-Cuot P, Carlier C, Martin P & Courvalin P (1987) Plasmid transfer by conjugation from *Escherichia coli* to Gram positive bacteria. FEMS Microbiol. Lett. 48, 289-294

[116] Valyasevi R, Kyle MM, Christie PJ & Steinkraus KH (1990) Plasmids in *Bacillus popilliae*. J. Invertebr. Pathol. 56, 286-288

[117] Vilas-Boas GFLT, Vilas-Boas LA, Lereclus D & Arantes OMN (1998) *Bacillus thuringiensis* conjugation under environmental conditions. FEMS Microbiol. Ecol. 25, 369-374

[118] Walter TM & Aronson AI (1991) Transduction of certain genes by autonomously replicating *Bacillus thuringiensis* phage. Appl. Environ. Microbiol. 57, 1000-1005

[119] Ward ES & Ellar DJ (1983) Assignment of the δ-endotoxin gene of *Bacillus thuringiensis* var. *israelensis* to a specific plasmid by curing analysis. FEBS Lett. 158, 45-49

[120] Wilcks A, Jayaswal N, Lereclus D & Andrup L (1998) Characterization of plasmid pAW63, a second self-transmissible plasmid in *Bacillus thuringiensis* subsp. *kurstaki* HD73. Microbiol. 144, 1263-1270

[121] Wiwat C, Panbangred W & Bhumiratana A (1980) Transfer of plasmids and chromosomal genes amongst subspecies of *Bacillus thuringiensis*. J. Ind. Microbiol. 6, 19-27

[122] Wiwat C, Panbangred W, Mongkolsuk S, Pantuwatana S & Bhumiratana A (1995) Inhibition of a conjugation-like gene transfer process in *Bacillus thuringiensis* subsp. *israelensis* by the anti-S-layer protein antibody. Curr. Microbiol. 30, 69-75

[123] Woodburn MA, Yousten AA & Hilu KH (1995) Random amplified polymorphic DNA fingerprinting of mosquito-pathogenic and nonpathogenic strains of *Bacillus sphaericus*. Int. J. Syst. Bacteriol. 45, 212-217

[124] Yousten AA (1984) Bacteriophage typing of mosquito pathogenic strains of *Bacillus sphaericus*. J. Invertebr. Pathol. 43, 124-125

[125] Yousten AA, de Barjac H, Hedrick J, Cosmao Dumanoir V & Myers P (1980) Comparison between bacteriophage typing and serotyping for the differenciation of *Bacillus sphaericus* strains. Ann. Microbiol. (Inst. Pasteur) 131B, 297-308

[126] Zahner V, Rabinovitch L, Suffys P & Momen H (1999) Genotypic diversity among *Brevibacillus laterosporus* strains. Appl. Environ. Microbiol. 65, 5182-5185

[127] Zhang J, Hodgman C, Krieger L, Schnetter W & Schairer HU (1997) Cloning and analysis of the first *cry* gene from *Bacillus popilliae*. J. Bacteriol. 179, 4336-4341

Table 1. Transposable elements from Bacillus thuringiensis and Clostridium bifermentans malaysia

Name	Family	Length (bp)	IR (bp)	DR (bp)	ORF (start-stop)	Bacillus thuringiensis serovar & strain	Acces. No.	Reference
IS231A	IS4	1656	20	10, 11, 12	478 (92-1528)	thuringiensis Berliner 1715 (p65kb)	X03397	[75]
IS231B	IS4	1643	7/20	ND	478 (92-1528)	thuringiensis Berliner 1715 (p65kb)	M16158	[87]
IS231C	IS4	1656	19/20	11	478 (92-1528)	thuringiensis Berliner 1715 (p65kb)	M16159	[87]
IS231D	IS4	1657	17/20	ND	478 (92-1528)	finitimus T02 001	X63383	[100]
IS231E	IS4	2075	16/17	ND	478 (511-1947)	finitimus T02 001	X63384	[100], Y Chen & J Mahillon unpublished
IS231F	IS4	1655	19/20	12	477 (108-1541)	israelensis 4Q2-72 (p137 kb)	X63385	[100]
IS231G	IS4	1649	20	ND	NONE	darmstadiensis 73-E-10-2	M93054	[105]
IS231H	IS4	> 817	ND	ND	241 (95-817) [P]	darmstadiensis 73-E-10-2	M93054 [P]	[105]
IS231V	IS4	1964	21/22	ND	ORF1: 244 (389-1121) ORF2: 245 (1072-1809)	israelensis 4Q2-72 (p137 kb)	M86926 [V]	[101]
IS231W	IS4	1964	21/22	ND	ORF1: 250 (389-1141) ORF2: 245 (1072-1809)	israelensis 4Q2-72 (p137 kb)	M86926	[101]
IS232A	IS21	2184	48/67	ND	ORF1: 431 (93-1388) ORF2: 250 (1378-2130)	thuringiensis Berliner 1715 (p65kb)	M38370	[91]
IS232B	IS21	2200	28/37	ND	ND	kurstaki HD-73 (p75kb)	M77344 [P]	[91]
IS232C	IS21	2200	28/37	ND	ND	kurstaki HD-73 (p75kb)	ND	[91]
IS233A	IS982	1028	25	8	302 (101-1009)	galleriae T05 001	ND	C Léonard, Y Chen, K Jendem & J Mahillon unpublished
IS233B	IS982	1026	25	ND	ND	medellin T30 001	ND	C Léonard, Y Chen, K Jendem & J Mahillon unpublished

Table 1 cont.

Name	Family	Length (bp)	IR (bp)	DR (bp)	ORF (start-stop)	Bacillus thuringiensis serovar & strain	Acces. No.	Reference
IS240A	IS6	861	16/17	ND	235 (92-799)	israelensis (p137 kb)	M23740	[31]
IS240B	IS6	861	16/17	ND	235 (92-799)	israelensis (p137 kb)	M23741	[31]
IS240F	IS6	806	16/17	ND	NONE	fukuokaensis 84I113 (p197kb)	Y09946	[35]
IS240J1	IS6	808	17	ND	228 (65-751)	jegathesan T28A001	ND	[32], M-L Rosso unpublished
IS240J2	IS6	809	18	ND	228 (67-753)	jegathesan T28A001	ND	[102, 103]
IS240M1	IS6	861	16/17	ND	235 (92-799)	medellin T30 001	AJ251978	A Delécluse unpublished
IS240M2	IS6	815	13/14	ND	204 (122-736) [P]	medellin T30 001	AJ251979	A Delécluse unpublished
IS240M3	IS6	862	15/16	ND	235 (91-798)	medellin T30 001	AJ251979	A Delécluse unpublished
IS240M4	IS6	1018	13/14	ND	187 (92-655) [P]	medellin T30 001	AJ251979	A Delécluse unpublished
ISBtl	IS3	999	15/17	ND	262 (101-889)	aizawai HD-229	L2910	[110]
ISBt2	IS3	> 187	ND	ND	45 [P]	YBT-226	S68409 [P]	[45]
Tn4430	Tn3	4149	38	5	TnpI: 284 (247-1101) TnpA: 987 (1120-4083)	thuringiensis H1.1	X07651	[84]
Tn5401	Tn3	4837	53	5	TnpI: 307 (764-1684) TnpA: 1006 (1756-4773)	morrisoni EG2158	U03554	[17]
TnBthl	Tn3	ND	ND	ND	TnpR: 194 TnpA: [P]	jegathesan T28A001	ND	M-L Rosso unpublished
TnCbil	Tn3	> 2994	ND	ND	TnpR: 195 TnpA: 605 [P]	Clostridium bifermentans malaysia	ND	V Juárez-Pérez unpublished

ND, not determined; p, plasmid; [P], partial sequence; [V], variant of IS231/W, displaying 99.5% DNA identity

Chapter 3.1

Pathogenesis of *Bacillus thuringiensis* toxins

Peter Lüthy[1] & Michael G. Wolfersberger[2]
[1]Institute of Microbiology, Swiss Federal Institute of Technology, 8092 Zurich, Switzerland, and [2]Biology Department, Temple University, Philadelphia, PA 19122, USA

Key words: δ-endotoxin, activation, gut epithelium, histopathology, origin of δ-endotoxin

Abstract: The mode of action, and hence the histopathological events of all the δ-endotoxins produced as secondary metabolites by the species *Bacillus thuringiensis* are identical. The protein particles produced during the sporulation process have to be ingested by the target insects. In the intestine they are dissolved and undergo activation by gut juice proteinases. Following receptor mediated binding pores are formed leading to the loss of normal membrane function. The gut epithelial cells swell. Microvilli and other cell organelles are distorted along with the formation of vacuoles. The pathogenesis, the site and mode of action, is unique for an insecticidal metabolite. The process involved in the pathogenesis explains the specificity and the high level of safety of *Bt* products. The possible origin of the δ-endotoxin, reflecting a close connection between the target insects and the bacterium is discussed.

1.   INTRODUCTION

Bacillus thuringiensis is recognised as an own species. The only justification to separate it from the closely related *B. cereus* is the formation of a proteinaceous insecticidal metabolite during the sporulation process and the importance of *B. thuringiensis* as a microbial insecticide.

The insecticidal metabolites of *B. thuringiensis* have become the most important alternative to chemical insecticides. *B. thuringiensis* preparations are still predominantly applied as conventional sprays in crop protection or against larval stages of mosquitoes and black flies.

J.-F. Charles et al. (eds.),
Entomopathogenic Bacteria: From Laboratory to Field Application, 167–180.
© 2000 *Kluwer Academic Publishers. Printed in the Netherlands.*

Since a few years major culture plants which have been transformed with genes encoding for δ-endotoxins, are on the market. The protection of corn, cotton and rice is aimed at pest insects causing damage to internal plant tissue.

Nearly half a century of research on the δ-endotoxin has been required to reach the present state of plant protection systems of these insecticides of microbial origin. There are several reasons for this slow progress. The initial studies on the mode of action, limited to histopathology, coincided with the climax on the euphoria on chemical pesticides. Therefore little attention has been given to the more complex situation encountered with the δ-endotoxins of *B. thuringiensis*. The microbial origin of the insecticidal metabolite along with its particulate form and the uncommon mode of action have kept the general interest within narrow limits.

Therefore only a small group of researchers was involved in histopathological studies, whereby many of the early results came out of the Insect Pathology Research Institute in Sault Sainte Marie, today incorporated into Natural Resources Canada, Canadian Forest Service, Great Lakes Forestry Centre.

The classical histopathology induced by the *B. thuringiensis* δ-endotoxins belongs to the past and this chapter will of necessity emphasise on a historical review. Therefore we will keep it short and limit it to the essential facts. On the other hand we have to be aware that pioneering work on the mode of action was a indispensable prerequisite for a successful shift into the molecular biology and genetics.

Studies on the histopathology of the δ-endotoxin were started at around 1960 with a climax during the decade of 1970. Afterwards, when an increased number of biochemical and genetic tools became available a steady transition to took place towards molecular mode of action studies. The number of scientists which became interested in the δ-endotoxin of *B. thuringiensis* expanded fast, followed by a steady increase in literature.

Several factors have added to the present importance of *B. thuringiensis*. These include the unique mode of action, the high specificity along with an outstanding safety record in comparison with chemical insecticides. Since the δ-endotoxins are polypeptides encoded by genes present on plasmids, characterisation at a molecular level and gene transfer to other organisms including plants has become easy routine.

The genetic analysis and the elucidation of the molecular mode of action have yielded a wealth of information. This, for example, has necessitated already a second revision of the nomenclature of the crystal proteins ([7]; see also Chapter 2.1 and the web site ; http://www.biols.susx.ac.uk/Home/Neil_Crickmore/Bt/toxins.html).   For persons not fully familiar with *B. thuringiensis*, it has become difficult to retain a clear and general picture. In this chapter we will try give an overview of the baseline factors, encompassing the δ-endotoxin of *B. thuringiensis*.

The term δ-endotoxin will be used for all forms of the insecticidal protein which accumulates within the sporulating cells of *B. thuringiensis* with the exception of the hemolytic / cytolytic Cyt1 and Cyt2 proteins which occur only in some *B. thuringiensis* subspecies such as *israelensis, morrisoni* strain PG-14, *medellin, jegathesan and kyushuensis*, which usely have a mosquitocidal activity. The terms inclusion bodies, crystals, crystal protein refer to the physical form in which the δ-endotoxin as ultimate active moiety is contained.

All the investigations on the δ-endotoxin have shown a uniform principle of the mode of action. Therefore the common features of the mode of action will be taken as a starting point. On the other hand we are faced with a big diversity in specific activity. These are due to differences in the δ-endotoxins and due to the target insects which have to offer the proper conditions for activation and interaction between toxin and the gut epithelial membrane.

2. FROM THE BACTERIAL INCLUSION TO THE ACTIVE POLYPEPTIDE

The polypeptides which are synthesised by the sporulating cell of *B. thuringiensis*, aggregate to microscopically visible inclusions. A given *B. thuringiensis* strain produces as a rule a single inclusion of a defined shape. The most common forms are bipyramidal, typical for the majority of lepidopteran active polypeptides. The strain tenebrionis (belonging to the subspecies *morrisoni*), producer of Cry3 proteins and active against some Coleoptera, forms one or two flat rhomboidal inclusions per cell. Spherical inclusion bodies which consist of four/five polypeptide species are produced by the subspecies *israelensis* as well as the PG-14 strain of the subspecies *morrisoni*. For example HD-1 strains assigned to the Cry2 group which exhibit dual activity against lepidopteran and dipteran insects form two inclusion bodies. Many *B. thuringiensis* strains produce more than one δ-endotoxin species incorporated into a single inclusion. *B. thuringiensis* strains which produce two separate inclusion bodies can be considered as an exception. In most cases micrographs reveal a regular crystalline-like assembly of the polypeptides.

The inclusions produced by the sporulating *B. thuringiensis* within the mother cell compartment are released into the medium upon disintegration of the sporangium. The inclusion bodies show a high degree of persistence and retain their insecticidal activity especially under dry conditions and ambient temperature for months. Commercial products in the form of wettable powders have therefore a shelf life comparable or even superior to chemical insecticides.

The insects have to ingest the inclusion bodies together with their food, a general disadvantage if compared with contact insecticides. The inclusions are dissolved in the alkaline and reducing environment of the gut juice. There are indications that the speed of dissolution of the crystal depends on the pH of the gut juice. For example, Yamvrias [27] working with *Ephestia kühniella*, an insect species with only a slightly alkaline gut pH, showed that the dissolution of the crystals and hence the insecticidal action was slow. On the other hand *Bombyx mori* with a gut juice pH at around 10 shows symptoms within minutes after uptake of crystals.

The polypeptides released from the inclusion bodies represent protoxins still bare of biological activity. The molecular weight of the protoxins differs from one polypeptide species to the other. The protoxins active against lepidopteran insect larvae are in the range of 130 kDa, those active against Coleoptera and Diptera have main protoxin components around 67 kDa [17, 19].

The next, most decisive step is the proteolytic activation of the protoxin. In the case of the lepidopteran active protoxin the activation involves serine proteinases, the main function of which is the digestion of food within the larval midgut. The serine proteinases truncate the protoxin at the C-terminus into a well defined polypeptide of about half the original size. Twenty-eight amino acids are removed from the N-terminus. The activated moiety shows considerable stability and further proteolytic degradation under loss of insecticidal activity continues at a reduced rate. The activated coleopteran toxic polypeptide has a size of 55 kDa. The *israelensis* activated Cry polypeptides ranged within 65 and 38 kDa, according to Federici [11].

Gut juice from different insect species, even belonging to the same order, will process a given protoxin differently. Müller-Jaquet [21] incubated protoxin of a *kurstaki* strain with gut juice of *Spodoptera littoralis*, *Heliothis virescens*, and *Pieris brassicae*, respectively. Two pronounced protein bands were obtained in the range of 65 kDa with *P. brassicae* gut juice. *H. virescens* gut juice yielded two main components in the same range but with numerous smaller polypeptide bands. Finally, the incubation of protoxin in gut juice of *S. littoralis* resulted in breakdown products smaller than 65 kDa. This explains the well known fact that the activity of *kurstaki* crystals is high against *P. brassicae* and low against *S. littoralis*.

Quantitative dissolution of the crystals and processing of the protoxin into the toxin are prerequisites for maximum insecticidal activity of a given crystal species. Müller-Jaquet [21] found big differences in insecticidal activity when she fed crystals, *in vitro* dissolved protoxin and activated toxin to larvae of *H. virescens*. Protoxin was in the average 3.5 times more toxic than intact crystals. This ratio increased to 6.0 when *in vitro* activated δ-endotoxin was assayed. More than tenfold differences were found between different crystal types, e.g. between *B. thuringiensis* subspecies.

Following proteolytic activation in the gut lumen, the toxin has to pass through the peritrophic membrane. According to a study by Wolfersberger *et al.* [27] with *Manduca sexta* the peritrophic membrane is no serious barrier for polypeptides smaller than 100 kDa.

The last step is the receptor mediated binding of the active polypeptide to the gut epithelium followed by toxin insertion which results in membrane permeabilisation (see Chapter 3.3). Early studies by Müller-Jaquet [21] have indicated that the specific activity is based on the degree of match between the toxin and the receptor. An updated review on the complexity of the crystal proteins from the DNA to the mode of action has been presented by Schnepf *et al.* [24].

3. REACTION OF INSECT LARVAE TO THE δ-ENDOTOXIN

Angus [1], 45 years ago, using larvae of *B. mori* and a strain of the *B. thuringiensis* subspecies *sotto*, was the first to prove that the crystal protein was the insecticidal agent. The initial reaction to an intoxication following uptake of crystals is the cessation of feeding. The stop of food uptake can take place within minutes in the case of larvae of *B. mori* or *P. brassicae*. If the amount of crystal protein is high then the stop of feeding is followed by excretion of gut content. General paralysis is the next step which can be demonstrated for example in bioassay with mosquito larvae fed with *B. thuringiensis israelensis* crystals. Within hours the mosquito larvae float motionless just below the water surface.

Although paralysed lepidopteran insects such as *P. brassicae* or *H. virescens* can stay alive for several days after ingestion of a lethal dose of *B. thuringiensis* crystal protein. The insects loose weight and start to shrink due to dehydration by loss of body fluid. Interestingly, quite often bacterial infection of the hemolymph starts only after the death of the insect larvae. This indicates that the gut epithelium although damaged continues to function as barrier against bacterial infection.

The δ-endotoxins are very potent insecticides if the insect and polypeptide species match. For example a third instar larvae of *P. brassicae*, weighing 200 to 300 mg, can be killed with 25 to 50 ng of crystal protein, encoded by *cry1A* genes. About 10 times less crystal protein of the subspecies *israelensis* (proteins encoded by *cry4*, *cry10* and *cry11* genes) is required to kill a 3rd instar larva of *Aedes aegypti*.

If the dose of crystals ingested is sublethal, then the insect larvae will be able to recover. The gut epithelium has a considerable potential to regenerate since the natural turn over of epithelial cells is high. Damaged gut cells are replaced by so-called stem cells generated at the basis of the gut epithelium. Uptake of sublethal amounts under field conditions can be due to restricted feeding activity, for example during cool weather periods. This is clear a disadvantage of *B. thuringiensis* preparations whose pest control activity is closely related to the feeding behaviour of the target insects.

4.   THE GUT EPITHELIUM AS THE TARGET TISSUE

In 1959 Angus & Heimpel [2] demonstrated with silkworm larvae (*B. mori*) as test insects that the gut epithelium was the site of action of the δ-endotoxin. They noticed under the light microscope a general disintegration of the gut epithelium within 30 to 45 min following toxin application. Numerous histopathological studies followed, making use of the advancing techniques in electron microscopy and the improvement in preparation techniques.

Our extended studies on the histopathology have been carried with *P. brassicae*. The site of action of the δ-endotoxin has been traced with gold labelled antibodies for visualisation in the electron microscope. The preferred binding sites of the δ-endotoxin were the tips of the microvilli. Hofmann [16] who has carried out comprehensive binding assays in our laboratory showed that binding of the δ-endotoxin was a high affinity saturable process which implied the presence of specific receptors. This has since been confirmed by many studies, reviewed for example by Schnepf *et al.* [24].

The δ-endotoxin causes a general disorder within the gut epithelial cells. The columnar and goblet cells are likewise affected. The most drastic histopathological change is the swelling of the columnar cells. The increase in volume of these cells is such that microvilli on the surface become shorter until they are finally fully integrated within the ballooning cells. In addition, the goblet cells swell filling out their cavity. Similar observations were done on other insects, such as mosquito larvae intoxicated by *B. thuringiensis israelensis* crystal proteins (Fig. 1).

The general cellular disorder upon exposure to δ-endotoxin which includes all the organelles has been first described for Lepidoptera [9, 20], then for mosquito larvae [5, 6]. The directly exposed microvilli loose their well defined internal fibrillar arrangement leaving only remnants of distorted membranes. In the interior of the gut epithelial cells, the appearance of vacuoles is one of the most prominent phenomenon of a δ-endotoxin poisoning (Fig. 2A and 2B). The formation of lytic vacuoles is particularly intense near the Golgi complexes. The intermembraneous space of the rough endoplasmatic reticulum becomes grossly enlarged. Subsequently, the connections between the membranes rupture, leading to a vesicular re-arrangement. Likewise the mitochondria swell continuously while the cristae gradually disintegrate. Vacuole formation does also occur along the membranes separating the individual cells of the gut epithelium. On the other hand it was shown that the intercellular connections such as the tight junctions, the desmosomes and the gap junctions remained morphologically intact for a long time.

We found that the intracellular histopathological changes take place very fast, within a time frame of five to ten minutes. Sub lethal doses of δ-endotoxin were applied in order to slow down the events as much as possible. *In vivo* studies did not reveal histopathological changes within the nuclei. Only the outer nuclear membrane which represents a part of the rough endoplasmatic reticulum becomes partly detached by the formation of vacuoles.

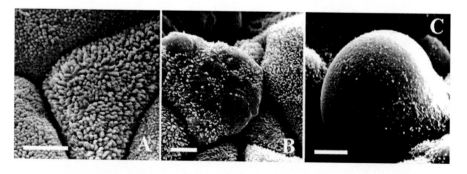

Figure 1. Midgut epithelium of *Aedes aegypti* larva (bar = 0.5μm). A. Healthly larva showing intact microvilli at the cell surface. B. Cell swelling and microvilli reduction after 30 min intoxication with δ-endotoxin. C. After 1 hour, microvilli disappear and the cell is about to lyse. (Micrographs: courtesy of J-F Charles [6]).

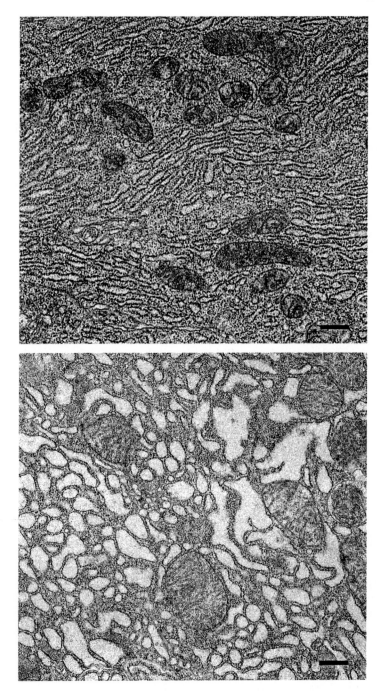

Figure 2. Section of columnar cells of *Pieris brassicae* (bar 0.1μm). Top: l healthy larva. Bottom: after 15 minutes force-feeding δ-endotoxin. Note the vacuolisation and the swollen appearance of the mitochondria.

The intact membrane lining the gut epithelium constitutes a highly selective barrier separating the digestive tract from the interior of the insect. It controls the active and passive transport of ions and metabolites. Rapid losses of the selective permeability under the influence of δ-endotoxin have been shown with ruthenium red, an inorganic stain with a molecular weight of 852, contrasting in the electron microscope [10]. Functional epithelial membranes are not permeable for ruthenium red. However, a few minutes following δ-endotoxin application ruthenium starts to seep through the membrane.

Early investigations with auto radiography carried out by Lüthy & Ebersold [20] showed that no appreciable amount of labelled δ-endotoxin could be traced within the gut epithelial cells. These results could only be understood later on when it had been established that the toxin binds to cell membrane receptors.

De Barjac [8] and de Barjac & Charles [5, 6] showed that the histopathological changes on the gut epithelia of *Aedes aegypti* induced by crystal protein of *B. thuringiensis israelensis* were similar to those found in Lepidoptera. This was the first indication that the mode of action of the δ-endotoxins could be uniform irrespective the *B. thuringiensis* subspecies and the order of the target insect.

The activity of the δ-endotoxin is limited to the gut epithelium. No histopathological changes occur in the hemocel beyond the basal membrane of the gut epithelium which seems to continue to be a barrier also for bacteria. In many instances the *P. brassicae* larvae succumbed to dehydration due to the loss of body fluid through the intestinal tract.

5.   *IN VITRO* STUDIES

Cytopathological changes induced by the δ-endotoxin were demonstrated by Murphy *et al.* in 1976 [22] in Lepidoptera cell cultures. It has to be added that high doses of activated δ-endotoxin were required and that only a part of the cells should symptoms. Geiser [12] working with the Cf-1 cell line derived from the spruce budworm, *Choristoneura fumiferana*, succeeded in reproducing the *in vivo* histopathological changes. In addition he found in his *in vitro* system that the nuclei were affected. The chromatin was dislocated towards the peripheral regions of the nuclei. Pronounced vacuolisation occurred between the inner and outer nuclear membrane. Drastic changes in the interior of the cell membranes were shown by Lüthy & Ebersold [20] with *in vitro* cultured Cf-1 cells.

Geiser carried out a number of experiments with non target organisms and cell lines. He did not find any activity when he exposed prokaryotes to the activated δ-endotoxin, such as *Bacillus subtilis, Staphylococcus aureus, Enterococcus faecalis, Escherichia coli* and *Pseudomonas aeruginosa*. Likewise *Candida tropicalis* and *Saccharomyces cerevisiae* did not react. The plasmalemna of *S. cerevisiae* remained morphologically intact. Cell cultures of chicken fibroblasts and even primary cells cultures of hemocytes derived from *P. brassicae* did not respond to the activated δ-endotoxin. Nishiitsutsji-Uwo *et al.* [23] were not able to show any toxic effect on mammalian cell lines. These *in vitro* studies indicate the high specificity of the δ-endotoxin which acts only when the corresponding receptors within the cell membrane are present.

Charles [4] and Laurent & Charles [18] confirmed the high specificity of δ-endotoxins derived from *B. thuringiensis thuringiensis* and *B. thuringiensis israelensis*. δ-endotoxin of both subspecies was assayed in cell cultures of *Aedes aegypti*. Activity was only found with the mosquito active *israelensis* strain. The cell cultures continued to develop normally in the presence of lepidopteran active *B. thuringiensis* toxin.

Concomitant with the studies on histopathology interest arouse in the changes of the flux of ions through the gut epithelial membrane during the early stages of δ-endotoxin poisoning. Harvey & Wolfersberger [14] determined that exposure to δ-endotoxin reduced the electrical resistance of isolated *Manduca sexta* larval midguts by about 55%, stimulated oxygen consumption by about 30%, and inhibited net transepithelial ion flux by about 60%. Unidirectional potassium flux measurements revealed that the effect on net ion flux was due a tripling of the passive flux from lumen to hemolymph with no significant change in the active flux from hemolymph to lumen. The results indicated that the toxin increased the potassium permeability of the epithelium but had no direct affect on the active potassium transport mechanism.

In 1980 Hanozet *et al.* [13] reported the preparation of functionally active brush border membrane vesicles (BBMV) from lepidopteran midgut. The cell free BBMV system has become an indispensable tool for studies of membrane-toxin interaction on the molecular level. Mode of action studies employing BBMVs have been reviewed by Wolfersberger [26].

6. ON THE ORIGIN OF THE δ-ENDOTOXIN: A HYPOTHESIS

Working on the mode of action of the δ-endotoxin during nearly two decades, from 1970 towards the end of the nineteen-eighties, we came upon increasing evidence in practice and in theory that the δ-endotoxins might be closely related to polypeptides of holometabolic invertebrates where they play a major role in the apoptosis of gut epithelial cells.

The larval midgut represents a dynamic tissue with a high turnover rate in columnar and goblet cells. Stem cells which are generated at the basis of the gut epithelium provide the source for the enlargement of the gut epithelium from one moult to the other and for the replacement of degenerating cells [3]. During metamorphosis from the larval to the pupal stage the gut epithelium is completely replaced by pupal cells. In the prepupal stage, stem cells form an entirely new protective layer lining the basal membrane. Subsequently the larval midgut cells undergo a programmed cell death or apoptosis.

The degenerative process of midgut cells during larval development and at the transition to the pupal stage is an active process mediated by lytic enzymes, present in vacuoles in goblet cells and prepupal stem cells.

In the course of our histopathological studies we regularly came across degenerating gut epithelial tissue of *P. brassicae* larvae which were used as control specimens. Electron microscopy revealed that the morphological changes in these cells very much resembled those treated with δ-endotoxin [20]. When midgut epithelial cells were checked at the end of the larval stage, we observed again changes in the cell organelles which were practically indiscernible from δ-endotoxin induced poisoning.

Since the histological changes of δ-endotoxin and the programmed cell death show a high degree of similarity it can not be excluded that the molecular principles are related. This would necessitate a horizontal gene transfer with some frequency from holometabolic insects to competent prokaryotes with *B. cereus* playing the dominant role. A possible mode of events for the horizontal gene transfer from the insect to the *B. cereus* could be as follows:

Cells of *B. cereus* which gain access into the hemocoel of an insect, e.g. through a lesion, easily proliferate and cause a septicaemia. According to own tests one single *B. cereus* cell is able to trigger a deadly infection in a larva of *P. brassicae*. Equipped with natural competence, a given *B. cereus* strain is transformed with insect DNA fragments containing the sequences for the lytic principle responsible for the programmed degeneration of midgut epithelial cells. Modification by *B. cereus* of the transformed insect DNA by restriction would be very likely the next step before incorporation into a plasmid. The transformed *B. cereus* plasmid would then be able to express the newly acquired gene in the mother cell compartment during the early sporulation process. Thus the transformed strain would have become a *B. thuringiensis* since δ-endotoxin formation is the only main difference between the two spore-formers. Subsequently if an insect larvae ingests sporulated *B. thuringiensis* cells along with the food, the bacterial inclusions are activated and lyse the unprotected gut epithelial cells prematurely. The above described process must be an event occurring with some regularity.

The above hypothesis could easily explain the high specificity of the δ-endotoxins and their diversity. It should be noticed that the first isolations of *B. thuringiensis* have been made from sericultures in Japan (*B. thuringiensis* subsp. *sotto*) and from flour moths (*Ephestia kühniella*) in Germany (*B. thuringiensis* subsp. *thuringiensis*). These are situations were insects are crowded. This should represent ideal environmental conditions for the transformation of *B. cereus* to *B. thuringiensis*. A similar situation applies to the first isolation of the subspecies *israelensis* in a small drying pond crowded with mosquito larvae in the Negev desert.

The above three subspecies produce δ-endotoxins with an extremely high activity against the original insect species. This is further strong evidence that a relationship has to exist between the δ-endotoxin and the insect. In the case of *B. thuringiensis israelensis* we may speculate that along with four δ-endotoxin genes, *B. cereus* has been transformed with the *cyt1A* gene. Genes for the expression of hemolytic factors are required by adult mosquitoes.

Admittedly the final proof for the above hypothesis is still missing. We were not able to find matching sequences between *cry* genes and insect genes. Little is known about the molecular events which take place on the level of the midgut epithelium between the larval and the pupal stage, followed by the signal given by ecdyson. Genes involved in the programmed cell death have, to our knowledge, have not yet been isolated and sequenced.

Using Ouchterlony we were able to find a precipitation line with gut extracts of *P. brassicae* larvae just before pupation and polyclonal antibodies prepared against Cry1 protein.

A similar situation is encountered in the case of *Bacillus sphaericus* where we find strains producing a binary mosquitocidal toxin which may act like the δ-endotoxin of *B. thuringiensis* (see Chapter 3.5). The preferred habitat of *B. sphaericus* are water bodies, rich in organic matter. This means that intimate contact between *B. sphaericus* and mosquito larvae, especially of the genus *Culex*, must exist.

Should the δ-endotoxin be an insect-own polypeptide, then isolation of the genes from holometabolic invertebrates could open the door for the directed development of additional δ-endotoxin based insecticides.

REFERENCES

[1] Angus TA (1954) A bacterial toxin paralyzing silkworm larvae. Nature 175, 545

[2] Angus TA & Heimpel AM (1959) Inhibition of feeding, and blood pH changes, in Lepidopterous larvae infected with crystal-forming bacteria. Can. Entomol. 91, 352-358

[3] Baldwin KM, Hakim RS, Loeb MJ & Sadrud-Din SY (1996) Midgut development, p. 31-54. *In* MJ Lehane & PF Billingsley (ed.), Biology of the Insect Midgut Chapman & Hall

[4] Charles J-F (1983) Action de la δ–endotoxine de *Bacillus thuringiensis* var. *israelensis* sur cultures de cellules de *Aedes aegypti* L. en microscopie électronique. Ann. Microbiol. (Inst. Pasteur), 134A, 365-381

[5] Charles J-F & de Barjac H (1981) Histopathologie de l'action de la δ–endotoxine de *Bacillus thuringiensis* var. *israelensis* sur les larves d'*Aedes aegypti* (Dipt.: Culicidae). Entomophaga, 26, 203-212

[6] Charles J-F & de Barjac H (1983) Action des cristaux de *Bacillus thuringiensis* var. *israelensis* sur l'intestin moyen des larves de *Aedes aegypti* L. en microscopie électronique. Ann. Microbiol. (Inst. Pasteur), 134A, 197-218

[7] Crickmore N, Zeigler DR, Feitelson J *et al.* (1998) Revision of the nomenclature for the *Bacillus thuringiensis* pesticidal crystal proteins. Microbiol. Mol. Rev. 62, 807-813

[8] de Barjac H (1978) Étude cytologique de l'action de *Bacillus thuringiensis var. israelensis* sur larves de Moustiques. C.R. Acad. Sc. Paris, t. 286, Série D, 1629-1632

[9] Ebersold HR, Lüthy P & Müller M (1977) Changes in the fine structure of the gut epithelium of *Pieris brassicae* induced by the delta-endotoxin of *Bacillus thuringiensis*. Bull. Soc. Ent. Suisse, 50, 269-276

[10] Ebersold HR, Lüthy P, Geiser P & Ettlinger L (1978) The action of the delta-endotoxin of *Bacillus thuringiensis*: An electron microscope study. Experientia 34, 1672

[11] Federici BA, Lüthy P & Ibarra JE (1990) Parasporal body of *Bacillus thuringiensis israelensis*: structure, protein composition, and toxicity, p. 16-44. *In* (de Barjac H & Sutherland DJ, ed), Bacterial control of mosquitoes and blackflies, Rutgers University Press, New Brunswick

[12] Geiser P (1979) Versuche in vitro zum Nachweis einer Wirkung des delta-Endotoxins von *Bacillus thuringiensis*. Dissertation ETH No. 6411, Swiss Federal Institute of Technology, 8092 Zurich, Switzerland

[13] Hanozet GM, Giordana B & Sacchi VF (1980) K^+-dependent phenylalanine uptake in membrane vesicles isolated from the midgut of *Philosamia cynthia* larvae. Biochem Biophysi. Acta 596, 481-486

[14] Harvey WR & Wolfersberger MG (1979) Mechanism of inhibition of active potassium transport in isolated midgut of *Manduca sexta* by *Bacillus thuringiensis*. J. Exp. Biol. 83, 293-304

[15] Heimpel AM & Angus TA (1959) The site of action of crystalliferous bacteria in Lepidoptera larvae. J. Insect Pathol. 1, 152-170

[16] Hofmann C (1988) The binding of *Bacillus thuringiensis* delta-endotoxin to cultured insect cells and to brush border membrane vesicles. Dissertation ETH No. 8498, Swiss Federal Institute of Technology, 8092 Zurich, Switzerland

[17] Huber HE, Lüthy P & Ebersold HR. (1981) The subunits of the parasporal crystal of *Bacillus thuringiensis*: size, linkage and toxicity. Arch. Microbiol. 129, 14-18

[18] Laurent P & Charles J-F (1984) Action comparée des cristaux solubilisés des sérotypes H-14 et H-1 de *Bacillus thuringiensis* sur des cultures de cellules de *Aedes aegypti*. Ann. Microbiol. (Inst.Pasteur) 135 A, 473-484

[19] Lereclus D, Delécluse A & Lecadet M-M (1993) Diversity of *Bacillus thuringiensis* Toxin and Genes, p. 37-69. *In* (Entwistle PF, Cory JS, Bailey MJ & Higgs S, ed), *Bacillus thuringiensis*, an environmental biopesticide: theory and practice, John Wiley & Sons, Chichester, New York, Brisbane, Toronto, Singapore

[20] Lüthy P & Ebersold HR (1981) *Bacillus thuringiensis* delta-endotoxin: histopathology and molecular mode of action, p. 235-267. *In* (Davidson EW, ed), Pathogenesis of invertebrate microbial diseases, Allanheld, Osmun & Co. Publishers, Totowa, New Jersey

[21] Müller-Jaquet F (1987) On the specificity of delta-endotoxins of *Bacillus thuringiensis*. Dissertation ETH No. 8474, Swiss Federal Institute of Technology, 8092 Zurich, Switzerland

[22] Murphy DW, Sohi SS & Fast PG (1976) *Bacillus thuringiensis* enzyme digested delta-endotoxin: Effect on cultures insect cells. Science 194, 954-956

[23] Nishiitsutsuji-Uwo J, Endo Y & Himeno M (1980) Effects of *Bacillus thuringiensis* delta-endotoxin on insect and mammalian cells in vitro. Appl. Ent. Zool. 15, 133-139

[24] Schnepf E, Crickmore N, Van Rie J *et al.* (1998) *Bacillus thuringiensis* and its pesticidal crystal proteins. Microbiol Mol. Biol. Rev. 62, 775-806

[25] Wolfersberger MG (1990) Specificity and mode of action of *Bacillus thuringiensis* insecticidal crystal proteins toxic to lepidopteran larvae: recent insights from studies utilizing midgut brush border membrane vesicles. Proc. Vth International. Colloquium on Invertebrate Pathology and Microbial Control, Adelaide, Australia

[26] Wolfersberger MG, Spaeth DD & Dow JAT (1986) Permeability of the peritrophic membrane of Tobacco Hornworm larval midgut. American Zoologist 26, 356

[27] Yamvrias C (1962) Contribution à l'étude du mode d'action de *Bacillus thuringiensis* vis-à-vis de la teigne de la farine, *Anagasta kühniella* Zell. (Lepidoptère). Entomophaga 7, 101-159

Chapter 3.2

Investigations of *Bacillus thuringiensis* Cry1 toxin receptor structure and function

Stephen F. Garczynski[1] & Michael J. Adang[1,2]
[1]*Affiliation Departments of Entomology, and* [2]*Biochemistry and Molecular Biology, University of Georgia, Athens, Georgia 30602, USA*

Key words: *Bacillus thuringiensis*, Cry1, δ-endotoxins, brush border membranes, aminopeptidase, cadherin, glycolipids

Abstract: *Bacillus thuringiensis* Cry1 proteins are highly toxic to some species of lepidopteran larvae. Cry proteins act on the brush border membrane through a multi-step process. The process includes Cry1 crystal solubilisation, activation by proteinases, attachment to receptors, then membrane insertion and permeation followed by cell lysis. Toxin binding to receptors is a pivotal event necessary for insect mortality. We review research that identified aminopeptidase N and cadherin-like proteins as Cry1 toxin receptors. Since enteric toxins recognise glycolipid receptors, we examined *Manduca sexta* brush border lipids as potential toxin receptors. *M. sexta* brush border membrane glycolipids lipids bound Cry1Ac toxin selectively on thin-layer chromatograph overlays. Because pore formation is necessary for toxicity, a functional receptor should catalyse Cry1-induced pore formation. Various data support aminopeptidases, cadherin-like proteins and glycolipids as "true" Cry1 receptors.

1.   CRY1 TOXIN BINDING TO BRUSH BORDER MEMBRANE VESICLES

Cry1 toxins bind to molecules in apical microvillar membranes, and form ion channels in the midguts of susceptible insect larvae [66]. Preparations of brush border membrane vesicles (BBMV) from the midguts of lepidopteran larvae provide an in vitro representation of the

J.-F. Charles et al. (eds.),
Entomopathogenic Bacteria: From Laboratory to Field Application, 181–197.
© 2000 *Kluwer Academic Publishers. Printed in the Netherlands.*

apical microvillar membrane [85]. The use of BBMV has facilitated study of the molecular interactions between Cry1 toxins and insect midgut epithelial cells. Hofmann *et al.* [30, 31] were first to use BBMV in competitive binding assays to characterise the interaction of Cry1 toxins with its target site from the apical membrane. They determined that Cry1 toxins bind saturably and with high affinity to BBMV isolated from the midguts of *Pieris brassicae* and *Manduca sexta* larvae, and toxin binding was correlated with insecticidal activity [30, 31]. Binding studies have been done to assess the interaction of several Cry1 toxins with BBMV prepared from a variety of lepidopteran larvae, and specific binding sites with affinity constants in the nM range have been reported [12, 19, 25, 64, 80, 81, 83].

Cry1 toxin binding to high-affinity sites present on BBMV has been positively correlated with insecticidal activity [12, 25, 30, 31, 80, 81]. However, the presence of binding sites does not ensure *in vivo* toxicity [19, 25, 50, 64, 81, 83], suggesting that post-binding events (membrane insertion and pore formation) are important for Cry1 toxicity. For example, Cry1Ac was not toxic to *Spodoptera frugiperda* larvae yet ^{125}I-Cry1Ac bound to BBMV prepared from the midguts isolated from this insect [25, 50].

The importance of binding sites in the insect midgut is best illustrated in the diamondback moth, *Plutella xylostella*, that have acquired resistance to Cry1 toxins. *P. xylostella* was the first insect to develop field resistance to *B. thuringiensis* toxins. A reduced number binding sites was found in a Cry1A resistant population of *P. xylostella* [19]. In further studies with a different population of *P. xylostella* larvae resistant to the Cry1A toxins, Tabashnik *et al.* [72] found no binding of Cry1Ac and that the resistance was reversible. The reversal of resistance was correlated with the return of Cry1 binding sites. Cry1Ac binds with high-affinity to BBMV isolated from susceptible and revertant strains of *P. xylostella*, while no binding was detected for resistant larvae [72]. The Cry1Ac binding protein in *P. xylostella* was identified as aminopeptidase N [53]. Analysis of BBMV isolated from susceptible and resistant strains of *P. xylostella* revealed that aminopeptidase N activity was the same, and there was no difference in Cry1Ac binding on ligand blots [53]. These results do not support aminopeptidase as a determinant of toxicity, and possible explanations as to how *P. xylostella* larvae could attain Cry1Ac resistance with a receptor present in BBMV were offered [53]. Potential mechanisms have been proposed for the presence of binding proteins on ligand blots, but not in soluble assays. These include epitope masking [46] and loss of binding based on GPI anchorage [53]. In epitope masking, toxin binding to receptor proteins may be blocked by other midgut molecules that are not recognised by Cry toxins [46]. Luo *et al.* [53] suggest that it is also possible that an endogenous phosphoinositol phospholipase C releases aminopeptidase from the apical membrane in resistant strains prior to Cry1A toxin pore formation. Another possibility is that BBMV molecules other than aminopeptidases may also serve as toxin binding determinants.

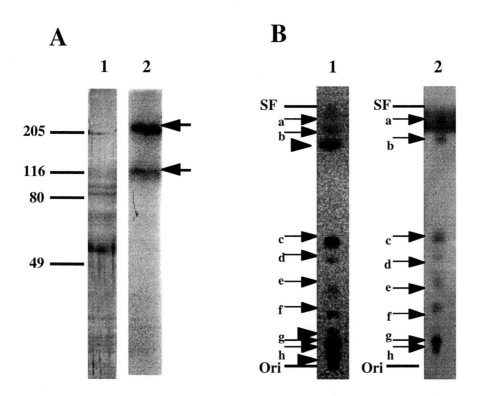

Figure 1. Detection of Cry1Ac binding molecules present in *Manduca sexta* BBMV by ligand blotting (Panel A) or TLC overlays. (A) *M. sexta* BBMV were separated by SDS-PAGE (10% acrylamide, 74:1 acrylamide:bisacrylamide) then transferred to polyvinylidene difluoride membranes. Lane 1: *M. sexta* BBMV molecules stained with India ink. Molecular size markers (kDa) are denoted with thick lines. Note: The band at 205 kDa is from the molecular size markers. Lane 2: Autoradiograph of a ^{125}I-Cry1Ac toxin overlay. Molecules bound by ^{125}I-Cry1Ac are indicated by the thick lines. (B) Organic extract of *M. sexta* BBMV separated by TLC in a solvent system containing chloroform:methanol:water (75:25:2). Lane 1: *M. sexta* BBMV glycolipids stained with orcinol spray reagent. Lane 2: Autoradiograph of a ^{125}I-Cry1Ac toxin overlay. Glycolipids bound by ^{125}I-Cry1Ac are indicated by the thin lines. The origin (Ori) and solvent front (SF) are indicated by the thick lines.

2.   CRY1 RECEPTOR DETECTION USING TOXIN OVERLAYS

The term receptor has been used to refer to a molecule on the surface of an insect cell to which a Cry protein binds [40]. This is a rather vague definition and does not imply any physiological function associated with these surface molecules or indicate an active role in Cry toxin action. In the literature, receptor and binding protein are often used interchangeably.

2.1   Ligand blotting

Visualisation of Cry1 binding to receptor glycoproteins on the plasma membrane of insect cell lines was done using a protein blotting technique [27, 41]. Protein blotting (also called ligand blotting; reviewed in [71]) is a technique that is analogous to Western blotting [75], except that Cry1 toxins are used instead of antibodies to identify proteins on blots. For ligand blots, BBMV molecules are first separated by SDS-PAGE, and then are transferred to a solid support (usually nitrocellulose or polyvinylidene difluoride). The membrane filters containing BBMV are incubated with a labeled Cry protein and then washed to remove any unbound toxin. Visualisation of receptor proteins present in BBMV can be done by autoradiography (for ^{125}I-labeled toxins) or colorimetric detection (for biotin or antibody based labeling). An example of a ligand blot is presented in Figure 1A.

Ligand blots have been used extensively to identify receptor proteins present in BBMV isolated from larval midguts. First reports of applying ligand blots to lepidopteran BBMV identified glycoproteins as putative Cry1A receptors in the larval midgut [25, 43, 60]. For Cry1Ac, a 120 kDa glycoprotein was detected in extracts of BBMV prepared from midguts isolated from *M. sexta* larvae [25, 43], while multiple proteins ranging from 60 to 155 kDa were identified from BBMV from *Heliothis virescens* [25, 60]. Putative receptor proteins have been identified for Cry1 toxins from BBMV isolated from a variety of insects [7, 25, 33, 43, 53, 56, 59, 60, 61, 64, 76, 78, 87].

2.2   Thin-layer Chromatography Overlays

Glycolipids serve as receptors for some protein toxins produced by mammalian pathogenic bacteria [20, 34, 47], and their carbohydrate epitopes serve as toxin recognition factors [34]. To detect toxin binding to glycolipids a thin-layer chromatography (TLC) overlay technique was developed [54]. The TLC overlay technique used to detect toxin binding to glycolipids is analogous to the ligand blot procedure used to identify receptor proteins in insect BBMV. For TLC overlays, BBMV lipids are separated on silica plates, incubated with a labeled toxin, and then washed to remove any unbound protein. As with ligand blots, Cry1 binding is detected by autoradiography (for ^{125}I-labeled toxins) or colorimetric detection (for biotin or antibody based labeling). An example of a TLC overlay is presented in Figure 1B. TLC overlays have been used to detect Cry1A toxin binding to insect glycolipids [10, 23] and will be discussed in detail in section 4.3.

3.   IDENTIFICATION OF CRY1 RECEPTOR PROTEINS

Some of the Cry1 receptor proteins detected on ligand blots of insect BBMV have been purified, characterised and cloned [11, 22, 26, 35, 36, 37, 38, 51, 52, 53, 57, 58, 59, 65, 76, 77, 78, 79, 87]. These proteins are identified as forms of aminopeptidase N [11, 26, 37, 51, 57 65, 78, 87] or as members of the cadherin superfamily of proteins [22, 36, 59, 77]. Aminopeptidases isolated from *Manduca sexta* BBMV have been identified as glycoprotein receptors of 106 kDa for Cry1C [51] and 115-120 kDa for Cry1Aa [57], Cry1Ab [11, 57], and Cry1Ac [37, 57, 65]. Aminopeptidases have been identified as Cry1A receptors by amino acid homology and enzyme activity in BBMV isolated from *Lymantria. dispar* [78, 79], *H. virescens* [26, 52], *P. xylostella* [11, 53], and *Bombyx mori* [87]. These results in total indicate an important role for aminopeptidases in Cry1A toxin action.

A 210 kDa glycoprotein has been identified by ligand blot analysis as the Cry1Ab receptor from BBMV prepared from the midguts of *M. sexta* larvae [56, 76]. Analysis of the cloned sequence showed the 210 kDa Cry1Ab receptor to have homologies with members of the cadherin superfamily of proteins [77]. Although initially detected with Cry1Ab, the 210-cadherin-like protein is also bound by Cry1Aa and Cry1Ac toxins [22, 35, 36]. Recently, a 175 kDa protein was identified as a Cry1Aa receptor in *B. mori* [59]. A cDNA clone corresponding to the 175 kDa Cry1Aa receptor from *B. mori* shows homology with members of the cadherin superfamily of proteins, and the deduced amino acid sequence is 69.5% identical with its *M. sexta* counterpart [58]. The identification of both aminopeptidases and cadherins as Cry1A receptors leads to some controversy in the literature.

4.   CHARACTERISTICS OF CRY1A BINDING MOLECULES IN *M. SEXTA* BBMV

4.1   The 120 kDa Aminopeptidase N

A 120 kDa protein present in *M. sexta* BBMV was first detected on protein blots as a putative binding molecule for Cry1Ac [25, 43]. Three years after this initial detection, the 120 kDa protein was identified as an aminopeptidase N [37, 65]. Aminopeptidase Ns are now identified as the major Cry1 binding proteins for *M. sexta, H. virescens, L. dispar, B. mori, P. xylostella* and *Trichoplusia ni* [26, 32, 46, 48, 51, 53, 78, 79, 87], indicating the potential importance of this midgut glycoprotein in toxin action.

Knowles *et al.* [44] suggested that the cell surface receptor for the Cry1A proteins was a glycoconjugate because insect cells could be protected from toxin effects by lectins and the sugar N-acetylgalactosamine (GalNAc). GalNAc competes with Cry1Ac in *in vitro* binding assays using *M. sexta* BBMV [25, 43]. Therefore, it was proposed that GalNAc is part of the carbohydrate epitope on *M. sexta* BBMV that Cry1Ac binds [25, 43]. Cry1Ac binds to carbohydrate structures associated with the 120 kDa APN from *M. sexta* BBMV, and GalNAc is part of that epitope [2, 6, 8, 25, 37, 43, 57]. Cry1Ac binding to the 120 kDa aminopeptidase N on ligand blots was not detected following periodate treatment of *M. sexta* BBMV extracts transferred to PVDF [23]. Mild periodate treatment has been successfully used to map carbohydrate epitopes for antibodies by chemically breaking the ring structure of sugars [86]. Elimination of Cry1Ac binding to aminopeptidase N following periodate treatment would indicate that sugar structure is important for toxin/receptor interaction on ligand blots.

Previous reports indicate that aminopeptidase N and alkaline phosphatase on the brush border of midgut cells in *Bombyx mori* larvae are attached to the membrane through a glycosyl-phosphatidylinositol (GPI) anchor [73, 74]. Consistent with reports for *B. mori*, release of aminopeptidase N and alkaline phosphatase activities from *M. sexta* BBMV by phosphatidylinositol-specific phospholipase C (PI-PLC) indicates the presence of GPI-anchors on these proteins [24, 38]. Ligand blots were used to detect a 115 kDa Cry1Ac binding protein presumably the PI-PLC-solubilised form of aminopeptidase N in *M. sexta* BBMV [24, 49, 57]. Aminopeptidase Ns identified as Cry1 binding proteins in BBMV from *M. sexta, H. virescens, L. dispar, P. xylostella, B. mori* and *T. ni* are all proposed to be inserted in the membrane by GPI-anchors [11, 26, 32,

38, 48, 51, 52, 57, 79, 87]. Because the GPI-anchor is a carbohydrate-rich structure at the membrane surface, it was speculated that it may be the epitope involved in Cry1Ac binding to aminopeptidase N from *M. sexta* [24].

It is unlikely that Cry1Ac binds to the lipid moiety of the GPI-anchor on *M. sexta* or *L. dispar* aminopeptidases because toxin still bound to these proteins after PI-PLC solubilisation [24, 49, 57, 79]. The glycan moiety of a GPI-anchor can be removed by cleaving the phosphate linkages with hydrofluoric acid [18]. Cry1Ac still binds to the 120 kDa APN on ligand blots after removal of the glycan core by hydrofluoric acid dephosphorylation, suggesting that another carbohydrate structure may contain the toxin binding epitope. Removal of the GPI-anchor was confirmed by detection (or lack of detection) of the cross-reacting determinant generated by PI-PLC cleavage [23]. If the GPI-anchor is not the carbohydrate structure Cry1Ac binds to on APN, then it would follow that the binding epitope may be contained on N- or O-linked sugars. Denolf *et al.* [11] reported that glycan structures sensitive to N-glycosidase F are involved in Cry1Ab binding to a *M. sexta* BBMV. In addition to GPI-anchor signal peptides, potential N- and O- glycosylation sites are present in the deduced amino acid sequences of cDNA clones encoding the different *M. sexta* aminopeptidases [11, 38]. However, the identity and structure of the carbohydrate epitopes that mediate Cry1Ab and Cry1Ac binding to *M. sexta* aminopeptidases remain unknown.

In addition to Cry1Ac, the 120 kDa aminopeptidase N from *M. sexta* has also been identified as the binding protein for Cry1Aa and Cry1Ab. On ligand blots, Cry1Ab and Cry1Ac both bind to 120 kDa proteins in *M. sexta* BBMV [9, 21, 62]. Furthermore, Cry1Aa, Cry1Ab and Cry1Ac toxins all bound specifically to the 115 kDa form of aminopeptidase N purified from *M. sexta* BBMV [57]. Ligand blots done with *M. sexta* BBMV separated on SDS-PAGE (10%, acrylamide:bis 74:1) do not support aminopeptidase N being a common binding protein for Cry1Ab and Cry1Ac toxins. Instead, Cry1Ab bound to a 106 kDa molecule, different from the 120 kDa aminopeptidase N bound by Cry1Ac and to a 210 kDa molecule discussed below [23]. Ligand blots using the 115 kDa aminopeptidase N from *M. sexta* produced similar results in that Cry1Ac, but not Cry1Ab, bound to this purified receptor protein [9]. In support of these results, cDNAs encoding two different APNs have been cloned from *M. sexta*, one corresponding to the 120 kDa Cry1Ac binding protein [38], the other to the 106 kDa Cry1Ab binding protein [11]. These ligand blot results indicate different forms of APN may serve as binding proteins for Cry1Ab and Cry1Ac in *M. sexta* BBMV. However, the BBMV proteins on ligand blots are denatured by the SDS-PAGE separation and may not reflect the interaction of the Cry1A toxins with native forms of APN.

4.2   The 210 kDa Cadherin-like Protein

A 210 kDa binding protein from *M. sexta* BBMV was detected on ligand blots probed with Cry1Ab [76]. This 210 kDa binding molecule was identified as a member of the cadherin-superfamily of proteins [77]. The 210 kDa cadherin-like protein from *M. sexta* BBMV has also been identified as the common binding site for Cry1Aa, Cry1Ab, and Cry1Ac toxins [21, 22, 35, 36, 56, 62, 77]. However, as discussed above, Cry1Ab also binds to a 106 kDa protein and Cry1Ac to the 120 kDa APN indicating that these toxins may interact with more than one molecule in *M. sexta* BBMV. Incubation of *M. sexta* BBMV molecules transferred to PVDF with certain detergents (NP-40, deoxycholate, and CHAPS) reduces Cry1Ab and Cry1Ac binding on ligand blots [23, 36]. It could be that the detergent treatments remove the toxin binding proteins from the PVDF, however that is unlikely due to the high protein retention rates of these membrane filters [82].

4.3   Insect Glycolipids

Previously, a mixture of Cry1A protoxins and activated-toxins (containing Cry1Aa, Cry1Ab, and Cry1Ac) prepared from *B. thuringiensis* var. *kurstaki* HD-1 were shown to bind to glycosphingolipids of dipteran and non-insect origin on TLC overlays [10]. The Cry1A protoxin mixture bound preferentially to glycosphingolipids containing a terminal α-Gal, while the activated toxins bound to glycolipids with terminal sugars consisting of α-Gal, β-Gal, β-GalNAc, and β-GlcNAc [10]. It was concluded that the activated toxins bound to glycolipids with a decreased specificity relative to protoxins [10], however these authors assumed the presence of a single toxin instead of the three Cry1A proteins present in HD-1 crystals.

The initial detection and characterisation of Cry1Ac binding to glycolipids present in *M. sexta* BBMV has been reported [23]. Detection of Cry1Ac binding to glycolipids from *M. sexta* BBMV was done using a TLC overlay technique [10, 54]. As described in section 2.2, the TLC overlay technique used to detect toxin binding to glycolipids is analogous to the ligand blot procedure used to identify Cry1Ac binding proteins in insect BBMV. When *M. sexta* BBMV lipids were separated in a solvent system composed of $CHCl_3$:methanol:H_2O (50:40:10 v/v/v), the TLC overlays revealed that Cry1Ac bound to six glycolipids that are detected with orcinol. This toxin interaction appears selective because Cry1Ac did not bind to all of the glycolipids stained by orcinol, and there was no detectable binding of denatured ^{125}I-Cry1Ac on TLC overlays. Furthermore, Cry1Ac binding to *M. sexta* glycolipids was reduced when a 1,000-fold excess of unlabeled toxin was added to the TLC overlay [23]. The glycolipid results above are consistent with those previously reported for competitive binding assays with Cry1Ac and *M. sexta* BBMV *in vitro* [25, 43, 81], and on ligand blots [25, 36]. This is the first indication that Cry1Ac binds to glycolipids found in BBMV isolated from a susceptible insect.

To characterise the glycolipids extracted from *M. sexta* BBMV that are bound by Cry1Ac, a more non-polar solvent system (CHCl$_3$:methanol:H$_2$O, 75:25:2 v/v/v) was used to achieve a better separation of these molecules. The autoradiogram of the TLC overlay showed that ^{125}I-Cry1Ac bound to eight molecules when the organic extract of *M. sexta* BBMV was separated using the non-polar solvent system (Figure 1B, lane 2). Orcinol spray reagent was used to confirm that Cry1Ac is binding to a subset of glycolipids present in the *M. sexta* BBMV extract (Figure 1B, lane 1). Visualisation of *M. sexta* BBMV glycolipids with molybdenum blue and antimony trichloride spray reagents show that not all of these molecules contain the same lipid moieties, indicating that toxin could possibly interact with the carbohydrate structures present. Soybean agglutinin, a lectin with a specificity for GalNAc and Gal, bound to the same glycolipids as Cry1Ac on TLC overlays [23]. This result would indicate that GalNAc or Gal are shared sugar constituents of the glycolipids in *M. sexta* BBMV that are bound by Cry1Ac. Interestingly, GalNAc and Gal were amongst the terminal sugars determined to be important for Cry1A toxin binding to dipteran glycolipids [10], consistent with the hypothesis that GalNAc is part of the carbohydrate epitope bound by Cry1Ac in *M. sexta* BBMV [6, 25, 37, 43]. As above, the structural features of the Cry1Ac epitopes on glycolipids are not yet determined, therefore, further characterisations of these binding molecules will be necessary.

5.   CRY1A TOXIN-INDUCED PORES

Knowles and Ellar [42] showed that Cry toxins affect membrane permeability in insect cell lines, and proposed that Cry toxins kill cells by forming non-specific pores leading to cell death by colloid osmotic lysis. Models for membrane insertion and disruption of ion gradients and osmotic balances of the midgut epithelial cells have been reviewed [15,

39, 40]. Cry1 toxin pores have been characterised in planar phospholipid bilayers [16, 68, 70], insect cell lines [67], hybrid phospholipid/BBMV [14, 84], and BBMV [3, 17, 29, 84]. Mutagenesis experiments assessing the role of the different domains of Cry proteins in membrane insertion and pore formation function has been recently reviewed [63, 66], and will not be discussed.

Cry1 toxins insert into artificial phospholipid bilayers at high doses (>1 µg/ml), creating pores that are permeable to ions and small molecules [14, 27, 70]. The concentration of Cry1Ac needed to create pores is drastically reduced when BBMV are incorporated into phospholipid bilayers [14, 55]. Carroll & Ellar [3] used a scattered light assay to show the specificity of Cry1Ac pore formation in BBMV isolated from the midguts of *M. sexta* larvae. Pore formation is also correlated with Cry1 toxicity in *Spodoptera exigua* and *S. frugiperda* [50]. Because pore formation has been correlated with toxicity, assays monitoring this post-binding event have been used to determine functionality of Cry1 toxin receptors.

6. FUNCTIONAL CRY1 RECEPTORS

A functional Cry1 receptor should catalyse the next step in toxin action, pore formation or, better yet, confer susceptibility to cells previously immune to the effects of Cry proteins. Aminopeptidases [13, 37, 65] and cadherin-like proteins [35, 77] have been identified as Cry1A receptors in the midgut of *M. sexta* larvae. When a partially purified mixture, containing aminopeptidase N and alkaline phosphatase (and possibly other components), from *M. sexta* BBMV were incorporated into phospholipid vesicles, Cry1Ac binding was increased, and Cry1Ac pore formation enhanced [65]. This result has been used to provide evidence that the 120 kDa aminopeptidase serves as a functional receptor for Cry1Ac toxin in *M. sexta* [reviewed in 1, 13]. Because the aminopeptidase mixture was not pure, it is possible that other molecules (including the 210 kDa cadherin-like protein or BBMV glycolipids) may have catalysed binding and pore formation. Further evidence to support aminopeptidase as a functional Cry1Ac receptor has been presented [48, 69]. As above, the samples tested were partially purified extracts of BBMV containing aminopeptidase. While these examples functional studies demonstrated a correlation between Cry1 toxin binding and channel formation, the exact molecules in the complex interacting with the toxin was not determined.

Cry1Ac also forms pores in liposomes prepared from protein-free extracts of *M. sexta* BBMV lipids. Cry1Ac pore formation in *M. sexta* BBMV has been monitored using scattered light assays [3, 4, 5]. Changes in scattered light were monitored on a flow cytometer to determine if Cry1Ac pores were formed in liposomes prepared from lipids extracted from *M. sexta* BBMV. Liposomes prepared from *M. sexta* BBMV lipids that were incubated with Cry1Ac showed a different scattered light response when KCl was added, indicating the presence of a toxin-induced ion channel [23]. The differences in scattered light response observed for Cry1Ac and KCl were similar to that seen when monensin, a known Na^+ ionophore, was incubated with the liposomes prepared from *M. sexta* BBMV lipids. This result indicates that Cry1Ac forms pores in liposomes prepared from *M. sexta* lipid extracts at toxin concentrations as low as 0.01 nM [23]. Enhanced uptake of $^{86}Rb^+$ has been used to the detect the ion channel activity of Cry1Ac [14]. The Cry1Ac pores observed by flow cytometry described above were confirmed by $^{86}Rb^+$-uptake experiments. Addition of Cry1Ac (1, 5, and 10 nM) to liposomes prepared from *M. sexta* lipid extracts increased $^{86}Rb^+$-uptake (22, 25, and 31%) over the control [23]. Overall, it appears that *M. sexta* BBMV lipids support the two known functions of Cry proteins, toxin binding and ion channel formation. However, the exact role of *M. sexta* BBMV lipids in Cry1Ac toxin action is not yet determined.

7. CONCLUDING REMARKS

Glycoproteins and glycolipids are both present in insect BBMV prepared from the apical membrane of midgut epithelial cells used for *in vitro* studies of Cry1 toxin action. Many laboratories world-wide are researching the interactions of Cry1 toxins with glycoproteins present in insect BBMV. Future research goals should include studies to determine the role that BBMV glycoproteins and glycolipids may have in Cry1 toxin action at the membrane surface.

ACKNOWLEDGEMENTS

Research described in this chapter conducted in Mike Adang's laboratory was supported by grants from the United States Department of Agriculture and the National Institutes of Health.

REFERENCES

[1] Adang MJ, Paskewitz SM, Garczynski SF & Sangadala S (1995) Identification and functional characterization of the *Bacillus thuringiensis* CryIA(c) δ-endotoxin receptor in *Manduca sexta*, p. 320-329. *In* Marshall Clark J (ed.), Molecular action of insecticides on ion channels, American Chemical Society

[2] Burton SL, Ellar DJ, Li J & Derbyshire DJ (1999) N-Acetylgalactosamine on the putative insect receptor aminopeptidase N is recognised by a site on the domain III lectin-like fold of a *Bacillus thuringiensis* insecticidal toxin. J. Mol. Biol. 287, 1011-1022

[3] Carroll J & Ellar DJ (1993) An analysis of *Bacillus thuringiensis* δ-endotoxin action on insect-midgut-membrane permeability using a light-scattering assay. Eur. J. Biochem. 214, 771-778

[4] Carroll J & Ellar DJ (1997) Analysis of the large aqueous pores produced by a *Bacillus thuringiensis* protein insecticide in *Manduca sexta* midgut-brush-border-membrane vesicles. Eur. J. Biochem. 245, 797-804

[5] Carroll J, Wolfersberger MG & Ellar DJ (1997) The *Bacillus thuringiensis* Cry1Ac toxin-induced permeability change in *Manduca sexta* midgut brush border membrane vesicles proceeds by more than one mechanism. J. Cell. Sci. 110, 3099-3104

[6] Cooper MA, Carroll J, Travis ER, Williams DH & Ellar DJ (1998) *Bacillus thuringiensis* Cry1Ac toxin interaction with *Manduca sexta* aminopeptidase N in a model membrane environment. Biochem. J. 333, 677-683

[7] Cowles EA, Yunovitz H, Charles J-F & Gill SS (1995) Comparison of toxin overlay and solid-phase binding assays to identify diverse CryIA(c) toxin-binding proteins in *Heliothis virescens* midgut. Appl. Environ. Microbiol. 61, 2738-2744

[8] de Maagd RA, Bakker PL, Masson L *et al.* (1999) Domain III of the *Bacillus thuringiensis* delta-endotoxin Cry1Ac is involved in binding to *Manduca sexta* brush border membranes and to its purified aminopeptidase N. Molec. Microbiol. 31, 463-471

[9] de Maagd RA, Van der Klei H, Bakker PL, Stiekema WJ & Bosch D (1996) Different domains of *Bacillus thuringiensis* δ-endotoxins can bind to insect midgut membrane proteins on ligand blots. Appl. Environ. Microbiol. 62, 2753-2757

[10] Dennis R D, Weigandt H, Haustein D, Knowles BH & Ellar DJ (1986) Thin layer chromatography overlay technique in the analysis of the binding of the solubilized protoxin of *Bacillus thuringiensis* var. *kurstaki* to an insect glycosphingolipid of known structure. Biomed. Chromatog. 1, 31-37

[11] Denolf P, Hendrickx K, Van Damme J *et al.* (1997) Cloning and characterization of *Manduca sexta* and *Plutella xylostella* midgut aminopeptidase N enzymes related to *Bacillus thuringiensis* toxin-binding proteins. Eur. J. Biochem. 248, 748-761

[12] Denolf P, Jansens S, Peferoen M *et al.* (1993) Two different *Bacillus thuringiensis* delta-endotoxin receptors in the midgut brush border membrane of the european corn borer, *Ostrinia nubilalis* (Hubner) (Lepidoptera: Pyralidae). Appl. Environ. Microbiol. 59, 1828-1837

[13] Ellar DJ (1994) Structure and mechanism of action of *Bacillus thuringiensis* endotoxins and their receptors. Biocontrol Science Tech. 4, 445-447

[14] English L, Readdy TL & Bastian AE (1991) Delta-endotoxin-induced leakage of $^{86}Rb^{+}$-K^{+} and H_2O from phospholipid vesicles is catalyzed by reconstituted midgut membrane. Insect Biochem. 21, 177-184

[15] English L & Slatin SL (1992) Mode of action of delta-endotoxins from *Bacillus thuringiensis*: A comparison with other bacterial toxins. Insect Biochem. Molec. Biol. 22, 1-7

[16] English L, Walters F, Von Tersch MA & Slatin S (1995) Modulation of δ-endotoxin ion channels, p. 302-307. *In* Marshall Clark J (ed.), Molecular action of insecticides on ion channels, American Chemical Society

[17] Escriche B, De Decker N, Van Rie J, Jansens S & Van Kerkhove E (1998) Changes in permeability of brush border membrane vesicles from *Spodoptera littoralis* midgut induced by insecticidal crystal proteins from *Bacillus thuringiensis*. Appl. Environ. Microbiol. 64, 1563-1565

[18] Ferguson MAJ (1992) Chemical and enzymic analysis of glycosyl-phosphatidylinositol anchors, p. 191-230. *In* Hooper NM & Turner AJ (ed.), Lipid modification of proteins: A practical approach, IRL Press

[19] Ferré J, Real MD, Van Rie J, Jansens S & Peferoen M (1991) Resistance to the *Bacillus thuringiensis* bioinsecticide in a field population of *Plutella xylostella* is due to a change in a midgut membrane receptor. Proc. Natl. Acad. Sci. USA 88, 5119-5123

[20] Fishman PH, Pacuszka T & Orlandi PA (1993). Gangliosides as receptors for bacterial enterotoxins. Adv. Lipid Res. 25, 165-187

[21] Flores H, Soberon X, Sanchez J & Bravo A (1997) Isolated domain II and III from the *Bacillus thuringiensis* Cry1Ab delta-endotoxin binds to lepidopteran midgut membranes. FEBS Lett. 414, 313-318

[22] Francis BR & Bulla LA (1997) Further characterization of BT-R_1, the cadherin-like receptor for Cry1Ab toxin in the tobacco hornworm (*Manduca sexta*) midguts. Insect Biochem. Molec. Biol. 27, 541-550

[23] Garczynski SF (1999) Detection and characterization of Cry1Ac binding molecules in brush border membrane vesicles isolated from the midguts of *Manduca sexta* larvae. Doctoral Dissertation. University of Georgia, Athens

[24] Garczynski SF & Adang MJ (1995) *Bacillus thuringiensis* CryIA(c) δ-endotoxin binding aminopeptidase in the *Manduca sexta* midgut has a glycosyl-phosphatidylinositol anchor. Insect Biochem. Mol. Biol. 25, 409-415

[25] Garczynski SF, Crim JW & Adang MJ (1991) Identification of a putative insect brush border membrane-binding molecules specific to *Bacillus thuringiensis* δ-endotoxin by protein blot analysis. Appl. Environ. Microbiol. 57, 2816-2820

[26] Gill S, Cowles EA & Francis V (1995) Identification, isolation, and cloning of a *Bacillus thuringiensis* CryIAc toxin-binding protein from the midgut of the lepidopteran insect *Heliothis virescens*. J. Biol. Chem. 270, 27277-27282

[27] Haider MZ & Ellar DJ (1987) Analysis of the molecular basis of insecticidal specificity of *Bacillus thuringiensis* crystal delta-endotoxin. Biochem J. 248, 197-201

[28] Haider MZ & Ellar DJ (1989) Mechanism of action of *Bacillus thuringiensis* δ-endotoxin: interaction with phospholipid vesicles. Biochim. Biophys. Acta 978, 216-222

[29] Hendrickx K, De Loof A & Van Mellaert H (1990) Effects of *Bacillus thuringiensis* delta-endotoxin on the permeability of brush border membrane vesicles from tobacco hornworm (*Manduca sexta*) midgut. Comp. Biochem. Physiol. 95C, 241-245

[30] Hofmann C, Luthy P, Hutter R & Pliska V (1988b) Binding of the delta-endotoxin from *Bacillus thuringiensis* to brush-border membrane vesicles of the cabbage butterfly (*Pieris brassicae*). Eur. J. Biochem. 173, 85-91

[31] Hofmann C, Vanderbruggen H, Höfte H *et al.* (1988) Specificity of *Bacillus thuringiensis* delta-endotoxins is correlated with the presence of high-affinity binding sites in the brush border membrane of target insect midguts. Proc. Natl. Acad. Sci. USA 85, 7844-7848

[32] Hua G, Tsukamoto K, Rasilo ML & Ikezawa H (1998) Molecular cloning of a GPI-anchored aminopeptidase N from *Bombyx mori* midgut: a putative receptor for *Bacillus thuringiensis* Cry1Aa toxin. Gene. 214, 177-185

[33] Indrasith LS, Hori H (1992) Isolation and partial characterization of binding proteins for immobilized delta endotoxin from solubilized brush border membrane vesicles of the silkworm, *Bombyx mori*, and the common cutworm, *Spodoptera litura*. Comp. Biochem. Physiol. 102B, 605-610

[34] Karlsson K-A, Angstrom J & Teneberg S (1991) On the characteristics of carbohydrate receptors for bacterial toxins: aspects of the analysis of binding epitopes, p. 435-444. *In* Alouf JE & Freer JH (ed.), Sourcebook of bacterial protein toxins, Academic Press

[35] Keeton TP, Bulla LA (1997) Ligand specificity and affinity of Bt-R$_1$, the *Bacillus thuringiensis* Cry1A toxin receptor from *Manduca sexta*, expressed in mammalian and insect cell cultures. Appl. Environ. Microbiol. 63, 3419-3425

[36] Keeton TP, Francis BR, Maaty WSA & Bulla LA (1998) Effects of midgut-protein-preparative and ligand binding procedures on the toxin binding characteristics of BT-R$_1$, a common high-affinity receptor in *Manduca sexta* for Cry1A *Bacillus thuringiensis* toxins. Appl. Environ. Microbiol. 63, 2158-2165

[37] Knight, PJK, Crickmore N & Ellar DJ (1994) The receptor for *Bacillus thuringiensis* CryIA(c) δ-endotoxin in the brush border membrane is aminopeptidase N. Molec. Microbiol. 11, 429-436

[38] Knight PJK, Knowles BH & Ellar DJ (1995) Molecular cloning of an insect aminopeptidase N that serves as a receptor for *Bacillus thuringiensis* CryIA(c) toxin. J. Biol. Chem. 270, 17765-17770

[39] Knowles BH (1994) Mechanism of action of *Bacillus thuringiensis* insecticidal δ-endotoxins. Adv. Insect Physiol. 24, 275-308

[40] Knowles BH & Dow JAT (1993) The crystal delta-endotoxins of *Bacillus thuringiensis*: models for their mechanism of action on the insect gut. Bioessays 15, 469-476

[41] Knowles BH & Ellar DJ (1986) Characterization and partial purification of a plasma membrane receptor for *Bacillus thuringiensis* var. *kurstaki* lepidopteran-specific δ-endotoxin. J. Cell Sci. 83, 89-101

[42] Knowles BH & Ellar DJ (1987) Colloid-osmotic lysis is a general feature of the mechanism of action of *Bacillus thuringiensis* delta-endotoxins with different insect specificities. Biochim. Biophys. Acta 924, 509-518

[43] Knowles BH, Knight PJ & Ellar DJ (1991) N-Acetylgalactosamine is part of the receptor in insect gut epithelia that recognizes an insecticidal protein from *Bacillus thuringiensis*. Proc. R. Soc. Lond. B 245, 31-35

[44] Knowles BH, Thomas WE & Ellar DJ (1984) Lectin-like binding of *Bacillus thuringiensis* var. *kurstaki* lepidopteran-specific toxin is an initial step in insecticidal action. FEBS Lett. 168, 197-202

[45] Lee MK, Rajamohan F, Gould F & Dean DH (1995) Resistance to Cry1A δ-endotoxins in a laboratory-selected *Heliothis virescens* strain is related to receptor alteration. Appl. Environ. Microbiol. 61, 3836-3842

[46] Lee MK, You TH, Young BA *et al.* (1996) Aminopeptidase N purified from gypsy moth brush border membrane vesicles is a specific receptor for *Bacillus thuringiensis* Cry IAc toxin. Appl. Environ. Microbiol. 62, 2845-2849

[47] Lingwood CA (1993) Verotoxins and their glycolipid receptors. Adv. Lip. Res. 25, 189-211

[48] Lorence A, Darszon A & Bravo A (1997) Aminopeptidase dependent pore formation of *Bacillus thuringiensis* Cry1Ac toxin on *Trichoplusia ni* membranes. FEBS Lett. 414, 303-307

[49] Lu Y-J & Adang MJ (1996) Conversion of *Bacillus thuringiensis* CryIAc-binding aminopeptidase to a soluble form by endogenous phosphatidylinositol phospholipase C. Insect Biochem. Molec. Biol. 26, 33-40

[50] Luo K, Banks D & Adang MJ (1999) Toxicity, binding, and permeability analyses of four *Bacillus thuringiensis* Cry1 δ-endotoxins using brush border membrane vesicles of *Spodoptera exigua* and *Spodoptera frugiperda*. Appl. Environ. Microbiol. 65, 457-464

[51] Luo K, Lu Y-j & Adang MJ (1996) A 106-kDa form of aminopeptidase is a receptor for *Bacillus thuringiensis* Cry1C δ-endotoxin in the brush border membrane of *Manduca sexta*. Insect Biochem. Molec. Biol. 26, 33-40

[52] Luo K, Sangadala S, Masson L *et al.* (1997) The *Heliothis virescens* 170 kDa aminopeptidase functions as "receptor A" by mediating specific *Bacillus thuringiensis* δ-endotoxin binding and pore formation. Insect Biochem. Molec. Biol. 27, 735-743

[53] Luo K, Tabashnik BE & Adang MJ (1997) Binding of *Bacillus thuringiensis* Cry1Ac toxin to aminopeptidase in susceptible and resistant diamondback moths (*Plutella xylostella*). Appl. Environ. Microbiol. 63, 1024-1027

[54] Magnani JF, Smith DF & Ginsburg V (1980) Detection of gangliosides that bind cholera toxin: direct binding of [125]-I-labeled toxin to thin-layer chromatograms. Anal. Biochem. 109, 399-402

[55] Martin FG & Wolfersberger MG (1995) *Bacillus thuringiensis* δ-endotoxin and larval *Manduca sexta* midgut brush-border membrane vesicles act synergistically to cause very large increases in the conductance of planar lipid bilayers. J. Exp.Biol. 198, 91-96

[56] Martinez-Ramirez AC, Gonzalez-Nebauer S, Escriche B & Real MD (1994) Ligand blot identification of a *Manduca sexta* midgut binding protein specific to three *Bacillus thuringiensis* CryIA-type ICPs. Biochem. Biophys. Res. Comm. 201, 782-787

[57] Masson L, Lu Y-J, Mazza A, Brousseau R & Adang MJ (1995) The CryIA(c) receptor purified from *Manduca sexta* displays multiple specificities. J. Biol. Chem. 270, 20309-20315

[58] Nagamatsu Y, Toda S, Koike T *et al.* (1998) Cloning, sequencing, and expression of the *Bombyx mori* receptor for *Bacillus thuringiensis* insecticidal CryIA(a) toxin. Biosci. Biotechnol. Biochem. 62, 727-734

[59] Nagamatsu Y, Toda S, Yamaguchi F *et al.* (1998) Identification of *Bombyx mori* midgut receptor for *Bacillus thuringiensis* insecticidal CryIA(a) toxin. Biosci. Biotechnol. Biochem. 62, 718-726

[60] Oddou P, Hartmann H & Geiser M (1991) Identification and characterization of *Heliothis virescens* midgut membrane proteins binding *Bacillus thuringiensis* δ-endotoxin. Eur. J. Biochem. 202, 673-680

[61] Oddou P, Hartmann H, Radecke F & Geiser M (1993) Immunologically unrelated *Heliothis* sp. and *Spodoptera* sp. midgut membrane-proteins bind *Bacillus thuringiensis* Cry1A(b) δ-endotoxin. Eur. J. Biochem. 212, 145-150

[62] Rajamohan F, Alcantara E, Lee MK *et al.* (1995) Single amino acid changes in domain II of *Bacillus thuringiensis* CryIAb δ-endotoxin affect irreversible binding to *Manduca sexta* midgut membrane vesicles. J. Bacteriol. 177, 2276-2282

[63] Rajamohan F, Lee MK & Dean DH (1998) *Bacillus thuringiensis* insecticidal proteins: molecular mode of action. Prog. Nucleic Acids Mol. Biol. 60, 1-27

[64] Sanchis V & Ellar DJ (1993) Identification and partial purification of a *Bacillus thuringiensis* CryIC δ-endotoxin binding protein from *Spodoptera litoralis* gut membranes. FEBS Letters 3, 264-268

[65] Sangadala S, Walters F, English LH & Adang MJ (1994) A mixture of *Manduca sexta* aminopeptidase and alkaline phosphatase enhances *Bacillus thuringiensis* insecticidal CryIA(c) toxin binding and $^{86}Rb^+$-K^+ leakage *in vitro*. J. Biol. Chem. 269, 10088-10092

[66] Schnepf E, Crickmore N, Van Rie J *et al.* (1998) *Bacillus thuringiensis* and its pesticidal crystal proteins. Microbiol. Mol. Biol. Rev. 62, 775-806

[67] Schwartz J, Garneau L, Masson L & Brousseau R (1991) Early response of cultured lepidopteran cells to exposure to δ-endotoxin from *Bacillus thuringiensis*: involvement of calcium and anionic channels. Biochim. Biophys. Acta. 1065, 250-260

[68] Schwartz JL, Garneau L, Savaria D *et al.* (1993) Lepidopteran-specific crystal toxins from *Bacillus thuringiensis* form cation- and anion-selective channels in planar lipid bilayers. J. Membr. Biol. 132, 53-62

[69] Schwartz JL, Lu Y-J, Sohnlein P *et al.* (1997) Ion channels formed in planar lipid bilayers by *Bacillus thuringiensis* toxins in the presence of *Manduca sexta* midgut receptors. FEBS Lett. 412, 270-276

[70] Slatin SL, Abrams CK & English L (1990) Delta-endotoxins form cation-selective channels in planar lipid bilayers. Biochem. Biophys. Res. Commun. 169, 765-772

[71] Soutar AK, Wade DP (1989) Ligand blotting, p. 55-76. *In* (Creighton TE, ed.), Protein function: a practical approach, IRL Press, Oxford

[72] Tabashnik BE, Finson N, Groeters FR *et al.* (1994) Reversal of resistance to *Bacillus thuringiensis* in *Plutella xylostella*. Proc. Natl. Acad. Sci. USA 91, 4120-4124

[73] Takesue, YK, Yokota K, Miyajima S, Taguchi R & Ikezawa H (1989) Membrane anchors of alkaline phosphatase and trehalase associated with the plasma membrane of larval midgut epithelial cells of the silkworm, *Bombyx mori*. J. Biochem. *105*, 998-1001

[74] Takesue S, Yokota K, Miyajima S *et al.* (1992) Partial release of aminopeptidase N from larval midgut cell membranes of the silkworm, *Bombyx mori*, by phosphatidylinositol-specific phospholipase C. Comp. Biochem. Physiol. 102B, 7-11

[75] Towbin H, Staehelin T & Gordon J (1979) Electrophoretic transfer of proteins from polyacrylamide gels to nitrocelluslose sheets: procedure and some applications. Proc. Natl. Acad. Sci. USA. 76, 4350-4354

[76] Vadlamudi RK, Ji TH & Bulla LA (1993) A specific binding protein from *Manduca sexta* for the insecticidal toxin of *Bacillus thuringiensis* subsp. *berliner*. J. Biol. Chem. 268, 12334-12340

[77] Vadlamudi R, Weber KE, Ji I, Ji TH & Bulla LA (1995) Cloning and expression of a receptor for an insecticidal toxin of *Bacillus thuringiensis*. J. Biol. Chem. 270, 5490-5494

[78] Valaitis AP, Lee MK, Rajamohan F & Dean DH (1995) Brush border membrane aminopeptidase-N in the midgut of the gypsy moth serves as the receptor for the CryIA(c) δ-endotoxin of *Bacillus thuringiensis*. Insect Biochem. Molec. Biol. 25, 1143-1151

[79] Valaitis AP, Mazza A, Brousseau R & Masson L (1997) Interaction analyses of *Bacillus thuringiensis* Cry1A toxins with two aminopeptidases from gypsy moth midgut brush border membranes. Insect Biochem. Molec. Biol. 27, 529-539

[80] Van Rie J, Jansens S, Höfte H, Degheele D & Van Mellaert H (1989) Specificity of *Bacillus thuringiensis* delta-endotoxins. Eur. J. Biochem. 186, 239-247

[81] Van Rie J, Jansens S, Höfte H, Degheele D & Van Mellaert H (1990) Receptors on the brush border membrane of the insect midgut as determinants of the specificity of *Bacillus thuringiensis* delta-endotoxins Appl. Environ. Microbiol. 56, 1378-1385

[82] Weitzhandler M, Kadlecek D, Avdalovic N *et al.* (1993) Monosaccharide and oligosaccharide analysis of proteins transferred to polyvinylidene fluoride membranes after sodium dodecyl sulfate polyacrylamide gel electrophoresis. J. Biol. Chem. 268, 5121-5130

[83] Wolfersberger MG (1990) The toxicity of two *Bacillus thuringiensis* δ-endotoxins to gypsy moth larvae is inversely related to the affinity of binding sites on midgut brush border membranes for the toxins. Experientia 46, 475-477

[84] Wolfersberger MG (1995) Permeability of *Bacillus thuringiensis* CryI toxin channels, In "Molecular action of insecticides on ion channels", J. Marshall Clark, ed. (Washington, D.C: American Chemical Society). pp. 294-301

[85] Wolfersberger MG, Luthy P, Maurer A *et al.* (1987) Preparation and partial characterisation of amino acid transporting brush border membrane vesicles from the larval midgut of the cabbage butterfly (*Pieris brassicae*). Comp. Biochem. Physiol. 86A, 301-308

[86] Woodward MR, Young WW & Bloodgood RA (1985) Detection of monoclonal antibodies specific for carbohydrate epitopes using periodate oxidation. J. Immunol. Methods 78, 143-153

[87] Yaoi K, Kadotani T, Kuwana H *et al.* (1997) Aminopeptidase N from *Bombyx mori* as a candidate for the receptor of *Bacillus thuringiensis* Cry1Aa toxin. Eur. J. Biochem. 246, 652-657

Chapter 3.3

Membrane permeabilisation by *Bacillus thuringiensis* toxins: protein insertion and pore formation

Jean-Louis Schwartz[1,2] & Raynald Laprade[2]

[1]*Biotechnology Research Institute, National Research Council of Canada, Montreal, and* [2]*Groupe de recherche en transport membranaire, Université de Montréal, Montreal, Quebec, Canada*

Key words: Ion channel, pore, bacterial toxin, *Bacillus thuringiensis*, lipid membrane, oligomerisation, structure-function, membrane permeabilisation

Abstract: Following recognition of specific binding sites at the surface of midgut target cell membranes, activated *Bacillus thuringiensis* (*Bt*) toxins act mainly by permeabilising the cells and disrupting vital ion and metabolite homeostasis. This chapter reviews our present knowledge on the various steps involved in the permeabilisation process at both structural and functional levels, and describes several biophysical and physiological approaches used for the study of membrane permeabilisation by *Bt* toxins.

1.   INTRODUCTION

The development of novel insecticides with high activity against important agricultural and forestry pests is critical to better protect our fibre and food supply while preserving the environment. The threat of insects becoming resistant to currently used products is intensifying the demand for alternative products with improved characteristics to strengthen the battery of available tools for integrated pest management. Among these, *Bacillus thuringiensis* (*Bt*) continues to represent the leading environment-friendly pest control agent. It has been used for several decades for the biological control of pest insects in forestry, agriculture and of human and animal disease vectors.

J.-F. Charles et al. (eds.),
Entomopathogenic Bacteria: From Laboratory to Field Application, 199–217.
© *2000 Kluwer Academic Publishers. Printed in the Netherlands.*

Bt toxins are powerful gut poisons which display a high level of specificity against a variety of lepidopteran, dipteran and coleopteran insects and several other invertebrates [75]. Present directions for *Bt* research and development focus on the construction of novel genes and engineered hosts to improve pest organism specificity. The development of the future generations of *Bt* products based on engineered microbial or plant constructs necessarily relies on a clear understanding of the toxin's mode of action at the molecular level. In the last decade, based on structural data on *Bt* protein, one of the few bacterial toxins with a known structure at atomic resolution [38, 57], several hypotheses on the functional role of various regions of *Bt* proteins were proposed. They were tested on whole toxins, toxin fragments, and site-directed mutated or hybrid constructs using a wide array of functional assays ranging from bioassays on live insects to experiments on natural membrane fractions and synthetic phospholipid membranes.

The purpose of this chapter is to review the current knowledge about the crucial steps of membrane insertion and pore formation by *Bt* toxins, to provide information on the techniques currently used and to discuss some of the related critical issues within the more general context of the mode of action of bacterial toxins. Other important topics in relation to the intoxicating mechanisms of *Bt* (reviewed in [30, 68, 73, 75]) have been deliberately left out although relevant results will be mentioned when appropriate. They include ion and aminoacid transport in the epithelial cells lining the gut lumen, toxin crystal processing in the gut, receptors, synergy or antagonism, and resistance. Moreover, this review does not cover studies that specifically deal with *Bt* spores (reviewed in [75]), Cyt toxins [56] or other *Bt* toxins like Vip [28].

2.   TOXICITY AT THE MOLECULAR LEVEL

Bt crystal ingestion results in insect death within several hours. The macroscopic mode of action involves a cascade of events which has been best described for Cry proteins that are primarily active against lepidopteran pests [48, 49, 68]. The *Bt* inclusion bodies contain the biologically inactive protoxins (~130 kDa) which are solubilised in the highly alkaline environment of the midgut lumen and cleaved by gut proteinases to 60-65 kDa activated proteins. The resulting toxins recognise specific sites located on the microvilli of the columnar cell brush border membrane. The next steps include pore formation in the membrane, ion and metabolite transport disruption, and cell lysis ultimately leading to insect death.

Table 3. The four steps of membrane insertion and channel/pore formation.

	Approach	Docking	Partition	Permeation
#1	T		Adsorption T	T^* or T_n^*
#2	Aggregation $T \Leftrightarrow T_n$		Adsorption T_n	T_n^*
#3	T	Binding/docking $T + R \Leftrightarrow TR$	Insertion T or TR	T^* or $(TR)^*$
#4	T	Binding/docking $T + R \Leftrightarrow TR$	Insertion $\Leftrightarrow$ Oligomerisation $T \Leftrightarrow T_n$	T_n^*
#5	T	Binding/docking $\Leftrightarrow$ Oligomerisation $T + R \Leftrightarrow TR \Leftrightarrow (TR)_n$	Insertion T_n or $(TR)_n$ or $T_n R$	T_n^* or $(TR)_n^*$ or $T_n^* R$
#6	Aggregation $T \Leftrightarrow T_n$	Binding/docking $T_n + R \Leftrightarrow T_n R$	Insertion T_n or $T_n R$	T_n^* or $(T_n R)^*$

T, monomer of toxin; R, receptor or docking molecule; subscripts indicate order of oligomerisation; pore constituents in their final conformation are marked with asterisks.

At the molecular level, the mode of action of *Bt* Cry toxins is not fully elucidated. Based on the fact that target cell death is mainly due to pore formation in the plasma membrane [48], and by analogy with the mechanism of action of other pore-forming bacterial toxins [36, 55], a number of kinetic schemes can be envisaged as a conceptual framework for the study of the molecular mechanisms responsible for *Bt* intoxication (Table 1). In principle, any of these schemes could take place depending on conditions prevailing near the cell membrane, on availability of high affinity binding molecules at the cell surface and on the physical and chemical properties of the target membrane itself. Moreover, interaction between toxin molecules and target cells may trigger intracellular signals which could either interfere with or promote toxin insertion and pore formation.

In Table 1 the toxin interacts either directly with the membrane as a monomer (scheme #1) or as a preformed aggregate/oligomer (scheme #2) or, alternatively, with a receptor/docking molecule, either alone (schemes #3, 4 and 5) or as a preformed aggregate/oligomer (scheme #6). In scheme #5, the receptor induces oligomerisation. The toxin then partitions into the membrane as a single molecule (schemes #1 and 3), as an oligomer made of several toxin molecules (schemes #2, 4, 5 and 6) or as a complex made of one (scheme #3) or more toxins and/or receptors (scheme #5 and 6). The functional permeating structure is made of a single toxin molecule (schemes #1 and 3) or, more likely, of several subunits (schemes #1, 2, 4, 5 and 6), possibly with some regions of the receptor itself (schemes #3, 5 and 6).

Bt toxin channel formation in receptor-free liposomes and planar lipid bilayers (PLBs) may be described by the simple mechanisms of schemes #1 and 2, while schemes #3, 4, 5 and 6 are more representative of the pore formation process in the insect gut or in midgut brush border membrane vesicles (BBMVs). The number of subunits of the oligomeric structures may increase with time, and insertion and oligomerisation could proceed in parallel, resulting in channels and pores of increasing size gradually becoming permeable to larger ions and molecules.

Table 4. Biophysical techniques for the determination of membrane permeabilisation by *Bt* Cry proteins[a]

	Electrophysiology					Swelling assays		Fluorescence assays			
	PLB	μE	PC	SC	TEP	LS	VM	FV_m	FP	FI	CV
Phospholipid membranes[b]	*[c],[A][d]										
BLMs											
Liposomes	*		*			[G]		[J]			
BBMVs	*, [B]		*			*, [H]		*, [K]	[L]		
Midguts Whole midguts		*, [C]	*	[E]	[F]					*	*
Epithelial sheaths		*	*							*	*, [O]
Cells			*				*			*, [M]	*, [P]
Target insects Cell lines		*, [D]					*, [I]			*, [N]	*, [Q]

[a] Abbreviations: *PLB*: planar lipid bilayers; *μE*: microelectrode; *PC*: patch-clamp; *SC*: short-circuit current; *TEP*: transepithelial voltage; *LS*: light-scattering assay; *VM*: videomicroscopy; *FVm*: membrane voltage; *FP*: fluorescent probe efflux; *FI*: intracellular pH, calcium, potassium or other ions; *CV*: cell viability, *BLMs*: black lipid membranes.
[b] Midgut membrane material like native lipids, native proteins or BBMVs may be reconstituted in BLMs or liposomes.
[c] Asterisks indicate techniques currently used in the authors' laboratories.
[d] Letters in brackets refer to the following references: *A*: [19, 26, 32, 38, 63, 70, 78-84, 91, 93]; *B* : [60, 62]; *C*: [67]; *D*: [65, 77]; *E*: [16, 17, 44, 52, 53, 58]; *F*: [54, 95]; *G*: [25, 43]; *H*: [13-15, 18, 27, 61, 96]; *I*: [72, 90]; *J*: [10, 34, 35]; *K*: [54, 59, 60, 87]; *L*: [14]; *M*: Fig. 3 herein; *N*: [23, 24, 41, 42, 65, 70, 72, 77, 88]; *O*: [8]; *P*: [5]; *Q*: [5, 64].

3. APPROACHES AND TECHNIQUES

Table 2, which provides relevant references, summarises the techniques that are currently available to extract, at increasing levels of biological complexity, steady-state and kinetic information on some of the above steps of *Bt* channel/pore formation, as well as some physical features of the permeabilising structures. Membrane permeabilisation by pore-forming toxins can be assessed by measuring the increased flux of solutes across the cell membrane. Since these fluxes take place via channels or pore structures made by the toxin in the membrane, the magnitude of the flux will depend on the number and the biophysical properties of these permeabilising structures. For each type of structure, the overall flux F_s across the cell membrane is given by: $F_s = NP_o f_s$, where N is the number of identical structures, P_o their probability of being open, and f_s the solute flux through each individual structure under a given electrochemical gradient. If several different types of structures are present, the resulting flux will be the sum of the fluxes generated by each individual structure type.

If the solute is a salt, ionic fluxes can be measured through the electric current they produce, using very sensitive electrophysiological techniques [74]. In the simplest system (PLB, Table 2), planar lipid bilayers are made with purified native lipids or synthetic lipids. Toxins are then usually incorporated in the bilayer via the aqueous phase bathing the membrane and electric potential differences are applied across the membrane. The analysis of the resulting current flowing through individual ion channels provides quantitative information on the channel biophysical properties, i.e. its conductance, selectivity, voltage-dependence and gating, thus providing the characteristic signature of each toxin in a receptor-free membrane environment (see Table 1, schemes #1 and 2). With this system, *Bt* Cry toxins in the 20-50 nM concentration range have been shown to readily insert and form ionic channels in phospholipid membranes. Indeed, all wild-type and modified *Bt* toxins that were tested in PLBs, including those which did not show insecticidal activity *in vivo*, induced channel activity [71]. The PLB technique can also be used for determining the channel properties of toxins in the presence of receptors by reconstituting purified receptor proteins in the membrane or by fusing target cell BBMVs with the bilayer.

Formation of ion channels by toxins and their properties in target cells can be studied with the patch-clamp technique (PC, Table 2) which mechanically and electrically isolates a microscopic portion of cellular membrane at the tip of a glass micropipette. The approach can be used on dissociated or cultured cells or cells from intact epithelia (as in cut-open midguts or isolated epithelial sheaths). Furthermore, the technique allows to investigate the effects of toxin exposure on endogenous channels. When target cells do not lend themselves to the patch-clamp technique, BBMV fusion with PLBs can be used as an alternative approach. Cellular metabolism regulation, of course, is lost under these conditions.

Alternatively, the overall permeabilising efficacy of toxins can be determined by their inhibitory effect on intracellular electrical potential, using the microelectrode technique (μE, Table 2) in which epithelial cells of isolated midguts are impaled with extremely fine capillary pipettes. The electric current mediated by toxin-formed ionic channels interferes with that of endogenous cellular ion channels and pumps, thereby reducing or, at high doses, totally abolishing the membrane potential and all ion gradients and currents. With this technique, the potency of different toxins at a given dose may be compared by measuring the time required to cause a 50% depolarisation of the apical midgut membrane. Alternatively, short-circuit current (SC, Table 2) or transepithelial potential (TEP, Table 2) measurements can be used on the isolated midgut as the measurable transducing parameters for the toxin effect. Likewise, these approaches provide a direct estimate of the rate at which the toxin forms functional permeabilising structures in the apical membrane of the midgut.

Movement of uncharged solutes like sugars and amino acids and size of the pores are best determined with swelling assays in which toxin-mediated increased influx of solute into cells, membrane vesicles or liposomes translates into an increased rate of swelling caused by the associated entry of water. Volume changes of isolated cells or BBMVs can be measured with the light-scattering technique (LS, Table 2) in which scattered light intensity is inversely related to cellular or vesicular size. Alternatively, swelling of isolated cells can be monitored by video microscopy (VM, Table 2). In general, BBMVs are preincubated with toxin in a low osmolarity solution and are then rapidly mixed with the solute solution. The permeabilising effect of the toxin is deduced from the increased rate of volume recovery following initial shrinking, as compared to that observed in the absence of toxin (Fig. 1, *left panel*). Using the rate of volume recovery as the measuring parameter for solute entry, the relative steady-state macroscopic permeability generated by different toxins or by the same toxin at different doses, can easily be determined. Alternatively, the rate of pore formation can be estimated from that of volume recovery following rapid and simultaneous mixing of BBMVs with toxin and solute (Fig. 1, *right panel*). Moreover, using

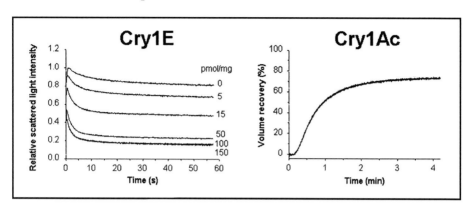

Figure 19. Manduca sexta midgut BBMV light scattering assay.

solutes of increasing size, different charge or both, the technique provides an estimate of the maximum size of channels or pores as well as their overall cation or anion selectivity. Finally, comparison of the volume recovery rates for different toxins using solutes of various molecular size allows to determine whether the permeabilising potency differences of the toxins are due to differences in pore number, size or selectivity.

Toxin-mediated fluxes can also be measured with various fluorescent probes by monitoring the movement of fluorescent molecules through the permeabilised membrane (FP, Table 2). Alternatively, fluorescent indicators can be used to measure changes in membrane voltage (FV_m, Table 2) or in the concentration of extracellular or intracellular ions like H^+, Na^+, Cl^-, K^+ and Ca^{2+} (FI, Table 2) in response to pore formation in cells or vesicle membranes. These techniques, applied either to cell and vesicle populations using cuvette fluoroscopy, or to single cells and small cell clusters using microspectrofluorescence, video-imaging or confocal microscopy, provide critical data on the macroscopic permeability, ion selectivity and formation rate of toxin pores [40]. Furthermore, they permit the monitoring of intracellular signalling related to the pathological state of the cells. Finally, the effects of toxins on nucleic membrane integrity and thus, cell viability (CV, Table 2), can be assessed by various fluorescent probes that specifically bind to DNA.

Using the methods described above, a wealth of biophysical information has been obtained. For example, *Bt* toxins form permeating structures in PLBs which clearly fall into the general class of ion channels [74]. At alkaline pH and under symmetrical 150 mM KCl conditions, they display conductances ranging from 10 to about 500 pS, low voltage-dependence (with some exceptions [26, 82]), cation selectivity and slow kinetics gating. In contrast, swelling assays demonstrate that molecules as large as sucrose can permeate toxin-mediated structures, suggesting that in cells or BBMV membranes, pore-like structures are actually formed. Swelling assays also show that functional pore formation takes place within seconds after toxin molecules are brought in the vicinity of the membrane.

4.   MICROENVIRONMENT OF TARGET CELLS

Several of the methods described in the preceding section are particularly suited to study the pore formation process that takes place when toxin molecules are located close to the cell membrane. In the insect midgut, several factors may influence the concentration and the physical state of toxin molecules. In particular, once the *Bt* protoxin has been processed in the insect gut, i.e. after solubilisation and proteolytic cleavage into the toxic moiety, the toxin molecule has to cross the peritrophic matrix (PM). The PM acts as a molecular sieve whose permeability is modulated by pH, ionic strength and calcium ions [85]. It has been shown that *Bt* toxins can bind to the PM [9]. Toxin molecules in the ectoperitrophic space, i.e. between the peritrophic matrix and the apical side of the midgut cells, may also get involved in a variety of subsequent events. They may experience aggregation or oligomerisation (Table 1, schemes #2 and 6). Their concentration may increase. They may dock to unspecified molecules in the neutral or weakly anionic fuzzy environment (glycocalix) of the columnar cell microvilli [51]. It cannot be excluded that the proteins find their way between gut cells and access targets located on the basal side of the gut (D. Baines, personal communication). What will actually happen will largely depend on the microenvironment found in the ectoperitrophic space [37]. In particular, aggregation or oligomerisation in solution are controlled by pH and ionic strength [29, 39]. Because of the lack of detailed information about the precise physical state of the ectoperitrophic region, which may be gel-like rather than an aqueous environment [85], the rules governing several properties of proteins in solution may not be directly applicable.

5.   ROLE OF THE RECEPTOR

Significant progress has recently been made in the area of toxin recognition molecules. *Bt* toxin molecules bind to midgut brush border membrane proteins that act as receptors and that are members of the aminopeptidase N or cadherin B families (see Chapter 3.2, and [75] for a review). Similar to what is observed with other bacterial toxins that also bind to specialised receptors at the cell surface [36], it is believed that receptor binding increases the concentration of toxin molecules [55]. This would explain the much lower doses necessary for *in vivo* intoxication compared to those required for *in vitro* activity in cell lines or synthetic membranes [76]. Indeed, when the 120 kDa aminopeptidase N protein, which acts as a receptor for Cry1Ac in *Manduca sexta*, was reconstituted in PLBs, a 500-fold lower concentration of Cry1Ac toxin was able to generate the same level of channel activity [80]. The presence of the receptor also affected the current-voltage relationship of the channels. It was no longer linear as is that of Cry1Ac alone, thus demonstrating that under these conditions, the conductance of Cry1Ac channels depends on their polarisation by membrane voltage. This suggests that interaction of Cry1Ac with its receptor induces structural modifications of the toxin channels (Table 1, schemes #3, 5 and 6). Similar observations on increased liposome permeability [25] and channel property changes were made with BBMVs fused to PLBs [60, 62].

It has been proposed that, similar to what happens in PLBs, receptor molecules play a role in lowering the energy barrier for the conformational changes needed for toxin insertion into target cell membranes [55], that they induce oligomerisation [2] (Table 1, schemes #4 and 5) and that they are included in the final architecture of the pore [62, 80] (Table 1, scheme #3, 5 and 6). Receptors may also be involved in secondary cellular responses responsible for promoting further toxin insertion and pore formation [64, 70] or alternatively, for affecting normal cell physiology either by activating defence mechanisms or by accelerating cell death.

6. CONFORMATIONAL CHANGES AND PORE STRUCTURE

The three-dimensional structures of three Cry proteins have recently been elucidated using X-ray diffraction: Cry3Aa, a coleopteran-specific toxin [57], Cry1Aa, a lepidopteran-specific toxin [38] and Cry2Aa, a protein which is active against both dipteran and lepidopteran pests [66]. There are three structural domains in the activated protein. The first domain is made of seven to eight α-helices, whereas domains 2 and 3 are made of β-sheets. Based on several mutagenesis studies (reviewed in [38, 75]) and on toxin structural data, it has been proposed that the α-helix-rich domain 1 is responsible for pore formation in PLBs and cell membranes. Both Cry3A toxin treated with Colorado potato beetle gut juice, cysteine or serine proteinases and a 39 kDa fragment produced by hydroxylamine cleavage of the toxin and corresponding mainly to its first domain, form channels of similar conductances in PLBs (Fig. 2) [31]. This data is consistent with earlier reports on N-terminal fragments of Cry1Ac [93] and Cry3Ba1 (formerly known as CryIIIB2) [91], supporting the concept of domain 1 being the pore-forming region in *Bt* toxins.

Domain 2 of Cry toxins is thought to be involved in binding site recognition and toxin specificity. The role of domain 3 is unclear. It may be involved in the modulation of specificity and permeabilisation, as demonstrated by recent studies showing that host range modification [20, 21] and augmented toxicity [72] can be achieved by domain swapping between different toxins, and that BBMV permeability [96] and PLB ionic currents [82] are affected by mutations in a highly conserved region of domain 3.

One of the most challenging aspects of the molecular mode of action of *Bt* toxins relates to the elucidation of how this water-soluble protein becomes a membrane protein and what is the final architecture of the lesion it makes in the target cell membrane. Like several other helix-rich bacterial toxins in solution [36, 50, 55], domain 1 of *Bt* toxins assumes an inside-out membrane protein structure which shelters the most hydrophobic regions by a barrel of more amphiphatic ones. Major conformational changes will be needed to trigger insertion of significant stretches of the molecule in the membrane [11]. Furthermore, as observed in pore-forming bacterial cytolysins, oligomerisation may be required to provide the energy to drive the membrane-spanning regions into the lipid bilayer [6].

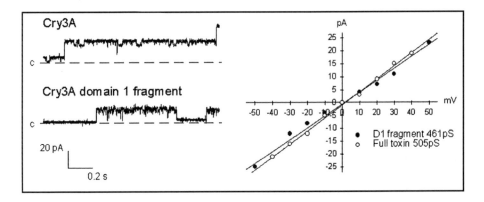

Figure 20. Channels formed by Cry3A toxin and Cry3A domain 1 fragment in PLBs.

Various models have been proposed that predict conformational changes in *Bt* toxins [33, 48]. Trypsin-activated Cry1Aa, which has no cysteine residue [38], was stabilised by means of strategically located disulfide bridges by making use of cysteines introduced by mutagenesis. Rupture of the solvent-exposed disulfide bonds by a reducing agent was examined in PLB experiments [79, 81]. It was shown that the ability of domain 1 to swing away from the rest of the toxin molecule was a necessary step for channel formation [79]. In the absence of receptor proteins, high pH and negative surface charges in the microenvironment of the lipid membrane may be responsible for the transition to the molten globule state [10, 29, 55, 89]. Interhelical proteolytic nicking observed in Cry3A toxin may also increase toxin flexibility [12]. Finally, in target cell membranes, binding to a receptor and possibly aggregation with nearby toxins may cause the protein to unfold, to oligomerise, or both [2, 55, 69, 93].

The insertion process appears to be initiated by hairpin $\alpha4/\alpha5$. Site-specific mutations in the helices of domain 1 of Cry toxins provided indirect evidence for an important role of helices 4 and 5 in toxicity [3, 86, 97]. PLB experiments on a set of double cysteine mutants designed for increased stability of various hairpins in domain 1 of Cry1Aa showed that flexibility of the $\alpha4/\alpha5$ hairpin was critical for membrane permeation [79, 81]. Data obtained with the entire activated toxin molecule confirmed observations made with synthetic $\alpha4$ and $\alpha5$ peptides [19, 35] and molecular dynamic modeling predictions [7]. Recent spectroscopic experiments on the interaction between phospholipid and synthetic Cry3A helices have provided further support to the $\alpha4/\alpha5$ hairpin model [33].

Helix 5 of the $\alpha4/\alpha5$ hairpin is highly conserved throughout the *Bt* family of Cry toxins. It is also the most hydrophobic helix [38, 57], making it a plausible candidate for being fully buried in the lipid membrane. On the other hand, helix 4 is more amphiphilic than $\alpha5$ and highly variable among Cry toxins. It is therefore quite possible, for both structural and functional reasons, that one side of helix 4 is facing the water-filled lumen of the channel. Such an arrangement of membrane-spanning hairpins from identical subunits is found in one general class of K^+ channels [22]. The topology of the Cry1Aa channel has been investigated with the substituted cysteine accessibility method [46], an approach that has been used to map the channel-lining residues of the diphtheria bacterial toxin [47]. Preliminary results indicate that $\alpha4$ inserts into lipid membranes, presumably as a hairpin with $\alpha5$, and lines the aqueous pore [63]. Based on this data and the results of the disulphide bond engineering study [79, 81], a model has been proposed in which domain 1 swings away from domains 2 and 3, and the most hydrophobic α4/α5 hairpin inserts transversally into the bilayer, while the other helices remain parallel to the membrane surface. The functional Cry1Aa ion channel will then assume a tetrameric structure in its minimal configuration. In this model, the membrane-inserted α4/α5 hairpins of four toxin molecules are thought to aggregate so that the lumen of the channel is lined by the hydrophilic faces of the α4 helices while the α5 components of the hairpins anchor the molecules in the lipid membrane.

All available data, including the presence of several subconducting states in PLBs [71, 78, 83], suggest that *Bt* toxin pores, like other bacterial toxins [36, 55, 69], are made by more than one toxin molecule either preassembled in solution [29, 39, 92] or as a result of oligomerisation in the membrane [2, 45]. However, there is presently no information on how many toxin molecules form the toxin pore, on the effect of time on the number and type of subunits or on the contribution of other proteins, like the receptor, to the pore architecture. For example, it has been shown that the pore formed by the *Clostridium perfringens* enterotoxin is a heterotrimer made of one toxin and two membrane proteins [69]. Furthermore, pores made by multiproteic *Bt* strains may be constituted by the heteromeric assembly of subunits from more than one protein, thus displaying altered efficiency compared to that of individual toxins, resulting in synergy or antagonism [75].

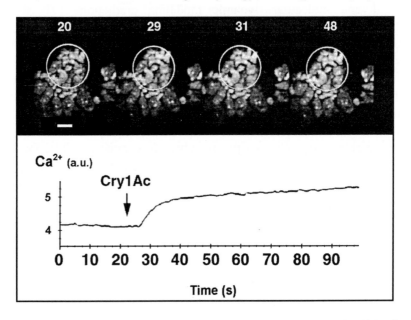

Figure 21. Intracellular calcium surge in *Lymantria dispar* midgut epithelial cells.

7.   POSTBINDING EVENTS, RESISTANCE

Several morphological and physiological effects of *Bt* Cry toxins have been reported in the literature, based on histological studies and metabolic measurements that were conducted on insect larval tissue and cells *in vitro* and, quite often, several hours after intoxication [9, 48, 49, 76]. Less information is available on the early events that may take place even before fully functional intoxicating pores are in place. These events may be responsible for long range cellular effects, similar to what is observed in a number of pore-forming bacterial cytolysins which, in addition to their cytocidal activity, trigger secondary reactions in cells under toxic attack [6]. For example, Ca^{2+} influx through *Escherichia coli* HlyA haemolysin toxin pores activates the arachidonic acid cascade, producing lipid mediators which profoundly affect intoxicated cells and their neighbours [6]. The concept of a *Bt*-mediated early surge in intracellular calcium is supported by several studies both in cell lines [65, 70, 77] and in midgut columnar cells (Fig. 3). In Cf1 cells, the calcium signal observed upon toxin exposure is largely due to the influx of calcium ions through toxin-made channels, but calcium is also mobilised from intracellular pools [70], supporting earlier results showing that toxicity is calcium-dependent and that it can be augmented by agents that modulate intracellular calcium mobilisation [64].

Postbinding events, including pore formation, may play a role in the mechanisms of insect resistance to *Bt* toxins (reviewed in [30, 68, 75]). Resistance has been ascribed to several prebinding processes such as modified gut proteinases, further toxin proteolysis, precipitation in the midgut environment and interaction with the gut peritrophic membrane, but it is currently believed that the major cause of resistance lies in modified binding-site properties. However, alteration of yet unidentified postbinding events may also be involved, including altered membrane permeabilisation or changes in the cascade of metabolic events connected to *Bt* toxin-mediated intracellular signalling.

Many challenging problems remain to be solved to provide a detailed description of how *Bt* toxins form pores in target cells and to identify the molecular determinants of their function [1]. The availability of physiologically competent cellular models derived from the insect gut will allow further investigation of the physiological interactions between *Bt* toxins and their target organisms [4, 5, 8, 94]. This will contribute to the elucidation of the events that follow binding of *Bt* toxins to their receptor and their role in toxicity, in particular in the process of toxin insertion and pore formation in target cell membranes.

ACKNOWLEDGEMENTS

We thank M. Juteau, M. Marsolais, A. Mazza, L. Potvin, G. Préfontaine for their expert technical assistance, and D. Baines, R. Brousseau, E. Franken, R. Frutos, G. Guihard, L. Masson, R. Monette, O. Peyronnet, J. Racapé, C. Rang, W. Schnetter, V. Vachon and M. Villalon for their participation in some of the studies reported here and for many stimulating discussions. The assistance of R. Cabot in helping to edit the manuscript is gratefully acknowledged. This work was supported in part by a research grant from the Natural Sciences and Engineering Council of Canada.

REFERENCES

[1] Andersen OS & Koeppe RE II (1992) Molecular determinants of channel function. Physiol. Rev. 72, S89-S158

[2] Aronson AI, Geng C & Wu L (1999) Aggregation of *Bacillus thuringiensis* Cry1A toxins upon binding to target insect larval midgut vesicles. Appl. Environ. Microbiol. 65, 2503-2507

[3] Aronson AI, Wu D & Zhang CL (1995) Mutagenesis of specificity and toxicity regions of a *Bacillus thuringiensis* protoxin gene. J. Bacteriol. 177, 4059-4065

[4] Baines D, Brownwright A & Schwartz JL (1994) Establishment of primary and continuous cultures of epithelial cells from larval lepidopteran midguts. J. Insect Physiol. 40, 347-357

[5] Baines D, Schwartz JL, Sohi S, Dedes J & Pang A (1997) Comparison of the response of midgut epithelial cells and cell lines from lepidopteran larvae to CryIA toxins from *Bacillus thuringiensis*. J. Insect Physiol. 43, 823-831

[6] Bhakdi S, Valeva A, Walev I, Zitzer A & Palmer M (1998) Pore-forming bacterial cytolysins. J. Appl. Microbiol. 84, 15S-25S

[7] Biggin PC & Sansom MSP (1996) Simulation of voltage-dependent interactions of α-helical peptides with lipid bilayers. Biophys. Chem. 60, 99-110

[8] Braun L & Keddie BA (1997) A new tissue technique for evaluating effects of *Bacillus thuringiensis* toxins on insect midgut epithelium. J. Invertebr. Pathol. 69, 92-104

[9] Bravo A, Jansens S & Peferoen M (1992) Immunocytochemical localization of *Bacillus thuringiensis* insecticidal crystal proteins in intoxicated insects. J. Invertebr. Pathol. 60, 237-246

[10] Butko P, Cournoyer M, Pusztai-Carey M & Surewicz WK (1994) Membrane interactions and surface hydrophobicity of *Bacillus thuringiensis* δ-endotoxin CryIC. FEBS Letters 340, 89-92

[11] Cabiaux V, Wolff C & Ruysschaert JM (1997) Interaction with a lipid membrane: a key step in bacterial toxins virulence. Int. J. Biol. Macromol. 21, 285-298

[12] Carroll J, Convents D, Van Damme J, Boets A, Van Rie J & Ellar DJ (1997) Intramolecular proteolytic cleavage of *Bacillus thuringiensis* Cry3A δ-endotoxin may facilitate its coleopteran toxicity. J. Invertebr. Pathol. 70, 41-49

[13] Carroll J & Ellar DJ (1993) An analysis of *Bacillus thuringiensis* δ–endotoxin action on insect-midgut-membrane permeability using a light-scattering assay. Eur. J. Biochem. 214, 771-778

[14] Carroll J & Ellar DJ (1997) Analysis of the large aqueous pores produced by a *Bacillus thuringiensis* protein insecticide in *Manduca sexta* midgut brush border membrane vesicles. Eur. J. Biochem. 245, 797-804

[15] Carroll J, Wolfersberger MG & Ellar DJ (1997) The *Bacillus thuringiensis* Cry1Ac toxin-induced permeability change in *Manduca sexta* midgut-brush-border-membrane vesicles proceeds by more than one mechanism. J. Cell Sci. 110, 3099-3104

[16] Chen XJ, Curtiss A, Alcantara E & Dean DH (1995) Mutations in domain I of *Bacillus thuringiensis* δ–endotoxin CryIAb reduce the irreversible binding of toxin to *Manduca sexta* brush border membrane vesicles. J. Biol. Chem. 270, 6412-6419

[17] Chen XJ, Lee MK & Dean DH (1993) Site-directed mutations in a highly conserved region of *Bacillus thuringiensis* δ–endotoxin affect inhibition of short circuit current across *Bombyx mori* midguts. Proc. Natl. Acad. Sci. USA 90, 9041-9045

[18] Cooper MA, Carroll J, Travis ER, Williams DH & Ellar DJ (1998) *Bacillus thuringiensis* Cry1Ac toxin interaction with *Manduca sexta* aminopeptidase N in a model membrane environment. Biochem. J. 333 , 677-683

[19] Cummings CE, Armstrong G, Hodgman TC & Ellar DJ (1994) Structural and functional studies of a synthetic peptide mimicking a proposed membrane inserting region of a *Bacillus thuringiensis* δ–endotoxin. Mol. Membr. Biol. 11, 87-92

[20] de Maagd RA, Bakker PL, Masson L *et al.* (1999) Domain III of the *Bacillus thuringiensis* δ–endotoxin Cry1Ac is involved in binding to *Manduca sexta* brush border membranes and to its purified aminopeptidase N. Mol. Microbiol. 31, 463-471

[21] de Maagd RA, Kwa MSG, Van der Klei H *et al.* (1996) Domain III substitution in *Bacillus thuringiensis* δ–endotoxin CryIA(b) results in superior toxicity for *Spodoptera exigua* and altered membrane protein recognition. Appl. Environ. Microbiol. 62, 1537-1543

[22] Doyle DA, Cabral JM, Pfuetzner RA *et al.* (1998) The structure of the potassium channel: molecular basis of K^+ conduction and selectivity. Science 280, 69-77

[23] English LH & Cantley LC (1984) Characterization of monovalent ion transport systems in an insect cell line (*Manduca sexta* embryonic cell line CHE). J. Cell. Physiol. 121, 125-132

[24] English LH & Cantley LC (1985) Delta endotoxin inhibits Rb^+ uptake, lowers cytoplasmic pH and inhibits a K^+-ATPase in *Manduca sexta* CHE cells. J. Membr. Biol. 85, 199-204

[25] English LH, Readdy TL & Bastian AE (1991) Delta-endotoxin-induced leakage of Rb^+ -K^+ and H_2O from phospholipid vesicles is catalyzed by reconstituted midgut membrane. Insect Biochem. 21, 177-184

[26] English LH, Robbins HL, Von Tersch MA *et al.* (1994) Mode of action of CryIIA: a *Bacillus thuringiensis* delta-endotoxin. Insect Biochem. Mol. Biol. 24, 1025-1035

[27] Escriche B, de Decker N, Van Rie J, Jansens S & Van Kerkhove E (1998) Changes in permeability of brush border membrane vesicles from *Spodoptera littoralis* midgut induced by insecticidal crystal proteins from *Bacillus thuringiensis*. Appl. Environ. Microbiol. 64, 1563-1565

[28] Estruch JJ, Warren GW, Mullins MA, Nye GJ, Craig JA & Koziel MG (1996) Vip3A, a novel *Bacillus thuringiensis* vegetative insecticidal protein with a wide spectrum of activities against lepidopteran insects. Proc. Natl. Acad. Sci. USA 93, 5389-5394

[29] Feng Q & Becktel WJ (1994) pH-Induced conformational transitions of Cry IA(a), CryIA(c), and Cry IIIA δ–endotoxins in *Bacillus thuringiensis*. Biochemistry 33, 8521-8526

[30] Ferré J, Escriche B, Bel Y & Van Rie J (1995) Biochemistry and genetics of insect resistance to *Bacillus thuringiensis* insecticidal crystal proteins. FEMS Microbiol. Letters 132, 1-7

[31] Franken E, Potvin L, Schwartz JL & Schnetter W (1996) Activation of the CryIIIA toxin of *Bacillus thuringiensis* subsp. *tenebrionis*: a comparative analysis. XXth Int. Congr. Entomol., Firenze, Italy, p.163

[32] Gazit E, Bach D, Kerr ID, Sansom MSP, Chejanovsky N & Shai Y (1994) The α5 segment of *Bacillus thuringiensis* δ–endotoxin: *in vitro* activity, ion channel formation and molecular modelling. Biochem. J. 304, 895-902

[33] Gazit E, La Rocca P, Sansom MSP & Shai Y (1998) The structure and organization within the membrane of the helices composing the pore-forming domain of *Bacillus thuringiensis* δ–endotoxin are consistent with an "umbrella-like" structure of the pore. Proc. Natl. Acad. Sci. USA 95, 12289-12294

[34] Gazit E & Shai Y (1993) Structural and functional characterization of the α5 segment of *Bacillus thuringiensis* δ–endotoxin. Biochemistry 32, 3429-3436

[35] Gazit E & Shai Y (1995) The assembly and organization of the α5 and α7 helices from the pore-forming domain of *Bacillus thuringiensis* δ–endotoxin – Relevance to a functional model. J. Biol. Chem. 270, 2571-2578

[36] Gouaux E (1997) Channel forming toxins: tales of transformation. Curr. Opinion Struct. Biol. 7, 566-573

[37] Gringorten JL, Crawford DN & Harvey WR (1993) High pH in the ectoperitrophic space of the larval lepidopteran midgut. J. Exp. Biol. 183, 353-359

[38] Grochulski P, Masson L, Borisova S *et al.* (1995) *Bacillus thuringiensis* CryIA(a) insecticidal toxin: crystal structure and channel formation. J. Mol. Biol. 254, 447-464

[39] Güereca L & Bravo A (1999) The oligomeric state of *Bacillus thuringiensis* Cry toxins in solution. Biochim. Biophys. Acta 1429, 342-350

[40] Guihard G, Falk S, Vachon V, Laprade R & Schwartz JL (1999) Real-time fluorimetric analysis of gramicidin D- and alamethicin-induced K^+ efflux from Sf9 and Cf1 cells. Biochemistry 38, 6164-6170

[41] Guihard G, Vachon V, Laprade R & Schwartz JL (1999) Kinetic properties of the channels formed by the *B. thuringiensis* insecticidal protein Cry1C in the plasma membrane of Sf9 cells. Submitted

[42] Guihard G, Vachon V, Yi J, Masson L *et al.* (1998) Cry1C toxin of *Bacillus thuringiensis*: pore formation and properties in live cells. Biophys. J. 74, A320

[43] Haider MZ & Ellar DJ (1989) Mechanism of action of *Bacillus thuringiensis* insecticidal δ–endotoxin: interaction with phospholipid vesicles. Biochim. Biophys. Acta 978, 216-222

[44] Harvey WR & Wolfersberger MG (1979) Mechanism of inhibition of active potassium transport in isolated midgut of *Manduca sexta* by *Bacillus thuringiensis* endotoxin. J. Exp. Biol. 83, 293-304

[45] Hodgman TC & Ellar DJ (1990) Models for the structure and function of the *Bacillus thuringiensis* δ–endotoxins determined by compilational analysis. DNA Sequence-J. DNA Sequencing and Mapping 1, 97-106

[46] Holmgren M, Liu Y, Xu Y & Yellen G (1996) On the use of thiol-modifying agents to determine channel topology. Neuropharmacol. 35, 797-804

[47] Huynh PD, Cui C, Zhan HJ *et al.* (1997) Probing the structure of the diphtheria toxin channel - reactivity in planar lipid bilayer membranes of cysteine-substituted mutant channels with methanethiosulfonate derivatives. J. Gen. Physiol. 10, 229-242

[48] Knowles BH (1994) Mechanism of action of *Bacillus thuringiensis* insecticidal δ-endotoxins. Adv. Insect Physiol. 24, 275-308

[49] Knowles BH & Dow JAT (1993) The crystal δ-endotoxins of *Bacillus thuringiensis*: models for their mechanism of action on the insect gut. BioEssays 15, 469-476

[50] Lacy DB & Stevens RC (1998) Unraveling the structures and modes of action of bacterial toxins. Curr. Opinion Struct. Biol. 8, 778-784

[51] Lane NJ, Dallai R & Ashhurst DE (1996) Structural macromolecules of the cell membranes and the extracellular matrices of the insect midgut. Lehane MJ, Billingsley PF, Editors. Biology of the Insect Midgut. London, U.K.: Chapman & Hall, pp. 115-150

[52] Lee MK, Curtiss A, Alcantara E & Dean DH (1996) Synergistic effect of the *Bacillus thuringiensis* toxins CryIAa and CryIAc on the gypsy moth, *Lymantria dispar*. Appl. Environ. Microbiol. 62, 583-586

[53] Lee MK, You TH, Young BA, Cotrill JA, Valaitis AP & Dean DH (1996) Aminopeptidase N purified from gypsy moth brush border membrane vesicles is a specific receptor for *Bacillus thuringiensis* CryIAc toxin. Appl. Environ. Microbiol. 62, 2845-2849

[54] Leonardi MG, Parenti P, Casartelli M & Giordana B (1997) *Bacillus thuringiensis* CryIAa δ-endotoxin affects the K^+ amino acid symport in *Bombyx mori* larval midgut. J. Membr. Biol. 159, 209-217

[55] Lesieur C, Vécsey-Sémjen B, Abrami L, Fivaz M & Van der Goot FG (1997) Membrane insertion: the strategies of toxins. Mol. Membr. Biol. 14, 45-64

[56] Li J (1996) Insecticidal δ-endotoxins from *Bacillus thuringiensis*. Parker MW, Editor. Protein Toxin Structure. Georgetown, TX: RG Landes, pp. 49-77

[57] Li J, Carroll J & Ellar DJ (1991) Crystal structure of insecticidal δ-endotoxin from *Bacillus thuringiensis* at 2.5 Å resolution. Nature 353, 815-821

[58] Liebig B, Stetson DL & Dean DH (1995) Quantification of the effect of *Bacillus thuringiensis* toxins on short-circuit current in the midgut of *Bombyx mori*. J. Insect Physiol. 41, 17-22

[59] Lorence A, Darszon A & Bravo A (1997) Aminopeptidase dependent pore formation of *Bacillus thuringiensis* Cry1Ac toxin on *Trichoplusia ni* membranes. FEBS Letters 414, 303-307

[60] Lorence A, Darszon A, Díaz C, Liévano A, Quintero R & Bravo A (1995) δ-Endotoxins induce cation channels in *Spodoptera frugiperda* brush border membranes in suspension and in planar lipid bilayers. FEBS Letters 360, 217-222

[61] Luo K, Banks D & Adang MJ (1999) Toxicity, binding, and permeability analyses of four *Bacillus thuringiensis* Cry1 δ-endotoxins using brush border membrane vesicles of *Spodoptera exigua* and *Spodoptera frugiperda*. Appl. Environ. Microbiol. 65, 457-464

[62] Martin FG & Wolfersberger MG (1995) *Bacillus thuringiensis* δ-endotoxin and larval *Manduca sexta* midgut brush-border membrane vesicles act synergistically to cause very large increases in the conductance of planar lipid bilayers. J. Exp. Biol. 198, 91-96

[63] Masson L, Tabashnik BE, Liu YB, Brousseau R & Schwartz JL (1999) Helix 4 of the *Bacillus thuringiensis* Cry1Aa toxin lines the lumen of the ion channel. Submitted

[64] Monette R, Potvin L, Baines D, Laprade R & Schwartz JL (1997) Interaction between calcium ions and *Bacillus thuringiensis* toxin activity against Sf9 cells (*Spodoptera frugiperda*, Lepidoptera). Appl. Environ. Microbiol. 63, 440-447

[65] Monette R, Savaria D, Masson L, Brousseau R & Schwartz JL (1994) Calcium-activated potassium channels in the UCR-SE-1a lepidopteran cell line from the beet armyworm (*Spodoptera exigua*). J. Insect Physiol. 40, 273-282

[66] Morse RJ, Powell G, Ramalingam V, Yamamoto T & Stroud RM (1998) Crystal structure of Cry2Aa from *Bacillus thuringiensis* at 2.2 Angstroms: structural basis of dual specificity. Proc. Annu. Meeting Soc. Invertebr. Pathol., Sapporo, Japan, pp. 1-2

[67] Peyronnet O, Vachon V, Brousseau R, Baines D, Schwartz JL & Laprade R (1997) Effect of *Bacillus thuringiensis* toxins on the membrane potential of lepidopteran insect midgut cells. Appl. Environ. Microbiol. 63, 1679-1684

[68] Pietrantonio PV & Gill SS (1996) *Bacillus thuringiensis* endotoxins: action on the insect midgut, p. 345-372. *In* Lehane MJ & Billingsley PF (ed.), Biology of the Insect Midgut, Chapman & Hall

[69] Popoff MR (1998) Interaction between bacterial toxins and intestinal cells. Toxicon 36, 665-685

[70] Potvin L, Laprade R & Schwartz JL (1998) Cry1Ac, a *Bacillus thuringiensis* toxin, triggers extracellular Ca^{2+} influx and Ca^{2+} release from intracellular stores in Cf1 cells (*Choristoneura fumiferana*, Lepidoptera). J. Exp. Biol. 201, 1851-1858

[71] Racapé J, Granger D, Noulin JF *et al.* (1997) Properties of the pores formed by parental and chimeric *Bacillus thuringiensis* insecticidal toxins in planar lipid bilayer membranes. Biophys. J. 72, A82

[72] Rang C, Vachon V & de Maagd RA *et al.* (1999) *Bacillus thuringiensis* insecticidal crystal proteins: interaction between functional domains. Appl. Environ. Microbiol. 65, 2918-2925

[73] Sacchi VF & Wolfersberger MG. 1996. Amino acid absorption, p. 265-292. *In* Lehane MJ & Billingsley PF (ed.), Biology of the Insect Midgut, Chapman & Hall, London

[74] Sakmann B & Neher E (1995) Single-Channel Recording, 2nd Edition, Plenum Press

[75] Schnepf E, Crickmore N, Van Rie J *et al.* (1998) *Bacillus thuringiensis* and its pesticidal crystal proteins. Microbiol. Mol. Biol. Rev. 62, 775-806

[76] Schwab GE & Culver P (1990) In vitro analyses of *Bacillus thuringiensis* δ–endotoxin action, p. 36-45. *In* Hickle LA & Fitch WL (ed.), Analytical Chemistry of *Bacillus thuringiensis*, American Chemical Society

[77] Schwartz JL, Garneau L, Masson L & Brousseau R (1991) Early response of cultured lepidopteran cells to exposure to δ–endotoxin from *Bacillus thuringiensis*: involvement of calcium and anionic channels. Biochim. Biophys. Acta 1065, 250-260

[78] Schwartz JL, Garneau L, Savaria D, Masson L, Brousseau R & Rousseau E (1993) Lepidopteran-specific crystal toxins from *Bacillus thuringiensis* form cation- and anion-selective channels in planar lipid bilayers. J. Membr. Biol. 132, 53-62

[79] Schwartz JL, Juteau M, Grochulski P *et al.* (1997) Restriction of intramolecular movements within the Cry1Aa toxin molecule of *Bacillus thuringiensis* through disulfide bond engineering. FEBS Letters 410, 397-402

[80] Schwartz JL, Lu YJ, Söhnlein P *et al.* (1997) Ion channels formed in planar lipid bilayers by *Bacillus thuringiensis* toxins in the presence of *Manduca sexta* midgut receptors. FEBS Letters 412, 270-276

[81] Schwartz JL & Masson L (2000) Structure-function analysis of cysteine-engineered entomopathic toxins, p. 101-114. *In* Holst O (ed.), Bacterial Toxins Methods & Protocols, Humana Press

[82] Schwartz JL, Potvin L, Chen XJ, Brousseau R, Laprade R & Dean DH (1997) Single site mutations in the conserved alternating arginine region affect ionic channels formed by CryIAa, a *Bacillus thuringiensis* toxin. Appl. Environ. Microbiol. 63, 3978-3984

[83] Slatin SL, Abrams CK & English L (1990) Delta-endotoxins form cation-selective channels in planar lipid bilayers. Biochem. Biophys. Res. Comm. 169, 765-772

[84] Smedley DP, Armstrong G, Ellar DJ (1997) Channel activity caused by a *Bacillus thuringiensis* δ–endotoxin preparation depends on the method of activation. Mol. Membr. Biol. 14, 13-18

[85] Tellam RL (1996) The peritrophic matrix, p. 86-114. *In* Lehane MJ & Billingsley PF (ed.), Biology of the Insect Midgut, Chapman & Hall

[86] Uawithya P, Tuntitippawan T, Katzenmeier G, Panyim S & Angsuthanasombat C (1998) Effects on larvicidal activity of single proline substitutions in α3 or α4 of the *Bacillus thuringiensis* Cry4B toxin. Biochem. Mol. Biol. Internat. 44, 825-32

[87] Uemura T, Ihara H, Wadano A & Himeno M (1992) Fluorometric assay of potential change of *Bombyx mori* midgut brush border membrane induced by δ–endotoxin from *Bacillus thuringiensis*. Biosci. Biotech. Biochem. 56, 1976-1979

[88] Vachon V, Paradis MJ, Marsolais M, Schwartz JL & Laprade R (1995) Ionic permeabilities induced by *Bacillus thuringiensis* in Sf9 cells. J. Membr. Biol. 148, 57-63

[89] Venugopal MG, Wolfersberger MG & Wallace BA (1992) Effects of pH on conformational properties related to the toxicity of *Bacillus thuringiensis* δ–endotoxin. Biochim. Biophys. Acta 1159, 185-192

[90] Villalon M, Vachon V, Brousseau R, Schwartz JL & Laprade R (1998) Video imaging analysis of the plasma membrane permeabilizing effects of *Bacillus thuringiensis* insecticidal toxins in Sf9 cells. Biochim. Biophys. Acta 1368, 27-34

[91] Von Tersch MA, Slatin SL, Kulesza CA & English LH (1994) Membrane-permeabilizing activities of *Bacillus thuringiensis* coleopteran-active toxin CryIIIB2 and CryIIIB2 domain I peptide. Appl. Environ. Microbiol. 60, 3711-3717

[92] Walters FS, Kulesza CA, Phillips AT & English LH (1994) A stable oligomer of *Bacillus thuringiensis* delta-endotoxin, CryIIIA. Insect Biochem. Mol. Biol. 24, 963-968

[93] Walters FS, Slatin SL, Kulesza CA & English LH (1993) Ion channel activity of N-terminal fragments from CryIA(c) delta-endotoxin. Biochem. Biophys. Res. Comm. 196, 921-926

[94] Wang SW & McCarthy WJ (1997) Cytolytic activity of *Bacillus thuringiensis* CryIC and CryIAc toxins to *Spodoptera* sp. midgut epithelial cells *in vitro*. In Vitro Cell. Dev. Biol. 33, 315-323

[95] Weltens R, Peferoen M, Steels P & Van Kerkhove E (1991) Electrophysiological investigation of the midgut of the Colorado potato beetle (*Leptinotarsa decemlineata*) : effect of a natural insecticide. Belg. J. Zool. 121, 53-54

[96] Wolfersberger MG, Chen XJ & Dean DH (1996) Site-directed mutations in the third domain of *Bacillus thuringiensis* δ–endotoxin CryIAa affect its ability to increase the permeability of *Bombyx mori* midgut brush border membrane vesicles. Appl. Environ. Microbiol. 62, 279-282

[97] Wu D & Aronson AI (1992) Localized mutagenesis defines regions of the *Bacillus thuringiensis* δ–endotoxin involved in toxicity and specificity. J. Biol. Chem. 267, 2311-2317

Chapter 3.4

Insect resistance to *Bacillus thuringiensis* insecticidal crystal proteins

Jeroen Van Rie[1] & Juan Ferré[2]

[1]*Aventis CropScience N.V., J. Plateaustraat 22, 9000 Gent, Belgium and* [2]*Department of Genetics, Faculty of Biology, Universitat de Valencia, 46100 Burjassot, Spain*

Key words: *Bacillus thuringiensis*, resistance

Abstract: Several insect species have developed resistance to insecticidal crystal proteins from *Bacillus thuringiensis*, either through laboratory selection, or under field conditions. In this chapter we review the current knowledge on the biochemical and genetic mechanisms of resistance to *B. thuringiensis*. This knowledge will be important in the design of appropriate tactics to manage the development of resistance in insect populations.

1. INTRODUCTION

Perhaps the most serious threat to the durability of insect control strategies involving the use of *Bacillus thuringiensis* (*Bt*) insecticidal crystal proteins (ICPs) is the potential of insect populations to develop resistance to *Bt* ICPs.

Indeed, insects have demonstrated their enormous genetic plasticity with over 500 insect species resistant to one or multiple insecticides [13]. In 1985, the first report on insect resistance to *Bt* was published [33]. Since then, many more such cases have been reported [5, 11, 43, 46]. Most of these strains were selected for resistance under laboratory conditions. Laboratory selection experiments do not predict if resistance will develop in the field or which resistance mechanisms will be selected, but can indicate the repertoire of resistance mechanisms available in a

J.-F. Charles et al. (eds.),
Entomopathogenic Bacteria: From Laboratory to Field Application, 219–236.

certain population and are essential to study inheritance of resistance
genes.

2.   BIOCHEMICAL BASIS OF RESISTANCE

In principle, the mechanism of insect resistance to *Bt* could be located
at each of the various steps in the mode of action of *Bt* ICPs
(solubilisation, proteolytic processing, passage through the peritrophic
membrane, receptor binding, pore formation, osmotic lysis of midgut
cells). Whereas different mechanisms have been observed in resistant
strains selected under laboratory conditions, only one major mechanism
has been reported so far for resistance developed under field conditions
(Table 1).

Table 1. Selected insect species and strains which have developed resistance to *Bt*

Species	Selecting agent[1]	Strain Name[2]	Resistance ratio[3]		Resistance mechanism[4]	Ref[5]
P. interpunctella	Dipel®	343-R	Dipel®	>250		[33]
			Cry1Ab	(877)	↓ binding	
			Cry1Ca	0.3		
P. interpunctella	Bte HD-198	198[r]	HD-198	32		[35]
			Cry1Ac	128	↓ protoxin activation	
P. xylostella	Btk	BL[§]	Dipel®	1		
			Cry1Ab	>200	↓ binding	[12]
			Cry1Ba	2		
			Cry1Ca	0.5		
P. xylostella	Btk		Cry1Aa	1.3		[3]
			Cry1Ab	236		
			Cry1Ac	1		
			Cry1Ba	1		
	Cry1Ab/	PHI	Cry1Aa	>1, n.d.	Unaltered binding	[53]
	Cry1Ac-1Ab		Cry1Ab	>1, n.d.	↓ binding	
			Cry1Ac	>1, n.d.	Unaltered binding	
			Cry1Ca	(1)		
			Cry1Fa	(1)		
			Cry1Ja	(1)		

Table 1 cont. Selected insect species and strains which have developed resistance to *Bt*

Species	Selecting agent [1]	Strain Name[2]	Resistance ratio[3]		Resistance mechanism[4]	Ref[5]
P. xylostella	Dipel®	NO-QA	Cry1Aa	>100	Altered binding	[51]
			Cry1Ab	>100	↓ binding	
			Cry1Ac	>100	↓ binding	
			Cry1Ba	3		
			Cry1Ca	2		
			Cry1Da	3		
			Cry1Fa	>100		
			Cry1Ia	3		
			Cry1Ja	>140		
P. xylostella	*Btk/Bta*	NO-95	Cry1Ca	22		[27]
			Btk HD-1	134		
			Bta	3		
P. xylostella	*Btk*	Loxa A	Javelin®	300		[57]
			Cry1Aa	>200		
			Cry1Ab	>200	↓ binding	
			Cry1Ac	>200		
			Cry1Ba	2.5		
			Cry1Ca	3.4		
			Cry1Da	1		
P. xylostella	*Btk/*	PEN	Cry1Aa	High	Altered binding	[53]
	Cry1Ac/		Cry1Ab	High	↓ binding	
	Cry1C		Cry1Ac	High	↓ binding	
			Cry1C	No		
			Cry1Fa	High		
			Cry1Ja	High		
P. xylostella	*Btk/Bta*	SERD3	Dipel®	330		[63]
			Florbac®	160		
			Cry1Ab	n.d.	↓ binding	
H. virescens	Cry1Ac	YHD2	Cry1Aa	n.d.	↓ binding	[16]
			Cry1Ab	>2,300	unaltered (?) binding	
			Cry1Ac	>10,000	unaltered (?) binding	
			Cry1Fa[+]	3,700		
			Cry1Ca[+]	2.5		
			Cry2Aa	25		

Table 1 cont. Selected insect species and strains which have developed resistance to *Bt*

Species	Selecting agent [1]	Strain Name[2]	Resistance ratio[3]		Resistance mechanism[4]	Ref[6]
H. virescens	CrylAb/	SEL	CrylAb*	71	Slightly altered binding	[30]
	Dipel®		CrylAc	16	Slightly altered binding	
			Dipel®	57		
H. virescens	CrylAc	CP73-3	CrylAb	13	unaltered (?) binding	[17]
			CrylAc	50	unaltered (?) binding	
			Cry2Aa	53		
S. exigua	CrylCa		CrylCa	850	↓ total binding ↑ non-specific binding	[36]
S. littoralis	CrylCa		CrylCa	>500		[38]

[1] *Bta* (*B. thuringiensis* var. *aizawai*) and *Btk* (*B. thuringiensis* var. *kurstaki*) refer to commercial formulations of *B. thuringiensis*. Dipel® and Javelin® are tradenames of formulations of *Btk*. Florbac® is a trade name for a commercial formulation of *Bta*. Bte = *B. thuringiensis* var. *entomocidus*.

[2] Name of the resistant strain as given in the reference paper, except when followed by [§]: those names are arbitrary names given by the authors of this paper.

[3] LC$_{50}$ (or LD$_{50}$) of resistant strain divided by LC$_{50}$ (or LD$_{50}$) of susceptible control strain. Values in parentheses are estimates. All values for Cry1 and Cry9 proteins refer to activated toxins, except for values followed by * or + which refer to protoxin or CellCap formulation respectively. n.d.: not determined.

[4] When available, the mechanism of resistance to the particular toxin is given. In case of binding, only results of binding experiments to native BBMVs or to tissue sections are given.

[5] Reference for the first paper describing the strain or containing most of the bioassay data is given; additional relevant references can be found throughout the text.

2.1   Altered proteolytic processing

A *Plodia interpunctella* strain (198^r), selected in the laboratory for resistance to *B. thuringiensis* var. *entomocidus* (*Bte*) HD-198, was 128-fold less susceptible to CrylAc as the sensitive strain. Midgut extracts from the resistant insects had significantly lower proteolytic activity towards several *p*-nitroanilide substrates than extracts from susceptible insects and had a reduced capacity to activate Cry1Ac protoxin [40]. A subsequent study demonstrated a genetic linkage between decreased susceptibility to Cry1Ac and the absence of a major gut proteinase [39].

Altered gut proteinase activity was reported not to be present in another *P. interpunctella* strain (343-R), selected for resistance to Dipel® (a *B. thuringiensis* var. *kurstaki* (*Btk*) HD-1 formulation) [22].

2.2   Binding site modification

Laboratory selection of a *P. interpunctella* colony, established from grain bins, with Dipel® resulted in a 250-fold resistance to this *Bt* formulation [33, 34]. Whereas the strain was highly resistant to Cry1Ab (present in Dipel®), it was even more susceptible to Cry1Ca (not present in Dipel®) than the susceptible control strain. Receptor binding studies using brush border membrane vesicles (BBMV) prepared from larval midguts showed a 50-fold reduction in binding affinity (K_d), but no change in the number of binding sites (R_t) for Cry1Ab in the resistant strain [60]. The K_d value for Cry1Ca, which binds to another high affinity site, was similar in both strains, but the R_t value was significantly higher (3-fold) in the resistant strain. These data provided evidence that resistance was due to an alteration only in the binding site for Cry1Ab. The increase in the number of binding sites for Cry1Ca could explain the higher susceptibility of the selected strain towards this crystal protein.

Cry1Ac bound to the brush border membrane (BBM) on midgut sections from susceptible *P. interpunctella* larvae, but not to sections from larvae of two resistant strains (Dipel®, HD-133[r]) selected for resistance to Dipel® and *B. thuringiensis* var. *aizawai* (*Bta*) HD-133, respectively [37]. Remarkably, Cry1Ac recognised a 80 kDa BBM protein in a ligand blot, suggesting a change in the accessibility of the ICP binding protein in the resistant strain.

Very high levels of resistance to Cry1Ac (over 10,000-fold) were obtained in a *H. virescens* colony by selection with Cry1Ac protoxin [16]. This strain (YHD2) was highly cross-resistant to Cry1Ab and Cry1Fa, only moderately resistant to Cry2Aa, and almost non-resistant to Cry1Ca and Cry1Ba. Binding of Cry1Aa to BBMV of resistant larvae was dramatically reduced but binding of Cry1Ab and Cry1Ac was apparently not reduced [25]. It had already been demonstrated that the Cry1A ICPs share some binding sites in this insect species. Thus, Cry1Ac and Cry1Ab also bind to the Cry1Aa binding site [59]. Consequently, it was proposed that the altered Cry1Aa binding site causes resistance to all three Cry1A ICPs and that the additional binding sites recognised by Cry1Ab and Cry1Ac may not be involved in toxicity. Cry1Fa shares a binding site with Cry1A toxins in *P. xylostella* [4], in *Spodoptera exigua* and in *S. frugiperda* [28]. Therefore, it is not unlikely that, also in *H. virescens*, Cry1Fa and Cry1A share a common receptor, explaining the observed cross-resistance.

The diamondback moth (*P. xylostella*), is the only insect species that has evolved high levels of resistance to *Bt* in the field. The first case of field resistance was reported from Hawaii. Laboratory selection of a population, that had developed about a 30-fold level of resistance under field conditions, using Dipel® increased resistance rapidly to over 800-fold [48]. The resulting strain (NO-QA) displayed high levels of (cross-) resistance to Cry1Aa, Cry1Ab, Cry1Ac, Cry1Fa, and Cry1Ja, whereas resistance to Cry1Ba, Cry1Bb, Cry1Ca, Cry1Da, Cry1Ia, and Cry2Aa was not considered significant [51]. Compilation of results from homologous and heterologous competition binding experiments using Cry1Aa, Cry1Ab, Cry1Ac, Cry1Ba, Cry1Ca and Cry1Fa in susceptible *P. xylostella* colonies [4] indicate the presence of (at least) 4 ICP binding sites: one (site 1) is recognised only by Cry1Aa, another (site 2) is shared among Cry1Aa, Cry1Ab, Cry1Ac and Cry1Fa, and two additional sites bind Cry1Ba and Cry1Ca respectively (site 3 and 4 respectively). Binding of Cry1Ac and Cry1Ab, but not Cry1Aa, to BBMV from resistant insects was strongly reduced as compared to binding in susceptible insects [4,47,53]. The Cry1Aa binding data can be understood in the light of the binding site model [4] where Cry1Aa competes for the Cry1Ab site, but also recognises an additional site. From the binding and cross-resistance data, it appears that the additional Cry1Aa binding site is not involved in toxicity. Surprisingly, binding of Cry1A ICPs to the resistant strain could be demonstrated in alternative binding assays (surface plasmon resonance based binding analysis [32], binding to tissue sections [8] or to purified aminopeptidase [29]), suggesting a difference in the accessibility of the common Cry1A binding site in the different assay systems. Binding of Cry1Ca was not altered in the NO-QA strain [47]. In conclusion, if we consider only sites 2, 3 and 4 to be functional ICP receptors and if we assume that Cry1Ja also binds to site 2, it seems that patterns of cross-resistance correspond to patterns of receptor specificity.

Similar to the strain from Hawaii, high levels of resistance to Cry1Aa, Cry1Ab, Cry1Ac, Cry1Fa and Cry1Ja were also observed in a *P. xylostella* strain (PEN) from Pennsylvania. Binding of Cry1Ab and Cry1Ac, but not of Cry1Aa, was dramatically reduced in this strain [53].

A diamondback moth colony (BL), derived from a field population in the Philippines regularly exposed to Dipel®, showed a more than 200-fold resistance to Cry1Ab but was still fully susceptible to Cry1Ba and Cry1Ca [12]. Binding of Cry1Ab, but not of Cry1Ba or Cry1Ca, was very strongly reduced in the resistant larvae. Lack of binding of Cry1Ab to the resistant strain was also confirmed by a histological study of resistant larvae intoxicated with Cry1Ab [6]. Furthermore, there was no damage to midgut epithelial cells from these larvae whereas cells from susceptible larvae showed clear binding and damage. In contrast to Cry1Ab, Cry1Ba bound to and damaged midgut cells in both resistant and susceptible larvae. These data clearly demonstrated a causal relationship between decreased binding and decreased susceptibility (resistance).

In another colony, sampled from the same location in the Philippines but about 3.5 years later, resistance was limited to Cry1Ab and did not extend to Cry1Aa, Cry1Ac or Cry1Ba [3]. Following additional selection with Cry1Ab and later with a hybrid Cry1A protoxin, this strain (PHI) was observed to be partially resistant to Cry1Aa, Cry1Ab, Cry1Ac and still susceptible to Cry1Ca, Cry1Fa and Cry1Ja [53]. Binding of Cry1Ab, but not of Cry1Aa and Cry1Ac, was dramatically reduced in the resistant strain [4, 53]. Thus, an alteration in site 2 may affect binding of Cry1Ab, without affecting Cry1Aa and Cry1Ac binding, suggesting that the binding epitopes of these three ICPs are not identical. Therefore, an additional mechanism of resistance, other than a reduction in binding, must be involved in this strain.

The mechanism of resistance was also studied in a *P. xylostella* colony (Loxa A) from Florida that was 1,640-fold resistant to Javelin® (a commercial formulation of *Btk* NRD12) in the second generation after the colony was collected from the field [57]. Resistance rapidly fell to about 300-fold in the absence of selection but remained stable at this level in subsequent generations. Insects with this level of stabilised resistance were more than 200-fold resistant to Cry1Aa, Cry1Ab, and Cry1Ac, but were still fully susceptible towards Cry1Ba, Cry1Ca, Cry1Da. Binding studies demonstrated a virtually complete lack of Cry1Ab binding in the resistant strain. Binding of Cry1Ba and Cry1Ca was not affected [57].

More recently, a *P. xylostella* colony (SERD3) from Malaysia resistant to both *Btk* HD-1 (330-fold) and *Bta* H7 (160-fold) has been reported [63]. Selection during 3 generations with *Btk* increased resistance to *Btk* but only marginally to *Bta* and vice versa. A large decrease in binding of Cry1Ab, but not of Cry1Aa, Cry1Ac, or Cry1Ca was observed in SERD3 larvae.

2.3   Slight alterations in binding and/or unknown mechanism

In a *H. virescens* colony selected on Cry1Ab protoxin, resistance to Cry1Ab was 20-fold after 14 generations [45] and further increased to 71-fold by 4 additional generations of selection with Dipel® [44]. Binding experiments showed marginally significant differences between the susceptible and the resistant (SEL) strain for binding of Cry1Ab and Cry1Ac (2 to 6-fold difference in K_d and R_t values) [30]. However, the differences were compensatory in nature, suggesting that the magnitude of the observed alterations in binding was insufficient to explain the resistance.

Selection of another colony of *H. virescens* with Cry1Ac resulted in a 50-fold resistance to Cry1Ac, 13-fold to Cry1Ab and 53-fold to Cry2Aa [17]. Cross-resistance in this strain (CP73-3) also extended to Cry1Aa, Cry1Ba, and Cry1Ca. No significant difference was found between the two strains in binding experiments with Cry1Ab and Cry1Ac. It should be pointed out that binding of Cry1Aa has not been studied in the above two *H. virescens* strains. Therefore, if the mechanism observed in strain YHD2 would also be present in these strains, it would have been overlooked.

Limited levels of resistance were obtained in a *S. exigua* strain upon 14 generations of selection with *E. coli* inclusion bodies containing Cry1Ca. Subsequent selection for 11 generations with activated Cry1Ca toxin resulted in significantly increased levels of resistance to Cry1Ca (850-fold) [36]. Binding experiments with Cry1Ca demonstrated a 5-fold decrease in K_d, but no change in R_t in the resistant strain. Furthermore, BBMV from the resistant larvae displayed a significantly higher level of non-specific binding of Cry1Ca. This strain was cross-resistant to Cry1Ab, Cry2Aa, Cry9C, and hybrid ICP G27.

Selection experiments on *Spodoptera littoralis* larvae using spore-Cry1Ca crystal mixtures generated a resistant colony with greater than 500-fold resistance to Cry1Ca [38]. The resistant strain was fully susceptible to Cry1Fa but exhibited limited cross-resistance to Cry1Ab, Cry1Da and *Bta* 7.29 and a somewhat higher level of cross-resistance to Cry1Ea.

Insect resistance in *P. xylostella* populations to *Btk* products has resulted in extensive use of *Bta*-based insecticides in certain locations. *Bta* strains typically contain, in addition to Cry1A toxins, Cry1Ca. Insects in two colonies (NO-93, NO-95) from one such location in Hawaii displayed up to 20-fold resistance to Cry1Ca [27]. These Cry1Ca-resistant colonies were only 2- to 4-fold less susceptible to *Bta*, somewhat less susceptible to Cry1Ab, and 50- to 130-fold less susceptible to *Btk* formulations when compared to a susceptible colony.

Decreased susceptibility to Dipel® was observed in all of the 5 colonies of *Ostrinia nubilalis* selected in the laboratory with this *Bt* formulation [21]. Upon 3 generations of selection, one colony displayed a 36-fold decrease in susceptibility and continued selection for 4 generations increased the resistance ratio to 73-fold. This rapid adaptation suggests that resistance genes in *O. nubilalis* may not be extremely rare.

Bt resistant insect colonies have also been reported for *Trichoplusia ni* [9], *Leptinotarsa decemlineata* [61], *Chrysomela scripta* [5] and *Culex quinquefasciatus* [62].

3.   GENETICS OF RESISTANCE

In insects, resistance is a pre-adaptive phenomenon that develops by selection of rare individuals that can survive a certain insecticide treatment of a population. A large number of genetic, biological, ecological and operational factors influence the rate of resistance development. Genetic factors include variability in natural populations, number of resistance genes and their frequency and mode of inheritance (number of alleles, degree of dominance and sex-linkage), and fitness costs associated with resistance. In the following sections we will review what is currently known about the genetic aspects of *Bt* resistance.

3.1   Natural variability

Two approaches have been used to obtain estimates of the variability regarding resistance genes for *Bt* in field populations and hence, estimates of the potential of these populations to evolve resistance. One approach involves the measurement of differences in susceptibility among and within populations. One such study was performed to measure variation in susceptibility to *Btk* HD-1 in nine field populations and two laboratory colonies of *Choristoneura fumiferana,* with no previous exposure to the bioinsecticide [58]. Although variation among populations was low, variation among families within same populations varied from 2.4- to 30.4-fold difference between the highest and lowest percent mortalities at a single dose. Similar observations were made in a study of variability in *Lymantria dispar* [42]. In *O. nubilalis*, the variation in susceptibility against *Btk* HD-1 among 5 colonies of insects collected from three different US states was also low (1.9-fold) [21]. A review by Tabashnik [46] lists variations in tolerance to *Bt* and *Bt* ICPs in 15 insect species. In all cases variation was low or minimal except in *P. xylostella* populations that had been repeatedly exposed to the bioinsecticide and in *P. interpunctella* (up to 42-fold), which presumably had been naturally exposed to the bacterium in grain bins [24].

A different approach to estimate variability for resistance genes in natural populations is to measure the heritability (h^2) in laboratory selection experiments on a sample of the population. The h^2 is the proportion of phenotypic variation for a given trait accounted for by additive genetic variation [10]. Tabashnik [46] estimated realised h^2 of resistance to *Bt* products and Cry1A toxins for 27 selection experiments. Compared to 8 other insect species, a relatively high h^2 was found in *P. interpunctella*, reflecting low phenotypic variation and high additive genetic variation, perhaps resulting from its relatively constant environmental conditions and periodic exposure to natural infestations of *Bt*.

3.2   Frequency of resistance genes

An indirect estimate of the frequency of resistance alleles is provided by laboratory selection experiments that have succeeded in selecting for resistance to *Bt*. In these cases at least one copy of the resistance allele had to be present at the start of selection. Given that successful selection experiments with *Bt* in Lepidoptera used a small sample of insects from the field population [16, 17, 35], the frequency of resistance alleles in those populations must have been relatively high (around 1 to 5 x 10^{-3}).

A direct approach to estimate the frequency of a major *Bt* resistance allele has been recently applied to field populations of *H. virescens* making use of a practically homozygous resistant strain for a recessive resistance allele [15]. Field collected males were individually mated to females of the resistant strain. By testing the F_1 and F_2 offspring from over 1,000 of those single pair matings the authors estimated a frequency of resistance alleles in the field sample of 1.5 x 10^{-3}. A similar approach has been applied to a susceptible laboratory colony of *P. xylostella* from Hawaii [52], resulting in an estimate of the frequency for recessive resistance alleles in the susceptible laboratory colony of 0.12. This is an extremely high frequency for a resistance allele unless it is held in a polymorphic state based on a primary physiological function unrelated to resistance [14].

Andow & Alstad [1] have recently proposed an efficient method to detect and estimate frequencies of resistance alleles in field populations based on an F_2 screening procedure. When this method was applied to a sample of 91 females from a field population of *O. nubilalis* the estimated frequency for Cry1Ab-resistance alleles was <0.013 [2].

3.3 Mode of inheritance of resistance

In five colonies of *P. interpunctella* that had evolved moderate to high levels of resistance to *Btk* upon laboratory selection, resistance was found to be autosomally inherited and partially to completely recessive [33, 34]. Backcross studies of strain 198^r, resistant to *Bte*, suggested that resistance was also recessive and linked to a major gut proteinase locus [39].

In three geographically distinct highly resistant colonies of *P. xylostella* [19, 55, 56] resistance to *Btk* formulations was found to be autosomally inherited and partially to nearly completely recessive. Results from backcross and F_2 progenies indicated that resistance was controlled by a single locus in the Loxa A colony and by one or a few loci in the other two colonies (ROO and NO-Q). Studies on the NO-QA strain indicated that resistance to Cry1Aa, Cry1Ab, Cry1Ac, and Cry1Fa was due to an autosomally inherited recessive allele and that there was at least one additional locus conferring resistance to Cry1Aa that segregated independently [52]. The same spectrum of resistance, dominance and binding results as those reported for NO-QA were found in the PEN strain [53]. Resistance in the F_1 progeny of crosses between NO-QA and PEN showed that the multi-toxin resistance mutation in both strains was allelic, suggesting that the same type of mutations are responsible for the resistance in both strains. In another *P. xylostella* colony (BL) [12], susceptibility of the F_1 progeny to Cry1Ab depended on the sex of the insects in the parental cross, yet sex linkage was discarded on the basis of the 1:1 sex ratio of the F_1 survivors at various doses of Cry1Ab [31]. Lack of binding of Cry1Ab, associated to resistance in the parental colony, segregated as an autosomal recessive trait [31]. Resistance to Cry1A toxins in the PHI strain was inherited autosomally and, at the test dose employed, resistance was partially recessive for Cry1Ab, but partially dominant for Cry1Aa and semi-dominant for Cry1Ac [53, 54]. Furthermore, crosses between NO-QA and PHI resistant insects showed that resistance to Cry1Ab in PHI was due to a different allelic mutation at the same locus that confers multi-toxin resistance in NO-QA and PEN strains [53, 54]. Cry1Ca resistance *P. xylostella* strain NO-95C was autosomally inherited, appearing recessive at high test doses and dominant at low test doses [26].

Studies on the inheritance of resistance in *H. virescens* have rendered different results depending on the selected strain used. In the SEL strain inheritance was found to be autosomally inherited, incompletely dominant, and controlled by several genetic factors [44]. In strain CP73-3 resistance to Cry1Ac was autosomally inherited, behaving as an additive trait at high concentrations of Cry1Ac and as a recessive to partially recessive trait at low concentrations of toxin [17]. For a third strain (YHD2), resistance appeared to be autosomally inherited, partially recessive to Cry1Ab and partially dominant to Cry2Aa [16]. Growth-inhibition tests of F_2 and backcross indicated that resistance to Cry1Ab was most likely due to a single locus (or a set of tightly linked loci).

Cry1Ca resistance in *S. littoralis* was partially recessive, whereas Cry3Aa resistance in *L. decemlineata* was partially dominant. In both cases resistance appeared to be autosomal and multifactorial [7, 41].

3.4   Number of resistance genes

In *P. xylostella,* different lines of evidence indicate the occurrence of more than one gene involved in resistance to *Bt*. In the Loxa A strain [56], resistance to Javelin® was extremely high when the insects were first brought to the laboratory (>1,500-fold) but then rapidly declined in the absence of selection, stabilising at a resistance ratio around 300-fold. A first selection attempt at generation 4 succeeded at increasing resistance to the initial levels (which also declined in the absence of selection), but a second selection at generation 8 failed to do so. The results suggest that the field population contained at least two genes conferring resistance to Javelin. Results from experiments of prolonged selection with a Hawaiian colony (NO-Q) and subsequent relaxation of selection of the resistant line and 6 isofemale lines suggested that resistance to Dipel® was not controlled only by one locus with two alleles [49]. As indicated in the above section, strain NO-QA had at least an additional locus for resistance to Cry1Aa segregating independently from the multi-toxin resistance locus [52]. In the NO-95C *P. xylostella* strain, resistance to Cry1Ab and Cry1C segregated independently [26]. Differences in susceptibility to various toxins of split broods from single-pair matings between the resistant PEN strain and a susceptible strain indicated that at least two independently segregating loci were involved in resistance in this strain [53, 54]. The strain PHI from the Philippines was shown to have a different mutation at the multi-toxin locus than the NO-QA and PEN strains [53]. Although the PHI strain is resistant to three Cry1A toxins, the multi-toxin mutation presumably confers resistance just to Cry1Ab, since binding to BBMV is reduced for this toxin but not for Cry1Aa or Cry1Ac [4, 53]. In a colony of *P.*

xylostella from Malaysia (SERD3), resistance to *Btk* segregated separately from resistance to *Bta* in laboratory selection experiments [63]. Reduced binding to BBMV was only found for Cry1Ab, but not for Cry1Aa, Cry1Ac or Cry1Ca, which suggests that more than one genetic mechanism is present.

In the YHD2 strain of *H. virescens*, there seems to be a major partially recessive gene conferring resistance to structurally related toxins (Cry1Aa, Cry1Ab, Cry1Ac, and Cry1Fa) by apparently modifying a membrane receptor [16, 25] and another gene (or genes) conferring moderate resistance to Cry2Aa, with resistance to Cry2Aa being partially dominant. Genetic analysis with marker loci revealed that linkage group 9 had a major contribution (as much as 80%) to the resistance to Cry1Ac in YHD2, that linkage group 11 made a smaller contribution, and that additional unlinked loci had minor effects on resistance [20].

3.5 Stability of resistance in the absence of selecting agent and fitness costs

In the vast majority of cases, resistance to *Bt* and individual ICPs has been found to be unstable. Instability is most likely to be caused by fitness costs associated with resistance genes or with other loci closely linked to these. *P. interpunctella* is one of the few cases where resistance did not decline in one strain (343-R) even after 29 generations on untreated diet, remaining at a level of around 49 to 87-fold resistance compared with a susceptible strain [34]. However, in another resistant strain with a much lower level of resistance (between 12 and 17-fold), resistance declined when selection pressure was discontinued, reaching a level of 4-fold after 29 generations. In a different study using 5 resistant strains and one susceptible strain, higher natural mortality was found in the resistant strains, though no difference in other fitness components was detected [23].

Other examples where resistance remained high and stable after discontinuing selection are found in *P. xylostella*. In a study with one selected strain (NO-Y) from Hawaii with very high levels of resistance to Dipel® (>3,500-fold) and 6 isofemale lines derived from it, resistance declined in most cases [49]. However, one isofemale line remained with extremely high resistance after 32 generations without exposure. In a study with colony NO-93, resistance to Dipel® (>20-fold) did not decline for 10 generations without exposure to the insecticide [26].

In *P. xylostella*, rapid decline in resistance to Dipel® has been found in three resistant laboratory-selected strains from Hawaii (with 90 to 2,800-fold resistance) [47, 50]. The decline was much slower in an

unselected colony from Hawaii with 22-fold resistance [48, 50].
Decreased fitness was found in the most resistant strain (NO-QA) as
compared with a susceptible colony [18, 47]. Resistance was also unstable
in three resistant greenhouse populations from Japan (with 220- to
700-fold resistance to *Btk*) [19]. In colony Loxa A, resistance to Javelin®
was unstable from the 2nd to the 3rd generation, but thereafter it reached
a plateau and remained practically stable for at least 7 generations [56].
In hybridised populations of susceptible insects and resistant insects (after
resistance had reached a plateau), resistance remained relatively stable
over 6 generations, suggesting that resistance was not causing detectable
reductions in fitness.

In *H. virescens* (SEL strain) [44], in the Cry1Ca resistant *S. littoralis*
strain [38] and in the Cry3Aa resistant *L. decemlineata* strain [41],
resistance declined when selection was discontinued. In the *H. virescens*
and *L. decemlineata* strain, some level of resistance remained stable in
the absence of the selecting agent.

4.   CONCLUSIONS

Two different biochemical mechanisms of insect resistance to *Bt* have
been observed so far: in one laboratory selected colony of *P.
interpunctella*, resistance was linked to altered proteolytic processing of
the protoxin, whereas in colonies of several insect species (*P.
interpunctella, H. virescens, P. xylostella*) that developed resistance under
field or laboratory conditions, resistance was due to modification of an
ICP binding site. In all cases of field developed resistance (*P. xylostella*)
in which resistance mechanisms were studied, resistance to one or more
ICPs could be correlated to an alteration of an ICP binding site. In some
colonies, additional mechanisms –still to be identified– seem to be
present.

In all cases of binding site modification, resistance seems to be due to
a recessive or partially recessive mutation in a major autosomal gene, and
the resistance levels reached are high. Cross-resistance appears to extend
only to ICPs sharing binding sites. In contrast, in many cases where
resistance is due to another modification, a more general cross-resistance
was observed.

On the basis of our current knowledge on resistance mechanisms, it is
proposed that resistance management strategies should consider
combinations of ICPs with different binding site specificity.

However, insect populations may contain more than one gene
conferring resistance to ICPs. If different genes controlling different

mechanisms of resistance could become coselected, the design of an optimal resistance management strategy may be more complex. The optimal strategy will, to a large extent, depend on the particular insect-crop interactions. It is clear that judicious deployment of *Bt* is desirable in order to maintain its usefulness as an effective and environmentally friendly insect control agent.

REFERENCES

[1] Andow DA & Alstad DN (1998) F_2 screening for rare resistance alleles. J. Econ. Entomol. 91, 572-578.

[2] Andow DA & Alstad DN, Pang YH, Bolin PC & Hutchison WD (1998) Using an F_2 screen to search for resistance alleles to *Bacillus thuringiensis* toxin in European corn borer (Lepidoptera: Crambidae). J. Econ. Entomol. 91, 579-584.

[3] Ballester V, Escriche B, Ménsua JL, Riethmacher GW & Ferré J (1994) Lack of cross-resistance to other *Bacillus thuringiensis* crystal proteins in a population of *Plutella xylostella* highly resistant to Cry1Ab. Biocontrol Sci. Technol. 4, 437-443.

[4] Ballester V, Granero F, Tabashnik BE, Malvar T & Ferré J (1999) Integrative model for binding of *Bacillus thuringiensis* toxins in susceptible and resistant larvae of the diamondback moth (*Plutella xylostella*). Appl. Environ. Microbiol. 65; 1413-1419.

[5] Bauer LS (1995) Resistance: a threat to the insecticidal crystal proteins of *Bacillus thuringiensis*. Florida Entomol. 78, 414-443.

[6] Bravo A, Jansens S & Peferoen M (1992) Immunocytochemical localization of *Bacillus thuringiensis* insecticidal crystal proteins in intoxicated insects. J. Invertebr. Pathol. 60, 237-246.

[7] Chaufaux J, Müller-Cohn J, Buisson C, Sanchis V, Lereclus D & Pasteur N (1997) Inheritance of resistance to the *Bacillus thuringiensis* CryIC toxin in *Spodoptera littoralis* (Lepidoptera: Noctuidae). J. Econ. Entomol. 90, 873-878.

[8] Escriche B, Tabashnik B, Finson N & Ferré J (1995) Imunohistochemical detection of binding of CryIA crystal proteins of *Bacillus thuringiensis* in highly resistant strains of *Plutella xylostella* (L.) from Hawaii. Biochem. Biophys. Res. Commun. 212, 388-395.

[9] Estada U & Ferré J (1994) Binding of insecticidal crystal proteins of *Bacillus thuringiensis* to the midgut brush border of the cabbage looper, *Trichoplusia ni* (Hübner) (Lepidoptera: Noctuidae), and selection for resistance to one of the crystal proteins. Appl. Environ. Microbiol. 60, 3840-3846.

[10] Falconer DS (1989) Introduction to Quantitative Genetics. New York: Longman.

[11] Ferré J, Escriche B, Bel Y & Van Rie J. (1995) Biochemistry and genetics of insect resistance to *Bacillus thuringiensis* insecticidal crystal proteins. FEMS Microbiol. Lett. 132, 1-7.

[12] Ferré J, Real MD, Van Rie J, Jansens S, Peferoen M (1991) Resistance to the *Bacillus thuringiensis* bioinsecticide in a field population of *Plutella xylostella* is due to a change in a midgut membrane receptor. Proc. Natl. Acad. Sci. USA. 88, 5119-5123.

[13] Georghiou GP Lagunes-Tejeda A (1991) The occurrence of resistance to pesticides in Arthropods. Food and Agriculture Organization of the United Nations, Rome.

[14] Gould F (1998) Sustainability of transgenic insecticidal cultivars: integrating pest genetics and ecology. Annu. Rev. Entomol. 43, 701-726.

[15] Gould F, Anderson A, Jones A et al. (1997) Initial frequency of alleles for resistance to *Bacillus thuringiensis* toxins in filed populations of *Heliothis virescens*. Proc. Natl. Acad. Sci. USA 94, 3519-3523.

[16] Gould F, Anderson A, Reynolds A, Bumgarner L & Moar W (1995) Selection and genetic analysis of a *Heliothis virescens* (Lepidoptera: Noctuidae) strain with high levels of resistance to *Bacillus thuringiensis* toxins. J. Econ. Entomol. 88, 1545-1559.

[17] Gould F, Martínez-Ramírez A, Anderson A, Ferré J, Silva FJ, Moar WJ (1992) Broad-spectrum resistance to *Bacillus thuringiensis* toxins in *Heliothis virescens*. Proc. Natl. Acad. Sci. USA 89, 7986-7990.

[18] Groeters FR, Tabashnik BE, Finson N & Johnson MW (1994) Fitness costs of resistance to *Bacillus thuringiensis* in the diamondback moth (*Plutella xylostella*). Evolution 48, 197-201.

[19] Hama H, Suzuki K & Tanaka H (1992) Inheritance and stability of resistance to *Bacillus thuringiensis* formulations of the diamondback moth, *Plutella xylostella* (Linnaeus) (Lepidoptera: Yponomeutidae). Appl. Entomol. Zool. 27, 355-362.

[20] Heckel DG, Gahan LC, Gould F & Anderson A (1997) Identification of a linkage group with a major effect on resistance to *Bacillus thuringiensis* Cry1Ac endotoxin in the tobacco budworm (Lepidoptera: Noctuidae). J. Econ. Entomol. 90, 75-86.

[21] Huang F, Higgins RA & Buschman LT (1997) Baseline susceptibility and changes in susceptibility to *Bacillus thuringiensis* subsp. *kurstaki* under selection pressure in European corn borer (Lepidoptera: Pyralidae). J. Econ. Entomol. 90, 1137-1143.

[22] Johnson DE, Brookhart GL, Kramer FJ, Barnett BD & McGaughey WH (1990) Resistance to *Bacillus thuringiensis* by the Indian meal moth, *Plodia interpunctella*: comparison of midgut proteinases from susceptible and resistant larvae. J. Invertebr. Pathol. 55, 235-244.

[23] Johnson DE & McGaughey WH (1996) Natural mortality among Indianmeal moth larvae with resistance to *Bacillus thuringiensis*. J. Invertebr. Pathol. 68, 170-172.

[24] Kinsinger RA, McGaughey WH (1979) Susceptibility of populations of Indianmeal moth and almond moth to *Bacillus thuringiensis* isolates (Lepidoptera: Pyralidae). J. Econ. Entomol. 72, 346-349.

[25] Lee MK, Rajamohan F, Gould F, Dean DH (1995) Resistance to *Bacillus thuringiensis* Cry1A δ-endotoxins in a laboratory-selected *Heliothis virescens* strain is related to receptor alteration. Appl. Environ. Microbiol. 61, 3836-3842.

[26] Liu YB & Tabashnik BE (1997) Inheritance of resistance to the *Bacillus thuringiensis* toxin Cry1C in the diamondback moth. Appl. Environ. Microbiol. 63, 2218-2223.

[27] Liu YB, Tabashnik BE & Pusztai-Carey M (1996) Field-evolved resistance to *Bacillus thuringiensis* toxin CryIC in diamondback moth (Lepidoptera: Plutellidae). J. Econ. Entomol. 89, 798-804.

[28] Luo K, Banks D & Adang MJ (1999) Toxicity, binding, and permeability analyses of four *Bacillus thuringiensis* Cry1 _-endotoxins using brush border membrane vesicles of *Spodoptera exigua* and *Spodoptera frugiperda*. Appl. Environ. Microbiol. 65, 457-464.

[29] Luo K, Tabashnik BE & Adang MJ (1997) Binding of *Bacillus thuringiensis* Cry1Ac toxin to aminopeptidase in susceptible and resistant diamondback moths (*Plutella xylostella*). Appl. Environ. Microbiol. 63, 1024-1027.

[30] MacIntosh SC, Stone TB, Jokerst RS & Fuchs RL (1991) Binding of *Bacillus thuringiensis* proteins to a laboratory-selected line of *Heliothis virescens*. Proc. Natl. Acad. Sci. USA 88, 8930-8933.

[31] Martínez-Ramírez AC, Escriche B, Real MD, Silva FJ & Ferré J (1995) Inheritance of resistance to a *Bacillus thuringiensis* toxin in a field population of diamondback moth (*Plutella xylostella*). Pestic. Sci. 43, 115-120.

[32] Masson L, Mazza A, Brousseau R & Tabashnik B (1995) Kinetics of *Bacillus thuringiensis* toxin binding with brush border membrane vesicles from susceptible and resistant larvae of *Plutella xylostella*. J. Biol. Chem. 270, 11887-11896.

[33] McGaughey WH (1985) Insect resistance to the biological insecticide *Bacillus thuringiensis*. Science 229, 193-195.

[34] McGaughey WH & Beeman RW (1988) Resistance to *Bacillus thuringiensis* in colonies of Indianmeal moth and almond moth (Lepidoptera: Pyralidae). J. Econ. Entomol. 81, 28-33.

[35] McGaughey WH & Johnson DE (1992) Indianmeal moth (Lepidoptera: Pyralidae) resistance to different strains and mixtures of *Bacillus thuringiensis*. J. Econ. Entomol. 85, 1594-1600.

[36] Moar WJ, Pusztai-Carey M, van Faassen H et al (1995) Development of *Bacillus thuringiensis* CryIC resistance by *Spodoptera exigua* (Hubner) (Lepidoptera: Noctuidae). Appl. Environ. Microbiol. 61, 2086-2092.

[37] Mohammed SI, Johnson DE & Aronson AI (1996) Altered binding of the Cry1Ac toxin to larval membranes but not to the toxin-binding protein in *Plodia interpunctella* selected for resistance to different *Bacillus thuringiensis* isolates. Appl. Environ. Entomol. 62, 4168-4173.

[38] Müller-Cohn J, Chaufaux J, Buisson C, Gilois N, Sanchis V & Lereclus D (1996) *Spodoptera littoralis* (Lepidoptera: Noctuidae) resistance to CryIC and cross-resistance to other *Bacillus thuringiensis* crystal toxins. J. Econ. Entomol. 89, 791-797.

[39] Oppert B, Kramer KJ, Beeman RW, Johnson D & McGaughey WH (1997) Proteinase-mediated insect resistance to *Bacillus thuringiensis* toxins. J. Biol. Chem. 272, 23473-23476.

[40] Oppert B, Kramer KJ, Johnson D, Upton S & McGaughey WH (1996) Luminal proteinases from *Plodia interpunctella* and the hydrolysis of *Bacillus thuringiensis* CryIA(c) protoxin. Insect Biochem. Molec. Biol. 26, 571-583.

[41] Rahardja U & Whalon ME (1995) Inheritance of resistance to *Bacillus thuringiensis* subsp. *tenebrionis* CryIIIA d-endotoxin in Colorado potato beetle (Coleoptera: Chrysomelidae). J. Econ. Entomol. 88, 21-26.

[42] Rossiter M, Yendol WG & Dubois NR (1990) Resistance to *Bacillus thuringiensis* in gypsy moth (Lepidoptera: Lymantriidae): genetic and environmental causes. J. Econ. Entomol. 83, 2211-2218.

[43] Schnepf E, Crickmore N, Van Rie J, *et al.* (1998) *Bacillus thuringiensis* and its pesticidal crystal proteins. Microbiol. Mol. Biol. Rev. 62, 775-806.

[44] Sims SR & Stone TB (1991) Genetic basis of tobacco budworm resistance to an engineered *Pseudomonas fluorescens* expressing the δ-endotoxin of *Bacillus thuringiensis kurstaki*. J. Invertebr. Pathol. 57, 206-210.

[45] Stone TB, Sims SR & Marrone PG (1989) Selection of tobacco buworm for resistance to a genetically engineered *Pseudomonas fluorescens* containing the _-endotoxin of *Bacillus thuringiensis* subsp. *kurstaki*. J. Invertebr. Pathol. 53, 228-234.

[46] Tabashnik BE (1994) Evolution of resistance to *Bacillus thuringiensis*. Annu. Rev. Entomol. 39, 47-79.

[47] Tabashnik BE, Finson N, Groeters F et al (1994) Reversal of resistance to *Bacillus thuringiensis* in *Plutella xylostella*. Proc. Natl Acad. Sci. 91, 4120-4124.

[48] Tabashnik BE, Finson N & Johnson MW (1991) Managing resistance to *Bacillus thuringiensis*: lessons from the diamondback moth (Lepidoptera: Plutellidae). J. Econ. Entomol. 84, 49-55.

[49] Tabashnik BE, Finson N, Johnson MW & Heckel D (1995) Prolonged selection affects stability of resistance to *Bacillus thuringiensis* in diamondback moth (Lepidoptera: Plutellidae). J. Econ. Entomol. 88, 219-224.

[50] Tabashnik BE, Groeters FR, Finson N & Johnson MW (1994) Instability of resistance to *Bacillus thuringiensis*. Biocontrol Sci. Technol. 4, 419-426.

[51] Tabashnik BE, Malvar T, Liu Y-B et al (1996) Cross-resistance of diamondback moth indicates altered interactions with domain II of *Bacillus thuringiensis* toxins. Appl. Environ. Microbiol. 62, 2839-2844.

[52] Tabashnik BE, Liu YB, Finson N, Masson L & Heckel DG (1997) One gene in diamondback moth confers resistance to four *Bacillus thuringiensis* toxins. Proc. Natl. Acad. Sci. USA 94, 1640-1644.

[53] Tabashnik BE, Liu YB, Malvar T, et al (1997) Global variation in the genetic and biochemical basis of diamondback moth resistance to *Bacillus thuringiensis*. Proc. Natl. Acad. Sci. USA 94, 12780-12785.

[54] Tabashnik BE, Liu YB, Malvar T, Heckel DG, Masson L & Ferré J (1998) Insect resistance to *Bacillus thuringiensis*: uniform or diverse? Phil. Trans. R. Soc. Lond. B 353, 1751-1756.

[55] Tabashnik BE, Schwartz JM, Finson N & Johnson MW (1992) Inheritance of resistance to *Bacillus thuringiensis* in diamondback moth (Lepidoptera: Plutellidae). J. Econ. Entomol. 85, 1046-1055.

[56] Tang JD, Gilboa S, Roush, RT & Shelton AM (1997) Inheritance, stability, and lack-of-fitness costs of field-selected resistance to *Bacillus thuringiensis* in diamondback moth (Lepdoptera: Plutellidae) from Florida. J. Econ. Entomol. 90, 732-741.

[57] Tang JD, Shelton AM, Van Rie J, et al (1996) Toxicity of *Bacillus thuringiensis* spore and crystal protein to resistant diamondback moth (*Plutella xylostella*). Appl. Environ. Microbiol. 62, 564-569.

[58] van Frankenhuyzen K, Nystrom CW & Tabashnik BE (1995) Variation in tolerance to *Bacillus thuringiensis* among and within populations of the spruce budworm (Lepidoptera: Tortricidae) in Ontario. J. Econ. Entomol. 88, 97-105.

[59] Van Rie J, Jansens S, Höfte H, Degheele D & Van Mellaert H (1989) Specificity of *Bacillus thuringiensis* δ-endotoxins - Importance of specific receptors on the brush border membranes of the mid-gut of target insects. Eur. J. Biochem. 186, 239-247.

[60] Van Rie J, McGaughey WH, Johnson DE, Barnett BD & Van Mellaert H (1990) Mechanism of insect resistance to the microbial insecticide *Bacillus thuringiensis*. Science 247, 72-74.

[61] Whalon ME, Miller DL, Hollingworth RM, Grafius EJ & Miller JR (1993) Selection of a Colorado potato beetle (Coleoptera: Chrysomelidae) strain resistant to *Bacillus thuringiensis*. J. Econ. Entomol. 86, 1516-1521.

[62] Wirth MC & Georghiou GP (1997) Cross-resistance among CryIV toxins of *Bacillus thuringiensis* subsp. *israelensis* in *Culex quinquefasciatus* (Diptera: Culicidae). J. Econ. Entomol. 90, 1471-1477.

[63] Wright DJ, Iqbal M, Granero F & Ferré J (1997) A change in a single midgut receptor in the diamondback moth (*Plutella xylostella*) is only in part responsible for field resistance to *Bacillus thuringiensis* subsp. *kurstaki* and *B. thuringiensis* subsp. *aizawai*. Appl. Environ. Microbiol. 63, 1814-1819.

Chapter 3.5

Mode of action of *Bacillus sphaericus* on mosquito larvae: incidence on resistance

Jean-François Charles[1], Maria Helena Silva-Filha[2] & Christina Nielsen-LeRoux[1]

[1]*Bactéries & Champignons Entomopathogènes, Institut Pasteur, 25 rue du Dr. Roux, 75724 Paris cedex 15, France, and* [2]*Centro de Pesquisas Aggeu Magalhães-FIOCRUZ, Av. Moraes Rêgo s/n 50670-420 Recife PE, Brazil*

Key words: *Bacillus sphaericus*, crystal Bin toxin, mosquito, midgut, receptor, α-glucosidase, resistance

Abstract: The larvicidal activity of *Bacillus sphaericus* is due mainly to the crystal toxin (Bin), which is composed of two polypeptides, BinA and BinB, acting in synergy. After ingestion of *B. sphaericus* by the larva, these proteins are released into the midgut, activated by midgut serine proteinases and, in susceptible mosquito species, they bind to a specific receptor present on apical midgut brush border membranes. The resulting damage to the midgut cells kills the mosquito. The receptor in *Culex pipiens* has been shown to be a 60kDa α-glucosidase, anchored in the epithelial cell membrane by a glycosyl-phosphatidylinositol anchor. Investigations with laboratory and field colonies of mosquitoes that have become highly resistant to the Bin toxin have shown that several mechanisms of resistance are involved, some affecting the toxin/receptor binding step, others unknown.

1.   INTRODUCTION

Bacteria were first used for the biological control of mosquitoes only after the discovery of the highly insecticidal bacterium *Bacillus thuringiensis* serovar *israelensis* [16, 18]. *Bacillus sphaericus* strains K and Q [23] were isolated in the 1960s but their larvicidal activities were so low that their use in mosquito control could never have been

J.-F. Charles et al. (eds.),
Entomopathogenic Bacteria: From Laboratory to Field Application, 237–252.
© 2000 *Kluwer Academic Publishers. Printed in the Netherlands.*

considered. The first strain of *Bacillus sphaericus* reported to be active against mosquitoes was strain SSII-1 [39]. However, it was only after the isolation in Indonesia, from dead mosquito larvae, of strain 1593, which had a much higher activity [40], that the potential of *B. sphaericus* as a biological control agent for mosquitoes was taken seriously [41].

Singer was the first to report that *B. sphaericus* may act by toxaemia rather than septicaemia, in the same way that *B. thuringiensis* acts on larvae of Lepidoptera [39]. After ingestion of the spore-crystal complex by mosquito larvae, the protein-crystal matrix rapidly dissolves in the lumen of the anterior stomach [5, 7, 44], due to the combined action of midgut proteinases and the high pH [6, 10]. The toxin is released from *B. sphaericus* crystals in all species, even those naturally resistant such as *Aedes aegypti*. Indeed, some studies have reported that differences in susceptibility to *B. sphaericus* are not due to differences in solubilisation and/or activation of the crystal toxin [2, 14, 26].

The crystal toxin (Bin toxin) is a hetero-dimer composed of 2 polypeptides, BinA and BinB, of approximately 42kDa and 51kDa, respectively, that act in synergy (see Chapter 2.3). Indeed, the two components must be present at an equivalent molar ration for the full expression of toxicity [4], whereas high doses of BinA alone may also be larvicidal [25] and may kill mosquito cells in culture [1].

2. CYTOLOGICAL AND PHYSIOLOGICAL EFFECTS

Unlike the toxins of *B. thuringiensis* serovar *israelensis*, *B. sphaericus* toxins do not completely destroy the midgut epithelium of the mosquito larva. Nevertheless, midgut alterations start as soon as 15 min after ingestion of the *B. sphaericus* spore-crystal complex [5, 11, 22, 42]. The extent of midgut damage in *Culex pipiens* is similar for ingestion of crystals of strain 1593 and 2297, but the symptoms of poisoning produced by these two strains differ in other species. Large vacuoles (and/or cytolysosomes) appear in *C. pipiens* midgut cells, whereas large area of low electron density appear in *Anopheles stephensi* midgut cells [5]. Mitochondria swelling generally occurs, as has been, described for *C. pipiens* and *A. stephensi*, and *A. aegypti* if treated with a very high dose of spore-crystals. The midgut cells, especially those of the posterior stomach and the gastric caecae, are the cells most severely damaged by the toxin, but Singh and Gill [42] also reported late damage to the in neural tissue and skeletal muscles.

Ultrastructural effects have been also reported *in vitro*, in cultured cells of *C. pipiens* within a few minutes of treatment with soluble and activated *B. sphaericus* Bin toxin. These changes involved mainly the swelling of the mitochondrial cristae and endoplasmic reticulum, followed by enlargement of vacuoles and condensation of the mitochondrial matrix [15]. These suggested that the toxin might exert its effects at the cell membrane itself. The electrophysiological effects of the native binary (Bin) toxin and its individual components, BinA and BinB, have been investigated in cultured *C. pipiens* cells using the patch clamp technique [9]. The authors reported a reduction in whole-cell membrane resistance, suggesting that the toxin creates pores or channels in the cell membrane.

In vitro assays using electrophysiological measurements in planar lipid bilayers and permeabilisation measurements in liposomes have confirmed that the Bin toxin forms channel in artificial phospholipid membranes, mainly due to the BinA component, although BinB may also, to a lesser extent, cause channel formation [31].

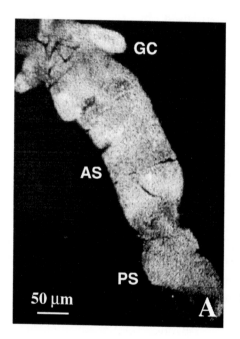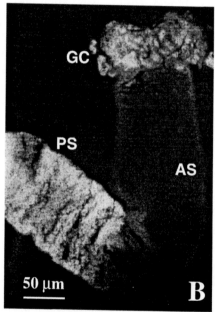

Figure 22. (A) Midgut of *C. quinquefasciatus* larva fed with a high dose of fluorescently labelled BinA, which bind over the entire midgut. (B) Larva fed with fluorescently labelled BinA and unlabelled BinB: binding of BinA is regional, restricted to the gastric caecae and posterior stomach. AS, anterior stomach, GC, gastric caecae, PS, posterior stomach. Micrographs kindly provided by Dr. C. Berry, Cardiff School of Biosciences [30].

The physiological effects of *B. sphaericus* Bin toxin *in vivo* have been little studied. Narasu & Gopinathan [24] were the noly one to report mitochondria isolated from *B. sphaericus*-treated *C. pipiens* larvae had a lower level of oxydative activity, and that activity of the larval choline acetyl transferase was inhibited in the presence of toxin.

3.   BINDING OF THE BINARY TOXIN TO A SPECIFIC RECEPTOR

Investigations using fluorescently-labelled Bin toxin have shown that the toxin binds mosquito midgut. Binding was not the same for larvae of different species fed with *B. sphaericus* [12]. A weak and variable pattern of binding to the midgut of *A. gambiae* and *A. stephensi* was observed, whereas a strong and regionalised binding was observed for *C. pipiens* [13]. *In vitro*, Davidson *et al.* [15] showed using *C. pipiens* cell lines that the bound toxin was apparently internalised by receptor-mediated endocytosis, probably involving a glycoprotein receptor containing N-acetyl-D-glucosamine. Oei *et al.* [30] confirmed that binding of the Bin toxin to *C. pipiens* midgut was regionalised, and that BinB bound to the same places in the midgut as the whole Bin toxin, whereas BinA bound uniformly but weakly to the whole midgut (Fig. 1).

Evidence that a specific receptor was involved in binding of the toxin to midgut apical cell membranes was provided by Nielsen-LeRoux & Charles [27], in experiments with radiolabelled Bin2 toxin (see Chapter 2.3) and isolated brush border membrane fractions (BBMF) of *C. pipiens*. Similar experiments with *Aedes aegypti*, a naturally resistant species, gave no specific binding [27], whereas a specific receptor was also shown to be present in *A. gambiae* and *A. stephensi* species [36]. *In vitro* binding assays seem to confirm that binding affinity is correlated with the *in vivo* susceptibility of larvae: the affinity of binding to BBMF is higher for *C. pipiens* larvae than for *A. gambiae*, and the affinity of BBMF binding is higher for *A. gambiae* than for *A. stephensi* larvae [36] (Fig. 2).

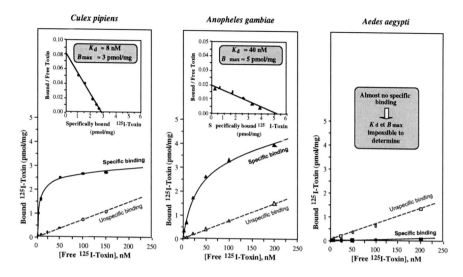

Figure 2. Comparative binding of ^{125}I-*B. sphaericus* Bin2 toxin to the BBMF of three mosquito species; *C. pipiens* (A), *A. gambiae* (B) and *A. aegypti* (C). Insets : same data in Scatchard plots, making possible the calculation of toxin/receptor dissociation constants (K_d) and binding site concentrations (B_{max} or R_t).

Kinetics studies of toxin binding showed that dissociation of the Bin toxin from BBMF was fast in *C. pipiens* and *A. gambiae*, but that association was different in these species [36]. Saturation of all the receptors was achieved in few hours in *C. pipiens*, whereas binding rate was much slower in *A. gambiae*. The slow and continuous binding observed for *A. gambiae* may correspond to toxin oligomerisation to the Bin molecules already bound, rather than binding to new sites. Indeed, oligomerisation of the Bin toxin has been reported in experiments conducted by Shanmugavelu *et al.* [34]. Bioassays with *C. quinquefasciatus* larvae, using binary toxins derived from single mutants of BinA and BinB, showed that substitution of alanine residues at some sites in both the BinB and the BinA polypeptides resulted in the total loss of larvicidal activity. However, mixing two non-toxic derivatives of the same peptide (*i.e.*, one mutated at the N-terminal end and the other at the C-terminal end of either the BinB or the BinA peptide), restored toxicity. These results indicate that mutated binary toxins can functionally complement each other by forming oligomers.

Another difference between *C. pipiens* and *A. gambiae* in toxin binding to BBMF involves the individual binding properties of each polypeptide (BinB and BinA) (Fig. 3). BinB competes for the binding site on BBMF from both species with an affinity similar to that for the whole

Bin toxin. In contrast, the binding affinity of BinA is much lower for BBMF from *C. pipiens* than for BBMF from *A. gambiae*. If BinB is added, maximal binding of BinA to *C. pipiens* BBMF is reached at a 1:1 molar ratio of each polypeptide (Fig. 3). For *A. gambiae*, the binding of either BinA or BinB greatly increases binding of the other polypeptide and maximal binding is achieved when the two proteins are present in a 1:1 molar ratio. These experiments also demonstrated the very strong affinity between the BinA and BinB proteins [8].

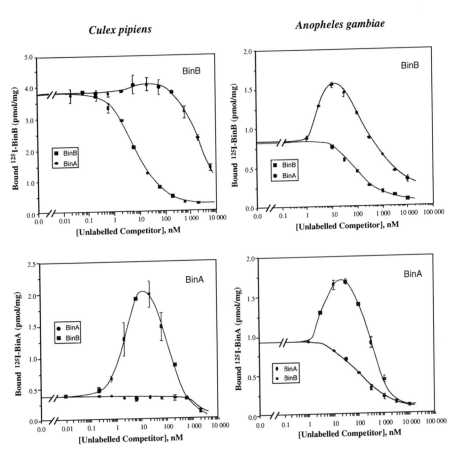

Figure 3. Competitive binding experiments with the *B. sphaericus* BinA and BinB polypeptides and BBMF from *C. pipiens* (left) and *A. gambiae* (right).
Top : competition experiments involving ^{125}I-BinB toxin and homologous BinB or heterologous BinA competitors. Bottom : competition experiments involving ^{125}I-BinA toxin and homologous BinA or heterologous BinB competitors.

4.   THE TOXIN RECEPTOR IN *CULEX PIPIENS*

The receptor was identified based on its affinity for the Bin toxin. *C. pipiens* and *A. aegypti* BBMF proteins released by phosphatidylinositol phospholipase C (PI-PLC) were radiolabelled and incubated with the Bin2 toxin immobilised on Sephadex beads (Bin-beads). An autoradiograph of the samples separated by SDS-PAGE showed that a 60 kDa protein from *C. pipiens* bound to Bin-beads, whereas no bands was detected for *A. aegypti* [37]. The molecule that serves as the receptor for *B. sphaericus* Bin toxin in the midgut of *C. pipiens* larvae was therefore identified as a 60 kDa protein attached to the cell membrane by a glycosyl-phosphatidylinositol (GPI) anchor [37]. In *A. gambiae*, the toxin receptor is also attached to the cellular epithelium by a GPI anchor, but no binding protein has yet been identified.

The 60 kDa protein was also detected by Western blotting using *C. pipiens* BBMF proteins solubilised with CHAPS detergent, after incubation with Bin-beads or BinB-beads, separation by SDS-PAGE, transfer and immunodetection using polyclonal antibodies againts soluble BBMF proteins. In both cases, binding to the 60 kDa protein binding is specific: the intensity of the signal is much lower if incubation is carried out in the presence of the competitor, free Bin toxin [37]. These experiments also confirm that BinB binds to the receptor as efficiently as does the whole Bin toxin, as previously reported [8].

4.1   Partial purification of the receptor

Fractionation of CHAPS extracts by ion exchange chromatography (Mono Q, Pharmacia) resulted in the elution of 4 peaks of proteins from the column. Affinity binding assays performed with pooled fractions for each peak showed that a 60 kDa protein eluted in peak I specifically bound Bin beads (Fig. 4). This protein was purified and subjected to amino acid microsequencing. It was found to be very similar in sequence to maltase-like proteins of the α-amylase family [37]. Sequence identity was found to maltases already described for dipteran insects including Mal1 from *A. aegypti* [20], Agm1 and Agm2 from *A. gambiae*, LvpD, LvpH and LvpL from *Drosophila melanogaster* [47], Mav1 and Mav2 from *D. virilis* [43].

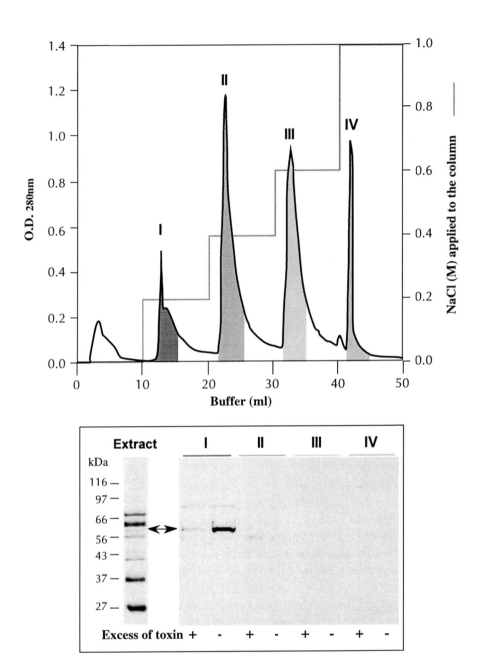

Figure 4. Anion exchange chromatography of CHAPS-solubilised BBMF proteins from *C. pipiens*. Grey areas correspond to fractions from peaks I to IV that were analysed by Western blotting after affinity binding to Bin-beads, carried out in the absence (-) or presence (+) of an excess of free Bin toxin.

Table 1. Maltase activities in PI-PLC-released proteins from *C. pipiens* BBMF preparations and evaluation of enzyme enrichment in BBMF and CHAPS-extracts.

Enzyme	Enzymes released from BBMF (ΔA_{405nm}/h)		Enzyme enrichment factor[1]	
	Control	PI-PLC treatment	BBMF	CHAPS Extracts
α-glucosidase	0.052	0.454	2.8	4.1
α-amylase	0.428	0.405	0.2	0.2

[1]Determined as the ratio of the specific enzyme activity in the sample to the specific enzyme activity in the starting material used in the first step of BBMF preparation.

The α-amylase family, *sensu lato*, comprises sugar hydrolases with about 20 different specificities and is widely distributed among plants, vertebrates, invertebrates and micro-organisms [21]. The level of identity for the amino acid sequences of these enzymes is low due to the diversity of sources butsome conserved domains have been identified such as the $(\alpha/\beta)_8$, "TIM barrel", an important marker of most enzymes of this family [21]. Amino acid microsequencing of the second fragment from the 60 kDa protein showed that 9 of 17 amino acids were involved in the $(\alpha/\beta)_8$ structure [37]. This protein has recently been cloned and sequenced by I. Darboux and D. Pauron (in preparation). It consists of 580 amino acids, consistent with the molecular weight of 60 kDa estimated in affinity binding assays, and its C terminal extremity contains a sequence typical of GPI anchor.

4.2 An α-glucosidase as the receptor for the Bin toxin

Microsequencing showed that the 60 kDa protein was an α-amylase, *sensu lato*. At least two different α-amylase activities were found in peak I which contained the 60 kDa protein: an α-amylase *sensu stricto* (EC 3.2.1.1) and an α-glucosidase (EC 3.2.1.20). Both activities were found in BBMF, in CHAPS extracts and in peak I eluted from the chromatography column [37]. The GPI anchoring of this enzyme, the enrichment of α-glucosidases in midgut BBMF preparations (Table 1) and α-amylase activity in the loose sense provided strong evidence that the receptor for the Bin toxin is an α-glucosidase rather than an α-amylase *sensu stricto*. It is known that α-glucosidases play a key physiological role in insect digestion and this is not the first case of enzymes acting as receptors for pathogens, as aminopeptidases were found to act as receptors of *B. thuringiensis* Cry1 toxins in the BBMF of some lepidopteran larvae (see Chapter 3.2). However, it is unclear whether the catalytic site and the binding site on this molecule are independent or related structures.

5.   TOXIN RECEPTOR INTERACTION IN *B. SPHAERICUS*-RESISTANT COLONIES OF *C. PIPIENS*

Previous work on susceptible *Culex* and *Anopheles* sp. larvae, has shown that receptor binding is a key step in the mode of action of the Bin toxin of *B. sphaericus*. For *Culex*, BinB is apparently the major binding component involved. This indicates that this heterodimeric toxin interacts like a single-site insecticide. The risk of the development of resistance is therefore high. This may not be the case for *Anopheles*, in which both components are able to interact in the receptor binding step [8].

The appearance of *C. pipiens* larvae resistant to *B. sphaericus* focused attention on the possible involvement of the receptor in the mechanisms of resistance. Some *Culex* colonies have developed various degrees of resistance under laboratory selection [17, 33] or when subjected to intensive field exposure [32, 35, 38, 46]. The only published work relating to resistance mechanisms and genetics involved studies with two colonies, GEO and SPHAE, both highly resistant to *B. sphaericus* [17, 38]. The laboratory-selected GEO colony was more than 100,000-times more resistant than controls, and the field-collected SPHAE colony was about 10,000-times more resistant. The mechanisms of resistance are different for these two colonies. In GEO, resistance is due to the failure of the toxin-receptor binding step: almost no specific binding was observed between the Bin toxin and the BBMF [28]. In contrast, the Bin toxin boundspecifically to BBMF from SPHAE larvae, despite the high level of *in vivo* resistance observed for the SPHAE colony [29].

The genetic basis of resistance has been investigated for both the SPHAE and GEO colonies. BBMF isolated from the F1 progeny produced by crossing mosquitoes from the resistant GEO colony with susceptible mosquitoes, showed specific binding with one class of binding sites, consistent with the recessive nature of resistance for both colonies, resulting in F1 susceptibility being close to that of the susceptible parental colony. The resistant backcross (F1xR) of the GEO colony indicated that there was a positive correlation between the presence of a plateau in the dose mortality response curve and there being two possible classes of binding site in the BBMF [28].

Genetic investigation has also indicated that resistance may be due to one major gene in each colony, clearly sex-linked for the SPHAE colony [29] whereas it may be autosomal for the GEO colony. The finding that the genes for resistance in the two colonies may be at two different loci provides further evidence for there being two different mechanisms of resistance.

Investigations of other resistant field colonies from various regions in China [46], Tunisia (N. Pasteur, personal communication), France (C. Chevillon, personal communication) and India (S. Poopatthi, personal communication) are in progress.

Once the 60 kDa, α-glucosidase receptor molecule was identified in the susceptible *C. pipiens* colony [37], investigations were carried out to determine whether this molecule was present in the BBMF of resistant colonies. Affinity binding studies with BBMF proteins, solubilised with PI-PLC or CHAPS, and Bin-beads, showed that the 60 kDa receptor was present in the SPHAE colony but not in the GEO colony, consistent with BBMF binding experiments (Silva-Filha *et al.*, unpublished data). Further investigation of the genes encoding these receptor proteins will show whether the lack of binding is due to the absence of a functional receptor molecule or to the absence of the molecule itself.

The resistance of the SPHAE colony does not involve the toxin-receptor binding step, so other mechanisms, as yet unknown, must be involved [29]. However, the proteinase processing of protoxin to give activated toxin is still functional [29] and binding kinetics (association and dissociation of the toxin with/from the receptor) are similar to those of the susceptible colony [36]. The mechanism of resistance in this colony may be at yet another level, possibly involving the receptor molecule, but at a step other than the binding step. There may, for example, be a structural change in the receptor, modifying presentation of the Bin toxin after binding, and rendering the toxic part of the molecule harmless.

The Bin toxin (BinA in particular) forms pores in artificial bilayers [31]. It is currently being investigated whether this is the case in BBMF too, and special attention is being paid to BBMV from the SPHAE colony because this may be the mechanism of resistance.

In case of resistance due to a lack of binding of the Bin2 binary toxin of the commercialy available 2362 and 1593 strains, we need to know whether the binary toxins of other strains, with amino acid sequences slightly different from that of Bin2 (see Chapter 2.3) could bind to the receptor. In aparticular, some natural *B. sphaericus* strains are able to kill resistant *Culex* larvae (see Chapter 6.2). Such investigations are ongoing, but it is likely that the four Bin toxin groups share the same receptor binding site, resulting in high risk of cross-resistance between Bin toxins.

The *B. sphaericus* strain LP1-G is active on *Culex* larvae that are resistant to strain 2362. This strain is of particular interest because: *1)* it produces large crystals composed of binary toxin, *2)* it has a lower level of activity than other Bin toxin-producing *B. sphaericus* strains and *3)* it is equally toxic against all *Culex* larvae whether susceptible or resistant to strains 2362 and 1593. The gene encoding the LP1-G Bin4 toxin was cloned and expressed. The resulting protein had no larvicidal activity, indicating that the LP1-G strain must contain other toxic elements [45].

6.   TOXIN STRUCTURE AND *IN VIVO/IN VITRO* ACTIVITY

The primary protein structure of the LP1-G Bin toxin is very similar to that of the other three types (1, 2 and 3, see Chapter 2.3). It differs from them, however, at position 93 of the 42kDa polypeptide (BinA4), which is a serine in LP1-G BinA4, whereas in all other types of BinA toxin from *B. sphaericus* (BinA1, BinA2 and BinA3) there is a leucine at this position [19, 45]. Reciprocal site-directed mutagenesis was performed to replace the serine at position 93 with a leucine in BinA4 from LP1-G and to replace the leucine with a serine in the BinA2 protein from strain 1593. Native and mutated genes were cloned and overexpressed in a *B. thuringiensis israelensis* crystal-minus (Cry⁻) strain. Inclusion bodies were purified and tested for activity against *C. pipiens* larvae. The mutated LP1-G Bin4 protein was toxic *in vivo*, which is not the case for the native LP1-G protein, and the reciprocal mutation in Bin2 (from strain 1593) led to a significant loss of toxicity. *In vitro* receptor-binding studies, saturation, heterologous and homologous competition experiments were therefore performed. Similar binding behaviour was observed for both native and mutated toxins. These results suggest that amino acid 93 of the BinA polypeptide is a key element for the optimal structural conformation of the BinA/BinB complex responsible for the toxicity of *B. sphaericus* Bin toxins [45]. These findings, along with those of previous studies showing that the amino acids in positions 99 and 104 of the BinA play an important role in both toxicity and specificity [3], suggest that despite the high level of identity observed for Bin toxin sequences overall, very small differences are sufficient to affect the structural integrity of the molecule, leading to a partial loss of funtionality, which may affect the steps after the initial toxin / receptor interaction.

7. CONCLUSIONS/PERSPECTIVES

Despite identification of the α-glucosidase as the midgut receptor of the *B. sphaericus* Bin toxin in *C. pipiens*, almost nothing is known about the receptor in other susceptible species such as *A. gambiae* and in naturally resistant species such as *A. aegypti*. For these species, as for the colonies of *C. pipiens* that are resistant to Bin toxins, knowledge of the sequences of the genes encoding these midgut α-glucosidases may improve our understanding of the reasons why these molecules function or do not function as Bin toxin receptors. The relationships between the physiological role of these molecules and their function as receptors should also be investigated. From a practical point of view, identification of the mutations responsible for resistance may also be useful for screening purpose and for monitoring resistance in *B. sphaericus*-treated *Culex* populations.

Investigations into the possible intracellular activity of the Bin toxin are necessary to elucidate further of the mode of action of these toxins, and its relationships to the still unknown mechanisms of resistance.

REFERENCES

[1] Baumann L & Baumann P (1991) Effects of components of the *Bacillus sphaericus* toxin on mosquito larvae and mosquito-derived tissue culture-grown cells. Curr. Microbiol. 23, 51-57

[2] Baumann P, Unterman BM, Baumann L *et al.* (1985) Purification of the larvicidal toxin of *Bacillus sphaericus* and evidence for high-molecular-weight precursors. J. Bacteriol. 163, 738-747

[3] Berry C, Hindley J, Ehrhardt AF *et al.* (1993) Genetic determinants of host ranges of *Bacillus sphaericus* mosquito larvicidal toxins. J. Bacteriol. 175, 510-518

[4] Broadwell AH, Baumann L & Baumann P (1990) Larvicidal properties of the 42 and 51 kilodalton *Bacillus sphaericus* proteins expressed in different bacterial hosts: evidence for a binary toxin. Curr. Microbiol. 21, 361-366

[5] Charles J-F (1987) Ultrastructural midgut events in Culicidae larvae fed with *Bacillus sphaericus* 2297 spore/crystal complex. Ann. Inst. Pasteur/Microbiol. 138, 471-484

[6] Charles J-F & de Barjac H (1981) Variation du pH de l'intestin moyen d'*Aedes aegypti* en relation avec l'intoxication par les cristaux de *Bacillus thuringiensis* var. *israelensis* (sérotype H 14). Bull. Soc. Path. Exot. 74, 91-95

[7] Charles J-F & Nicolas L (1986) Recycling of *Bacillus sphaericus* 2362 in mosquito larvae: a laboratory study. Ann. Microbiol. (Inst. Pasteur) 137B, 101-111

[8] Charles J-F, Silva-Filha MH, Nielsen-LeRoux C, Humphreys M & Berry C (1997) Binding of the 51- and 42-kDa individual components from the *Bacillus sphaericus* crystal toxin on mosquito larval midgut membranes from *Culex* and *Anopheles* sp. (Diptera: Culicidae). FEMS Microbiol. Lett. 156, 153-159

[9] Cokmus C, Davidson EW & Cooper K (1997) Electrophysiological effects of *Bacillus sphaericus* binary toxin on cultured mosquito cells. J. Invertebr. Pathol. 69, 197-204

[10] Dadd RH (1975) Alkalinity within the midgut of mosquito larvae with alkaline-active digestive enzymes. J. Insect Physiol. 21, 1847-1853

[11] Davidson EW (1981) A review of the pathology of bacilli infecting mosquitoes, including an ultrastructural study of larvae fed *Bacillus sphaericus* 1593 spores. Dev. Industr. Microbiol. 22, 69-81

[12] Davidson EW (1988) Binding of the *Bacillus sphaericus* (Eubacteriales: Bacillaceae) toxin to midgut cells of mosquito (Diptera: Culicidae) larvae: relationship to host range. J. Med. Entomol. 25, 151-157

[13] Davidson EW (1989) Variation in binding of *Bacillus sphaericus* toxin and wheat germ agglutinin to larval midgut cells of six species of mosquitoes. J. Invertebr. Pathol. 53, 251-259

[14] Davidson EW, Bieger AL, Meyer M & Shellabarge RC (1987) Enzymatic activation of the *Bacillus sphaericus* mosquito larvicidal toxin. J. Invertebr. Pathol. 50, 40-44

[15] Davidson EW, Shellabarger C, Meyer M & Bieber AL (1987) Binding of the *Bacillus sphaericus* mosquito larvicidal toxin to cultured insect cells. Can. J. Microbiol. 33, 982-989

[16] de Barjac H (1978) Une nouvelle variété de *Bacillus thuringiensis* très toxique pour les moustiques: *Bacillus thuringiensis* var. *israelensis*, sérotype 14. C. R. Acad. Sc. Paris, sér. D 286, 797-800

[17] Georghiou GP, Malik JI, Wirth M & Sainato K (1992) Characterization of resistance of *Culex quinquefasciatus* to the insecticidal toxins of *Bacillus sphaericus* (strain 2362). in: University of California, Mosquito Control Research, Annual Report 1992 (J. Coast & L. Chase, eds), University of California Press, Berkeley pp. 34-35

[18] Goldberg LJ & Margalit J (1977) A bacterial spore demonstrating rapid larvicidal activity against *Anopheles sergentii*, *Uranotaenia unguiculata*, *Culex univitattus*, *Aedes aegypti* and *Culex pipiens*. Mosq. News 37, 355-358

[19] Humphreys MJ & Berry C (1998) Variants of the *Bacillus sphaericus* binary toxins: implications for differential toxicity strains. J. Invertebr. Pathol. 71, 184-185

[20] James AA, Blackmer K & Racioppi JV (1989) A salivary gland-specific, maltase-like gene of the vector mosquito, *Aedes aegypti*. Gene 75, 73-83

[21] Janecek S (1997) An α-amylase family: molecular biology and evolution. Prog. Biophys. Molec. Biol. 67, 67-97

[22] Karch S & Coz J (1983) Histopathologie de *Culex pipiens* Linné (Diptera, Culicidae) soumis à l'activité larvicide de *Bacillus sphaericus* 1593-4. Cah. ORSTOM, sér. Ent. méd. Parasitol. XXI, 225-230

[23] Kellen W, Clark T, Lindergren J *et al.* (1965) *Bacillus sphaericus* Neide as a pathogen of mosquitoes. J. Invertebr. Pathol. 7, 442-448

[24] Narasu ML & Gopinathan KP (1988) Effect of *Bacillus sphaericus* 1593 toxin on choline acetyl transferase and mitochondrial oxidative activities of the mosquito larvae. Indian J. Biochem. Biophys. 25, 253-256

[25] Nicolas L, Charles J-F & de Barjac H (1993) *Clostridium bifermentans* serovar *malysia*: characterization of putative mosquito larvicidal proteins. FEMS Microbiol Lett 113, 23-28

[26] Nicolas L, Lecroisey A & Charles J-F (1990) Role of the gut proteinases from mosquito larvae in the mechanism of action and the specificity of the *Bacillus sphaericus* toxin. Can. J. Microbiol. 36, 804-807

[27] Nielsen-LeRoux C & Charles J-F (1992) Binding of *Bacillus sphaericus* binary toxin to a specific receptor on midgut brush-border membranes from mosquito larvae. Eur. J. Biochem. 210, 585-590

[28] Nielsen-LeRoux C, Charles J-F, Thiéry I & Georghiou GP (1995) Resistance in a laboratory population of *Culex quinquefasciatus* (Diptera: Culicidae) to *Bacillus sphaericus* binary toxin is due to a change in the receptor on midgut brush-border membranes. Eur. J. Biochem. 228, 206-210

[29] Nielsen-LeRoux C, Pasquier F, Charles J-F *et al.* (1997) Resistance to *Bacillus sphaericus* involves different mechanisms in *Culex pipiens* mosquito larvae (Diptera: Culicidae). J. Med. Entomol. 34, 321-327

[30] Oei C, Hindley J & Berry C (1992) Binding of purified *Bacillus sphaericus* binary toxin and its deletion derivates to *Culex quinquefasciatus* gut: elucidation of functional binding domains. J. Gen. Microbiol. 138, 1515-1526

[31] Potvin L, Charles J-F, Berry C *et al.* (2000) Permeabilisation of model lipid membranes by *Bacillus sphaericus* mosquitocydal binary toxin and its individual components. *(in preparation)*

[32] Rao DR, Mani TR, Rajendran R *et al.* (1994) Development of a high level of resistance to *Bacillus sphaericus* in a field population of *Culex quinquefasciatus* from Kochi, India. J. Am. Mosq. Contr. Assoc. 11, 1-5

[33] Rodcharoen J & Mulla MS (1994) Resistance development in *Culex quinquefasciatus* (Diptera : Culicidae) to the microbial agent *Bacillus sphaericus*. J. Econ. Entomol. 87, 1133-1140

[34] Shanmugavelu M, Rajamohan F, Kathirvel M *et al.* (1998) Functional complementation of nontoxic mutant binary toxins of *Bacillus sphaericus* 1593M generated by site-directed mutagenesis. Appl. Environ. Microbiol. 64, 756-759

[35] Silva-Filha M-H, Regis L, Nielsen-LeRoux C & Charles J-F (1995) Low-level resistance to *Bacillus sphaericus* in a field-treated population of *Culex quinquefasciatus* (Diptera: Culicidae). J. Econ. Entomol. 88, 525-530

[36] Silva-Filha MH, Nielsen-LeRoux C & Charles J-F (1997) Binding kinetics of *Bacillus sphaericus* binary toxin to midgut brush border membranes of *Anopheles* and *Culex* sp. larvae. Eur. J. Biochem. 247, 754-761

[37] Silva-Filha MH, Nielsen-LeRoux C & Charles J-F (1999) Identification of the receptor of *Bacillus sphaericus* crystal toxin in the brush border membrane of the mosquito *Culex pipiens* (Diptera: Culicidae). Insect Biochem. Molec. Biol. 29, 711-721

[38] Sinègre G, Babinot M, Quermel J-M & Gavon B (1994) First field occurrence of *Culex pipiens* resistance to *Bacillus sphaericus* in southern France. Abstr. VIIIth Eur. Meet. Society of Vector Ecology, Barcelona, Sept. 5-8 p. 17

[39] Singer S (1973) Insecticidal activity of recent bacterial isolates and their toxins against mosquito larvae. Nature 244, 110-111

[40] Singer S (1974) Entomogenous bacilli against mosquito larvae. Dev. Industr. Microbiol. 15, 187-194

[41] Singer S (1977) Isolation and development of bacterial pathogens in vectors. *in* :"Biological regulation of vectors", DHEW Publication No. (NIH) 77-1180, Bethesda pp. 3-18

[42] Singh GJP & Gill SS (1988) An electron microscope study of the toxic action of *Bacillus sphaericus* in *Culex quinquefasciatus* larvae. J. Invertebr. Pathol. 52, 237-247

[43] Vieira CP, Vieira J & Hartl DL (1997) The evolution of small gene clusters: evidence for an independent origin of the maltase gene cluster in *Drosophila virilis* and *Drosophila melanogaster*. Mol. Biol. Evol. 14, 985-93

[44] Yousten AA & Davidson EW (1982) Ultrastructural analysis of spores and parasporal crystals formed by *Bacillus sphaericus* 2297. Appl. Environ. Microbiol. 44, 1449-1455

[45] Yuan Z, Rang C, Nielsen-LeRoux C *et al.* (2000) Identification of a toxicity determinant for the *Bacillus sphaericus* crystal toxin. Eur. J. Biochem. *(in preparation)*

[46] Yuan Z, Zhang Y, Cai Q & Liu E (2000) High-level field resistance to *Bacillus sphaericus* C3-41 in *Culex quinquefasciatus* from southern China. Biocontrol. Sci. & Technol. *(in press)*

[47] Zheng L, Whang LH, Kumar V & Kafatos FC (1995) Two genes encoding midgut-specific maltase-like polypeptides from *Anopheles gambiae*. Exp. Parasitol. 81, 272-283

Chapter 4

Safety and ecotoxicology of entomopathogenic bacteria

Lawrence A. Lacey[1] & Joel P. Siegel[2]

[1]*Fruit and Vegetable Research Unit, Yakima Agricultural Research Laboratory, USDA-ARS, 5230 Konnowac Pass Road, Wapato, WA 98951, USA, and* [2]*Horticultural Crops Research Laboratory, USDA-ARS, 2021 S. Peach Ave., Fresno, CA 93727, USA*

Key words: *Bacillus thuringiensis, Bacillus sphaericus, Bacillus popilliae,* nontarget organisms, safety, environmental impact

Abstract: *Bacillus* entomopathogens, especially *Bacillus thuringiensis*, have been used extensively for control of insect pests in crops, forests, and the aquatic environment. Their safety for vertebrates and nontarget invertebrates has been thoroughly documented in a myriad of studies. Their short term effects on nontarget organisms that are unrelated to target insects is negligible. However, the effect of repeated applications on most ecosystems is relatively unknown. It is highly probable that any regular disruption of large insect communities, due to chemical or microbial insecticides or natural disaster, could have long term deleterious effects on higher trophic levels and ecosystem structure. The more diversified the food web, the less likely that complete or partial removal of a single species will result in catastrophic consequences. The more species a given intervention affects, the greater the likelihood of altering ecosystem structure. The safety and environmental impact of entomopathogenic bacteria should be evaluated in light of the risk for nontarget organisms in comparison with other interventions and the effect no treatment at all will have on an ecosystem.

1.   INTRODUCTION

Entomopathogenic spore forming bacteria, most notably *Bacillus thuringiensis* (*Bt*), are the most widely used microbial pest control agents (MPCA). The broad spectrum of susceptible hosts, production on

J.-F. Charles et al. (eds.),
Entomopathogenic Bacteria: From Laboratory to Field Application, 253–273.
© 2000 Kluwer Academic Publishers. Printed in the Netherlands.

artificial media and ease of application using conventional equipment have resulted in widespread use against several insect pests in crops, forest and aquatic habitats.

In order to register a *Bacillus* species as a microbial insecticide, a series of studies must be conducted that assess the toxicity and infectivity of the candidate organism to a designated group of invertebrate and vertebrate nontarget organisms (NTOs). The emphasis of these studies has traditionally been direct effects, typically assessed in one month laboratory studies or one season field studies. Recently, concerns have been raised about long term indirect effects on NTOs when the pest species controlled by the *Bacillus* microbial insecticide becomes unavailable as a source of food. Additionally, questions have been raised about the vulnerability of endangered species of NTO to these insecticides. In the following pages we will highlight studies that address all of these safety issues and attempt to place these data in perspective.

The safety of *Bacillus* entomopathogens for NTOs has been addressed by a number of researchers over the past 50 years. The literature before 1989 on their direct effects on specific NTOs was reviewed in several chapters in *Safety of Microbial Insecticides* [49]. Research on the safety and environmental impact of entomopathogenic *Bacillus* spp. that has been conducted since 1989 will be emphasised in this chapter.

2.   DIRECT EFFECTS OF *BACILLUS* ENTOMOPATHOGENS ON INVERTEBRATE NTOs

2.1   *Bacillus thuringiensis (Bt)*

Here we address varieties of *Bt* that do not produce the β-exotoxin. Because of its toxicity to numerous NTOs including vertebrates, all commercial formulations intended for use in crops, forests and aquatic systems no longer contain β-exotoxin. The reader interested in β-exotoxin should refer to the reviews of Sebesta *et al.* [76] and Melin & Cozzi [56]. Varieties of *Bt*, the insecticidal activity of which is based on Cry toxins (also known as δ-endotoxins), are now commercially available for use against a wide variety of insect pests including species of Lepidoptera, Coleoptera and Diptera. The individual Cry toxins are for the most part active against single orders of insect pests and may affect one to several families within an order. There are exceptions such as the Cry2 toxins which are active against certain families of Diptera and Lepidoptera. The specificity of toxins is determined by the molecular configuration of the toxin and the physiology of the host midgut and presence of toxin receptors on the midgut epithelium [20, 75].

The order with the broadest spectrum of families affected by *Bt* toxins, most notably Cry1 toxins, is the Lepidoptera. Discovery and development of hundreds of lepidopteran-active isolates and subsequent genetic manipulation of some has resulted in production of highly efficacious biopesticides for control of several lepidopteran pests of crops and forest. The vast majority of non-lepidopteran NTOs are not directly affected by exposure to commercial products or purified Cry toxins used for control of lepidopteran pests [43, 53, 56, 83, 84]. There are some exceptions that will be presented in subsequent sections of the chapter.

Concerns over the impact of *Bt* products on nontarget Lepidoptera is predominantly focused on indigenous species found mainly in forest habitats or habitats peripheral to agroecosystems. Laboratory bioassays are the starting point to determine possible untoward effects on nontarget Lepidoptera, but they may not always accurately reflect the level of impact in nature. However, phenomena that may be difficult to quantify in the field may be more readily assessed in the lab. For example, Peacock *et al.* [70] evaluated the effects of two formulations of *Bt* on 42 species of native Lepidoptera and demonstrated differential susceptibility due to species and larval age. They also showed that larvae surviving sublethal dosages of *Bt* were likely to reach adulthood.

Measurement of direct effects in field situations is the most reliable method for determining the impact of a *Bt* spray programme on nontarget Lepidoptera. Several researchers have reported on the susceptibility of nontarget Lepidoptera in forest that had been treated with *Bt* [25, 39, 60, 73, 90, 95]. Beneficial and endangered lepidopteran species are among those reported at risk [5, 25, 38]. In addition to effects within treatment zones, Whaley *et al.* [95] demonstrated that drift of aerially applied *Bt* subsp. *kurstaki* (*Btk*) for gypsy moth control, killed nontarget Lepidoptera as much as 3000 m from the application site.

Effects of *Bt* isolates with activity towards Lepidoptera have negligible effects on insects in aquatic habitats at operational dosages [17]. Kreutzweiser *et al.* [42] observed no significant effect of high concentrations of *Btk* on drift and mortality of Ephemeroptera, Plecoptera, and Trichoptera. However, they observed 30% mortality in the plecopteran, *Taeniopteryx nivalis*, exposed to the massive concentration of 600 iu/ml for 24 hours. Mortality of species from pristine lotic habitats may be due to the effects of turbidity and formulation components rather than to *Bt* toxins.

Predators that are exposed to prey that were fed lepdiopteran-active *Bt* have not been shown to be susceptible to *Bt* toxins [7, 100], but *Chrysoperla carnea* that were fed directly on purified Cry1Ab toxin in diet, responded with 57% mortality compared to 30% control mortality [29]. Several additional studies on the effect of *Bt* on predators and other NTOs are summarised by Melin and Cozzi [56] and the USDA Forest Service [89].

There has been less field testing of beetle active isolates containing Cry3 toxins against NTOs. Field trials of *Bt* subsp. tenebrionis (*Btt*) in combination with the predatory bug, *Perillus bioculatus*, for control of Colorado potato beetle, *Leptinotarsa decemlineata*, demonstrated compatibility between the bacterium and predator [9, 32]. In another study, five weekly applications of low and high label rates of a genetically engineered isolate of *Bt* for control of *L. decemlineata*, resulted in fair to good control of the beetle with no detectable effects on NTOs including predatory Hemiptera [46]. In contrast, few or no predatory Hemiptera were observed in plots treated with the systemic carbamate insecticide, aldicarb [46]. Giroux *et al.* [21] reported on the negative effects of beetle-active *Btt* on duration of development of *Coleomegilla maculata* (Coccinellidae) larvae, but concluded it did not cause mortality. Langenbruch [51] reported on the lack of untoward effects of *Btt* on other predators in the potato agroecosystem.

Isolates of *Bt* with Cry4 toxins (*e.g. Bt* subsp. *israelensis* [*Bti*] and others) are highly active against mosquitoes and black flies [48] and have been shown to kill dipteran larvae in closely related families in the sub-order Nematocera, such as certain chironomid, tipulid and blepharocerid species [2, 8, 26, 47, 61, 97]. Numerous bioassays and field trials of *Bti* against NTOs other than Nematocera have demonstrated that the vast majority are not directly affected by *Bti* toxins [8, 27, 36, 45, 47, 58, 97]. Filter feeding species are the most likely to capture and concentrate the parasporal crystals of *Bti* and formulation components that may be harmful. Wipfli & Merritt [97] produced mortality in the filter feeding mayfly, *Arthroplea bipunctata*, at 500 times the concentration required for black fly control. Increased drift of black flies and NTOs, such as species of Ephemeroptera, Plecoptera, Trichoptera, and Blepharoceridae following treatment with *Bti* has been reported by some authors ostensibly due in part to increased turbidity and formulation components [2, 12, 47, 97].

Under most conditions the majority of predators of mosquitoes and black flies are not susceptible to *Bti*. However, predaceous mosquito larvae in the genus *Toxorhynchites* are susceptible to *Bti* when fed on prey larvae that have been exposed to the bacterium [44, 52]. Because of the lack of direct deleterious effects in other predatory insects, *Bti* is an ideal microbial insecticide for use in integrated control programmes. Although its larvicidal activity is short lived in most habitats [48], undisturbed predators can continue suppression of target insects [1, 47, 66].

2.2 Direct effects of other *Bacillus* entomopathogens

Relatively few studies have been conducted on the effects of *Bacillus sphaericus* and *B. popilliae* on nontarget invertebrates. Both organisms are considerably more specific in their host spectra than *Bt*.

Bacillus sphaericus is an entomopathogen of mosquitoes, but has a markedly narrower mosquito host range than *Bti* and does not appear to directly affect nontarget fauna including chironomids and other Nematocera [1, 47, 63, 91,101]. The bacterium is attractive because of elevated activity against *Culex* and *Psorophora* species and its greater persistence in organically enriched larval habitats [62, 67]. Recycling of *B. sphaericus* in larval cadavers has been reported or suspected, further extending its activity and persistence in the environment [11,40, 67]. Spores may be returned from inaccessible substrates by feeding activity of NTOs which are not harmed by the bacterium [40, 101].

The efficacy of *B. sphaericus* against certain mosquito species, persistence in larval habitats and compatibility with predators has provided extended control in certain circumstances [1, 47].

Isolates of *B. popilliae* are specific pathogens of the Scarabaeidae with no demonstrated effects on NTOs [16, 69]. Their lethal activity is based on septicemia in the host and not on the production of toxin [6, 16] as is the case with *B. thuringiensis* and *B. sphaericus*. Although spores of *B. popilliae* may persist for protracted periods of time in the soil, they only germinate, grow and sporulate in nature within scarab hosts.

3.   INDIRECT EFFECTS OF *BT* ON NONTARGET INVERTEBRATES

The indirect effect of *Bt* on nontargets can be broadly divided into two categories: immediate impact and longer term impact. The implications of longer term impact will be addressed under section 6 of this chapter. The immediate indirect effects are most often observed in insects that prey upon or parasitise targeted insects.

Parasitoids are most commonly affected by premature death of the host before development can be completed [3, 4]. Brooks [4] reviewed the literature on host-parasitoid-pathogen interactions and presented several examples of parasitoids that were unable to complete development due to death of their hosts caused by *Bt* and *B. popilliae*. The effects of *Bt* treatment of host insects on survival and percentage parasitism depends on the host, timing of applications and dosage of *Bt*. Nealis & van Frankenhuyzen [65] noted that spruce budworm larvae, *Choristoneura fumiferana*, that are parasitised by *Apanteles fumiferanae* (Braconidae) are more likely to survive exposure to *Bt* because they feed less. However they did observe a 50-60% reduction in parasitoid populations when the host was treated with *Bt* as third instars. They observed better parasitoid survival when late fourth instars were treated and concluded that *Bt* would complement the beneficial effects of *A. fumiferanae*. A benefit to ingestion of sublethal dosages by host insects is an extended period of development and increased exposure to parasitoids [94].

The effects of host removal on the survival of predators will depend on the specificity of the predator, and the availability of other prey. Studies conducted in aquatic habitats demonstrate some changes in feeding habits of two species of Plecoptera. *Acroneuria lycorias*, a generalist predator, preferred live larvae, but after treatment of prey populations (simuliids) with *Bti* will feed on dead larvae and may exploit other food sources [59]. The total prey ingested by *A. lycorias*, however, declined after treatment of streams with *Bti* [96]. The detrivore, *Prostoia completa*, prefers dead larvae and was not affected by *Bti* treatments [59].

Synergistic and antagonistic activity between *Bt* and other entomopathogens has been reported. Koppenhöfer & Kaya [41] demonstrated synergistic activity between *Bt* subsp. *japonensis* and the entomopathogenic nematodes, *Heterorhabditis bacteriophora* and *Steinernema glaseri* for control of white grubs. A decrease in the incidence of nucleopolyhedrovirus infections has been reported in forests treated with *Btk* [93, 99].

4.   EFFECTS OF *BACILLUS* ENTOMOPATHOGENS ON VERTEBRATES

Vertebrate safety testing traditionally refers to a series of tests designed to evaluate the infectivity and pathogenicity of a candidate MPCA. Initially, tests evaluating infectivity and pathogenicity were additions to the standard protocols used to evaluate the toxicity of chemicals, but considerable evolution of these tests has occurred over the past 40 years. An example of this process is the elimination of long term (two year) feeding studies from the evaluation of MPCAs, because these tests were designed to assess carcinogenicity and this is not applicable. Infectivity is a concern unique to the evaluation of the safety of entomopathogenic bacilli.

Current MPCA tests are typically short term (< one month), and evaluate infectivity using a high dose of the MPCA and include an invasive route of exposure such as intravenous or intraperitoneal injection [79, 80]. If mortality occurs, it must be judged in the context of the test administered. For example, mortality following intracerebral injection of rats with 2 x 10^8 colony forming units (cfu) of *Btk* is not surprising but would be cause for concern if it occurred after ingestion [79, 80]. A finding of acceptable risk does not mean that under every circumstance a product will never prove harmful. Burges [6] stated that "Registration of a chemical is essentially a statement of usage in which the risks are acceptable. The same must apply to biological agents". Even when products have successfully cleared these hurdles, new questions can arise based on the changing public perception of risk. Questions have been raised periodically concerning the susceptibility of immunosuppressed individuals to *Bt* products [18].

What is the proper way to design safety tests to address this issue, or is it even necessary? There has been considerable debate about the value of safety tests employing immunosuppressed animals. Those opposing this type of safety testing contended that immunosuppressed individuals would succumb to a variety of opportunistic agents before they would become infected by an entomopathogen and furthermore, interpreting data from immunosuppressed animal studies is problematic [6]. In contrast, Shadduck [77] advocated a philosophy of testing known as maximum challenge, which included the use of immunosuppressed or immune deficient animals. Shadduck noted however, that a disadvantage of this approach is that a potentially useful organism may be unfairly labelled as unsafe based on a single test and emphasised that hazard evaluations must be based on a series of tests. A recent study illustrates the difficulty in interpreting test results using immunosuppressed animals [28]. This study reported that *Bt* subsp *konkukian* was infectious when injected subcutaneously (10^7 cfu) in cyclophosphamide-injected mice. However, the mice were only followed for two days after injection and the alleged infection did not become systemic. This study underscores the caveat that a single test cannot be used to determine the hazard of an entomopathogen.

4.1   Direct effects of *Bt* on mammals

There have been thousands of research papers published on *Bt*, but there are relatively few published studies on vertebrate safety. That does not mean that research has not been conducted, but rather that these data are proprietary. Initially, one of the main issues raised about the safety of *Bt* was its close relationship to *B. anthracis*. Some feared that it would somehow mutate and become a human pathogen although Steinhaus [88] eloquently rebutted these concerns. More recent questions have centered on the relationship between *Bt* and *B. cereus*. *B. cereus* has been recognised as the causal agent of an increasing number of cases of food poisoning and as a source of ocular infections [14, 37]. Other researchers have reported that various isolates belonging to several *Bt* serotypes produced *B. cereus* enterotoxins [10]. *Bt* production of enterotoxins has been rebutted by studies that have raised questions about the specificity of the *in vitro* test used to detect enterotoxins [78]. Additionally, no evidence of mammalian toxicity has been found in the numerous laboratory safety studies conducted on *Bt* insecticides; many of these tests were designed to assess the presence of toxins with mammalian activity [15]. No evidence of human infection has been found in epidemiologic studies following mass *Bt* forest spraying campaigns [18, 22, 68]. Finally, at this point *Bt* products have been used for decades and numerous people have been exposed; there has been ample opportunity for any negative effects to be recognised.

As early as 1958, Thuricide®, a *Bt* subsp. *thuringiensis* insecticide, was granted an exemption from tolerance by the United States Food and Drug administration based on a series of human and animal studies [19]. These studies included serial passage through mice, intraperitoneal injection in guinea pigs, inhalation studies in mice and human volunteers, and short term feeding studies using human volunteers. Various isolates and subspecies of *Bt* were also tested in long term feeding studies, using a daily dose of 10^9 spores per rat per day for 730 days [33] and a daily dose of 10^{12} spores per sheep per day for 150 days [23]. In highly invasive tests, *Btk* and *Bti* were injected into rats intracerebrally with inocula containing as many as 10^6 cfu; no mortality resulted [79]. In contrast, subcutaneous injection of as little as four spores of *B. anthracis* can kill mice [50].

Determining the infectivity of entomopathogenic bacilli is complicated by their biology. Inocula typically used in safety tests contain a mixture of spores and vegetative cells; commercial products may contain both spores and vegetative cells as well. The spores can remain viable in tissue for periods longer than six weeks [79, 80]. This ability to remain viable without multiplying is referred to as persistence

[81]. Persistence may cause confusion, if researchers or clinicians regard simple recovery of *Bt* following exposure as synonymous with infection. In assessing safety, it is more useful to regard infection as established when recovery of a MPCA is linked to tissue damage.

There are three well-documented reports associating human infection with *Bt* (see also Chapter 1.2). In the first case, a farmer was accidentally splashed in the face with a commercial preparation of *Btk*. An ocular ulcer subsequently developed and *Bt* was recovered [74]. In the second case, a laboratory worker accidentally stuck himself with a needle used to resuspend a cell spore crystal pellet of *Bti* and *Acinetobacter calcoaceticus* var *anitratus* [92]. In the final case, a French soldier stepped on a land mine and suffered a traumatic injury to his leg. Twenty-four hours after the blast *Bt* subsp. *konkukian* was recovered from multiple abscesses [28]. When the first two reports are examined critically, one cannot definitively state that *Bt* caused infection. In the first case, viable spores may have persisted in the conjunctival *"cul de sac"* and been recovered when the eye was swabbed. In the second case, the laboratory worker was exposed to both *Bt* and another bacterium, as well as metabolites in the culture medium. *Bt* and *A. calcoaceticus* var. *anitratus* were cultured from the wound, so it is impossible to state that *Bt* alone caused the infection. The only case where *Bt* was clearly the cause of infection was the French soldier. A land mine blast is certainly a worst case scenario and the *Bt* serotype recovered is not used commercially. Repeated human exposure by this route is unlikely. When one takes into account the tens of thousands of humans exposed to *Bt* products over the past 40 years, we submit that this single clear-cut case of human infection underscores the mammalian safety of *Bt*.

4.2   Direct effects of *Bt* on other vertebrates

In the United States, as part of the testing necessary for registration, *Bt* products were administered orally to mallard ducks (*Anas platyrhynchos*) and Northern bobwhite quail (*Colinus virginianus*). The exposure period was five days and the total dose was as high as 1×10^{12} cfu/kg; there were no adverse effects reported. Three species of fish were also tested during the registration process, sheepshead minnow (*Cyprionodon variegatus*), steelhead trout (*Oncorhynchus mykiss*) and bluegill sunfish (*Lepomis macrochiurus*). These species were exposed to *Bt* in concentrations as high as 2.87×10^{10} cfu/L in a 30-day static renewal test; test solutions were renewed twice weekly. There was no evidence of pathogenicity or infectivity (bacterial recovery 100 times the administered dose). In one study, there was significant mortality among steelhead trout exposed to *Bt*. The mortality was attributed to the extreme turbidity of the water in the test group. The fish could not see their food, and in turn attacked each other (WHO, personal communication). Data published in refereed journals on direct effects of *Bt* insecticides support the conclusions of these industry studies. Starlings (*Sturnus vulgaris*), white crowned sparrows (*Zonotrichia leucephrys*) and slate-colored junco (*Junco hyemalis*) fed 7.5 x 10^8 spores of *Bt* experienced no mortality [85]. Caged rock bass (*Ambloplites rupestris*) exposed to *Bt* over a three-day period experienced no mortality [58]. In contrast Snarski [87] reported that larval and juvenile fathead minnows (*Pimephales promelas*) exposed to 2.0 x 10^6 cfu/ml of *Bti* died within 24 hours. However, the mortality was due to dissolved oxygen depletion by formulation components. Wipfli *et al*. [98] were also able to kill the embryos of brook trout (*Salvelinus fontinalis*), brown trout (*Salmo trutta*) and steelhead trout with *Bti*. Although mortality occurred, it was only achieved using levels 12,000 times the recommended dose rate. Mortality was attributed to the formulation components. In conclusion, there is no evidence from industry and academic studies that *Bt* insecticides are infectious or pathogenic to birds and fish. Formulated *Bt* products can kill fish indirectly by depleting oxygen levels or making it difficult to find food, but to do so must be applied at a level that is many thousand times the recommended label rate.

4.3   Direct effects of *B. sphaericus* on mammals and other vertebrates

B. sphaericus was subjected to the same infectivity and pathogenicity studies as *Bt* for registration in the United States. These data are unpublished, but the tests included inhalation, oral and intraperitoneal exposure to at least 10^6 cfu per test animal. Nontarget vertebrate studies included fish and birds. Additional studies were conducted by researchers funded by the WHO and included intraocular, intracerebral, subcutaneous, oral, intraperitoneal and aerosol exposure [80, 81]. There was no evidence of infectivity or pathogenicity in these studies. Many of these tests emphasised intracerebral injection (as many as 10^7 cfu) because there were reports in the literature associating *B. sphaericus* with fatal human central nervous system infections. It is noteworthy that in all cases, when these human isolates were injected in experimental animals the isolates were uninfectious. The most well documented human isolate of *B. sphaericus* was in fact misidentified; nevertheless, these cases are periodically cited as cause for concern [82]. In conclusion, entomopathogenic isolates of *B. sphaericus* were noninfectious and nonpathogenic in laboratory animal studies that included worse case exposure scenarios such as intraocular and intracerebral injection.

4.4   Direct effects of *B. popilliae* on mammals and other vertebrates

The mammalian safety studies on *B. popilliae* are summarised by Obenchain & Ellis [69]. Test animals included mice, rats, guinea pigs, rabbits, monkeys, starlings, and chickens. Doses as high as 10^8 spores were used in these studies, and there was no evidence of infectivity or pathogenicity. Heimpel [24] reported that a Maryland researcher ate a spoonful of spore dust to demonstrate its safety. There are no published studies on the infectivity and pathogenicity of *B. popilliae* to fish presumably because of its specificity as well the fact that *B. popilliae* runoff into aquatic systems is minimal.

5.   INDIRECT EFFECTS OF *BACILLUS* ENTOMOPATHOGENS ON VERTEBRATES

5.1   Indirect effects of *Bt* on mammals and other vertebrates

It is far more difficult to calculate the indirect effects of *Bacillus* based insecticides than direct effects because these determinations must be made in the field. Natural population fluctuations may be confounded with the effect of the microbial insecticide, and the proper time scale for observation may encompass many years. Additionally, one can argue about what is the proper control to include in these studies. Should the control plots be untreated, or is more appropriate to use as the control plots treated with currently used chemical insecticides? In the case of a forest defoliator such as the gypsy moth (*Lymantria dispar*), should the control plots be defoliated? These issues must be addressed when interpreting data on indirect effects.

Numerous published studies of the indirect effects of *Bt* on small mammals and birds concluded that any effects were minor. *Btk*, fenitrothion and aminocarb were applied aerially for control of jack pine budworm and there were no significant differences in abundance of small mammal populations that could be attributed to *Bt* [35]. Spraying *Btk* in forests significantly reduced the proportion of caterpillars brought to the nests by chestnut-backed chickadees (*Parus rufescens*) but reproductive success and nestling growth was not affected (WHO, personal communication). Nagy & Smith [64] studied the effect of *Btk* aerial application on hooded warblers (*Wilsonia citrina*) and reported that overall, reduction in Lepidoptera larvae due to spraying had minimal effect, although differences in feeding rates occurred for small clutches. Rodenhouse & Holmes [72] studied food reductions in black-throated blue warblers (*Dendroica caerulescens*) using *Bt*, and found that when caterpillar abundance was reduced, the warblers made significantly fewer nesting attempts and that diets of hatchlings included fewer caterpillars. However, clutch size, hatching success, and the number of fledglings/nest did not differ between treated and control sites. Finally, Holmes [30] studied the reproduction and behavior of Tennessee warblers (*Vermivora peregrina*) in forests treated with *Btk* and tebufenozide. Nestling growth and survival were unaffected by either insecticide, although females in the plots treated with tebufenozide spent more time foraging.

5.2   Indirect effects of *B. sphaericus* and *B. popilliae* on mammals and other vertebrates

There are no studies available on the indirect effects of *B. sphaericus*. This is due in part to its recent commercialisation, and since it is used to control mosquitoes there is less concern about its impact on forests. If studies are conducted, they will presumably focus on assessing any deleterious effects on fish caused by prey reduction. There are no published studies of the indirect effects of *B. popilliae* as well. In the case of *B. popilliae*, it was registered before current tests were mandated, and it was then grandfathered into existing legislation when the rules changed. As stated above, since *B. popilliae* is not used to treat forests or in aquatic habitats, there have been few concerns expressed about effects on vertebrates.

6.   LONG TERM IMPACT OF *BACILLUS* PATHOGENS USED AS MPCAs

There is a paucity of studies that have assessed the long term impact of *Bacillus* entomopathogens on ecosystem community structure. Of the three pathogens addressed in this chapter, *B. thuringiensis* is the best studied in this regard. Naturally occurring *Bt* is found throughout the world in a number of habitats, yet relatively little is known regarding its role in ecosystems. There are three main theories on the role of *Bt* in nature [55]. Varieties of *Bt* have been referred to as entomopathogens, soil organisms, or saprophytic inhabitants of the phylloplane. Although widespread in soil habitats [54], they usually exist in low numbers in soils relative to other soil bacilli such as *Bacillus cereus* and do not germinate and grow in the soil habitat as readily as *B. cereus*. Diverse varieties have also been isolated from leaf surfaces [86] and grain dust [13]. Natural infections in insects are common in certain protected habitats such as grain silos, but epizootics caused by *Bt* are rare. Although spores can persist in soil and aquatic habitats [57, 71], parasporal inclusions are rapidly denatured in the field. However, in protected field settings, insecticidal activity may persist for up to 30 days [39]. Applications of *Bt* bioinsecticides to agroecosystems and other habitats usually do not result in a build up of spores in the environment. A steady decline in the viability of spores is observed, especially those exposed to sunlight [34]. As a microbial control agent, *Bt* is always inundatively applied to infestations of insects and results in rapid kill of target insects usually without recycling. As discussed above, their short term effects on NTOs that are unrelated to target insects is minimal. Although *Bt* has been used extensively in a variety of habitats, its long term effects on most ecosystems is relatively unknown. Assessment of the long term environmental impact of *Bt* in agroecosystems may be difficult due the short term nature of such systems and the effect of other agricultural practices on community structure. Stable environments such as forests and permanent aquatic habitats provide the best systems in which to monitor the impact of repeated *Bt* treatments.

Large scale control programmes that utilise *Bt* for suppression or eradication of pests such as gypsy moth provide an opportunity to monitor long term effects of the bacterium. USDA Forest Service [89] contends that permanent changes in nontarget populations do not appear likely in gypsy moth suppression projects. Suppression treatments normally consist of a single application of *Btk* in the spring when target foliage averages 45% expansion. However, eradication treatments may include 2-3 applications on a yearly basis. Miller [60] observed reductions in species richness in the guild of leaf-feeding Lepidoptera in forest that was treated over a three year period. Other authors also report a decline in species richness and diversity in Lepidoptera in forests after *Btk* treatments [73, 90]. Risks could be highest for univoltine species especially when populations of sensitive species are clumped in a restricted habitat within the treatment zone [60]. Factors that will contribute to the long term impact on individual susceptible species include voltinism, phenology and distribution of the insect, location within the habitat, and the number and frequency of *Bt* applications.

Sample *et al.* [73] concluded that the long term impact of gypsy moth reduction could benefit some native species.

Routine treatment of mosquito breeding sites with *Bti* in several programmes around the world is on the increase. Mosquito control efforts in the Rhine Valley of Germany rely exclusively on applications of *Bti* and to a lesser extend, *B. sphaericus*. Becker (see Chapter 6.2) reports no long term deleterious effect on NTOs that are monitored as part of the Rhine Valley programme. However, long term monitoring of a wetland ecosystem in the United States indicated that initial regular application of *Bti* for control of mosquito larvae did not result in short term changes [26, 27], but after the wetlands were treated with the bacterium for 2-3 years, species diversity and richness declined significantly [26].

Lotic habitats have also been periodically or regularly treated with *Bti* for control of black fly larvae in Africa, Brazil and North America. The longest ongoing use of the bacterium in rivers has been in the Onchocerciasis Control Programme in West Africa where it is used as an intervention during the dry season [31]. Although alternation with conventional chemical larvicides during the wet season precludes long term assessment of the individual impact of *Bti*, Dejoux & Elouard [12] contend that there is no evidence of long term deleterious effects on ecosystems receiving weekly applications during the dry season. However, the structure of the invertebrate community in the Maraoué River in Ivory Coast after one year of treatment with *Bti* was different in some respects from community structure before treatment and after treatment with temephos and chlorphoxim [12]. In the United States, Molloy [61] observed very little effect on NTOs after multiple applications of *Bti* in small streams. The most sensitive nontarget species, a filter feeding chironomid, responded with an average of 23% mortality to a concentration of *Bti* that killed 98% of simuliid larvae. Wipfli & Merritt [96] observed that reduction of simuliid larvae with *Bti* indirectly and differentially affected predators. Specialist predators in black fly-poor environments were most affected, whereas generalist predators were least affected.

7.   CONCLUSION

It is highly probable that any regular disruption of large insect communities, due to chemical or microbial insecticides or natural disaster, could have long term deleterious effects on higher trophic levels and ecosystem structure. The more diversified the food web, the less likely that complete or partial removal of a single species will result in catastrophic consequences. The more species a given intervention affects, the greater the likelihood of altering ecosystem structure. The safety and environmental impact of entomopathogenic bacteria should be evaluated in light of the risk for NTOs in comparison with other interventions and the effect no treatment at all will have on an ecosystem. Major defoliation of a forest by a pest insect such as the gypsy moth, for example, may have broader and more intense long term negative effects on the ecosystem than periodic removal of gypsy moths and affected nontarget Lepidoptera. If the complete removal of an introduced pest has deleterious effects on predators that have grown dependent on the pest, but an ecosystem reverts to its original state, should that be considered catastrophic?

ACKNOWLEDGEMENTS

We thank Heather Headrick and Dwight Brown for help in assembling and processing the bibliography. David Horton and Don Hostetter reviewed the manuscript and provided constructive comments. We are also grateful to the several colleagues who furnished literature and information.

REFERENCES

[1] Aly C & Mulla MS (1987) Effect of two microbial insecticides on aquatic predators of mosquitoes. J. Appl. Entomol. 103, 113-118

[2] Back C, Boisvert J, Lacoursiere JO & Charpentier G (1985) High-dosage treatment of a Quebec stream with *Bacillus thuringiensis* serovar. *israelensis*: efficacy against black fly larvae (Diptera: Simuliidae) and impact on non-target insects. Can. Entomol. 117, 1523-1534

[3] Blumberg D, Navon A, Keren S, Goldenberg S & Ferkovich SM (1997) Interactions among *Helicoverpa armigera* (Lepidoptera: Noctuidae), its larval endoparasitoid *Microplitis croceipes* (Hymenoptera: Braconidae), and *Bacillus thuringiensis*. J. Econ. Entomol. 90, 1181-1186

[4] Brooks WM (1993) Host - parasitoid - pathogen interactions, p. 231-272. *In* Beckage NE, Thompson SN & Federici BA (ed.), Parasites and Pathogens of Insects, Vol 2: Pathogens, Academic Press

[5] Brower LP (1986) Comment: the potential impact of DIPEL spraying on the monarch butterfly over-wintering phenomenon. Atala 14, 17-19

[6] Burges HD (1981) Safety, safety testing and quality control of microbial pesticides, p. 738-767. *In* Burges HD (ed.), Microbial control of pests and plant diseases 1970-1980, Academic Press

[7] Cameron EA & Reeves RM (1990) Carabidae (Coloeptera) associated with gypsy moth, (*Lymantria dispar* (L.) Lepidoptera: Lymantriidae), populations subjected to *Bacillus thuringiensis* Berliner treatments in Pennsylvania. Can. Entomol. 122, 123-129

[8] Charbonneau CS, Drobney RD & Rabeni CF (1994) Effects of *Bacillus thuringiensis* var. *israelensis* on nontarget benthic organisms in a lentic habitat and factors affecting the efficacy of the larvicide. Env. Toxicol. Chem. 13, 267-279

[9] Cloutier C & Jean C (1998) Synergism between natural enemies and biopesticides: a test case using the stinkbug *Perillus bioculatus* (Hemiptera: Pentatomidae) and *Bacillus thuringiensis tenebrionis* against Colorado potato beetle (Coleoptera: Chrysomelidae). J. Econ. Entomol. 91, 1096-1108

[10] Damgaard PH, Larsen HD, Hansen BM, Bresciani J & Jorgensen K (1996) Enterotoxin-producing strains of *Bacillus thuringiensis* isolated from food. Lett. Appl. Microbiol. 23, 146-150

[11] Davidson EW, Urbina M, Payne J *et al.* (1984) Fate of *Bacillus sphaericus* 1593 and 2362 spores used as larvicides in the aquatic environment. Appl. Environ. Microbiol. 47, 125-129

[12] Dejoux C & Elouard J-M (1990) Potential impact of microbial insecticides on the freshwater environment, with special reference to the WHO/UNDP/World Bank, Onchocerciasis Control Programme, p. 66-83. *In* Laird M, Lacey LA & Davidson EW (ed.), Safety of Microbial Insecticides, CRC Press

[13] DeLucca AJ, Palmgren MS & Ciegler A (1982) *Bacillus thuringiensis* in grain elevator dusts. Can. J. Microbiol. 28, 452-456

[14] Drobniewski FA (1993) *Bacillus cereus* and related species. Clin. Microbiol. Rev. 6, 324-338

[15] Drobniewski FA (1994) The safety of *Bacillus* species as insect vector control agents. J. Appl. Bacteriol. 76, 101-109

[16] Dutky SR (1963) The milky diseases, p. 75-115. *In* Steinhaus EA (ed.), Insect Pathology: An Advanced Treatise, Vol. 2, Academic Press

[17] Eidt DC (1985) Toxicity of *Bacillus thuringiensis* var. *kurstaki* to aquatic insects. Can. Entomol. 117, 829-837

[18] Elliott LJ, Sokolow R, Heumann M & Elefant SL (1988) An exposure characterization of a large scale application of a biological insecticide, *Bacillus thuringiensis*. Appl. Ind. Hyg. 3, 119-122

[19] Fisher R & Rosner L (1959) Toxicology of the microbial insecticide, Thuricide. Agr. Food Chem. 7, 686-688

[20] Gill SS, Cowles EA & Pietrantonio PV (1992) The mode of action of *Bacillus thuringiensis* endotoxins. Annu. Rev. Entomol. 37, 615-636

[21] Giroux S, Coderre D, Vincent C & Cote JC (1994) Effects of *Bacillus thuringiensis* var. *san diego* on predation effectiveness, development and mortality of *Coleomegilla maculata lengi* (Col.:Coccinellidae) larvae. Entomophaga. 39, 61-69

[22] Green M, Heumann M, Sokolow R *et al.* (1990) Public health implications of the microbial pesticide *Bacillus thuringiensis*: an epidemiological study, Oregon, 1985-1986. Am. J. Publ. Health 80, 848-852

[23] Hadley WM, Burchiel SW, McDowell TD *et al.* (1987) Five-month oral (diet) toxicity/infectivity study of *Bacillus thuringiensis* insecticides in sheep. Fund. Appl. Toxicol. 8, 236-242

[24] Heimpel AM (1971) Safety of insect pathogens for man and vertebrates, p. 469-489. *In* (Burges, HD & Hussey NW, ed.), Microbial Control of insects and mites, Academic Press, New York

[25] Herms CP, McCullough DG, Bauer LS *et al.* (1997) Susceptibility of the endangered Karner blue butterfly (Lepidoptera: Lycaenidae) to *Bacillus thuringiensis* var. *kurstaki* used for gypsy moth suppression on Michigan. Great Lakes Entomol. 30, 125-141

[26] Hershey AE, Lima AR, Niemi GL & Regal RR (1998) Effects of *Bacillus thuringiensis israelensis* (BTI) and methoprene on nontarget macroinvertebrates in Minnesota wetlands. Ecol. Appl. 8, 41-60.

[27] Hershey AE, Shannon L, Axler R, Ernst C & Michelson P (1995) Effects of methoprene and *Bti* (*Bacillus thuringiensis* var. *israelensis*) on non-target insects. Hydrobiologia 308, 219-227

[28] Hernandez E, Ramisse F, Ducoureau JP, Cruel T & Cavallo JD (1998) *Bacillus thuringiensis* subsp. *konkukian* (Serotype H34) superinfection: case report and experimental evidence of pathogenicity in immunosuppressed mice. J. Clin. Microbiol. 36, 2138-2139

[29] Hilbeck A, Baumgartner M, Fried PM & Bigler F (1998) Effects of transgenic *Bacillus thuringiensis* corn-fed prey on mortality and development time of immature *Chrysoperla carnea* (Neuroptera: Chrysopidae). Environ. Entomol. 27, 480-487

[30] Holmes SB (1998) Reproduction and nest behavior of Tennessee warblers *Vermivora peregrina* in forests treated with Lepidoptera-specific insecticides. J. Appl. Ecol. 35, 185-194

[31] Hougard J-M, Yaméogo L, Skékétéli A, Boatin B & Dadzie KY (1997) Twenty-two years of blackfly control in the Onchocerciasis Control Porgramme in West Africa. Parasitol. Today 13, 425-431

[32] Hough-Goldstein J & Keil CB (1991) Prospects for integrated control of the Colorado potato beetle (Coleoptera: Chrysomelidae) using *Perillus bioculatus* (Hemiptera: Pentatomidae) and various pesticides. J. Econ. Entomol. 84, 1645-1651.

[33] Ignoffo CM (1973) Effects of entomopathogens on vertebrates. Ann. NY Academ. Sci. 217, 141-164

[34] Ignoffo CM (1992) Environmental factors affecting persistence of entomopathogens. Florida Entomol. 75, 516-525

[35] Innes DG & Bendell JF (1989) The effects of small-mammal populations of aerial applications of *Bacillus thuringiensis*, fenitrothion, and Matacil® used against jack pine budworm in Ontario. Can. J. Zool 67, 1318-1323

[36] Jackson JK, Sweeney BW, Bott TL, Newbold JD & Kaplan LA (1994) Transport of *Bacillus thuringiensis* var. *israelensis* and its effect on drift and benthic densities of nontarget macroinvertebrates in the Susquehanna River, Northern Pennsylvania. Can. J. Fish. Aquar. Sci. 51, 295-314

[37] Jackson SG, Goodbrand RB, Ahmed R & Kasatiya S (1995) *Bacillus cereus* and *Bacillus thuringiensis* isolated in a gastroenteritis outbreak investigation. Lett. Appl. Microbiol. 21, 103-105

[38] James RR, Miller JC & Lighthart B (1993) *Bacillus thuringiensis* var. *kurstaki* afects a beneficial insect, the cinnabar moth (Lepidoptera: Arctiidae). J. Econ. Entomol. 86, 334-339

[39] Johnson KS, Scriber JM, Nitao JK & Smitley DR (1995) Toxicity of *Bacillus thuringiensis* var. *kurstaki* to three nontarget Lepidoptera in field studies. Environ. Entomol. 24, 288-297

[40] Karch S, Monteny N, Jullien JL, Sinègre G & Coz J (1990) Control of *Culex pipiens* by *Bacillus sphaericus* and role of nontarget arthropods in its recycling. J. Amer. Mosq. Control Assoc. 6, 47-54

[41] Koppenhöfer AM & Kaya HK (1997) Additive and synergistic interaction between entomopathogenic nematodes and *Bacillus thuringiensis* for scarab grub control. Biol. Contr. 8, 131-137

[42] Kreutzweiser DP, Holmes SB, Capell SS & Eichenberg DC (1992) Lethal and sublethal effects of *Bacillus thuringiensis* var. *kurstaki* on aquatic insects in laboratory bioassays and outdoor stream channels. Bull. Environ. Contam. Toxicol. 49, 252-258

[43] Krieg A & Langenbruch GA (1981) Susceptibility of arthropod species to *Bacillus thuringiensis*. pp. 837-896 *In:* Burges HD, Ed., *Microbial control of pests and plant diseases 1970-1980.* Academic Press, London.

[44] Lacey LA & Dame DA (1982) The effect of *Bacillus thuringiensis* var. *israelensis* on *Toxorhynchites rutilus rutilus* in the presence and absence of prey. J. Med. Entomol 19, 593-596

[45] Lacey LA, Escaffre H, Philippon B, Sékétéli A & Guillet P (1982) Large river treatment with *Bacillus thuringiensis* (H-14) for the control of *Simulium damnosum* in the Onchocerciasis Control Programme. Z. Tropenmed. Parasitol. 33, 97-101

[46] Lacey LA, Horton DR, Chauvin RL & Stocker JM (1999) Comparative efficacy of *Beauveria bassiana*, *Bacillus thuringiensis*, and aldicarb for control of Colorado potato beetle in an irrigated desert agroecosystem and their effects on biodiversity. Ent. Exp. Appl. *submitted*

[47] Lacey LA & Mulla MS (1990) Safety of *Bacillus thuringiensis* (H-14) and *Bacillus sphaericus* to non-target organisms in the aquatic environment, p. 169-188. *In* Laird M, Lacey LA & Davidson EW (ed.), Safety of Microbial Insecticides, CRC Press

[48] Lacey LA & Undeen AH (1986) Microbial control of black flies and mosquitoes. Ann. Rev. Entomol. 31, 265-296.

[49] Laird M, Lacey LA & Davidson EW (Eds.) 1990. Safety of Microbial Insecticides, CRC Press, Boca Raton, 259 pp.

[50] Lamanna C & Jones L (1963) Lethality for mice of vegetative and spore forms of *Bacillus cereus* and *Bacillus cereus*-like insect pathogens injected intraperitoneally and subcutaneously. J. Bact. 85, 532-535

[51] Langenbruch GA (1992) Experiences with *Bacillus thuringiensis* subsp. *tenebrionis* in controlling the Colorado potato beetle. Mitt. der Deutschen Gesellschaft. 8, 193-195

[52] Larget I & Charles JF (1982) Etude de l'activité larvicide de *Bacillus thuringiensis* variété *israelensis* sur les larves de Toxorhynchitinae. Bull. Soc. Path. Exp. 75, 121-130

[53] MacIntosh SC, Stone TB, Sims SR *et al.* (1990) Specificity and efficacy of purified *Bacillus thuringiensis* proteins against agronomically important insects. J. Invertebr. Pathol. 56, 258-266

[54] Martin PAW & Travers RS (1989) Worldwide abundance and distribution of *Bacillus thuringiensis* isolates. Appl. Environ. Microbiol. 55, 2437-2442

[55] Meadows MP (1993) *Bacillus thuringiensis* in the environment: ecology and risk assessment, p. 193-220. *In* Entwistle PF, Cory JS, Bailey MJ & Higgs H (ed.), *Bacillus thuringiensis*, An Envrionmenntal Biopesticide: Theory and Practice, John Wiley & Sons

[56] Melin BE & Cozzi EM (1990) Safety to nontarget invertebrates of lepidopteran strains of *Bacillus thuringiensis* and their β−exotoxins. pp. 149-167. *In* Laird M, Lacey LA & Davidson EW (ed.), Safety of Microbial Insecticides, CRC Press

[57] Menon AS & DeMestral J (1985) Survival of *Bacillus thuringiensis* var. *kurstaki* in waters. Water, Air, Soil Poll. 25, 265-274

[58] Merritt RW, Walker ED, Wilzbach MA, Cummins KW & Morgan WT (1989) A broad evaluation of B.t.i. for black fly (Diptera: Simuliidae) control in a Michigan river: efficacy, carry and nontarget effects on invertebrates and fish. J. Am. Mosq. Cont. Assoc. 5, 397-415

[59] Merritt RW, Wipfli MS & Wotton RS (1991) Changes in feeding habits of selected nontarget aquatic insects in response to live and *Bacillus thuringiensis* var. *israelensis* de Barjac-killed black fly larvae (Diptera: Simuliidae). Can. Entomol. 123, 179-185

[60] Miller JC (1990) Field assessment of the effects of a microbial pest control agent on nontarget Lepidoptera. Am. Entomol. 36, 135-139

[61] Molloy DP (1992) Impact of the black fly (Diptera: Simuliidae) control agent *Bacillus thuringiensis* var. *israelensis* on chironomids (Diptera: Chironomidae) and other non target insects: results of ten field trials. J. Am. Mosq. Control Assoc. 8, 24-31

[62] Mulla MS, Axelrod H, Darwazeh HA & Matanmi BA (1988) Efficacy and longevity of *Bacillus sphaericus* 2362 formulations for control of mosquito larvae in dairy wastewater lagoons J. Am. Mosq. Control Assoc. 4, 448-452

[63] Mulla MS, Darwazeh HA, Davidson EW, Dulmage HT & Singer S (1984) Larvicidal activity and field efficacy of *Bacillus sphaericus* strains against mosquito larvae and their safety to nontarget organisms. Mosq. News 44, 336-342

[64] Nagy LR & Smith KG (1997) Effects of insecticide-induced reduction in lepidopteran larvae on reproductive success of hooded warblers. The Auk 114, 619-627

[65] Nealis V & van Frankenhuyzen K (1990) Interactions between *Bacillus thuringiensis* Berliner and *Apanteles fumiferanae* Vier. (Hymenoptera: Braconidae), a parasitoid of the spruce budworm, *Choristoneura fumiferana* (Clem.) (Lepidoptera: Tortricidae). Can. Entomol. 122, 585-594

[66] Neri-Barbosa JF, Quiros-Martinez H, Rodriguez-Tovar ML, Tejada LO & Badii MH (1997) Use of Bactimos briquets (B. t. i. formulation) combined with the backswimmer *Notonecta irrorata* (Hemiptera: Notonectidae) for control of mosquito larvae. J. Amer. Mosq. Control Assoc. 13, 87-89

[67] Nicolas L, Dossou-Yovo J & Hougard J-M (1987) Persistence and recycling of *Bacillus sphaericus* 2362 spores in *Culex quinquefasciatus* breeding sites in West Africa. Appl. Microbiol. Biotech. 25, 341-345

[68] Noble M A, Riben PD & Cook GJ (1992) Microbiological and epidemiological surveillance programme to monitor the health effects of Foray 48B BTK spray. Ministry of Forests, Province of British Columbia.

[69] Obenchain FD & Ellis B-J (1990) Safety considerations in the use of *Bacillus popilliae*, the milky disease pathogen of Scarabaeidae. pp. 189-201. *In* Laird M, Lacey LA & Davidson EW (ed.), Safety of Microbial Insecticides, CRC Press

[70] Peacock JW, Schweitzer DF, Carter JL & Dubois NR (1998) Laboratory assessment of the effects of *Bacillus thuringiensis* on native Lepidoptera. Environ. Entomol. 27, 450-457

[71] Petras SF & Casida LE, Jr (1985) Survival of *Bacillus thuringiensis* spores in soil. Appl. Environ. Microbiol. 50, 1496-1501

[72] Rodenhouse NL & Holmes RT (1992) Results of experimental and natural food reductions for breeding black-throated blue warblers. Ecology 73, 357-372

[73] Sample BE, Butler L, Zivkovich C, Whitmore RC & Reardon R (1996) Effects of *Bacillus thuringiensis* Berliner var. *kurstaki* and defoliation by the gypsy moth [*Lymantria dispar* L. (Lepidoptera: Lymantriidae)] on native arthropods in West Virginia. Can. Entomol. 128, 573-592

[74] Samples JR & Buettner H (1983) Ocular infection caused by a biological insecticide. J. Inf. Dis. 148, 613-614

[75] Schnepf E, Crickmore N, Van Rie J *et al.* (1998) *Bacillus thuringiensis* and its pesticidal proteins. Microbiol. Mol. Biol. Rev. 62, 775-806

[76] Sebesta K, Fargas J, Horska K & Vankova J (1981) Thuringiensin, the Beta-exotoxin of *Bacillus thuringiensis*, p. 249-281. *In* Burges HD (ed.), Microbial control of pests and plant diseases 1970-1980, Academic Press

[77] Shadduck JA (1983) Some considerations on the safety evaluation of nonviral microbial pesticides. Bull. WHO 61, 117-128

[78] Shinagawa K (1990) Purification and characterization of *Bacillus cereus* enterotoxin and its application to diagnosis, p. 181-193. *In* Pohland AE & Dowell VR, Jr. (ed.), Toxin in foods and feeds, Plenum Press

[79] Siegel JP & Shadduck JA (1990) Mammalian safety of *Bacillus thuringiensis israelensis*, p. 202-217. *In* de Barjac H & Sutherland DJ (ed.), Bacterial control of mosquitoes & black flies: biochemistry, genetics & applications of *Bacillus thuringiensis israelensis* and *Bacillus sphaericus*, Rutgers University Press

[80] Siegel JP & Shadduck JA (1990) Mammalian safety of *Bacillus sphaericus*. pp. 321-331. *In* de Barjac H & Sutherland DJ (ed.), Bacterial control of mosquitoes & black flies: biochemistry, genetics & applications of *Bacillus thuringiensis israelensis* and *Bacillus sphaericus*, Rutgers University Press

[81] Siegel JP & Shadduck JA (1990) Clearance of *Bacillus sphaericus* and *Bacillus thuringiensis* subsp. *israelensis* from mammals. J. Econ. Entomol. 83, 347-355.

[82] Siegel JP, Smith AR & Novak RJ (1997) Cellular fatty acid analysis of a human isolate alleged to be *Bacillus sphaericus* and *Bacillus sphaericus* isolated from a mosquito larvicide. J. Appl. Environ. Microbiol. 63, 1006-1010

[83] Sims SR (1995) *Bacillus thuringiensis* var. *kurstaki* [CryIA (c)] protein expressed in transgenic cotton: effects of beneficial and other non-target insects. Southwest. Entomol. 20, 493-500

[84] Sims SR (1997) Host activity spectrum of the cryIIa *Bacillus thuringiensis* subsp. *kurstaki* protein: effects on Lepidoptera, Diptera, and non-target arthropods. Southwest. Entomol. 22, 395-404

[85] Smirnoff WA & MacLeod CF (1961) Study of the survival of *Bacillus thuringiensis* var. *thuringiensis* Berliner in the digestive tracts and in feces of a small mammal and birds. J. Insect Pathol. 3, 266-270

[86] Smith RA & Couche GA (1991) The phylloplane as a source of *Bacillus thuringiensis* variants. Appl. Environ. Microbiol. 57, 311-315

[87] Snarski VM (1990) Interactions between *Bacillus thuringiensis* subsp. *israelensis* and fathead minnows, *Pimephales promelas* Rafinesque, under laboratory conditions. Appl. Environ. Microbiol. 56, 2618-2622

[88] Steinhaus EA (1959) On the improbability of *Bacillus thuringiensis* Berliner mutating to forms pathogenic for vertebrates. J. Econ. Entomol. 52, 506-508

[89] USDA Forest Service (1995) Gypsy moth Management in the United States, Final Environmental Impact Statement. (5 Volumes)

[90] Wagner DL, Peacock JW, Carter JL & Talley SE (1996) Field assessment of *Bacillus thuringiensis* on nontarget Lepidoptera. Environ. Entomol. 25, 1444-1454

[91] Walton WE & Mulla MS (1991) Integrated control of *Culex tarsalis* larvae using *Bacillus sphaericus* and *Gambusia affinis*: Effects on mosquitoes and nontarget organisms in field mesocosms. Bull. Soc. Vector Ecol. 16, 203-221

[92] Warren RE, Rubenstein D, Ellar DJ, Kramer JM & Gilbert RJ (1984) *Bacillus thuringiensis* var. *israelensis*: protoxin activation and safety. Lancet 1 (8378), 678-679

[93] Webb RE, Shapiro M, Podgwaite JD *et al.* (1989) Effect of aerial spraying with Dimilin, Dipel, or Gypchek on two natural enemies of the gypsy moth (Lepidoptera: Lymantriidae). J. Econ. Entomol. 82, 1695-1701

[94] Weseloh RM, Andreadis TG, Moore REB *et al.* (1983) Field confirmation of a mechanism causing synergism between *Bacillus thuringiensis* and the gypsy moth parasitoid, *Apanteles melanoscelus*. J. Invertebr. Pathol. 41, 99-103

[95] Whaley WH, Anhold J & Schaalje GB (1998) Canyon drift and dispersion of *Bacillus thuringiensis* and its effects on select nontarget lepidopterans in Utah. Environ. Entomol. 27, 539-548.

[96] Wipfli MS & Merritt RW (1994) Disturbance to a stream food web by a bacterial larvicide specific to black flies: feeding responses of predatory macroinvertebrates. Freshwater Biol. 32, 91-103

[97] Wipfli MS & Merritt RW (1994) Effects of *Bacillus thuringiensis* var. *israelensis* on nontarget benthic insects through direct and indirect exposure. J. N. Am. Benthol. Soc. 13, 190-205

[98] Wipfli MS, Merritt RW & Taylor WW (1994) Low toxicity of the black fly larvicide *Bacillus thuringiensis* var. *israelensis* to early stages of brook trout (*Salvelinus fontinalis*), brown trout (*Salmo trutta*), and steelhead trout (*Oncorhynchus mykiss*) following direct and indirect exposure. Can. J. Fish. Aquat. Sci. 51, 1451-1458

[99] Woods SA & Elkinkton JS (1988) Effects of *Bacillus thuringiensis* treatments of the occurrence of nuclear polyhedrosis virus in gypsy moth (Lepidoptera: Lymnatriidae) populations. J. Econ. Entomol. 81, 1706-1714

[100] Yousten AA (1973) Affect of the *Bacillus thuringiensis* δ–endotoxin on an insect predator which has consumed intoxicated cabbage looper larvae. J. Invertebr. Pathol. 21, 312-314

[101] Yousten AA, Benfield EF, Campbell RP, Foss SS & Genthner FJ (1991) Fate of *Bacillus sphaericus* 2362 spores following ingestion by nontarget invertebrates. J. Invertebr. Pathol. 58, 427-435

Chapter 5.1

Is *Bacillus thuringiensis* standardisation still possible?
Update and improvement of Bt titration over 20 years

Ole Skovmand[1], Isabelle Thiéry[2] & Gary Benzon[3]
[1]Intelligent Insect Control, 80 rue Paul Ramart, Montpellier 34070, France,
[2]Laboratoire des Bactéries et Champignons Entomopathogènes, Institut Pasteur, Paris, France and [3]Benzon Research, Carlisle, Pennsylvania, USA

Key words: *Bacillus thuringiensis*, potency, standardisation, immunoassay, larvae, Lepidoptera, Coleoptera, Culicidae

Abstract: The aim of this review is to clarify the situation surrounding titration of bacterial formulations against *Bacillus* reference standards and to discuss the importance of standard procedures. Secondly, with the presence of numerous *B. thuringiensis* Cry toxins, we examine a possible remedy for the absence of classical reference standards. We propose a new way of titrating *Bt* products combining insect based bioassays and quantitative determination of toxin protein with biochemical methods. Further, we suggest improvement in the standard protocols used for bioassays and the statistical methods applied.

1.   INTRODUCTION

The potency of a *Bacillus thuringiensis* (*Bt*) product indicates the dose-dependent, lethal activity of the product compared to that of an accepted standard *Bt*. The activity is a result of the interaction of several components of the bacterium of which δ-endotoxins are often very important. Several other toxins and molecules act in concert with δ-endotoxins [29]. In the vegetative phase, some *Bt* strains produce a strong synergism for δ-endotoxins and a toxic phospholipase [20], and an elimination of some of these enzymes reduced toxicity of *Bt* [58]. These and possible other toxins and enzymes are only available when the spore

J.-F. Charles et al. (eds.),
Entomopathogenic Bacteria: From Laboratory to Field Application, 275–295.
© *2000 Kluwer Academic Publishers. Printed in the Netherlands.*

is alive in the product. A synergising effect of the spore itself was noted by [11, 21]. Insects have defences against bacteria including bactericides called attacines and cecropines. *Bt* is to some extent able to destroy this system by excreting a metallo-proteinase [29, 43]. The target of most *Bt* products is the insect larva. As the development of the immune system of the insect larva is dependent on the instar, toxicity depends on insect species as well as larval instar. Accordingly, toxicity is a complex process and not simply dependent on δ-endotoxin concentration.

In a potency bioassay, mortality is measured against dose or concentration of the *Bt* product and data are usually transformed into log dose-probit mortality to obtain straight lines that are more easily compared [23]. The ratio of LC_{50} values of the standard and of the product determines the potency of the product, expressed in International Toxic Units (ITU) *Aedes aegypti* / mg product, on a test-insect using a simple formula:

LC_{50} (ppm) reference standard x titre standard (ITU/mg) / LC_{50} (ppm) product

but behind this formula are many preconditions that are often not met as discussed below.

Potency determination allows comparisons between laboratories that use the same standard, despite minor differences in methods of testing or rearing, since these differences should have the same linear influence on the dose response of the product and of the standard. Internationally accepted reference standard *Bt*'s and test protocols can help make potency determinations comparable world-wide. The potency bioassay can be used for research laboratories to compare industrial and experimental formulations and by industry and applicators to check for variability between batches and stability during storage.

Despite the increasing demand for documented quality control in research and production, we believe that standardisation methods for *Bt* products are not progressing. This is partly due to the ease of protein assay methods used to determine concentration of δ-endotoxins, but also due to problems inherent in the bioassays themselves. International test protocols are often not followed, and for many new *Bt* strains they do not exist. The increasing numbers of *Bt* strains used for insect control – whether of natural origins or genetically modified – suggest the need for just as many international standards, but this is not practical. In this chapter we propose a way to compare *Bt* products across laboratories and *Bt* strains by defining certain standard test methods and protocols.

2. HISTORY OF *BT* STANDARDISATION

About 50 years ago, the appearance of the first commercial products based on *B. thuringiensis* created a need for methods to compare products in various countries. The first protocols were based on spore counts, but because number of spores does not reflect the number of crystals present in bacteria, this method was unreliable and resulted in some major product failures in the field. With the development of commercial *Bt* products in France, [7, 9] developed a titration method based on comparison of the LC_{50}'s of a product and a standard material. This method led to rapid improvements not only in product reliability, but also in increased biological activity of products since the new technique became an instrument for industry fermentation and formulation development.

The evolution of various Lepidopteran *Bt* reference standards has been reviewed in [6, 52] and that of Dipteran reference standards in [51]. Table 1 (page 294) summarises all the reference standard powders. Presently only 2 reference standards for mosquitocidal products are still stable and available on request to the Laboratoire des Bactéries et Champignons Entomopathogènes, Institut Pasteur, Paris, France.

During this evolution, it became clear that more specific test protocols were needed to obtain comparable results between laboratories. In several instances where an identical product was tested at several laboratories, there were significant differences between potency determinations [5, 6, 47]. Beegle suggested restrictions on methods of testing and rearing the larvae to reduce this variation, but these suggestions were not translated into new protocols. With the appearance of new strains for lepidopteran control, attempts were made to establish standard materials for *Bt aizawai* and *Bt entomocidus,* but these standards were never recognised internationally. With the discovery of strains active on mosquito larvae, *Bt israelensis* (serotype H-14) and *B. sphaericus*, new mosquitocidal products appeared. World Health Organization (WHO) [14, 39] and US [19] titration methods were proposed. The WHO recommended method was based on the protocol proposed by [14] and recommend the use of particular mosquito strains (Table 1).

The international standard materials developed for Lepidopteran-active *Bt*'s have been depleted, and no standard has been made for Coleopteran-active *Bt*'s. When *Bt morrisoni* (tenebrionis 256-82 strain) was used commercially by Novo Nordisk and Mycogen in the early 1990s, there were no attempts to make a reference standard for product comparison, and only the producers themselves had methods capable of determining whether a product met label specifications. As of 1998 more than 100 *B. thuringiensis* toxins had been identified and the number is continually increasing.

The development of chemical and immunochemical assays for protein determination provided the industry with tools that were faster and more accurate than the bioassay as long as the protein quality did not change. Such methods are discussed below under *in vitro* test. These methods made it possible for industry to evaluate fermentation developments much more precisely, but registration authorities still demanded bioassays for potency determinations for a time. Today, where natural and genetically engineered strains are used commercially (Table 2), several products are sold on a percentage protein basis, but assay methods used cannot assure that the protein components are active.

3.   STANDARD PROCEDURES

There are three main components in a bioassay: a protocol, a reference standard, and a test insect. Each of these components must be agreed upon internationally to make the titration valuable outside the laboratory or company. We will treat each component separately below.

3.1   Standard protocols for agricultural pests

Susceptibility of insects may fluctuate from day-to-day even when strenuous efforts are made to standardise rearing and testing conditions [47]. Variations between laboratories can be especially great since they may use different strains of insect species, or have different methods and equipment for rearing, testing, incubation, determining death, etc. To compensate for these variations at least partially, the test product(s) and the reference standard material are tested simultaneously. It is assumed that variations in insect susceptibility will affect the LC_{50}'s of the products and the reference standard equally, and thus will be equalised by potency calculations.

Standard protocols were developed for *Bt kurstaki* and *Bt israelensis* during a period where the complexity of toxins in these strains was not known, though it was observed that various lepidopteran active *Bt* strains had different activities against various species [18]. The strains commercialised in Europe and the US were all based on these two serotypes until the appearance of new *Bt* products containing *Bt morrisoni* toxic towards Coleopteran larvae, or more recently *Bt aizawai*, toxic towards *Spodoptera* sp.

Some protocols developed for standardisation are discussed below. These and protocols used for other types of bioassays are presented as practical laboratory guidelines in [6, 19, 33, 39]. Problems have to be solved regarding the *Bt* products which contain other single or multiple toxins, either naturally or by genetic manipulation, with different spectra of activity against various insect targets. A list of *Bt* strains and products are given in Table 2.

3.1.1 Diet incorporation protocol

The protocol for a lepidopteran test was based on *Bt kurstaki* HD-1 [17]. This variety of *kurstaki* is still the most widely used *Bt* strain. When the last HD-1 standard was produced and the potency was established, the standard protocol was rewritten to be more specific [5].This protocol defined 4-day-old larvae (late 2nd instar) of *Trichoplusia ni* as the test insect, fed on an agar-based diet in which quantities of *Bt* are uniformly incorporated. During the test, the larvae are kept with the diet at 25-27°C, L:D 14:10. Mortality is scored after 4 days. The potency of a product is defined as the mean of 3 valid bioassays on 3 different days, each confirmed by a one sided analysis of variance and a Chi-square test. The mean and Coefficient of Variation (CV) of the three tests is calculated. If the CV is greater than 20%, 2 additional replicates are performed. The one value that initially caused the CV to be greater than 20% is omitted, and the calculations are repeated incorporating the new results. If the CV is still beyond 20%, perform three additional tests and discard two, if necessary. Outlayers are removed by rules of thumb. An alternative to this procedure is proposed below.

Table 2. Examples of *Bt* products active on lepidopteran and / or beetle larvae.

Product	Toxin expressed (Cry)										
	1Aa	1Ab	1Ac	1Ba	1Ca	1Da	1G	2Aa	2Ab	2Ac	3A
Bt kurstaki HD-1 based products:											
Biobit®, Dipel®, Foray®, Cutlass®, Thuricide®	+	+	+	-	-	-	-	+	+	-	-
Bt aizawai based products:											
Florbac®	+	+	-	-	+	+	+	-	+	-	-
Transconjugant *Bt* Agree®	+	-	-	-	+	+	+	-	+	-	-
Transconjugant *Bt* Foil®	-	-	+	-	-	-	-	-	-	-	-
Bt tenebrionis based products:											
Novodor®, Trident®	-	-	-	-	-	-	-	-	-	-	+

The diet incorporation method requires rather large quantities of *B. thuringiensis* material. This is normally not a problem for quality control of production lots, but for screening formulations made in small quantities or to evaluate new *Bt* strains, methods requiring less material are preferable. Several have been developed and are discussed in [33]. For instance, the coloured droplet test uses small, semitransparent larvae that drink small droplets of *Bt* suspensions containing a dye [53].

In general, the results obtained with the diet incorporation technique and other methods do not compare well. Only the diet incorporation test has been used by industry for quality control [6]. However, none of the protocols set criteria for larval rearing before the test, and this may be the reason for significant variations between laboratories that are otherwise using the same methods.

3.1.2   Other factors influencing test results

Several factors that are not specified in the standard test protocols may influence the potency determination. Before the USDA protocol of 1982 was developed, antibiotics were added to the rearing or testing diet in some laboratories. Since spore activity or the presence of other live bacteria has a great influence on LC_{50}, results from laboratories using antibiotics were different from those not using them [6].

The duration of insect exposure to contaminated diet also influences results. The USDA protocol states 4 days for the *Trichoplusia ni* test. Laboratories may use different periods when other insect species are being used. For example, 5 to 7 days of exposure is used for *Spodoptera littoralis*, which is harder to kill with any known *Bt* strain than is *T. ni*. The slope of the concentration-mortality curve declines with longer duration of exposure, and as a result, the LC_{50} declines and fiducial limits expand. This was demonstrated in a study with *Plutella xylostella* [49]. Furthermore, control mortality increases with time interval. Therefore, more consistent results may be obtained by using a shorter exposure period with higher concentrations.

3.2   Standard protocols for mosquito larvae

Two standard protocols were developed by Institut Pasteur for the WHO [15] and by the United States Agricultural Department (USDA) for the Environmental Protection Agency (EPA), the US registration authority [34]. The latter defines more test criteria than the former, but none of them defined criteria for rearing except that Institut Pasteur recently added such information to the test protocol provided with the standards. IPS82 is the most commonly used standard which has maintained its activity after 17 years in storage [51]. The standard developed by USDA, HD-968-S-1983, [19] was never accepted internationally, but was periodically used by some producers who used the potency unit ITU *Aedes aegypti* (AA) per mg. This standard was calibrated against the former IPS78 standard by 6 American laboratories. However, tests of products against IPS82 and HD-968-S-1983 resulted in quite different potency values. Therefore, the use of IPS82 is recommended to avoid confusion. For *B. sphaericus*, the only standard, SPH88, is from Institut Pasteur (see Table 1). Unit values for *Bti* (on *Aedes aegypti* larvae) and *B. sphaericus* (on *Culex pipiens pipiens*) are not related.

The Institut Pasteur protocol requires the use of early 4th instars of mosquito larvae for the bioassay. Dilutions of a homogenised product are added to plastic cups containing larvae and mortality is scored after 24 h (48 h for *B sphaericus*). It is important to use a dilution factor that reflects the slope of the dose-mortality curve to obtain 5 or more mortality scores between 0 and 100 %, preferably between 5 and 95 %.

We advocate the use of strict criteria for "alive" *vs.* "dead" and avoidance of criteria such as "about to die"; "totally dead and cannot move even when prodded" is much less dependent on the assay reader. For tests with *Bti*, the standard insect is *Aedes aegypti* Bora-Bora strain, which can be obtained on request from Institut Pasteur. For *B. sphaericus*, *Culex pipiens pipiens* (Montpellier strain) is specified. *Culex p. quinquefasciatus* is not recommended because it is usually 10 times more susceptible than the former. *Culex* larvae are fed at the beginning of the test to minimise mortality by starvation or cannibalism.

An inter-laboratory test of a liquid *Bti* product against IPS82 showed that sample preparation and mosquito rearing methods influence LC_{50} values differentially, causing significant inter-laboratory differences in determined potencies [47]. The rearing method influenced the LC_{50}, and the sample preparation influenced the slope of the dose-mortality curves [45]. Sample preparation especially influenced results with powdered products such as IPS82. Moreover, intense feeding increased growth rate of the larvae and resulted in increased LC_{50} values.

3.3   Bioassay protocols for Coleoptera

International protocols and standards were never defined for *Bt morrisoni* (tenebrionis 256-82 strain). The primary target for *Bt* tenebrionis was the Colorado potato beetle, *Leptinotarsa decemlineata*, but many other chrysomelid beetles are susceptible [28]. [22] proposed a standardised method and a standard (5653, see Table 1) for quantifying activity using 2nd instar larvae of *L. decemlineata* on excised potato leaves dipped in *Bt* solutions, 20 larvae per concentration. Leaflets and larvae are kept in aerated boxes. Mortality is evaluated after 96 h of incubation at 25°C, photoperiod 16:8 h. Reduced variation in test results was obtained by using an advanced spray technique instead of dipping. An alternative bioassay method was proposed using sucrose droplets containing bacterial toxin fed to third instars [16]. Bioassay using artificial diet and corn was tested on *Diabrotica* sp. [31]. We have seen no reports in which coleopteran active *Bt* is tested by diet incorporation.

3.4   Sample preparation

Initial weighed quantities of a sample should reflect the homogeneity and availability of the sample. For the international standard, which is highly homogenous but available in very small quantities, 50 to 100 mg weighs are appropriate. For industrial products, more material is normally available so that larger quantities can be used to insure a representative sample. Primary powders and standards are made without detergents or other dispersing agents, and particles may re-agglomerate after homogenisation. Therefore, a non-ionic detergent should be added before homogenisation.

The initial homogenisation and suspension of product in water can be made with a glass bead mill (method of Institut Pasteur and recommended in the WHO protocol), a sonicator, a rod homogeniser, or a Waring blender [47]. The last agreed modification of the USDA *T. ni* test protocol recommended the use of a bath sonicator for 5 min at 3ml/watt sonicator loading [6].

Measurements of particle size after various periods of homogenisation showed that intense homogenisation leads to decreased median particle size of powder products, but did not change particle size of liquid products because it was already minimised [45]. Further homogenisation, and especially sonication, breaks up the *Bti* and some complex *Bt kurstaki* crystals [34]. A sample preparation technique should be fixed for various product types to obtain a single cell/single crystal state and thereby overcome differences due to formulations. For *Bti* powders, this will lead to increase in slope of dose-mortality curves and thus better defined LC_{50} values. The interlaboratory study comparing a fluid product with IPS82 lyophilised powder thus showed that sample preparation caused larger variations in LC_{50} of IPS82 than in LC_{50} of the fluid product [47]. It might be argued that such sample preparation induces a weaker correlation between field results and bioassays, but our concern is the determination of biological activity, and not field performance that is also dependent on inert ingredients, application methods, and biotope.

Unfortunately, the same relationship does not exist for *Bt kurstaki* products. Here, liquid samples and intensely homogenised powder samples have lower slopes of dose-mortality curves. Thus, even when all products are homogenised to single cell level, bioassays become less accurate. Lower slopes are often due to a slower toxic action. It has been demonstrated (Skovmand, unpublished results) that liquid *Bt kurstaki* products and very intensely homogenised powder products give lower slopes of concentration mortality curves than those obtained with lightly homogenised powder products.

Since all standards are based on powders, the potency determination of liquid products (in principle always having small particle size) will depend on the intensity of the powder homogenisation.

3.5   Statistics

The statistical procedures recommended in the USDA protocols were not based on sound statistics and the criteria for determining and discarding non-valid data are somewhat arbitrary. A test should be discarded if the *F* value is too low or Chi-square value is too high. The number of replicates needed for determining the potency of a product type should be calculated by observing the CV of repeated tests of one sample of the product. Given that the registration authorities accept that the "true" potency of a product is within 20 % of the determined value, the minimum number of repetitions required for a certain product type can be calculated iteratively from a student *t*-test. Only bioassay values outside 2 times the standard deviation of the mean should be omitted. Using the label value ± 20% as acceptance criteria and just 3 bioassays for a test, CV should be lower than 8 % [44]. A review of the data used for titrating the values of all standards of *Btk* and *Bti* show that this criterion was never met. Indeed, CV's of the mean potencies of reference standard through the years is often around 20%, even under the most standardised conditions of rearing, test method, personnel, etc. [51].

3.6   Choice of test insect

The first titration protocol used *Ephestia kühniella* as test insect for *Bt thuringiensis* and *Bt kurstaki* products, whereas later *Bt kurstaki* protocols used *T. ni* larvae (Table 1). When HD-1-S-1980 was calibrated against the two other standards, the results were not comparable for some laboratories [19]. Thiéry *et al.* [50] showed that *B. sphaericus* products gave large titre variations when titrated on either *Culex pipiens pipiens* or *C. p. quinquefasciatus,* and even between *C. p. quinquefasciatus* from different geographical areas. Accordingly, to reduce interlaboratory variation, insect species and probably even insect strains need to be included in protocols and provided from reference laboratories. To our knowledge, only the WHO protocol administered by Institut Pasteur recommends and distributes on request the standard IPS82 and the specified mosquito, *A. aegypti,* Bora-Bora strain. Nevertheless, the inter-laboratory evaluation of a standard and a product where 4 of 6 laboratories used this strain, one of the 4 obtained potency estimates of the product significantly different from those of the other 3 [47]. Differences in sample preparation and rearing procedures were found to contribute to the discrepancy.

The larval stage used for the assay is important. Neonate and 4-day old larvae of *T. ni* gave different results when 6 different *Bt kurstaki* products were tested against the HD-1 standard at 5 laboratories and when 4 products were tested in two laboratories [6].

The number of larvae per test unit is important for mosquito larvae as well as for *T. ni* larvae. Groups of 20 *S. littoralis* larvae in a petri dish were 10 times more susceptible than larvae tested singly (I. Thiéry, unpublished data). Using one compartment or cell size and increasing the number of larvae from 1 to 5, the LC_{50} decreased with larval density and control mortality increased. No significant difference in LC_{50} was found between 1 and 2 larvae per compartment, but the CV for potency values increased from 18 % to nearly 30 % (O. Skovmand, unpublished data).

Table 3. Comparison of photoimmunoassay method to insect bioassay on fresh and heat-treated *Bt kurstaki* products. Detection of active ingredient.

Treatment	Analytical Method		
	Immunoassay	Bioassay LC_{50}	Bioassay LC_{90}
Non heated	1.0^1	38.0^2 (1.0^1)	$137^2 (1.0^1)$
Stored 3 months at 5°C	1.0	37.9 (1.0)	168 (0.8)
Stored 3 months at 25°C	1.0	57.1 (0.7)	295 (0.5)
Stored 3 months in warehouse	1.0	41.6 (0.9)	215 (0.6)

[1] Relative values

[2] Calculated concentration of *Bt kurstaki* in diet causing 50 and 90% mortality of *T. ni.* The immunoassay is not susceptible to minor changes in the active ingredients of the products. The faster increase of LC_{90} is due to a decline of the slope of the dose mortality curve, a change in quality of the product that only the insect-bioassay can reveal (O. Skovmand, unpublished results).

3.7   Choice of diet

Most protocols do not define the insect diet (or plant species for foliar tests), but the last modification of the USDA protocol for a *T. ni* bioassay specified Brownsville diet without antibiotics during the test [6]. An inter-laboratory test of a *Bti* product showed that feeding for the rearing period influenced the level of LC_{50} for *A. aegypti* larvae [47] and suggested that food quality and/or quantity should be fixed in an improved rearing protocol.

4.   *IN VITRO* ASSAYS

Insect cell lines and midgut brush border membrane vesicles have been used for screening of activated *Bt* toxins. Immunoassays, High Pressure Liquid Chromatography (HPLC) and quantitative protein determinations have been used to quantify *Bt* toxins in products and fermentations. The method of the first immunoassay, though results correlated to bioassays, was nearly as slow. Andrews *et al.* [2] and Smith & Ulrich [48] reduced assay time to 4 hours. An even faster immunoassay method (Photo Immuno Assay) reported by Skovmand & Sterndorff [46] had a very high correlation with repetitive bioassays. It was recognised by EPA and the Canadian registration authority as a method to replace bioassays for potency determinations of freshly made products provided that the correlation with bioassays can be demonstrated. The *in vitro* test lacks the ability to distinguish between active and non active Cry protein as reported by Beegle [6] and also seen in Table 3. Further, the immuno-polyclonal methods have problems with batch variation of antibodies and cross reaction between various toxins (H. Sterndorf, personal communication). This problem is not solved with the use of monoclonals unless they are raised against non-homologous parts of the toxins. Comparison of I_{50} (inhibition ELISA factor) and LC_{50} of various formulations did not show an acceptable degree of correlation for this method to be used as a possible replacement for bioassay (I. Thiéry & Schaefer, unpublished data).

The protein dosage methods to determine total protein suffers from variability in results due to dosing method. When protein determination follows Bradford's or Lowry's methods, inter-laboratory variations depend on protein standard (ovalbumin, bovine serum albumin, or purified preparation of the protein being assayed) and solubilisation method. In order to obtain accurate results, these parameters must be fixed.

Analysis by SDS-PAGE may allow evaluation of protein quality, but any changes in molecular weight (protein degradation) do not necessarily indicate loss of activity. Moreover, identical protein profiles do not insure equal activity.

The 25 kDa hemolytic part of the delta endotoxin in *Bti* plays an important role in activity against *Simulium* spp., but is only poorly related to activity in mosquito larvae. A hemolytic assay was developed to quantify this 25 kDa toxin, but it correlated poorly to *Simulium* activity [41] and Delécluse personal communication).

Unless an *in vitro* method can measure the presence of each individual toxin, it will not be able to determine if a toxin is no longer expressed. The method reported by Abbott Laboratories (oral communication, Society of Invertebrate Pathology annual meeting, Cordoba, 1996) claimed to quantify each toxin of Dipel®.

In conclusion, the *in vitro* test that is correlated to bioassays can be used for product quality control provided that the product does not change qualitatively. If the product changes qualitatively, e.g. due to improper storage conditions, the methods are unreliable.

5.   FUTURE OF BIOASSAYS

The bioassay is likely to remain the first choice for testing new formulations and reliably screening new *Bt* isolates if PCR and HPLC methods are used for finding new Cry toxins [32]. On the other hand, its role in titration of products for label information is waning. In a review of the situation in the 1980's, [19] quoted Churchill, "That those who are not willing to learn from history are doomed to repeat it's failures". Before the bioassay became standard practice, spore counts were often used for standardisation. With the increasing knowledge of the role of δ-endotoxins in *Bt* mode of action, the role of the germinated spore and the other toxins and proteinases of the vegetative cells have been increasingly ignored. This may be due to the tendency to draw a parallel between chemical pesticides and δ-endotoxins for registration purposes. But this is a false parallel. Percent Cry protein of, e.g. *Bt galleriae* consists of several endotoxins of which only one or a few may be important for the target insect. Comparison of percentage toxin between two products with different toxin composition is therefore meaningless. Some investigations have shown synergistic effect between some toxins, e.g. the 4 *Bti* Cry toxins are more active together than individually, and various cross reactions between the different toxins have been demonstrated in a few cases. None of this information is indicated by protein concentration measures. An insecticide label provides application rates for different group of insects, but these rates are usually given as a wide range to be adjusted according to field conditions. The label recommendations take into account the effect of formulation ingredients (detergents etc.) and are thus not entirely related to the bacterial active ingredient content that may degrade much faster during storage (Table 3). The end user, and to a certain degree the registration and enforcement authorities, have few opportunities to compare and evaluate products. Accordingly, we feel that labels should continue to provide information on biological activity as measured in bioassays.

Due to the small existing quantities or lack of internationally accepted reference standards, most *Bt* producers perform quality control bioassays using their own internal reference standards to check the potency, reproducibility, and storage stability of their spore-crystal formulations. Unless internal reference standards are exhaustively calibrated against international standards, such potency information cannot be used for the comparison of products between companies.

6.   SUGGESTION FOR NEW TYPE OF STANDARD

Ideally, a bioassay is based on identical toxins in the standard and the measured products producing the same dose-response in the test organism. In 1998, more than 100 *B. thuringiensis* toxins have been identified and the number is continuously increasing, particularly with the development of modified toxins. Furthermore, most natural *Bt* strains produce several toxins and their effect may be additive or synergistic. Clearly, this complicates potency standardisation. Below, we try to unify two purposes of the bioassay: (1) verify that a (stored) product remains at label potency; (2) allow users to compare products.

As seen in Table 4 (page 295), *T. ni* larvae are susceptible to most lepidopteran active δ-endotoxins. *T. ni* larvae may thus measure a general level of potency of most products. If toxins are inactivated due to improper or lenghtly storage, *T. ni* bioassay will reveal this. This solves the problem that *in vitro* assays so far cannot.

For comparison of products, we propose that a new type of standard and a triple-species bioassay be developed. The new lepidopteran standard should be produced by mixing, in a 1:1:1 ratio, 3 *Bt* strains (produced separately) containing a broad range of toxins including toxins active against major groups of pest larvae. *Spodoptera* species and *Mamestra* are not particularly susceptible to Cry1A toxins (Table 4), but more so to Cry1C and Cry1E. *Heliothis/Helicoverpa* are most susceptible to Cry1C and moderately susceptible to Cry1A. *T. ni* and *P. brassicae* and the forest pests *Lymantria dispar* and *Choristoneura fumiferana* seem to be susceptible to most of the Cry1 toxins. The major lepidopteran market for non-boring and non-digging larvae, *i.e.* species targeted with foliar spraying, fall into these three major groups of species.

Bioassays using at least two or three test insect species should then be performed using specified insect strains and methods of rearing. The registration of a product would then be based on a plasmid/toxin profile as well as a general potency determined by bioassay against *T. ni* using an internationally recognised standard containing the mixture of strains as suggested above, plus a more specific potency for *Spodoptera* (covering at least *Spodoptora* and *Mamestra*) and *Heliothis/Helicoverpa*. The *T. ni* activity will also apply to other species with very broad susceptibility.

We believe that a solution utilising a combination of quantitative protein analysis and bioassays is better than either method alone. The bioassays should be used for registration, initial and periodic calibration of the biochemical assay, and for evaluation of stored products.

Mosquito standardisation does not present a similar problem as only *Bti* and *B. sphaericus* are commercially used and all the commercial strains contain the same toxins. But the current problem with lepidopteran *Bt* strains may re-occur if strains like *Bt jegathesan* and *Bt medellin* become used commercially.

6.1   Suggestion for improvements in bioassay protocols

Several interlaboratory surveys have shown that differences in rearing and testing are major sources for variation in results [19] and [6, 45]. These have led to suggestions for improvements, but only a few have been incorporated into generally accepted protocols. Today, Good Laboratory Practices (GLP) and ISO 9002 systems are implemented for quality control. GLP is designed, among other things, to assure proper laboratory accountability, worker training, and proper use of instruments, test materials, standards, and protocols. ISO 9002 is somewhat similar but actually provides standards and standard protocols. Therefore, the current industrial environment is more open to stricter standardisation than before. For the lepidopteran tests, the stricter protocol suggested by [6] should be adopted. The protocol adds specific constraints of the older published protocol of Beegle [5]. We suggest further specifications as follows: (1) The diet used in the bioassay must be the same as that used for rearing. (2) The rearing temperature must be specified. (3) During bioassay, one larvae per cavity or container is used, and the cavity size range should be specified. (4) As long as the result of homogenisation is known, it is not necessary to define the equipment used. Instruments are available to assess particle size. (5) The statistical protocol should be corrected as described previously.

For the mosquito assay, a stricter protocol should include rearing conditions and strain of mosquito used [47, 51]. The rearing protocol needs to define the amount of diet used relative to the density of mosquito larvae or, more generally, the rearing conditions should be such that 4th instars are obtained in a defined number of days, *e.g.* 5-6, at 25°C. Concerning *A. aegypti*, no food is added to rearing containers on the test day and no food is added in the test cups since it will compete with the product. For *B. sphaericus*, food may be added after 24 hours since it has been shown that lethal concentrations are ingested within the first hours of the test [35]. Sample preparation should reduce particle size to a few microns, the size of a single or a few spores and crystals together thus avoiding the influence of particle size on the potency value [45].

For an assay using Coleoptera, we still need artificial diets to replace the leaf assays to reduce both laboratory and inter-laboratory variations. The main target for *Bt morrisoni* (256-82 strain tenebrionis) was *Leptinotarsa decemlineata*, but this species cannot be reared in many countries due to quarantine restrictions. Perhaps a closely related species such as *Phaedon cochleariae* could be used.

6.2   Suggestion for a new type of registration specification

The registration of a lepidopteran active product should continue to include a plasmid/toxin profile (but with standardised methods) and add a general potency set by *T. ni* using an internationally recognised standard containing a mixture of toxin groups as suggested above, and a more specific potency for *Spodoptera* (covering at least *Spodoptera* and *Mamestra*) and a *Heliothis/Helicoverpa* activity. Recognised bioassays for all *Bt* types should be carried out in accordance with international recognised protocols. Initiatives for such international cooperation was previously undertaken by USDA and WHO. Today the initiative might come from an international forum for biopesticide registration.

ACKNOWLEDGEMENTS

We thank Wendy Gelertner (Pace Consulting) for critical reading of the manuscript.

REFERENCES

[1] Adang MJ, Staver MJ, Rocheleau TA *et al.* (1985) Characterized full-length and truncated plasmid clones of the crystal protein of *Bacillus thuringiensis* subsp. *kurstaki* HD-73 and their toxicity to *Manduca sexta*. Gene 36, 289-300

[2] Andrews REJ, Bibilos MM & Bulla LAJ (1985) Protease activation of the entomocidal protoxin of *Bacillus thuringiensis* subsp. *kurstaki*. Appl. Environ. Microbiol. 50, 734-742

[3] Aranda E, Sanchez J, Peferoen M, Guĕreca L & Bravo A (1996) Interactions of *Bacillus thuringiensis* crystal proteins with the midgut epithelial cells of *Spodoptera frugiperda* (Lepidoptera: Noctuidae). J. Invertebr. Pathol. 68, 203-212

[4] Bai C, Degheele D, Jansens S & Lambert B (1993) Activity of insecticidal crystal proteins and strains of *Bacillus thuringiensis* against *Spodoptera exempta* (Walker). J. Invertebr. Pathol. 62, 211-215

[5] Beegle CC (1981) Basic differences between *Bacillus thuringiensis* var. *israelensis* and Lepidopterous active varieties. XIVth Ann. Meeting Soc. Invertebr. Pathol., Bozeman, Montana

[6] Beegle CC (1990) Bioassay methods for quantification of *Bacillus thuringiensis* δ-endotoxin, p. 14-21. *In* Hickle LA & Fitch WL (ed.), Analytical Chemistry of *Bacillus thuringiensis*, ACS Symposium series, American Chemical Society, Washington DC 1990

[7] Bonnefoi A, Burgerjon A & Grison P (1958) Titrage biologique des préparations de spores de *Bacillus thuringiensis*. C.R. Acad. Sci. 27, 1418-1420

[8] Bourgouin C, Larget-Thiéry I & de Barjac H (1984) Efficacy of dry powders from *Bacillus sphaericus*: RB80, a potent reference preparation for biological titration. J. Invertebr. Pathol. 44, 146-150

[9] Burgerjon A & Dulmage H (1977) Industrial and International standardization of microbial pesticides - I. *Bacillus thuringiensis*. Entomophaga 22, 121-129

[10] Burgerjon A & Yamvrias C (1959) Méthode de titrage des préparations à base de *Bacillus thuringiensis* Berliner avec *Anagasta kühniella* Zell. C. R. Acad. Sci., Paris, 249, 2871-2872

[11] Burges HD (1976) Techniques for the bioassay of *Bacillus thuringiensis* with *Galleria mellonella*. Entomologia Exp. Appl. 19, 243-254

[12] Chakrabati SK, Mandaokar A, Kumar PA & Sharma RP (1998) Efficacy of Leipdopteran specific δ-endotoxins of *Bacillus thuringiensis* against *Helicoverpa armigera*. J. Invertebr. Pathol. 72, 336-337

[13] Chambers JA, Jelen A, Gilbert MP *et al.* (1991) Isolation and characterization of a novel insecticidal crystal protein gene from *Bacillus thuringiensis* subsp. *aizawai*. J. Bacteriol. 173, 3966-3976

[14] de Barjac H & Larget I (1979) Proposals for the adoption for standardized bioassay method for the evaluation of insecticidal formulations derived from serotype H14 of *Bacillus thuringiensis* var. *israelensis*. Mimeogr. Doc., WHO/VBC/79-744, 15 pp.

[15] de Barjac H & Larget-Thiéry I (1984) Characteristics of IPS82 as standard for biological assay of *Bacillus thuringiensis* H-14 preparations. Mimeogr. Doc., WHO/VBC/84.892, 10pp.

[16] De Leon T & Ibarra JE (1995) Alternative bioassay technique to measure activity of Cry III proteins of *Bacillus thuringiensis*. J. Econ. Entomol. 88, 1596-1601

[17] Dulmage H, Boening OP, Rehnborg CS & Hansen GD (1971) A proposed standardized bioassay for formulations of *Bacillus thuringiensis* based on the International Units. J. Invertebr. Pathol. 18, 240-245

[18] Dulmage HT (1981) Insecticidal activity of isolates of *Bacillus thuringiensis* and their potential for pest control. Microbial Control of Pests and Plant Diseases, p. 193-222. *In* Burges HD (ed.), Academic Press

[19] Dulmage HT, Mc Laughlin RE, Lacey LA *et al.* (1985) HD-968-S-1983, a proposed U.S. standard for bioassays of preparations of *Bacillus thuringiensis* subsp. *israelensis*-H14. Bull. Entomol. Soc. Am. 31, 31-34

[20] Faust RM, Reichelderfer CF & Thorne CB (1981) Plasmid-bacteriophage-recombinant DNA in genetic manipulations of entomopathogenic bacteria, p 225-254. *In* Panopoulos NJ (ed), Genetic Engineering in the Plant Sciences, Prager Publications, Inc.

[21] Federici BA & Wu D (1994) Synergism of insecticidal activity in *Bacillus thuringiensis*. *In* Ackhurst R (ed), Proc. II *Bacillus thuringiensis* meeting, Canberra 1994

[22] Ferro DN & Gelerntner WD (1989) Toxicity of a new strain of *Bacillus thuringiensis* to Colorado Potato Bettle (Coleoptera: Chrysomelidae). J. Econ. Entomol. 82, 750-755

[23] Finney DJ (1971) Probit analysis. A statistical treatment of the sigmoid response curve. University Press, Cambridge

[24] Frankenhuyzen K (van), Gringorten JL, Mile RE *et al.* (1991) Specificity of activiated CryIA proteins from *Bacillus thuringiensis* subsp. *kurstaki* HD-1 for defoliating forest lepidoptera. Appl. Environ. Microbiol. 57, 1650-1655

[25] Ge AZ, Shivarova NI & Dean DH (1989) Location of the *Bombyx mori* specificity domain on a *Bacillus thuringiensis* δ-endotoxin protein. Proc. Natl. Acad. Sci. USA 86, 4037-4041

[26] Granero F, Ballester V & Ferré J (1996) *Bacillus thuringiensis* crystal proteins Cry1Ab and Cry1Fa share a high affinity binding site in *Plutella xylostella* (L.). Biochem. Biophys. Res. Commun. 224, 779-783

[27] Höfte H & Whiteley HR (1989) Insecticidal crystal proteins of *Bacillus thuringiensis*. Microbiol. Rev. 53, 242-255

[28] Krieg A, Huger AM, Langenbruch GA & Schnetter W (1983) *Bacillus thuringiensis* subsp. *tenebrionis*: ein neuer, gegenüber Larven von coleopteran wirksamer Pathotyp. Z. Ang. Entomol. 96, 500-508

[29] Lövgren A, Zhang M, Engström A, Dalhammar G & Landén R (1990) Molecular characterization of immune inhibitor A, a secreted virulence protease from *Bacillus thuringiensis*. Mol. Microbiol. 4, 2137-2146

[30] MacIntosh SC, Stone TB, Sims SR *et al.* (1990) Specificity and efficacy of purified *Bacillus thuringiensis* proteins against agronomically important insects. J. Invertebr. Pathol. 56, 258-266

[31] Marrone PG, Ferri FD, Mosley TR & Meinke LJ (1985) Improvements in laboratory rearing of the southern corn rootworm, *Diabrotica undecimpunctata howardi* Barber (Coleoptera: Chrysomelidae), on an artificial diet and corn. J. Econ. Entomol. 78, 290-293

[32] Masson L, Erlandson M, Puztai-Carey M *et al.* (1998) A holistic approach for determining the entomopathogenic potential of *Bacillus thuringiensis* strains. Appl. Environ. Microbiol. 64 64, 4782-88

[33] Mc Guire MR, Galan-Wong LJ & Tamez-Guerra P (1997) Bacteria : Bioassay of *Bacillus thuringiensis* against lepidopteran larvae, p. 91-99. *In* Lacey L (ed.), Manual of Techniques in Insect Pathology, Academic Press, San Diego, London

[34] Mc Laughlin RE, Dulmage HT, Alls R *et al.* (1984) A U.S. standard bioassay for the potency assessment of *Bacillus thuringiensis* serotype H-14 against mosquito larvae. Bull. Entomol. Soc. Am. 30, 26-29

[35] Mian LS & Mulla MS (1983) Effect of proteolytic enzymes on the activity of *Bacillus sphaericus* against *Aedes aegypti* and *Culex quinquefasciatus* (Diptera: Culicidae).

[36] Moar WJ, Masson L, Brousseau R & Trumble JT (1990) Toxicity to *Spodoptera exigua* and *Trichoplusia ni* of individual P1 protoxins and sporulated cultures of *Bacillus thuringiensis* subsp. *kurstaki* HD-1 and NRD-12. Appl. Environ. Microbiol. 56, 2480-2483

[37] Moar WJ, Trumble JT, Hice RH & Backman PA (1994) Insecticidal activity of the CryIIA protein from NRD-12 isolate of *Bacillus thuringiensis* subsp. *kurstaki* expressed in *Escherichia coli* and *Bacillus thuringiensis* and in a leaf-colonizing strain of *Bacillus cereus*. Appl. Environ. Microbiol. 60, 896-902

[38] Padidam M (1992) The insecticidal crystal protein CryIA(c) from *Bacillus thuringiensis* is highly toxic for *Heliothis armigera*. J. Invertebr. Pathol. 59, 109-111

[39] Rishikesh N & G. Q (1983) Introduction à une méthode normalisée pour l'evaluation de l'activité des produits à base de *Bacillus thuringiensis*, sérotype H-14. Bull. Org. Mond. Santé 61, 99-103

[40] Salama HS, Foda MS & Sharaby A (1989) A proposed new *Bacillus thuringiensis* standard for bioassay of bacterial insecticides versus *Spodoptora* spp. Trop. Pest Management 35, 326-30

[41] Sandler N, Zomper R, Keynan A & Margalit J (1985) *Bacillus thuringiensis* var. *israelensis* crystal hemolysis as a possible basis for an assay of larval toxicity. Appl. Microbiol. Biotechnol. 23, 47-53

[42] Sasaki J, Asano S, Iizuka T *et al* (1996) Insecticidal activity of the protein encoded by the Cry V gene of *Bacillus thuringinsis kurstaki* INA-02. Curr. Microbiol. 32, 195-200

[43] Sidén I, Dalhammar G, Telander B, Boman HG & Somerville H (1979) Virulence factors in *Bacillus thuringiensis*: purification and properties of a protein inhibitor of immunity in insects. J. Gen. Microbiol. 114, 45-52

[44] Skovmand O (1992) *Trichoplusia ni* for QC assays - QC of the test strain, method developments and laboratory safety, Abstract in Symposium on "Insect Behaviour - its influence on the development of insect rearing technology for research and pest management", XIX International Congress of Entomology, Beijing, China

[45] Skovmand O, Hoegh D, Pedersen HS & Rasmussen T (1997) Parameters influencing the potency of *Bacillus thuringiensis* var *israelensis* products. J. Econ. Entomol. 90, 361-369

[46] Skovmand O & Sterndorff HB (1994) Comparison of bioassays and photo immun assays on *Bacillus thuringiensis var. kurstaki* products. *In* Proc. VIth Int. Colloquium on Invertebrate Patholology and Microbial Control, Montpellier, France 2, 58-59

[47] Skovmand O, Thiéry I, Benzon G *et al.* (1998) Potency of products based on *Bacillus thuringiensis* var. *israelensis* : inter laboratory variations. J. Am. Mosq. Control Assoc. 14, 298-304

[48] Smith RA & Ulrich JT (1983) Enzyme-linked immunosorbent assay for quantitative detection of *Bacillus thuringiensis* crystal protein. Appl. Environm. Microbiol. 45, 586-590

[49] Tabashnik BE, Finson N, Johnson MW & Moar WJ (1993) Resistance to toxins from *Bacillus thuringiensis* subsp. *kurstaki* causes minimal cross-resistance to *B. thuringiensis* subsp. *aizawai* in the diamondback moth (Lepidoptera, Plutellidae). Appl. Environ. Microbiol. 59, 1332-1335

[50] Thiéry I, Baldet T, Barbazan P *et al.* (1997) International indoor and outdoor evaluation of *Bacillus sphaericus* products : complexity of standardizing outdoor protocols. J. Am. Mosq. Control Assoc. 13, 218-226

[51] Thiéry I & Hamon S (1998) Bacterial control of mosquito larvae: investigation of stability of *B.t.i.* and *B. sphaericus* standard powders. J. Am. Mosq. Control Ass. 14, 472-476

[52] Tompkins G, Engler R, Mendelsohn M & Hutton P (1990) Historical aspects of the quantification of the active ingredient percentage for *Bacillus thuringiensis* products, p. 9-13. *In* Hickle LA & Fitch WL (ed.), Analytical Chemistry of *Bacillus thuringiensis*, American Chemical Society, Washington DC

[53] Van Beek NAM & Hughes PR (1986) Determination by fluorescence spectroscopy of the volume ingested by neonate lepidopterous larvae. J. Invertebr. Pathol. 48, 249-251

[54] Visser B, Munsterman E, Stoker A & Dirkse WG (1990) A novel *Bacillus thuringiensis* gene encoding a *Spodoptera exigua*-specific crystal protein. J. Bacteriol. 172, 6783-6788

[55] Visser B, Van der Salm T, Van den Brink W & Folkers G (1988) Genes from *Bacillus thuringiensis entomocidus* 60.5 coding for insect-specific crystal proteins. Mol. Gen. Genet. 212, 219-224

[56] Von Tersch MA, Robbins HL, Jany CS & Johnson TB (1991) Insecticidal toxins from *Bacillus thuringiensis* subsp. *kenyae*: gene cloning and characterization and comparison with *B. thuringiensis* subsp. *kurstaki* CryA(c) toxins. Appl. Environ. Microbiol. 57, 349-358

[57] Wu D, Cao XL, Bay YY & Aronson AI (1991) Sequence of an operon containing a novel δ-endotoxin gene from *Bacillus thuringiensis*. FEMS Microbiol. Lett. 81, 31-36

[58] Zhang MY, Lövgren A, Low MG & Landen R (1993) Characterization of an avirulent pleiotropic mutant of the insect pathogen *Bacillus thuringiensis*: reduced expression of flagellin and phospholipases. Infect. Immun. 61, 4947-4954

Table 1. Reference standard powders of B. thuringiensis and B. sphaericus

Powders	Strain	Potency* (ITU/mg)	Test-insect	Stability	Stock	Presently used	Reference
B. thuringiensis							
E-61	Bt H1 type-strain	1,000	Ephestia kühniella	stable	no	no	[10]
HD1-S-1971	Bt H3a,3b,3c HD-1	18,000	Trichoplusia ni	stable	no	no	[17]
HD1-S-1980**	Bt H3a,3b,3c HD-1	16,000	Trichoplusia ni	unstable	no	yes/no	[6]
635- S-1987	Bt H6	10,000	Spodoptera sp.	?	?	?	[40]
IPS78	Bt H14 ONR-60A	1,000	Aedes aegypti (Bora-Bora strain)	unstable	yes	no	[14]
IPS80	Bt H14 ONR-60A	10,000	Aedes aegypti (Bora-Bora strain)	stable	yes	no	[51]
IPS82	Bt H14 1884	15,000	Aedes aegypti (Bora-Bora strain)	stable	yes	yes	[8]
HD-968-S-1983	Bt H14	4,740	Aedes aegypti	stable	no	no	[15]
5653	Bt morrisoni san diego	50,000	Leptinotarsa decemlineata	?	no	no	[19]
B. sphaericus							
RB80	Bs H5a,5b 1593	1,000	Culex pipiens pipiens (Montpellier strain)	unstable	yes	no	[22]
SPH84	Bs H25 2297	1,500	Culex pipiens pipiens (Montpellier strain)	stable	yes	no	[51]
SPH88	Bs H5a,5b 2362	1,700	Culex pipiens pipiens (Montpellier strain)	stable	yes	yes	[51]

* Potency is expressed in international toxic units/mg product. Potency of 1,000 ITU/mg has been arbitrary assigned, titer of the other reference standards were then assigned against the first one. For example IPS80, IPS82 and HD-968-S-1983 against IPS78

** This reference standard is no longer available, but the internal standard (previously titrated against the HD1-S-1980) used by Abbott Laboratories Inc. is accessible from that company

Table 4. Lepidopteran species susceptibility to various Cry toxins.

Species	1Aa	1Ab	1Ac	1Ba	1C	1D	1E	1F	2Aa	2Ab	2Ac	5A	References
Bombyx mori	+	±	-	-	+	+	+					-	[25], [36]
Choristoneura fumiferana	+	+	+	+	+	+	+	+					[42]
Heliothis virescens	+	±	+	-	-	-	±	+	+	+	+		[1], [13], [27], [37], [55], [56]
H. armigera	±	+	+	-	-	-	-	-	+				[12], [38]
Lymantria dispar	+	+	±	-	±	+	-	+	+	+			[24], [56], [57]
Manduca sexta	+	+	+	-	-	+	-	+	+	+	+		[1], [27], [55], [57]
Mamestra brassica	±	±	-	-	±	-	-						[27], [54]
Plutella xylostella	+	+	+	+	+	-	-	+				+	[26], [27], [56]
Pieris brassicae	+	+	+	+	±	-	-	+	+				[27], [55], [54]
Spodoptera exigua	±	±	-	-	+	+	±	+	-				[12], [55], [56], [54]
S. exempta	±	+	-	±	+	+	+						[4]
S. frugiperda/littoralis	-	-	-	-	+	-						+	[3], [38], [42]
S. littoralis	-	-	-	-	+	-	+						[27], [55]
Trichoplusia ni	+	+	+	+	+	+	+	+	+	+	+	+	[37], [56]

The list is limited to *B. thuringiensis* toxins referred by several publications.
+ indicates susceptibility, - very little or no susceptibility, ± moderate susceptibility or conflicting informations, sometimes because of minor variations in aminoacid sequences that did not lead to splitting a toxin-type into (further) subtypes.

Chapter 5.2

Industrial fermentation and formulation of entomopathogenic bacteria

Terry L. Couch
Becker Microbial Products, 9464 NW 11th st., Plantation, FL 33322, USA

Key words: Entomopathogenic   bacteria,   Fermentation,   formulation,   *Bacillus thuringiensi*s, *Bacillus sphaericus*

Abstract: The entomopathogenic bacteria have been used commercially as microbial pesticides for decades. Their success is the result of continued improvement in fermentation and formulation technology and to a lesser extent to genetic manipulation of the toxins produced by these organisms. This chapter discusses current fermentation and formulation techniques used in industry.

1.   INTRODUCTION

Because of the industrial proprietary aspects associated with mass production and formulation of entomopathogenic bacteria, little published literature is available which deals directly with this subject. In recent years several excellent reviews addressing this topic have been published (Beegle *et al.* [2] and Lisansky *et al.* [13]). These publications specifically address *Bacillus thuringiensis* and *B. sphaericus* but the principals outlined can be applied to all entomopathogenic bacteria produced commercially by deep tank fermentation. Dr. D Burges [4] also published an excellent text on formulation of microbial pesticides. Which is one of the most comprehensive ever published on this important topic.

Unfortunately, there is little published literature on commercial production of *B. sphaericus*. Literature on this topic is primarily associated with laboratory scale fermentation and not commercial production. There is almost no information on other entomopathogenic

J.-F. Charles et al. (eds.),
Entomopathogenic Bacteria: From Laboratory to Field Application, 297–316.
© 2000 *Kluwer Academic Publishers. Printed in the Netherlands.*

bacteria. Hence, this presentation will be concerned primarily with the truly commercial entomopathogenic bacteria species, *B. thuringiensis* (*Bt*) and *B. sphaericus*. However, the basic methodology described in the production section of this chapter is applicable to other species.

Since the author has been involved in all aspects of the commercial development of entomopathogenic bacteria, the emphasis of this chapter will be to present what actually is done during production and formulation of these agents. No attempt will be made to review the entire spectrum of the literature, which describe media components and methodology not used in actual industrial processes. Chapter 5.3 of this book provides an excellent review of solid state fermentation techniques.

2.   CULTURE SELECTION

Since this chapter deals with the industrial production and formulation of entomopathogenic bacteria, the assumption is being made that the most active, stable isolate of the appropriate species has already been selected by the commercial producer. This selection has often been made easier because numerous strains of entomopathogenic bacteria have been maintained by several large culture collections. Major collections for *Bacillus thuringiensis* are; Institut Pasteur, Paris, France; United States Department of Agriculture (USDA) laboratory, Peoria, Illinois, USA; Institut fur Biologische Schadlingsbekampfung, Darmstadt, Germany; Kyushu University, Fukuoka, Japan and Huazhong Agricultural University, Wuhan, People's Republic of China.

A bacterial strain is selected for industrialisation based on several important criteria. These are insecticidal spectrum (market available for sales); potency per unit volume of fermenter broth; fermentation media requirements, ease of production; genetic stability; and storage stability. To optimise the fermentation, formulation and insecticidal activity a significant amount of research and development is involved. This includes both laboratory and field research. These efforts include fermentation medium optimisation, insect and biochemical assay development, determination of the best formulation for the insects affected by the particular isolate, shelf life studies, and spray ability of the formulations.

3. LABORATORY TECHNIQUES FOR CULTURE MAINTENANCE

The selected bacterial culture must be maintained and stored under conditions that will have no effect on its growth or potency characteristics. Therefore, proper storage of the isolate is of paramount importance. Since the development of sophisticated methods for determining the genetic make-up of entomopathogenic bacteria, the extent of sub-culturing the primary isolate is of great concern. These bacteria, especially *B. thuringiensis*, can lose and exchange plasmids during culturing [10]. Therefore, manufacturers try to limit the number of passes of the bacterium from the initial flask culture to the fermentation tank.

Once the pure culture has been determined to be free of phage and other bacterial contaminants, it is grown in a shake flask using Trypticase Soy Broth or Lemco Broth and incubated at 30°C for 48–72 hours [13]. The culture is checked to insure the presence of spores and crystals. Aliquots are taken and transferred to liquid nitrogen vials and stored under liquid nitrogen. A sample of the broth is diluted serially and cultured on nutrient agar plates or other appropriate medium. These are incubated at 30°C for up to five days to ensure the absence of phage or bacterial contaminants. A second sample is tested for the presence of β-exotoxin. If the culture is found to be pure and free of β-exotoxin, the vials can be used for production of the seed culture for large-scale fermentation.

An alternate storage technique is to lyophilise the broth culture and store the lyophils at minus 28°C until needed for production. Usually sufficient vials are made to permit at least six months of production before additional seed cultures are required. All efforts must be made to minimise the number of passes from the parent culture to the fermenter.

4. FERMENTATION INOCULUM PREPARATION

In the preparation of the culture for the fermentation, one of the lyophils or nunc vials is aseptically transferred to several two litre flasks containing Trypticase Phosphate Broth for *B. thuringiensis* or nutrient broth supplemented with 0.05% yeast extract for *B. sphaericus* [2]. Flasks are placed on a rotary shaker in an incubator and incubated for 24 hours at 28°C. These are then checked for purity. If found to be satisfactory, two flasks (700 ml) are generally used to inoculate the seed fermenter. The seed fermenter is usually sized so that it is one to five percent of the volume of the commercial fermenter.

The seed fermenter is charged with the same medium used in the main fermentation tank. The selection and type of medium is discussed below. The seed fermenter has the same fermentation controls (pH, dissolved oxygen, sugar concentration) as the main fermentation vessel. When the culture has reached peak log phase in the vegetative cycle it is pumped into the main fermenter.

Table 1. Typical deep tank fermentation ingredients and concentrations

Ingredient	Concentration (g/l)
Soya Flour	20 — 40
Cotton Seed Flour	14 — 30
Potato Protein	15 — 40
Corn Steep Liquor or Solids	15 — 30
Glucose	10 —30
Peptone	2 — 5
Corn Syrup	20 — 45
Molasses	1.0 — 18.6
Glycerol	2.0 — 10
Corn Starch	10 — 15
Yeast Extract	2.0
KH_2PO_4	1.0
K_2HPO_4	1.0
$FeSO_4$	0.02
$FeSO_4.7H_2O$	0.0005 — 0.02
$MgSO_4.7H_2O$	0.3
$MnSO_4.H_2O$	0.02
$ZnSO_4.7H_2O$	0.02
$(NH_4)_2SO_4$	2
$CuSO_4.5H_2O$	0.005
$CaCO_3$	1.0 — 1.5
PPG2000	2.0 — 5.0
Hodag FD62 anti-foam	3.0
SAG 5693 anti-foam	0.5 — 1.25
Dow Corning AF anti-foam	0.1

References: [2, 13, 16]

5.   FERMENTATION MEDIUM SELECTION

Hundreds of articles have been written on the requirements for fermentation of *Bacillus thuringiensis*, and only a few on *B. sphaericus*, and *Serratia entomophila* [11]. Although no specific medium was referenced for *S. entomophila*, fermentation media components available tend to be similar regardless of the bacterium involved. These always involve carbon and nitrogen sources and trace minerals. Typical fermentation media ingredients used for production of most entomopathogenic bacteria are listed in Table 1.

The choice of media ingredients is critical for successful commercial production of an entomopathogenic bacterium. For purposes of this discussion commercial fermentation is defined as deep tank liquid fermentation using batch or fed batch techniques. Continuous fermentation does not work and to the author's knowledge is never used. A commercial fermentation also means that the size of the batch is never less than 30,000 litres. Anything smaller than this is a pilot plant or laboratory scale production and will not be discussed here.

The ingredients chosen for use in the production of the bacterium is of extreme importance and considerable research is done to ensure a medium is selected which produces the most activity per unit volume in the fermenter. For example, unlike *B. thuringiensis, B. sphaericus* can not use carbohydrates as a carbon source. The media are all balanced for carbon and nitrogen content and are all fortified with trace minerals and other additives that have been determined to add to the potency of the fermentation broth. As Table 1 clearly demonstrates the media components used in commercial production are undefined and relatively inexpensive. Major undefined components can be produced with specifications that do not vary significantly from lot to lot. The percent available glucose or in the case of soy flour, protein, can be determined for each lot of the respective ingredient. The trace materials can be from defined (analytical) sources. Corn steep liquor and solids are hard to define and often it is not known why this component increases the insecticidal activity of the beer.

Besides yield of insecticidal activity per unit volume of the fermentation beer, the most important criteria for media selection is price. It must be inexpensive and often excessive cost of the medium can essentially preclude a bacterium from commercialisation.

Each company involved in commercial production of entomopathogenic bacteria has its own proprietary mix of ingredients but all are combinations of those listed in Table 1. These are kept proprietary and are never revealed through patent or process protection. To do so would be to jeopardise their exclusivity.

6.   FERMENTATION PROCESS

The actual fermentation process parameters monitored and the parameter limits are also closely guarded proprietary company secrets. However, Lisansky *et al.* [13] have done an excellent job describing these.

Typically the parameters monitored during the fermentation process are temperature, dissolved oxygen (DO), pH, sugar concentration, and oxygen uptake rate (OUR). Because these bacteria tend to be voracious users of oxygen, control of the DO is extremely important. Care must be taken to prevent this value from falling below 20%. pH should be controlled between 6.8 and 7.2. Some producers do not have sophisticated pH controls on their fermenters and therefore they use buffered fermentation media. However, it has been found by some investigators that not controlling pH during the fermentation of *B. thuringiensis* subspecies *israelensis* (*Bti*) [19] and *B. sphaericus* [22] resulted in an increase in insecticidal activity when compared to fermenter cultures subjected to pH control. This is an area where there is some dispute but major producers use some form of pH control during fermentation.

Typically fed batch systems require monitoring of sugar levels. These must be controlled to ensure concentrations do not go below 2.0 g / l. Control of the fermentation temperature is also extremely important and is never permitted to vary from 30°C plus or minus 2°C. Higher temperatures result in suppression of endotoxin formation [16]. Lower temperatures <25°C slow the fermentation cycle and add to production costs. Although there is a great body of literature available on effects of oxygen (air feed 0.1 to 1.0 VVM) and glucose concentration (0.5–8.0 g/l) on the quality of bacteria growth [8, 17] the research was invariably done in flasks or small laboratory fermenters. Results in these types of systems rarely if ever transfer to commercial fermenters and current production systems rarely use less than one VVM or 18 g / l of glucose in those systems where glucose is the primary carbon source.

The fermentation cycle from seed tank to harvest is amazingly uniform for deep tank fermentation of *B. thuringiensis*. This time generally ranges from 62–92 hours but does not include fermenter clean up or harvest of the spores and δ-endotoxin. Times for other *Bacillus* species and *Serratia* may vary.

The end of the fermentation cycle for *B. thuringiensis* and *B. sphaericus* is easy to determine and to predict. This is facilitated by the morphological changes in the bacterium as the culture matures. For *B. thuringiensis* some intracellular spore and crystal formation can be seen after only 18 hours. The culture is typically mature at 28–32 hours. Mature spore and crystals are evident at this time in the *B. thuringiensis* vegetative cell. When approximately 80% of the culture is lysed the fermentation beer is immediately cooled to 4°C and the pH of the broth adjusted to 4.5. By the time the cooling stage is complete the culture is completely lysed and the beer contains only free spores and crystals.

7.   RECOVERY OF ENTOMOPATHOGENIC BACTERIA

The final fermentation beer containing the active isolate will generally contain 6–8% solids. 1–3% of these will be spores and in the case of *B. thuringiensis*, spores and crystals. The remainder will consist of soluble and insoluble carbohydrates and proteins.

To be commercial the active ingredient must be efficiently and inexpensively recovered. The methods used to recover the spore crystal complex are centrifugation and micro-filtration or a combination of the two.

Continuous centrifugation at >8,000 x *g* is the most common and least expensive method for recovery of the biomass. A flocculant may be added to the fermentation beer prior to centrifugation to optimise recovery of the solids. If the solids content of the fermenter beer is 4–6% this method is effective and the fermenter beer can be concentrated to produce slurry containing 15–30% solids [13]. If the cell mass is less dense centrifugation may be not be particularly effective since longer residence times in the centrifuge are required to recover the solids. This makes recovery of large fermenter volumes time consuming and costly. When centrifugation is used as the primary concentration methodology some product loss usually 10–15% must be accepted. The smaller crystals of *Bti* are especially subject to loss.

An alternative to centrifugation is micro-filtration. Membranes with pore sizes of 0.1–0.2 microns are typically used. If the filtration area of micro-filters is properly sized a fermenter of >70,000 litres can be recovered in 30 hours. Micro-filtration recovers 100% of the active solids including the small crystals of *Bti*. During recovery the solids are continuously washed with at least one volume of water to each volume of broth. This has an added benefit of removing any undesirable metabolites.

Micro-filtration equipment is more expensive to acquire and operate when compared to continuous centrifugation. A less expensive alternative is to combine the two methods. Centrifugation is used to do an initial recovery of the fermentation solids. The centrifugate is not discarded but is recovered in an appropriate storage tank. This is then concentrated with an properly sized micro-filter. Using this method reduces the size of the filtration system required and speeds the recovery time. It also recovers 100% of the active solids.

Because entomopathogenic bacterial slurries consist of living spores and heat labile crystal proteins or combinations thereof, care must be taken to preserve these from exposure to high temperatures, alkalinity and the introduction of exogenous bacteria, fungi, or yeasts. Extreme care must be taken to adequately cool the fermentation beer prior to recovery and to keep it cool while being processed. All tankage, piping and recovery equipment must be capable of being sanitised thoroughly. This is important to prevent the introduction of microbial contaminants into the recovered slurry. In addition to cooling and pH adjustment an anti-microbial is often added to the chilled beer to help ensure against an incidental contaminant introduced into the recovered product from growing.

8.   FORMULATION

Once the biomass is recovered it must be formulated. For purposes of this discussion the following definition of a formulation will be used. A basic commercial formulation constitutes the form and contents of the insecticide as supplied by the manufacturer to distributor and ultimately to end-user: a tank mix formulation is the commercial formulation plus spray vehicle; e. g. water, oil, etc., added by the end user and applicator [5]. The methodology involved depends on whether the product is to be an aqueous suspension, wettable powder, water dispersible granule, oil emulsifiable suspension, corncob granule, sand granule, pellet, briquette, dust or donut. Since bacteria are discreet, insoluble particles, they can not be formulated into soluble powders.

Formulation is the process used to convert a technical slurry or powder containing the active ingredient produced by the bacterium into a useful, end use insecticide compatible with existing ground and aerial application systems. The following general criteria must be satisfied by all formulations to be commercially successful. [1, 2, 5, 6, 13, 16]:

a) Insecticidal activity must be maintained and/or augmented during the formulation process.

b) The formulations must have an acceptable shelf life under the range of storage conditions typically found at end user locations. A reasonable shelf life is 18–36 months under ambient conditions averaging 25°C. Exposure to temperatures above 30°C for prolonged periods needs to be minimised with these products.
c) The formulations must be easily mixed and applied under field conditions.
d) The formulations must be designed to ensure adequate distribution of the active ingredient and to enhance deposition properties. This maximises the presence of the active ingredient on the target application site. Formulations of *Bti* often include dispersing agents designed to distribute and prolong the presence of the crystal protein within the feeding horizon of mosquito and blackfly larvae.
e) The formulation must be stable to environmental conditions and should be formulated so that the active ingredient is reasonably resistant to wash off and ultra violet light.
f) All formulations must be economical to make since these must have a competitive price in the market place.

The obvious delineation of commercial formulations is into liquid and solid products. Downstream processing to provide the technical grade active ingredient for these formulations differs.

8.1   Liquid formulations

The aqueous liquid suspension is the least expensive to produce when compared to dry formulations. The aqueous suspension is also inherently less stable than dry formulations. This type of product is more susceptible to the effects of high temperature and the introduction of exogenous contaminants.

An aqueous suspension is formulated directly from the recovered slurry following the concentration step. The concentration technique is always designed to preserve the viability and virulence of the bacterium and/or its toxin. The concentrated suspension is stabilised with the appropriate microbialstatic agent, anti-evaporant, and/or dispersant. This system is always a proprietary mix and is unique to each commercial manufacturer of these insecticides. Table 2 lists some of the typical ingredients used to formulate aqueous suspensions of entomopathogenic bacteria. This list is by no means exhaustive but represents the more common ingredients used to stabilise, suspend and facilitate mixing and application of these products. The formulation must be constructed so no special handling is required by the end user to optimise field applications and it must be compatible with existing ground and aerial application systems.

This blend of ingredients is selected not only to facilitate application of the aqueous suspension but also to ensure biological stability of the active ingredient. The formulation must have an adequate shelf life under reasonable storage conditions for a liquid. A shelf life of at least 18 months must be achievable when the product is stored at temperatures below 30°C. Recently a draft guideline was suggested by a World Health Organization (WHO) informal consultation in which it was recommended manufacturers meet the following minimum shelf life standards for formulations of *Bti* and *B. sphaericus* [21]. The recommendations are: (a) no more than 10% loss in biopotency below the labelled potency value when stored at 5°C for two years; and (b) no more than 10% loss in biopotency below the labelled potency value when stored at 20 to 25°C for one year. Such storage stability test shall be performed using representative product samples and the biopotency shall be assessed using the WHO standard method using mosquito bioassays. The informal Consultation noted that the variability of the potency test may be as great as 30% and must be included as a factor in determining potency loss.

Table 2. Typical formulation components used to prepare liquid suspensions of entomopathogenic bacteria.

Formulation components	Concentration (g/kg)	Function
Surfynol TGE	5	Dispersant
Bevaloid 211	30	Dispersant
Veegum	2.25	Suspending Agent
Xanthan Gum	0.3	Suspending Agent
Potassium Sorbate	0.2–0.5	Fungistat
Disodium Hydrogen Phosphate	16.5	Buffer / Stabiliser
Citric Acid	8.8	Buffer / Stabiliser
1, 2-benzisothiazolin-3-one	5.0–10.0	Bacteriastat / fungicide
Methyl Paraben	1.0–3.0	Bacteriastat / fungicide
Propylene Glycol	10.3	Paraben Stabiliser
Sodium Chloride	10–50	Preservative
Propionic Acid	30–50	Preservative
Methanol	50	Preservative
Glycerol	50	Thickener
Carboxymethylcellulose	1.5	Thickener
Tween 80	30	Dispersant
Sorbitol	30–450	Anti-evaporant
H_2SO_4	pH 4.5–5.2	Bacteriastat

References: [2, 4, 5, 13]

Although these recommendations are good target values for all bacterial formulations, manufacturers generally warrant that their aqueous formulations will work at labelled rates for two seasons if stored according to label instructions.

Once the aqueous suspensions are formulated and prior to packaging, the products are screened through a 100 mesh (149 µm) screen. This ensures that the formulations will not plug screens or nozzles in typical ground and aerial application systems. They are then packaged in appropriate containers for shipping.

8.2   Solid formulations

8.2.1   Wettable Powders, Water Dispersible Granules, and Dusts.

These formulations are essentially all wettable powders (WP). The major differences among the formulations listed involve the method of application and ease of mixing. Dusts are usually spread without mixing with water and have a lower concentration of active ingredient. A water dispersible granule formulation (WDG) or dry flowable is an agglomerated powder specifically formulated to disperse and wet quickly when mixed with water.

Wettable powder formulations of entomopathogenic bacteria are the most common products in the market place. However, water dispersible granules are slowly replacing wettable powders because of ease of handling and freedom from dust.

These formulations all begin from the production of the dry, technical grade, active ingredient (TGAI) from the concentrated slurry of the organism. This concentrated slurry is obtained as described earlier in this chapter. The TGAI is almost always produced by spray drying since other drying technologies tend to be too expensive.

The type of spray drier used is not important and stationary and rotary nozzles may be used [2, 13]. The conditions under which the powder is dried are extremely important. Spray driers are simple machines and consist of a device that permits introduction of liquid spray droplets into a heated cyclone of air. The water component of the slurry evaporates and powder is collected at the bottom of the drier. It is immediately removed to a collection hopper.

The temperatures used to dry the entomopathogenic bacteria must be strictly controlled or the bacterial spore can be damaged and the protein crystal destroyed. The typical inlet temperatures must be in the range of 180–215°C and the outlet temperature in the range of 70–80°C. These produce a powder with moisture content in the range of five to eight percent. An attempt to produce a powder with a moisture content under five percent often causes a significant decrease in insecticidal activity [2, 13].

The exact temperature suitable for the strain of bacteria being dried must be determined through experimentation since there are strain differences. As a general rule care must be taken to not exceed a moisture level in the TGAI of eight percent. A moisture content of six to eight percent promotes the longest shelf life of any other formulation of *B. thuringiensis* [13]. Some producers add a protectant to the slurry before drying. These are usually arabic gum or lactose but lactose is the most common. It is also a common practice to add a detergent to the slurry to add in the dispersion of the spray dried powder.

Once the TGAI is produced the particle size distribution of this material must be standardised. The easiest way to do this is by employing some type of milling device such as an air mill or hammer mill. The particle size of the TGAI should not exceed 50 microns.

Once the TGAI is produced its insecticidal activity (potency) must be determined before formulation into the final product.

To produce a wettable powder the TGAI is mixed with an appropriate diluent, wetting agent, dispersing agent and sometimes a proprietary mix of phagostimulants. Some of the most common additives are listed in Table 3. Once the wettable powder is mixed with its carriers and other formulation ingredients it is again air milled. Physical testing is then done to ensure the wetting time, dispersibility and particle size (<50 µm) are within product specifications. A sample is then sent for bioassay to ensure potency meets label claim.

The resulting wettable powder is then packaged into the appropriate size container. An extremely important requirement for packaging this type of formulation is it must be totally impervious to moisture. If the moisture content of the final formulation is in excess of five percent, caking, loss of wettability and dispersibility properties can result. In addition, high moisture content and caking is often associated with potency loss [5].

Water dispersible granules (WDG) are essentially agglomerated wettable powders containing a binding agent such as gypsum to facilitate the agglomeration process. The granulation takes place using a pan granulator or fluid bed drier or a combination thereof. The resulting formulation contains many of the common ingredients listed in Table 3 plus a binding agent. The potency requirements, mixing properties, dispersibility and suspensibilty parameters, packaging specifications and moisture content are the same as the wettable powders. Since the formulation is dry and easily stored, a three year shelf life is easily obtained. The WDG formulation is also referred to as a dry flowable (DF). It is dust free and thus minimises inhalation exposure of the mixing crews to the bacterium.

Dusts are the least common formulation used to distribute entomopathogenic bacteria. They typically contain less insecticidal activity per unit weight and are more difficult to apply. This formulation is usually reserved for the home and garden market.

Table 3. Common additives used to formulate wettable powders, water dispersible granules, granules, pellets, briquettes, dusts and donuts

Dry Diluents and Carriers	Surfactants
Barden Clay	Atplus 300
Attaclay	Triton B 1965
Silica Powder	Al-1246
1095 marble Dust	Tween 80
Silica sand	Tween 20
Continental Clay	Surfynol S485
Talc	Morwet EFW
Microcell	Triton X-35
Pyrax	Triton X-45
Lactose	Atlox 848
Corn Flour	Atlox 849
Soy Flour	Witconol H-31A
Rice Flour	Galoryl 546
Cotton Seed Flour	Rhodocal BX-78
Sipernat 22	Plurfac A-24
Agsorb	Surfynol TGE
Diluex	Bevaloid 211
Celite	Triton X-100
Plaster of Paris	
Powdered Cork	
Corn Cobs	
Biodac	
Gypsum	

References: [2, 4, 5, 13].
This list is by no means exhaustive but is representative of the components in current commercial entomopathogenic bacterial formulations.

8.2.2   Granule, Briquette, Pellet and Donut Formulations.

These are speciality formulations designed to improve availability of the formulation to the target insect. Granules formulated on sand have a very small particle size and are generally formulated on site. This type of product is generally formulated by the end user in the field using simple equipment such as a cement mixer. The powdered TGAI is added to sand and bound to it with a mineral oil. This dense formulation is excellent for penetration of foliage covering mosquito-breeding habitats and is almost exclusively used for this application.

The most common granule formulation until the advent of genetically manipulated corn containing the *B. thuringiensis* toxin gene was the corn grit granule. This granule is produced by coating defatted corn grit with *Bt* and binding the TGAI to the corn grit with an edible mineral oil, corn oil or soybean oil. If vegetable oils are used these typically are subject to more rapid degradation and an edible mineral oil is recommended. The particle size of this formulation is in the 10/14-mesh range (U. S. Standard Sieve Classification). Today the market for this formulation is essentially non-existent in corn. A similar formulation but with phagostimulants added such as molasses and cottonseed flour has also been produced by several manufacturers to service this market.

By far the largest market served by granular formulations of entomopathogenic bacteria is in mosquito control. Corncob formulations are produced to service this market. The formulations consist of *Bti* TGAI powder bound to a corncob or cellulose based granule. The most common carrier of the two is corncob. Two mesh sizes of cob are used 10/14, 5/8 (U. S. Standard Sieve Classification). The TGAI is bound to the granule using a food grade, white paraffin oil. Care must be taken to balance the amount of powder and oil to prevent loss of the TGAI from the surface of the granule during application and storage. The biological stability of these formulations is excellent and if stored covered and kept dry, shelf life can easily exceed two years at ambient conditions.

The advantages of the corn-cob granule are[1] it easily penetrates the over story of foliage often found in mosquito breeding areas; and [2] the formulation floats and slowly releases its active ingredient into the mosquito feeding horizon.

There have always been attempts to formulate granules containing *Bti* that are dense, contain higher concentrations of the TGAI and which slowly release the active ingredient into the mosquito feeding zone. Levy *et al.* [12] and Sjögren [18] report success with several different types of granule formulations. Levy used selected granule, briquette and pellet formulations containing proprietary delivery systems and coating systems to deliver the active ingredient to mosquito larvae. This system can contain feeding attractants and extenders to keep the TGAI available to mosquito larvae for a longer exposure period.

Sjögren developed a heavy, storable, quick-release granule of *Bti*, LarvXSG®. The granule is heavy and easily penetrates any foliage present in the mosquito breeding sites. The TGAI is released between 0.1–72 hours. Claims are made that a reduced concentration of LarvXSG® can effect control similar to higher rates of conventional corncob granules.

B. sphaericus can also be formulated on corn-cobs and other floating carriers. These granule formulations were more effective than *B. sphaericus* liquid concentrates [4].

Bti granules formulated with a cellulose carrier sold under the brand name Biodac have also been effective for mosquito control. However, its use is limited and abatement districts prefer corncob granules.

Entomopathogenic bacteria formulated as pellets or briquettes have had limited use. The market for these remains small. These products are made using corncob technology or extrusion presses.

In both cases heat must be controlled to prevent degradation of the TGAI. These formulations are used primarily in the mosquito market and typically contain a higher concentration of TGAI than the corncob granule (400 ITU/mg vs. 200 ITU/mg).

The donut formulation is a relatively new formulation and the market is growing. This formulation is used primarily in mosquito control. The technology is patented and Summit Chemicals is the sole producer. The formulation is in the shape of a donut and is formulated to slowly disintegrate over time in the mosquito habitat. Company literature says that control of larvae can be sustained up to one month with a single treatment of this product form. There is no similar product form for *B. sphaericus* or agricultural isolates of *Bt*.

8.3 Oil emulsifiable suspensions

This type of liquid formulation is unique enough to be described separately from other liquid products. It consists of the TGAI powder suspended in oil. The oils used may be either mineral or vegetable. Unlike aqueous liquid suspensions, oil formulations have biological stability profiles that are identical to those for dry powders. These formulations have excellent deposition characteristics since droplets do not evaporate during application.

There are two types of oil formulation. The first type does not contain an oil emulsifier and therefore, cleanup of spray equipment must be done with oil instead of water. This non-emulsifiable oil formulation is compatible only with oil carriers or other organic carriers. They are never mixed with water. The second type of formulation is an oil emulsifiable suspension. This type of formulation is miscible with water and oil carriers. However, care must be taken when mixing with water carriers to prevent formation of an invert emulsion during tank mixing.

An invert emulsion is usually formed when the quantity of the oil flowable formulation added to the tank mix exceeds that of water.

These formulations contain oil, gelling and suspending agents, and emulsifiers. Burges & Jones [4] and Couch & Ignoffo [5] prepared a list of these. In reality the registered recipes used by each company are proprietary and ingredients are kept secret. The author has never seen a recipe for a registered formulation published in a book or journal article.

The key component of this type of formulation is mineral oil. This is always aliphatic with no aromatic component to ensure the product will not be phytotoxic. The powder, which must be very uniform and of small size (<50 microns), is suspended in the oil using gelling and suspending agents. The product must be of reasonable viscosity (<1200 centapoise) to aid handling and pumping and the formulation must be constructed to prevent settling of the powder from the oil. If settling occurs the formulation becomes useless. This is the most popular formulation in areas that are semi-arid and where the oil formulation of the bacteria is applied in combination with chemical pesticides also formulated in an oil carrier.

In this section, little time was spent describing specific formulations based on *Bt* toxins grown in alternate bacterial species such as *Pseudomonas* sp.. These are novel formulations produced by Mycogen / Dow and referred to as CellCap® or Mcap® products. The *Bt* δ-endotoxin is enclosed in the *Pseudomonas* cell and is protected by the cell wall [9]. The *Pseudomonas* sp. cells are killed and then recovered and concentrated by conventional means. The resulting slurry concentrate is dried or it can be formulated directly into an aqueous liquid preparation.

CellCap® TGAI can be formulated in the same way as *Bt* formulations containing the spores and crystals. However, the aqueous, liquid slurry produced from CellCap® technology is easier to stabilise and has a better shelf life than *Bt* formulations containing viable spores. CellCap® formulations of this type are claimed to have better field persistence than conventional *Bt* formulations.

There also has been no discussion of transgenic plants expressing the *Bt* toxin. However, the plant can be considered a formulation in the strictest sense. Unfortunately, the regulatory agencies responsible for regulating sales of these products do not define then as such. Therefore, labelling requirements applicable to other insecticide formulations and active ingredients do not apply to plants (see Chapter 5.3).

A discussion of formulations of new biological insecticides derived from the insecticidal toxins of *Photorhabdus luminescens* has been excluded as well. *P. luminescens* is a bacterium carried by a nematode. The toxins produced by the bacterium are apparently broad spectrum but little information is available describing commercial production or formulation.

9.   QUALITY CONTROL REQUIREMENTS

The quality control requirements of formulations of entomopathogenic bacteria are straightforward. The formulations must meet the labelled potency requirement and physical specifications for each product form.

Biological and physical specifications are set by the commercial manufacturers to ensure the products mix, suspend, and disperse to ensure optimal field performance. For *Bti* and *B. sphaericus* the WHO has developed a set of draft guidelines covering physical and biological specifications [21]. Although developed for mosquito larvicides these requirements apply to all formulations of entomopathogenic bacteria.

The insect assay procedures to determine potency of formulations have been used for many years to standardise agricultural isolates of *Bt* [3, 7, 15]. The standard mosquito bioassay has been in existence for at least 20 years [14, 20]. These methods depend on using a standard technique combined with a standard of known potency. The methods are effective for establishing potency and determining degradation patterns of formulations.

New formulations of entomopathogenic bacteria containing transconjugates having crystal proteins containing toxins from more than one *Bt* serotype present different challenges when determining insecticidal activity. Manufacturers of this type of product often use biochemical techniques to determine the actual insecticidal protein concentration. The US EPA has accepted these methods based on HPLC) and / or ELISA techniques. Several products describe their active ingredients as percent lepidopteran or dipteran toxin protein instead of in International Toxic Units or International Units. Chapter 5.1 discusses standardisation in more detail.

Each lot of TGAI used in formulations of entomopathogenic bacteria must be tested using a mouse safety test. Five mice are injected subcutaneously with 10^6 spores and observed for seven days. This test screens for possible infectious contaminants and must be conducted on each production batch of the TGAI whether liquid or powder.

Tests for exogenous microbial contaminants in the TGAI and formulated products are also conducted although these are not currently required by the various regulatory agencies. Background levels of coliforms, *Staphylococcus aureus*, *Salmonella* sp., faecal streptococci, yeasts and molds are determined according to the internationally recognised specification for animal feeds developed by the Protein Advisory Group of the United Nations (Protein Advisory Group Bulletin, Vol. IV No. 3 Sept 1974- section 5, page 20 [2]).

Acceptable levels of these organisms are:

- Coliforms   <10/g
- *Staphylococcus aureus*   absent in 1 g
- *Salmonella*   absent in 10 g
- Faecal streptococci   <1 x 10^4 / g
- Viable yeasts and molds   <100 / g

The World Health Organization will publish new guidelines covering exogenous contaminants in bacterial insecticides in the first quarter of 2000.

Control of exogenous contamination begins with the fermentation and maintenance of sanitary conditions during recovery and mixing of the formulations. Control of exogenous contaminants in the end use product is extremely important in light of the negative press received lately for several *Bt* isolates.

10.   CONCLUSION

This chapter provides a concise overview of the methodology and problems associated with formulation of entomopathogenic bacteria. There are many available in the market place and these have been successful. However, formulation research is never finished. Manufacturers spend continuously on additional research and development to extend the shelf life of these products and to increase and synergism field performance. New formulations are in the pipeline especially for bacteria applied to the aquatic environment. Products with more carry, persistence and residual activity are being developed.

ACKNOWLEDGEMENTS

Thanks to Ole Skovmand for critical review of the manuscript.

REFERENCES

[1] Angus TA & Lüthy P (1971) Formulation of microbial insecticides, p. 623-628. *In* Burges HD & Hussey NW (ed.), Microbial Control of Insects and Mites, Academic Press

[2] Beegle CC, Rose RI & Ziniu Y (1990) Mass Production of *Bacillus thuringiensis* and *B. sphaericus* for Microbial Control of Insect Pests, p. 195-216, *In* Maramorosch KB (ed.), Biotechnology for Biological Control of Pests and Vectors, CRC Press

[3] Beegle CC, Couch TL, Alls RT, Versoi PL & Lee BL (1986) Standardization of HD-1-S-1980, U. S. standard for assay of ledidopterous - Active *Bacillus thuringiensis.* Bull. Entomol. Soc. Am., 32-44

[4] Burges HD & Jones KA (1998) Formulation of Bacteria, Viruses and Protozoa to Control Insects, p. 31-127. *In* Burges HD (ed.), Formulation of Microbial Pesticides, Kluwer Academic Publishers

[5] Couch TL & Ignoffo CM (1981) Formulation of insect pathogens, Chapter 34, p. 621-634, *In* Burges HD (ed.), Microbial Control of Pests and Plant Diseases 1970-1980, Academic Press

[6] Couch TL (1978) Formulations of Microbial Insecticides; conventional formulations, Misc. Publ. Entomol. Soc. Am. 10, 3-10

[7] Dulmage HD, Boening OP, Rehnborg CS & Hansen GD (1971) A proposed standardized bioassay for formulation of *Bacillus thuringiensis* based on the international unit. J. Invertebr. Pathol. 18, 240-245

[8] Foda MS, Salama HS & Selim M (1985) Factors affecting growth physiology of *Bacillus thuringiensis*, Appl. Microbiol. Biotechnol. 22, 50-54

[9] Gelernter W & Schwab GE (1993) Transgenic bacteria, viruses, algae and other microorganisms as *Bacillus thuringiensis* toxin delivery systems,, p. 89-104. *In* Entwistle PF, Cory JS, Bailey MJ & Higgins S (ed.), *Bacillus thuringiensis*, an environmental biopesticide: Theory and Practice, Wiley

[10] Gonzales JM Jr, Brown BJ & Carlton BC (1982) Transfer of *Bacillus thuringiensis* plasmids coding for δ−endotoxin among strains of *B. thuringiensis* and *B. cereus.* Proc. Nat. Acad. Sci., USA 79, 6951-6955

[11] Jackson TA, Pearson JF, O'Callagham M, Mahanty HK & Willocks MJ (1992) Pathogen to Product, Development of *Serratia entomophila* (Enterobacteriaceae) as a Commercial Biological Control Agent for the New Zealand Grass Grub (*Costelytra zealandica*), p. 191-198. *In* Jackson TA & Glare TR (ed.), Use of Pathogens *in* Scarab Pest Management. Intercept Ltd.

[12] Levy R, Nichols MA & Opp WR (1997) Targeted delivery of pesticides from Matricap® compositions, p. 63-93. *In* Gr oss GR, Hopkins MJ & Collins HM (ed.), Pesticide Formulation and Applications Systems, vol.17, ASTM STP 1328 Am. Soc. Testing and Materials

[13] Lisansky SG, Quinlan R &Tassoni G (1993) The *Bacillus thuringiensis* Production Handbook, CPL Press, United Kingdom

[14] McLaughlin RE, Dulmage HT, Alls R *et al.* (1984) U. S. Standard bioassay for the potency assessment of *Bacillus thuringiensis* serotype H-14 against mosquito larvae, Bull. Entomol. Soc. Am. 30, 26-29

[15] Riethmuller U & Langenbrush GA (1989) Two bioassay methods to test the efficacy of *Bacillus thuringiensis* subspec. tenebrionis against larvae of the Colorado Potato Beetle (*Leptinotarsa decemlineata*), Entomophaga 34, 237-244

[16] Rowe GE, Argyrios M (1987) Bioprocess Developments in the Production of Bioinsecticides by *Bacillus thuringiensis*, CRC Critical Reviews in Biotechnology, 6, 87-127

[17] Scherer P, Lüthy P & Trumpi B (1973) Production of δ–endotoxin by *B. thuringiensis* as a function of glucose concentration, Appl. Microbiol. 25, 644-646

[18] Sjögren RD (1996) Insecticide composite time release particle. US Patent 5484600

[19] Smith RA (1982) Effect of strain and medium variation on mosquito toxin production by *Bacillus thuringiensis* serovar. *israelensis*, Can. J. Microbiol. 28, 1089-1092

[20] World Health Organization (1983) Mosquito Bioassay Method for *Bacillus thuringiensis* subsp. *israelensis* (H-14), de Barjac H , WHO Report TDR, Annex 5, VEC-SWG 81.

[21] World Health Organization Communicable Disease Prevention and Control, WHO Pesticide Evaluation Scheme (WHOPES) (1999) Draft guideline specifications for bacterial larvicides for public health use, WHO/CDS/CPC/WHOPES/99.2, p. 27

[22] Yousten AA & Wallis DA (1987) batch and continuous culture of the mosquito larval toxin of *Bacillus sphaericus* 2362, J. Ind. Microbiol. 277-283

Chapter 5.3

Rural production of *Bacillus thuringiensis* by solid state fermentation

Eduardo Aranda, Argelia Lorence & Ma. del Refugio Trejo
Centro de Investigación en Biotecnología (CEIB), Universidad Autónoma del Estado de Morelos (UAEM). Av. Universidad 1001, Col. Chamilpa. C.P. 62210, Cuernavaca, Morelos, México

Key words: Sustainable agriculture, *Bacillus thuringiensis*, solid state fermentation

Abstract: Through sustainable agriculture, mankind may find a way for a better use of natural resources, cultivating commodities with lower inputs of energy. Sustainable production of biopesticides may become an important component of pest control strategies, recycling agricultural by-products and using low value carbon sources as substrates in developing countries. *Bacillus thuringiensis* (*Bt*) production by solid state fermentation (SSF) is already practised in rural communities in China and Cuba, where currency savings are limiting the use of expensive and contaminating chemical insecticides. In this work we propose the development of a cost-effective technological package of *Bt* production by SSF using native *Bt* strains and available local wastes or by-products directed to small farmers. Our proposal seeks to link the development of biocontrols at academic research level with field application.

1.   INTRODUCTION

Although enough food is produced world-wide, most of it is concentrated in developed nations, while developing countries are not self-sufficient. By the year 2005, human population will rise to approximately 8 thousand million, with a big percentage allocated only for developing nations (2,500 millions); food production, then, should be increased at least two-fold by the same year in those areas. At that time, 1 thousand million people will be considered malnourished [3]. It is because of this that a great effort must be given to the development of

J.-F. Charles et al. (eds.),
Entomopathogenic Bacteria: From Laboratory to Field Application, 317–332.
© 2000 *Kluwer Academic Publishers. Printed in the Netherlands.*

technologies to have efficient agricultural systems based on local sustainability. Many governments around the world have tried to establish research programs dedicated towards this end. Sustainable food production should use methods compatible with the environment and the biodiversity, but preserving natural resources for future generations. Applied Integrated Pest Management (IPM) strategies were considered to positively impact food production, although practical results obtained today, are not always consistent with theory.

At the present time, agricultural production uses a high input of chemical products (fertilisers, pesticides, and herbicides) to increase productivity and quality of food, and economical profits to farmers. However, a great percentage of those chemicals used for pest control are highly toxic and presently forbidden by several international agencies [13, 14]. On the other hand, sustainable agriculture, while promoting enough food production, also seeks to apply alternative strategies for better environmental protection. Because the majority of rural communities in developing countries face poverty and lack of opportunities to develop, it becomes unsuitable to keep using conventional methodologies (costly and ineffective at long term) for pest control that irreversibly alter the ecosystems, affecting also health conditions of farmers and consumers.

2.   STRATEGIES FOR INSECT CONTROL

2.1   Insect pest control: conventional *vs.* non-conventional

The use of synthetic chemical pesticides has posed great hazards world-wide for both, farmers and consumers. Environmental pollution in water, air, and soil, along with the development of resistance in target pests [18], and the negative impact in human health [29], are the main troubles. During the past 10 years, farmers and general public have become aware of the importance of those strategies related to biological control of insect pests in agriculture, forestry, and home dwellings [4, 12, 45]. Approximately, 400 species of fungi and 40 species of bacteria are known to attack insects; some of these microorganisms can be produced under field conditions to prepare biocontrol agents. Microbial insecticides, made of toxins and/or spores, have many advantages over chemicals and are widely used. The most successful microbial insecticide in the past 40 years is *Bacillus thuringiensis* (*Bt*), which has been used extensively for the control of agricultural pests and flies and mosquitoes that are vectors of diseases [16], but it still is at a low percentage when compared to chemical pesticides [23].

Bt is a ubiquitous Gram-positive, spore-forming bacterium that synthesises one or two parasporal crystals during the stationary phase of its life cycle. The mixture of spores and crystals has been used since 1960's as the active ingredient of commercial biopesticides for the control of certain insect pests. From 1904 to date, *Bt* strains toxic to species among the orders of Lepidoptera, Diptera, Coleoptera, Hymenoptera, Homoptera, Orthoptera, Mallophaga and against nematodes, mites, and protozoa have been described ([43] and see Chapter 2.1). The genes encoding the toxic proteins of the crystals (*cry* genes family) have more than 100 members and some are being used as a source for transgenic expression to provide pest resistance in plants.

The mechanism of action of the *Bt* Cry proteins (also called δ-endotoxins), involves solubilisation of the crystal in the insect midgut, proteolytic processing of the protoxin by midgut proteinases, binding of the Cry toxins to specific high affinity toxin binding proteins on the surface of midgut columnar cells, insertion of at least part of the toxin into the apical membrane, and presumably oligomerisation of the Cry protein to create ionic channels which lyse the cell and eventually lead to the death of the insect (for a review see [34] and section 3 of this book).

2.2 *Bt*: a natural strategy for insect pest control

Because *Bt* is a bacterial species that does not require relatively complicated nutritional media, the most common technology to produce spores-crystals of *Bt* is submerged fermentation in aerated and agitated bioreactors. A complete review of this technology is presented in the Chapter 5.2. The main industry related with production of *Bt* using submerged fermentation is distributed in United States and Europe. Companies as Abbott and Novartis (previously Ciba and Sandoz) have been the leader companies since 1990, and in 1996 both companies dominated 70% of the world market [23]. Of this total market, more than half of it is used in the United States and Canada; 18% in Eastern Countries and only 10% in China, and 8% in Latin America. The lack of knowledge of this biological control agent, plus the cost of *Bt*-based products are determinant factors with which small farmers are limited to make use of them in developing countries.

Besides this situation, most of the commercial *Bt* products are based on few strains which are non-native to many countries, users of these products; it is a common feature that these strains are few or non active against local pest problems. Additionally, *Bt* products are not directed to small local markets. It has been observed that the profile of *cry* genes from *Bt* around the world is quite specific for each region [7, 8]. For example, in Latin American countries, like Mexico, there exists a very diverse geography and patterns of microclimates, important conditions for the diversity of insects and *Bt* strains [8]. From this point of view, it would be a great advantage to design the production of *Bt* based on local, and "adapted" strains.

In this paper, we explore the possibilities of producing *Bt* by means of solid state fermentation (SSF) at rural scale in developing countries, using agricultural by-products and low-cost carbon sources with the consequent savings in money for producers [11, 28, 41, 46]. Recent applications of SSF processes have demonstrated effective *Bt* production, a process that can be adapted for small-scale rural production. Although China is mentioned as pioneer in the production of *Bt* [41], Brazil and Cuba have been developing their own technology for the rural production of *Bt* and other entomopathogens using agricultural by-products (Brazil [10, 11], Cuba [33, 37, 38, 47]).

3.   SOLID STATE FERMENTATION (SSF)

3.1   Recent applications of SSF

SSF has gained renewed interest because it can be applied to produce primary and secondary metabolites (enzymes, alkaloids, antibiotics, carboxylic acids, and biopesticides). SSF systems involve growth of microorganisms on moist solid substrate in the absence of free water [20, 22, 26, 36, 39], and is performed on natural organic substrates or inert supports. When using organic substrates, the moist solid substrates (polymeric in nature and insoluble in water), act as sources of carbon, nitrogen, minerals, water and other nutrients, and also provide support for microbial growth. In the inert matrices, the solids function as support of microorganisms and wide changes in the medium used for impregnating it are feasible [42, 48]. Natural or synthetic materials can be used as support or substrate to the SSF processes. Solid materials used in SSF are water insoluble; so in this condition, water is absorbed onto substrate particles which can be used by microorganisms for growth and metabolic activity. Bacteria and yeasts grow on the surface of solid material while fungal mycelia penetrate into the particles of the matrix [32, 44]. The microbial growth in SSF is influenced by several factors. Among the physical ones, availability of substrates to microbes, film effects and mass effects are important [22]. Physical morphology, mainly porosity and particle size of solid material, governs the available surface area to the organism. On the other hand, the chemical nature of substrate (as degree of polymerisation and organic matter content) plays an important role in SSF. In general, several agricultural or agroindustrial residues, which are cellulosic or starchy in nature have been used in SSF processes.

Recently, several research groups have realised the vast economical and practical advantages of SSF. Several factors seem to support this new approach to SSF applications: a) in this type of fermentation the substrate may be put on a tray, in a plastic bag or in a flask; b) most of microorganisms studied in this processes have the property to invade solid substrates and transform complex biochemicals in simple nutritional compounds, and; c) from the engineering point of view, such features help to develop non sterile, as well as water and energy saving fermentations [39].

Process parameters are very important factors in bioreactor design of SSF [22, 26, 31]. Kinetics studies of SSF processes have not received much attention as compared to submerged fermentations. SSF processes have difficulties associated with the measurement of various factors, due to the heterogeneity of the medium which is structurally and nutritionally complex. In addition, the control of fermentation parameters is also difficult. Several types of bioreactors have been used in batch or continuos mode in SSF. Laboratory studies are usually carried out in flasks, jars, glass column reactor or Roux bottles [39]. Most large scale SSF employ either tray-type or drum bioreactors. Tray bioreactors represent the simplest system of fermenters and may be built of wooden, plastic, or metallic materials. The bottom is perforated in such way that it holds substrate and allows aeration of the undersurface of the substrate. Drum bioreactors basically consist of a drum-shaped reactor equipped with a rotating device and provided with an air inlet and outlet. These systems employ forced aeration [31, 32].

3.2   SSF as an important alternative of an integrated ecological pest management program

The recent applications of SSF processes have demonstrated effective *Bt* production, which can be developed for small scale local production. The most important considerations to obtain an economic mixture of spores and δ-endotoxins production by SSF are: the fermentation media has to be as cheap as possible and it has to support the endotoxin production; the selection of constituents of fermentation media should be available in the region where the work is conducted; consistency of the composition of the by-product, and low cost of the local production processes [10, 11]. In developing countries, the local *Bt* production would result in considerable monetary savings, reduction of imports of chemical and biological insecticides, and also reduction of environmental deterioration.

In submerged fermentation, the most basic media to produce *Bt* are chemically defined (with diverse sources of nitrogen, carbohydrates, and inorganic materials). However, defined media are expensive for a large-scale production. Some authors have considered that defined media can be substituted for less expensive substrates (agricultural products or by-products). Such substrates may contain themselves sources of organic nitrogen and carbon, as well as inorganic components for bacterial growth. Availability and costs of different raw materials used for *Bt* fermentation media, will depend on the local production of such materials. The agricultural wastes and by-products, in culture media, have been found to affect the growth pattern and the toxicity of several *Bt* subspecies: *kurstaki, entomocidus, israelensis, thuringiensis* and *aizawai* among others. For example, for *Bt* var. *aizawai* (HD-133) the most efficient substrates for the production of spore-crystal biomass and endotoxin potency are cotton seed meal, defatted soy flour and corn gluten meal. The carbohydrate/nitrogen ratio of these additives ranged from 0.3 to 0.5 and the glutamic acid content of their proteins from 9.2 to 16.0% (glutamic acid is the most abundant amino acid constituent of *Bt* crystals). For particular strains in specific regions, an evaluation should be performed [25].

3.3   International experiences on the production of *Bt* by SSF

As cited above, the wider use of *Bt* based products in developing countries has been restricted for economical reasons. However, several authors recognises that *Bt* can be cheaply and easily produced on a wide of low value, organic substrates (mainly agricultural and industrial by-products) for local use by SSF [10, 11, 24, 25, 40].

Bt production by SSF has been investigated since 1970´s until now. On a first work, a solid medium consisting in groundnut cake and tamarind kernel powder was used; it yielded a product with a high level of sporulation and toxic to larvae of the almond moth (*Ephestia cautella*) [27]. Scientists of several countries have been studying different alternatives of *Bt* production by SSF in order to generate technologies that could be used by small farmers. Table 1 presents a summary of this work.

Table 1. Summary of *Bt* production using agricultural and animal wastes, as well as diverse sources of carbon and nitrogen.

Country [Ref.]	*Bt* Variety	Media	Results
AFRICA			
Egypt[1][40]	*kurstaki*	Fish meal, bovine blood, residues from slaughterhouses and fodder yeast	Laboratory
Nigeria[1][28]	*israelensis* IPS78	Groundnut cake, cow peas (2 varieties) and bambara beans	Laboratory
ASIA			
China [40]	*kurstaki*/HD-1, *galleriae*, *wuhanensis*, *dendrolimus*	Peanut bran, soybean meal, wheat bran, corn meal and defatted cotton seed cake	Industrial production[2]
Malaysia[1][21]		Fish meal and soybean waste	Laboratory
Thailand [46]	*israelensis* H-14	Wheat bran, and rice husk	Small scale
LATIN AMERICA			
Brazil [11]	*kurstaki* HD-1 *israelensis* H-14	Solid residue from pulp and paper industry and rice	Laboratory
Cuba [33, 37, 38, 47]	Berliner LBT-13 Berliner LBT-24	Banana wastes, yeast, corn meal and rice powder	Industrial production[2]
Mexico [This work]	Mexican *Bt* strain	Sugarcane pith bagasse, corn meal and soybean meal	Laboratory
Peru (P. Ventilosa, personal communication)	*israelensis*	Cassava wastes and whole coconut	Small scale

[1]Although these countries employ by-products for *Bt* production, this is done in submerged fermentations.
[2]It is considered industrial because of the size of the production, which ranges in the scale of tons.

Undoubtedly China is the country where *Bt* production by SSF is more widely used, being Wuhan the main site for *Bt* production using both, submerged fermentation and SSF. In the pilot plant at Hubei Academy of Agricultural Sciences, *Bt* production has grown from 23 tons in 1983 to 900 tons in 1990. *Bt* is now widely used in 30 provinces (approximately 16×10^6 ha) for the control of several insect pests of agriculture, forest and flies and mosquitoes vectors of human diseases. A portion of *Bt* production is exported to Thailand and South East Asia.

In China the *Bt* production in solid media is achieved using pure cultures. After 24 h growing in test tubes the *Bt* culture reaches a "solid-white state" and is then incorporated in a larger volume of media in litre bottles. Spore count is assessed after 24 h. This inoculum is then mixed with piles of peanut bran and soybean meal, dampened and put in shallow trays. It is ready to use in 30 h. Alternatively, on some communes, this procedure is achieved in 2.5-3 days in trays or soil pits [40].

It was reported the production of *Bt israelensis* H-14 by SSF in Thailand at village or small scale as mosquito larvicide especially for *Aedes* and *Anopheles* species. Innoculum is made from waste soybean extract in shake flasks (50 g waste soybean, 0.5 g monosodium glutamate, 2 g glucose, 0.5 g K_2HPO_4 per litre of water pH 7.6). The solid medium for *Bt* production contains 100 g wheat bran, 27.34 g rice husk, 6.45 g glucose, 0.165 g NaCl, 0.05 g $CaCl_2$ and 0.134 g $(NH_4)_2SO_4$. Materials are placed in appropriate trays and covered with foil or glass and sterilised for 30 minutes. Following, the innoculum and medium are mixed at a ratio of 2:5 and moisture adjusted to 70%. It is incubated at 30°C for 3 days. The final product is used to prepare three kinds of formulations: larvicidal floating briquette, dry granules and concentrated liquid. Dry granules are made drying at 40-45°C for 2 days until the moisture reached 5%. Larvicidal floating briquette are made by mixing oven dried product with polyvinyl glue and using a hydraulic press to make a doughnut shape briquette. Concentrated liquid can be prepared by extracting the dry or wet solid state culture with sterile water repeatedly until the resulting extract is brownish and cloudy. Viable cell count of solid final products is 9-10x10^{10} cfu/g dry weight and 10^8 spores per ml for concentrated liquid [46].

Brazil and Cuba are the Latin American countries where *Bt* production by SSF is more developed. In the case of Brazil it is reported that there are many groups producing *Bt* in laboratory scale by means of SSF and submerged fermentation [10]. Intensive work has been done in order to select the components of the medium based on availability of the agroindustrial solid by-product in the region where the work is conducted, the consistency of the composition of the by-product and its low-cost. For example *Bt kurstaki* is very efficiently produced by SSF using solid residues from pulp and paper industry. Final product has 10^{10} cfu/g dry matter after 168 h of incubation with periodic humidification and it is very active against soybean caterpillar (*Anticarsia gemmatalis*). Low-cost type rice has been used for *Bt israelensis* production by SSF where small plastic bags are used as reactors. After 112 h of incubation, a product with 2.5×10^8 cfu/g dry matter is obtained. It is necessary to use 143 mg of the whole biomass (100 h fermentation, dried in an oven at 60°C) to achieve 100% kill of *Culex* sp. larvae when spread over an area of 1 m^2 of clear water [11].

Cubans produce different *Bt* strains by SSF (Berliner LBT-3, LBT-13, LBT-24), using very diverse substrates including banana wastes [33], yeast [37], corn meal and rice powder [47]. *Bt* production is performed in most of the provinces of Cuba; however, no publications are available about the details of the SSF process.

In Peru two approaches have been used to produce *Bt israelensis*. Either piles of cassava or whole coconuts (minced) are inoculated and incubated in trays during 96 h. In the case of coconuts (already produced directly in the community), the average *Bti* spores is 2.2×10^5/ml; the average larvicidal effect on *Anopheles albimanus* was a 100% after 12 h with a concentration of 2.2×10^3 spores/ml [49].

Bt production by SSF has not been restricted to academic areas. There is a patent of *Bt* production by SSF. In this process, innoculum is grown in shake flask using a dextrose-corn steep liquor-yeast autolysed medium. The production media contains 545 g wheat bran, 380 g expended perlite, 62 g soybean meal, 36 g dextrose, 3.6 g lime, 0.9 g NaCl, 0.29 g $CaCl_2$ and 160 ml water. The medium is partially sterilised by passing steam for one hour. The inoculated media is placed in a bin in which air volume per fermenter volume per minute (VVM) is passed at 0.4 to 0.6 for 3 h at 30-34°C and at 1.2 VVM thereafter for 36 h to yield a product containing 4% moisture. The product is then ground and passed through a 80 mesh screen. A batch containing 250 kg of inoculated bran yields $3-17 \times 10^9$ spores/g of dry matter. The product may be in the form of dust, wettable powder or granular formulation. It is reported that Nutrile Products Inc. use this technology to make a product named Biotrol containing *Bt* var. *kurstaki* HD-1 by SSF [24].

3.4 Local *Bt* production in Mexico by SSF: an alternative proposal

Unfortunately, commercial agriculture in Mexico is heavily dependent on inputs of pesticides; most, if not all, small farmers use pesticides and few of them are really concerned about the hazards of pesticide use. Illiteracy and lack of training are part of the abuse of pesticides [2]. However, public concern about chemicals is growing and more producers are now moving towards biological control strategies.

Among the entomopathogens investigated in Mexico for biocontrol purposes, *Bt* has been widely studied at least by four different research groups in CINVESTAV-Irapuato (IPN) [5], CINVESTAV-México (IPN) [15], UANL, Monterey [17] and UNAM, Cuernavaca [7]. These studies are concerned with the screening of soil for *Bt* strains with novel insecticidal activities, characterisation of *cry* genes, study of the mechanisms of action of Cry toxins and *Bt* production by submerged fermentation.

In this work, studies were made to explore the feasibility of producing *Bt* by SSF, and to evaluate the best system for rural or local production. Based on the results obtained in our laboratory and on previous tests with SSF techniques for bacteria and fungi (*Bt* native strains from Morelos soil, [19]); the fungal pathogen *Isaria* sp. (isolated from *Gynaikothrips ficorum*, [1]), it is possible to develop the SSF as an alternative for small farmers. First, selection of medium component was made, targeting agroindustrial solid by-products available in the region. Three regions within Mexico have been selected: Sinaloa in the North, Morelos in the central plateau, and Chiapas in the South. In the state of Morelos, the most important by-products are sugarcane pith bagasse and rice bran. Preliminary experiments have been performed using selected residues as substrate to support fermentation in column type reactors [35, 42, 48], keeping constant humidity (70%) by means of water saturated air. In our work, the substrate-support tested to produce *Bt* was sugarcane pith bagasse mixed with corn meal or soybean meal as an additional carbon source. Some of the results obtained in our lab show promising yields of spores (2.9 X 10^9/g dry weight) within 48 h of SSF process.

Toxicity of final product was evaluated on the cabbage looper (*Trichoplusia ni*). Spores and crystals were 100% toxic to first instar larvae after 48 to 72 h exposure on contaminated discs of cabbage leaves (5 µg/ml of total protein as measured by the Bradford method [6]). The bioassay indicates the potential of the product obtained in the SSF process. At present, we are working in the selection of specific substrates for each region, and the design of the best process for fermentation in field conditions.

At the same time, we are working in the adequacy of the process to local needs, because we know that one of the limiting factors in the countryside is the first approach to people, so they gain credibility and engage in active participation in projects. For example, a team of experts has proved that farmers and villagers of Sinaloa may be trained in irrigation, seed selection, farming practices, production of green lacewings (Insecta: Neuroptera), and use of local plant species with insecticidal properties to improve quality and quantity of produce in family plots [30].

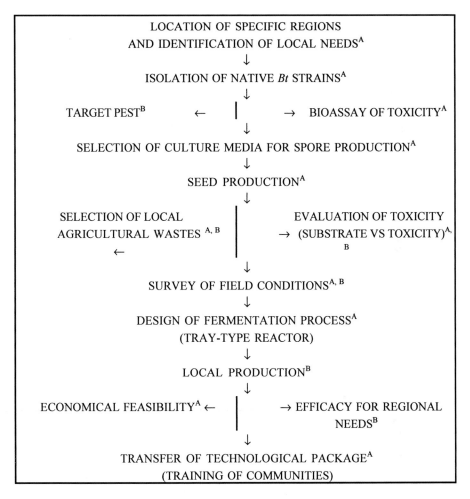

Figure 1. Schematic diagram of the transfer of a technological package to produce native strains of *Bt* by means of SSF. [A]Activity developed by the institution, [B]Activity developed by farmers and/or villagers.

The possibility of producing fungi or *Bt* using local inexpensive material by SSF, suits quite adequately the expectations of small farmers in developing countries. We propose to develop a technological package starting with the selection of native strains and local substrates trough the training of people (Figure 1).

To achieve a rural production system of *Bt*, the best process for field conditions is the tray-type reactor. A schematic drawing of this type of reactor is shown in Figure 2 [9], and will be adapted according to local materials and conditions as well as ideas from mushroom producers.

Usually, aluminium trays are arranged one above the other keeping suitable air-space among them. The fermentation is carried out in a chamber where a semi-controlled humidity is maintained. Temperature of the fermenting substrate is controlled by circulating warm or cool air as necessary. Pandey [31] discussed some applications of tray bioreactors in different countries.

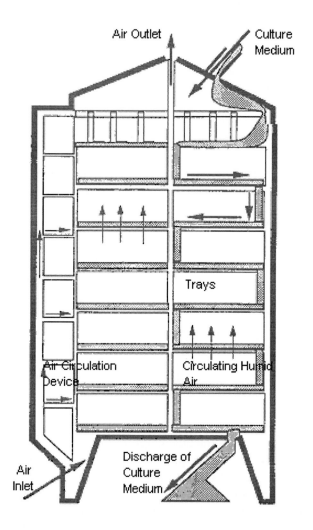

Figure 2. Example of tray-type reactor as shown in [49]. This system can be adapted for rural production of *Bt*, using shallow plastic trays (2 cm depth).

4.   CONCLUDING REMARKS

In Latin America, most small farmers depend on an agriculture of subsistence, farming on commodities that they use to tackle immediate needs: to feed themselves and to maintain their farm animals. Minimal excess is sold in local markets. Agriculture in the majority of rural areas, in general, is lagging far behind the technologies already applied in developed countries to increase yields. Illiteracy, poverty, and lack of training are the major drawbacks to achieve good programs of food production and insect pest control. The SSF process for *Bt* production may become an important component of integral food production in rural communities because all the advantages that offers as: process simplicity, reduced energy inputs, easiness of product recovering, and revaluation of local agricultural wastes or by-products. This strategy may lead to a true sustainability. The technological package for *Bt* rural production by SSF, based on native strains proposed here seeks to link the development of biocontrols of local pests at academic research level with field application, taking advantage of experiences already established.

ACKNOWLEDGEMENTS

Experimental work was partially supported by Dirección General de Investigación y Posgrado, Universidad Autónoma del Estado de Morelos (México), PFI-1997. We thank Solcamiri Hernández for her help in the SSF process, Dr. Alejandra Bravo (Instituto de Biotecnología, UNAM) for discussion, and Dr. Paulina Balbás and Dr. Sevastianos Roussos for critical review of the manuscript.

REFERENCES

[1] Abarca P, Hernández J, Montiel E, Martínez E, Trejo MR & Aranda E (1998) Producción de cepas nativas de hongos entomopatógenos por medio de fermentación en estado sólido. IX Encuentro Regional de Investigadores en Flora y Fauna del Centro-Sur de la República Mexicana. Cuernavaca, p. 29
[2] Bejarano F (1995) Pesticide proliferation in Mexico, p. 87-96. *In* Dinham B (ed.), The Pesticides Trust, The Pesticide Trail. The impact of trade controls on reducing pesticide hazards in developing countries, Eurolink Centre, London
[3] BID/PNUD (1991) Nuestra propia agenda. Comisión de Desarrollo y Medio Ambiente de América Latina y el Caribe, Banco Interamericano de Desarrollo
[4] Bishop DHL (1994) Biopesticides. Curr. Opinion Biotechnol. 5, 307-311

[5] Bohorova N, Maciel AM, Brito RM, Aguilar L, Ibarra JE & Hoisington D (1996) Selection and characterization of Mexican strains of *Bacillus thuringiensis* active against four major lepidopteran maize pests. Entomophaga 41, 153-165

[6] Bradford M (1976) A rapid and sensitive method for the quantitation of microgram quantities of protein utilizing the principle of protein-dye binding. Anal. Biochem. 72, 248-254

[7] Bravo A, Ortíz M, Ortíz A *et al.* (1996) Búsqueda y construcción de nuevas proteínas insecticidas de *Bacillus thuringiensis*, p. 375-379. *In* Galindo E (ed.), Fronteras en Biotecnología y Bioingeniería, Sociedad Mexicana de Biotecnología y Bioingeniería, Mexico

[8] Bravo A, Sarabia S, López L, *et al.* (1998) Characterization of *cry* genes in a Mexican *Bacillus thuringiensis* strain collection. Appl. Environ. Microbiol. 64, 4965-4972

[9] Cannel E & Moo Young M (1980) Solid-state fermentation systems. Process Biochem. June/July, 2-7

[10] Capalbo DMF (1995) *Bacillus thuringiensis*: fermentation process and risk assessment. A short review. Mem. Inst. Oswaldo Cruz 90, 135-138

[11] Capalbo DMF & Moraes IO (1997) Use of agro-industrial residues for bioinsecticidal endotoxin production by *Bacillus thuringiensis* var. *israelensis* or *kurstaki* in solid state fermentation, p. 475-478. In Roussos S, Lonsane BK, Raimbault M & Viniegra-González G (ed.), Advances in Solid State Fermentation, Kluwer Academic Publishers, Dordrecht

[12] Cate J (1990) Biological control of pests and diseases: integrating a diverse heritage, p. 23-43. *In* Baker RR & Dunn PE (ed.), New directions in biological control: alternatives for suppressing of agricultural pests and diseases, Alan R. Liss Inc., New York

[13] Dinham B (1993) The Pesticide Hazard. A Global Health and Environmental Audit. Zed Books Ltd., London and New Jersey

[14] Dinham B (1995) The Pesticides Trust, The Pesticide Trail. Dinham B (ed.), The impact of trade controls on reducing pesticide hazards in developing countries, Eurolink Centre, London

[15] Farrera RR, Guevara FP & De la Torre M (1998) Carbon:nitrogen ratio interacts with initial concentration of total solids on insecticidal crystal protein CryIA(c) and spore production in *Bacillus thuringiensis* HD-73. Appl. Microbiol. Biotechnol., 49:758-765

[16] Feitelson J, Payne J & Kim L (1992) *Bacillus thuringiensis*: insects and beyond. Bio/technology 10, 271-275

[17] Galán-Wong L.J., Rodríguez-Padilla C & Luna-Olvera HH (1996) Avances recientes en la biotecnología de *Bacillus thuringiensis*. Universidad Autónoma de Nuevo León, Monterrey

[18] Heitefus R (1989) Consequences of the use of chemical plant protection agents, p. 190-216. *In* Heitefus R (ed.), Crop and plant protection: the practical foundation, Ellis Horwood Limited, New York

[19] Hernández S, Aranda E & Trejo MR (1998) Producción de cepas nativas de *Bacillus thuringiensis* por medio de fermentación en estado sólido. XI Semana de la Investigación Escolar. Universidad Autónoma del Estado de Morelos

[20] Hesseltine C W (1977) Solid-state fermentations. Process Biochem. 12, 24-27

[21] Lee HL & Seleena P (1991) Fermentation of a Malaysian *Bacillus thuringiensis* serotype H-14 isolate, a mosquito microbial control agent utilizing local wastes. Southeast Asian J. Trop. Med. Public Health, 22, 108-112

[22] Lonsane BK, Saucedo-Castañeda G, Raimbault M *et al.* (1992) Scale-up strategies for solid state fermentation systems. Process Biochem. 27, 259-273

[23] Lorence A (1996) Biopesticidas. Cuadernos de Vigilancia Tecnológica. Solleiro J and Castañón R (ed.), CAMBIOTECH, Mexico

[24] Misra MC (1991). Production of microbial insecticides by solid state fermentation. Short Term Course on Solid State Fermentation. Central Food Technological Research Institute, Mysore, p. 23.1-23.5

[25] Morris ON, Kanagaratnam P & Converse V (1997) Suitability of 30 agricultural products and by-products as nutrient sources for laboratory production of *Bacillus thuringiensis* subsp. *aizawai* (HD133). J. Invertebr. Pathol. 70, 113-120

[26] Mudgett R E (1986). Solid-state fermentations, p. 66-84. *In* Demain AL & Solomon NA (ed.), American Society and Microbiology, Manual of Industrial Microbiology and Biotechnology, Washington

[27] Nagamma MV, Ragunathan AN & Majumder SK (1972) A new medium for *Bacillus thuringiensis* Berliner. J. Appl. Bacteriol. 35, 367-370

[28] Obeta J and Okafor N (1984) Medium for the production of primary powder of *Bacillus thuringiensis* subsp. *israelensis*. Appl. Environ. Microbiol. 47, 863-867

[29] Ortega J, Espinoza F & López L (1994) El control de los riesgos para la salud generados por los plaguicidas organofosforados en México: retos ante el tratado de libre comercio. Salud Pública de México 36, 624-632

[30] Padilla E, Peña G & Aranda E (1997) Control integral de plagas: hacia la constitución de un equilibrio. Instituto Mexicano de Tecnología del Agua, 1.

[31] Pandey A (1991) Aspects of fementor design for solid-state fermentations. Process Biochem. 26, 355-361

[32] Pandey A (1992) Recent process developments in solid-state fermentations. Process Biochem. 27, 109-117

[33] Pérez J (1994) Una nueva alternativa en los componentes del medio de cultivo para la producción del biopreparado *Bacillus thuringiensis* Berliner cepa L*BT*-13. IX Forum Ciencia y Técnica INISAV, II Encuentro Nacional de Bioplaguicidas, II-EXPOCREE, La Habana, p. 52

[34] Pietrantonio PV & Gill SS (1996) *Bacillus thuringiensis* endotoxins: action on the insect midgut, p.345-372. *In* Lehane MJ & Billingsley PF (ed.), Biology of the Insect Midgut, Chapman and Hall, London

[35] Raimbault M (1980) Fermentation en milieu solide. Croissance de champignons filamenteux sur substrat amylacé. PhD. Thesis, Université Paul Sabatier de Toulouse, Paris

[36] Raimbault M & Alazard D (1980) Culture method to study growth in solid fermentation. Eur. J. Appl. Microbiol. Biotechnol. 9, 199-209

[37] Rodríguez A, Castellanos L, Gómez M, Hernández OH & López C (1994) Estudios sobre *Bacillus thuringiensis* Ber cepa LBT-24 en la Provincia de Cienfuegos. IX Forum Ciencia y Técnica INISAV, II Encuentro Nacional de Bioplaguicidas, II-EXPOCREE, La Habana, p. 49

[38] Rodríguez-Hernández R (1994) Sustitución de la sosa caústica por la cal en la producción de *Bacillus thuringiensis*. IX Forum Ciencia y Técnica INISAV, II Encuentro Nacional de Bioplaguicidas, II-EXPOCREE, La Habana, p. 32

[39] Roussos S, Lonsane BK, Raimbault M & Viniegra-Gonzalez G (1997) Advances in solid state fermentation. Proceedings of the 2nd International Symposium on in solid state fermentation FMS-95, Kluwer Academic Publishers, Dortrecht

[40] Salama HS, Foda MS, Dulmage HT & El-Sharaby A (1983) Novel fermentation media for production of δ-endotoxins from *Bacillus thuringiensis*. J. Invertebr. Pathol. 41, 8-19

[41] Salama HS & Morris ON (1993) The use of *Bacillus thuringiensis* in developing countries, p. 237-252. *In* Entwistle PF, Cory JS, Brailey MJ & Higgs S (ed.), *Bacillus thuringiensis*, An Environmental Biopesticide: Theory and Practice, J Wiley, London

[42] Saucedo-Castañeda G, Trejo-Hernandez MR, Lonsane BK et al (1994) On-line monitoring and control system for CO_2 and O_2 concentrations in aerobic and anaerobic solid state fermentations. Process Biochem. 29, 13-24

[43] Schnepf E, Crickmore N, Van Rie J *et al.* (1998) *Bacillus thuringiensis* and its pesticidal crystal proteins. Microbiol. Mol. Biol. Rev. 62, 775-806

[44] Shankaranand VS, Ramesh MV & Lonsane BK (1992) Idiosyncrasies of solid-state fermentation systems in the biosynthesis of metabolites by some bacterial and fungal cultures. Process Biochem. 27, 33-36

[45] Smith S (1990) The broad perspectives on biological control, p. 45-48. *In* Baker RR and Dunn PE (ed.), New directions in biological control: alternatives for suppressing of agricultural pests and diseases, Alan R. Liss Inc., New York

[46] Suyanandana P, Potacharoen W, Aranpairojana V *et al.* (1996) The production of *Bacillus thuringiensis* by solid state fermentation for public health and agricultural application. The Second Pacific Rim Conference on Biotechnology of *Bacillus thuringiensis* and Its Impact to the Environment, Chiang Mai, Proceedings, p. 549-558

[47] Toro M & Hinojosa D (1994) Utilización de la harina de maíz para la reproducción del bioinsecticida *Bacillus thuringiensis*. IX Forum Ciencia y Técnica INISAV, II Encuentro Nacional de Bioplaguicidas, II-EXPOCREE, La Habana, p. 32

[48] Trejo-Hernandez MR, Lonsane BK, Raimbault M & Roussos S (1993) Spectra of ergot alkaloids produced by *Claviceps purpurea* 1029c in solid-state fermentation system: influence of the composition of liquid medium used for impregnating sugar-cane pith bagasse. Process Biochem. 28, 23-27

[49] Ventosilla P, Reyes E, Velez J, Guerra H & Novak P (1998) Community biological control of malaria using *Bacillus thuringiensis* var. *israelensis*-coconut in Piura, Peru. J. Am. Mosq. Control Assoc. 3, 220-210

Chapter 5.4

Registration of biopesticides

Gary N. Libman[1] & Susan C. MacIntosh[2]

[1]Ecogen Inc. Company 39 Sage Hill DrivePlacitas, NM 87043USA, [2]AgrEvo USA
Company,7200 Hickman Road, Suite 202,Des Moines, IA 50322, USA

Key words: *Bacillus thuringiensis*, transgenic plants, Biopesticides, microbial
              pesticides, product registrations, regulatory, food safety, environmental
              safety, reduced risk, recombinants, transconjugates

Abstract: Strains of *Bacillus thuringiensis* and other microbial pesticides have been
              shown to be inherently safer than most synthetic chemical pesticides due
              to their specific mode of action, narrow specificity and rapid degradation in
              the environment. There has been a rapid growth in registrations of these
              microbial pesticides because of a global effort to reduce the barriers to
              registration. Naturally occurring *Bt* strains, including transconjugates,
              along with genetically modified *Bt* organisms and plant pesticides
              containing *Bt* genes, have an excellent track record of effectiveness for
              specific target pests along with their intrinsic qualities of being
              environmentally benign. This chapter discusses the efforts of the United
              States Environmental Protection Agency, the European Union and other
              major world-wide regulatory bodies to induce increased use of these
              biopesticides by reducing regulatory requirements and positioning these
              products into Integrated Pest Management Programs which result in
              reduced amounts of synthetic chemical pesticides in the world. Guidelines
              for registration requirements of Microbial pesticides are added as Annex.

1.   INTRODUCTION

There has been a significant rise in registrations of Biopesticides
globally. This reflects the fact that Biopesticides are being used on a
much larger scale than in the past. They are used as stand-alone

J.-F. Charles et al. (eds.),
Entomopathogenic Bacteria: From Laboratory to Field Application, 333–353.
© 2000 *Kluwer Academic Publishers. Printed in the Netherlands.*

pesticides, or as a combination with other pest control agents within an Integrated Pest Management (IPM) program. This market shift towards Biopesticide use is directly related to the reduction or elimination of pesticides with a more toxic mode of action on humans, wildlife or the environment. The naturally occurring *Bt* strains (including transconjugates), along with the genetically modified *Bt* organisms and plant pesticides containing *Bt* genes, have a proven track record of effectiveness for specific target pests as well as being environmentally benign.

From a regulatory and registration perspective, most countries in the world have adopted the United States definitions of Biopesticides (below). This chapter will summarise different incentives in key countries and will delineate the procedures necessary for registration and world-wide harmonisation of Biopesticides products. The evolving regulatory environment for plant pesticides is also explored.

There has been a global effort to reduce the barriers to registration and use of Biopesticides and other reduced-risk pesticides. An OECD/FAO (Organisation for Economic and Co-operative Development/ United Nation Food and Agricultural Organisation) workshop on IPM, which had a strong focus on Biopesticides particularly *Bt*'s, took place in 1998, and was hosted by the Swiss Federal Office of Agriculture. In this meeting, which one of the authors attended, there were representatives from 25 OECD and FAO countries, as well as the European Union (EU), United Nations Environment Programme (UNEP) and participants from industry, researchers, growers, etc. The approximately 100 workshop participants reviewed ways to increase IPM and Biopesticide adoption, and reviewed what governments and other IPM stakeholders can do to enhance IPM adoption. The significant message is that Biopesticides when used as a component of IPM programs can decrease greatly the use of conventional synthetic pesticides while crop yields remain high.

2.   WHAT ARE BIOPESTICIDES?

The US Environmental Protection Agency's (EPA) Division of Biological and Pollution Prevention (BPPD) separates the term "Biopesticides" into three broad categories:

- Microbial pest control agents
- Biochemical pesticides
- Genetically enhanced organisms (GEO's)

Microbial pest control agents are those which contain a live microorganism or are derived from live microorganisms. The microorganisms that have been used as active ingredients in pesticides world-wide are:

1. bacteria
2. fungi – yeasts and molds
3. viruses
4. nematodes
5. protozoans

A major part of the registration process of these microorganisms is related to the mode of action of the organism. Some microorganisms have a mode of action similar to synthetic chemical toxins. *Bacillus thuringiensis* strains are an example of this mode of action. The active ingredients of *Bt*'s are linked to δ-endotoxins and in at least one case an exotoxin produced by the organism. Different strains of *Bt*'s produce different endotoxin proteins. The strain of *Bt*, which is used more than any other is the *kurstaki* subspecies (*Btk*) but other strains have been or are still being utilised i.e. *aizawai, israelensis,* tenebrionis, etc. *Bt*'s in nature generally contain several different genes that endow different insecticidal activities. Additionally, there are registered *Bt*'s which are the byproduct, either in nature or *in vitro,* of a natural mating process called conjugal transfer. These transconjugates can be very effective in insect control.

The separate category of "biochemicals" is utilised in the United States and several other countries, particularly in Asia Pacific and Latin America. However, neither the EU, nor Canada, nor many other countries, separate out biochemicals from the synthetic chemical category.

In the United States, the term biochemicals refers to naturally occurring substances, which have a non-toxic mode of action and which are generally used at relatively low use rates (<50 grams/acre). Examples of biochemicals are plant growth regulators such as gibberellic acid, auxins, ethylene, etc. Where the category exists, data requirements tend to be more extensive than the microbials, but less than the synthetic chemicals.

2.1   Plant pesticides

Plant pesticides incorporate the insect control agent directly into the plant cells using a variety of transformation techniques. By expressing the insecticidal agent within the plant, insect pests can be controlled without concerns of insect age, application timing, application coverage or premature environmental degradation. Genetically enhanced (GE) plant technology has been extensively studied in field tests since 1986 in over 45 different countries according to ISAAA (International Service for the Acquisition of Agri-biotech Applications) briefs. The first transgenic crop (virus-resistant tobacco) was commercially introduced during the early 1990s in China, but by 1996, the US accounted for more than half of the total global acreage. The grower acceptance of these new agricultural products has far exceeded the initial expectations. It is estimated that transgenic crops make up more than half of the soybean crop, 40% of the cotton crop and a third of the corn crop in the U.S. for the 1999-growing season. *Btt* potatoes continue to gain acceptance in the U.S. and *Bt* cotton has been limited to 150,000 HA in Australia.

To date, all pesticidal plants contain protein toxins from *Bt,* but other insect control agents are being tested. In North America, pesticidal plants are regulated under existing legal and regulatory frameworks. In the EU, Japan and other countries, new regulations have been or are under development to ensure safety.

3.   REGISTRATION OF PRODUCTS CONTAINING *BACILLUS THURINGIENSIS* TOXINS AS THE ACTIVE INGREDIENT

There has been an effort towards harmonisation of registration requirements of microbials and pesticidal plants globally. This is particularly the case in the United States, Canada and Mexico (NAFTA, North American Free trade Association) as well as the EU and other developed countries. Attachment I is a draft matrix from OECD (Organisation for Economic Co-operation and Development), which delineates required regulatory information and conditionally required information for registration of microbials in the EU, Canada and the USA. The sections below will summarise briefly the regulatory approaches in a few key countries.

3.1   United States

Generally speaking, the US EPA approach to registration of microbials and plant pesticides is very different from that for conventional chemicals. The data requirements for microbials revolve around a multi-tiered approach of toxicity and pathogenicity testing, using high doses for the initial tiered tests. Above and beyond product characterisation and identity data, risk assessments are made of potential toxicity and pathogenicity to humans and selected non-target organisms, including aquatic invertebrates, fish and birds, pollinators and plants. An extensive literature review of the particular microorganism is generally submitted to the Agency as well.

3.1.1 Microbial pesticides

With the exception of public health pests (e.g. for mosquitoes and black flies controlled by *Bt israelensis*), efficacy data are not required to be sent into the US EPA, although it is incumbent on all companies to generate these data and present them if asked. Additionally, in the United States, federal registration is just one milestone in a continuum of regulatory activities in that registrants are also required to register the active ingredients and formulations in the individual states where there is interest in marketing the products. Each state has its own specific requirements, and some of the states, e.g. California, Florida, New York, etc., may ask for efficacy data above and beyond what is required for federal registration.

Regulatory oversight of biopesticides within the US EPA is overseen by the Biopesticides and Pollution Prevention Division (BPPD), which was established in 1994 in the Office of Pesticide Programs. This Division promotes the use of safer pesticides, including Biopesticides, as components of IPM programs. All pesticides in the U.S. are regulated by the Federal Insecticide, Fungicide, and Rodenticide Act (FIFRA) as amended by the Food Quality Protection Act (FQPA) of 1996. Additionally, the Federal Food, Drug and Cosmetic Act (FFDCA) provide for the establishment of tolerances or exemptions from the requirement of a tolerance for levels of pesticides in food or feed. Note: all *Bt*'s that have been registered to-date have received an exemption from the requirement of a tolerance.

Data requirements for *Bt* registration are outlined in the Code of Federal Regulations Title 40 Part 158 (40 CFR Part 158). Additionally, the US EPA has published Pesticide Assessment Guidelines for Microbial Pest Control Agents and Biochemicals. These guidelines and protocols are published under the Pesticide Testing Guidelines Subdivision M. Subdivision M is a nonregulatory companion to 40 CFR Part 158 and contains the protocols that may be used to perform testing on *Bt*'s.

3.1.2   Plant pesticides

Plant pesticides are regulated by all three U.S. agencies – USDA, FDA and EPA. The USDA regulates all field trials performed during the product development phase for all genetically enhanced crops and grants non-regulated status after a full review of all aspects related to environmental safety. Specifically, the USDA determines if the modified crop could be a plant pest threat upon uncontrolled environmental release. The FDA ensures that the genetically enhanced crop is nutritionally equivalent to the conventional crop variety. They confer with the EPA on toxicology and allergenicity issues when necessary. The EPA is considered the lead agency for regulating the introduced pesticidal agent, but the agency does not review non-pesticidal genetically enhanced crops, such as virus or herbicide resistance.

Like the genetically enhanced microorganisms, pesticidal plants also require a detailed molecular analysis of the genetic changes introduced into the crop. The registration process is focused on a specific transformation event, which is transferred in a stable manner to all subsequent progeny. Simple insertion events, those with 1 or 2 copies of the introduced gene(s), are preferred, since they are relatively easier to characterise. The stability of the event over several generations is proven with both a molecular analysis, using Southern blots, and field studies for measuring Mendelian inheritance. Large plots are utilised to gain sufficient statistical evidence. All three U.S. agencies independently review the molecular characterisation data.

Regulatory bodies around the world have embraced the concept of "substantial equivalence" to assess the nutritional quality of genetically enhanced crops, including pesticidal plants. For this analysis, the GE crop is compared to the non-transformed or natural crop for various nutritional parameters. A full compositional analysis (e.g. proximates, minerals, vitamins, amino acid content, etc.) is performed on all edible portions of the crop. For corn, this includes not only the grain, used for both food and animal feed purposes, but also the whole corn plant used for animal feeding. Antinutrients (e.g. phytic acid – corn, trypsin inhibitors – soybean, etc.) are also measured to ensure that the transformation process did not increase the levels naturally found in the crop. The structure of the introduced protein is also compared to global protein databases for any similarities to known toxins and allergens.

The levels of the newly introduced pesticidal agents are measured in detail for the pesticidal crop to gain a full understanding of the exposure limits. The amount of the pesticide is measured in all plant parts across the growing season and in different genetic backgrounds. Using *Bt* corn as an example, field measurements are made on the *Bt* protein levels in root, stalk, pith, leaf, tassel, pollen, silk, kernel, and cob. Samples are taken during different field stages, the green vegetative, fodder, forage and at harvest to ensure that adequate levels of the *Bt* protein are present, even when protein production declines towards harvest. Since hundreds of different corn hybrids are utilised around the world with widely varying characteristics, the *Bt* protein levels are further quantitated in several different genetic backgrounds to make sure that the level is consistent, regardless of the corn germplasm. Finally, the crop is processed in a typical manner and these fractions are tested for the absence or presence of the *Bt* protein. For corn, more than a dozen different fractions are produced via wet and dry milling processes. If the crop is directly consumed, e.g. sweet corn, then samples after boiling, and processing for freezing and canning will be tested for the *Bt* protein. Although, this data is reviewed by the FDA, these studies carry even more importance to meet the proposed labelling rules under consideration in the EU and Japan.

Some data requirements are very similar to the microbial biopesticides, with special focus on the exposure aspects. For instance, inhalation testing is not necessary since the active ingredient is contained within the plant cell, but data on degradation of the insecticidal protein in soil is required. The same set of non-target organisms, listed above for microbial biopesticides, are also tested with the active ingredients from plant pesticides to measure any potential adverse effects. Typically, the non-target organisms are tested with plant material that would best mimic the usual exposure route. For example, aquatic invertebrates are tested using pollen, while soil invertebrates are tested with plant powder incorporated into soil. Finally, large-scale field studies are continued for several years after market introduction to conclude the risk assessment for non-target organisms.

Since the insect control agent may survive in food products and by-products, a toxicology and allergenicity assessment is made. As with synthetic pesticides, a tiered approach is taken, first using acute high dose toxicology studies. If a level of toxicity is measured in the acute testing, more extensive sub-chronic and chronic studies are then triggered. For allergenicity assessment, the pesticidal crop and protein are both evaluated. Over 90% of allergenic food reactions stem from eight food groups: shellfish, egg, fish, milk, peanuts, soy bean, tree nuts and wheat. Therefore, if either the recipient host crop or the introduced protein is from one of these sources, an inherent allergenicity study is undertaken. Other characteristics of known allergens are compared to the introduced protein(s), such as protein size, stability in simulated gastric fluids and to heat, protein modifications made by the plant, sequence similarity and prevalence in food. The allergenic risk assessment is the sum total of all available data – the "weight of evidence" approach. Unfortunately, there are no validated animal models to test for potential food allergy, although research continues to validate such a model in several laboratories around the world.

3.1.3   Insect resistance management requirement

The most remarkable regulatory aspect is the requirement by the EPA for the registrant to submit a detailed insect resistance management plan (IRM). Most plant pesticidal registrations are time limited to ensure that a robust IRM plan is in place and that ongoing monitoring and grower compliance efforts continue for the lifetime of the product. The elements of the IRM plan include the biology of the pest(s), potential strategies and their deployment options, product fit with integrated pest management practices, field monitoring for pest susceptibility, communication /education for growers to implement the plan, and proof of compliance. All pesticidal crops currently utilise the high dose-refuge strategy, where the level of *Bt* protein within the plant is at a sufficient level to kill any resistant insects that are heterozygous for the resistance trait. The refuge fosters the development of a population of susceptible insects that can dilute the genes from the rare resistant insects that may emerge from the *Bt* crop. Rotation of different *Bt* proteins is also encouraged, where available, along with destruction of crop residue. Ultimately the best strategy is a pyramid strategy, where two pesticidal ingredients with unique modes of action, are combined within the same plant. These dual gene plants are currently being field tested and should be commercially available in just a few years. As new research information becomes available, the IRM plans may be altered to reflect this new information - taking into account the relative risk.

3.1.4   Registered products

By mid-1999, the US EPA has registered over 50 viable and at least 5 non-viable microbial pesticides, and 7 plant pesticides. The review of a complete dossier for a new microbial pesticide is 12 to 18 months, as compared to 3 years for a conventional synthetic chemical pesticide. Table 1 is a listing of registered microbial pesticides in the United States.

Table 1. Listing of registered microbial pesticides in the United States (to September 1999).

Microbial pesticide	Species
Bacteria	*Bacillus thuringiensis* Berliner
	Bt subsp. *aizaiwai* — two strains
	Bt subsp. *israelensis* — two strains
	Bt subsp. *kurstaki* — nine strains
	Bt subsp. tenebrionis
	Burkkholderia cepacia type Wisconsin — two strains
	Pseudomonas aureofaciens — two strains
	Pseudomonas syringae — three strains
	Streptomyces griseoviridis
Fungi	*Ampelomyces quisqualis*
	Beauveria bassiana — two strains
	Candida oleophila
	Colletotrichum gloeosporioides
	Gliochadium catenulatum
	Gliocladium virens
	Lagenidium giganteum
	Metarhizium anisopliae
	Paecilomyces fumorsoroseus
	Puccinia canaliculate
	Trichoderma harzianum — two strains
	Trichoderma polysporum
Viruses	*Anagrapha falcifera* NPV
	Autographa californica NPV
	Cydia pomonella Granulosis virus
	Douglas fir trussock moth NPV
	Gypsy moth NPV
	Helicoverpa zea NPV
	Spodoptera exigua NPV
Plant Pesticides	Several different *cry* gene markers for expression in corn, cotton and potatoes

The organic farming community has requested national uniform certification standards to be overseen by the USDA. In a counter intuitive move, genetically modified plants, regardless of their modified genes, have been prohibited from crops meeting the organic certification standards. The use of biopesticides is allowed.

3.2   Canada

The Pest Management Regulatory Agency (PMRA) was established in Canada in 1995, and is responsible for pest management regulation under the Canadian Pest Control Practices Act (PCPA). Before making a registration decision regarding a new pest control agent, whether it's a microbial or a conventional chemical, the PMRA conducts the appropriate assessment of the risks and value of the product specific to its proposed use. Within the PMRA, the Alternative Strategies and Regulatory Affairs Division, develops policies, programs and projects related to sustainable pest management and coordinates international activities – harmonisation programs. Additionally this Division develops IPM partnership projects with stakeholders.

General information regarding registration of *Bt*'s is provided in the Canadian Registration Handbook for Pest Control Products under the Pest Control Products Act and Regulations. When compared to the US EPA, the PMRA tends to give fewer waivers for residue testing, and there are more data required for toxicological testing and non-target studies. Scientific-based rationales must accompany all waiver requests.

One significant Canadian efficacy requirement centres on "Ecozones" which are large ecologically distinctive geographical areas – with similar landforms, water, soil, climate, flora, etc. Thus, efficacy data must be provided for each ecozone for which registration is sought.

The Canadian Food Inspection Agency and Health Canada regulate pesticidal plants. The data requirements are almost identical to those of the U.S. Data must be collected on crops grown in Canada in addition to data collected in other countries. As with microbial biopesticides, Canada and the U.S. are moving towards a bilateral agreement to harmonise the regulatory procedures for pesticidal plants.

3.3   The European Union

On July 15, 1991, the Council of Ministers of the EU adopted a "Council Directive concerning the placing of plant protection agents on the market", (91/414EEC). This Directive became effective in all EU countries in 1993.

The provisions of the Directive are:

- A positive list of active ingredients (AI's) as Annex I. New active ingredients since July, 1993 are being evaluated by a new procedure for inclusion in the positive list.

- National registration is required for formulations of the AI's
- Mutual recognition is required by Member States of registrations granted by others, provided conditions (climatic, agricultural, environmental) are comparable
- Three year provisional approvals of formulations may be granted by Member States
- AI's in use before 1993 will be reviewed over the next 12 years
- Data protection will be harmonised throughout the EU to 10 years for new AI's
- Packaging and labelling will be harmonised
- Procedures for efficacy trials will be harmonised
- Uniform principles for data requirements and evaluation will be established

Data requirements for *Bt*'s and other microbial agents are included in Annex I, and are comparable to those requirements in Canada and the United States.

Pesticidal plants are regulated in the EU under the 90/220 directive, which addresses both environmental and feed issues. A 90/220 application can be entered for import purposes or for growing in the EU. A revision of the directive 90/220 is underway, but the expectation is that it will not be concluded until the 2001 – 2002 time frame. Assessing the safe use for human nutrition is required under the EC Regulation Directive 258/97 for Novel Foods and Novel Foods Ingredients. A detailed labelling policy for foods derived from genetic modifications is in the final stages of EU approval.

The overall EU regulatory process is complex and only a couple of *Bt* crops have successfully completed the entire process. The data requirements are stricter, in some cases, than the U.S. system, with an emphasis on a certainty of no harm. The review also uses the broad consideration of substantial equivalence, which is described above, of the pesticidal plant to the non-transgenic (conventional counterpart) crop variety. This thorough process has taken at least 2 – 3 years for completion. Many consumers in EU countries are opposed to genetically enhanced organisms of any kind, providing a difficult political environment. While the EU has operated under a *de facto* moratorium since mid-1998, North America (NA) continues to embrace the technology with enthusiasm. Many reasons can explain the differences in the attitudes of the general public in the two regions. The EU commission is a relatively new political organisation with less public trust than that observed in NA. Several food incidents, e.g. mad-cow's disease, dioxin-tainted meat, etc., have undermined the public confidence in regulatory and governmental bodies. NA has a strong agricultural culture, often dubbed "the breadbasket of the world", which has reduced economic subsidies to growers over the last few years thus moving to a market-driven agricultural system. High production agriculture has been adopted to increase crop yields, reduce the amount of land utilised for agriculture, while preserving wildlife areas. Agricultural biotechnology products are but one of many management components (e.g. fertiliser, synthetic pesticides, optimised equipment and technology, etc.) available to the grower to reduce input costs, improve yields and protect the environment. In contrast, agriculture in the EU remains highly subsidised and agricultural exports are less important contributors to the overall economy. Thus, the incentive for highly cost efficient agricultural production is different from in the U.S.

3.4   Rest of the World (ROW)

Most other countries are taking the lead from the US, Canada and the EU. Specific microbial guidelines have been issued or are being reviewed in other key countries. Specifically:

3.4.1   Japan

In 1994 data requirements were updated to include requirements for *Bt*'s and other microbials. Studies are underway at the Japanese MAFF to review and finalise tiered approaches to determine the safety profile of *Bt*'s. Because of the concern regarding effect of *Bt*'s on silkworms, additional studies are required in Japan.

Both the Japanese MAFF and the Ministry of Health and Welfare (MHW) regulate pesticidal plants. A tiered series of environmental testing, including a field trial in Japan, is required even if the desire is simply an import clearance for a traded commodity. The remaining data requirements are similar to that of the US EPA.

3.4.2   Australia

The National Registration Authority (NRA) has provided tiered requirements for registration of *Bt*'s and other microbials. Similar to the US and Canada, scientific-based waivers can be submitted and will be reviewed on a case-by-case basis. Australia's risk assessment is based on the principle that the product under review will be efficacious under the conditions of use, and not present any unacceptable hazard to use. A formal system for import clearances of pesticidal plants was initiated in 1998. Australian farmers have embraced *Bt* cotton.

3.4.3   Other countries

Most of the other countries of the world, not mentioned above, have varying degrees of requirements for microbial and plant pesticide registrations. Most, i.e. Brazil, Argentina, Taiwan, Korea, Eastern European countries, etc. have data requirements similar to the United States. Some, however, have one set of requirements for all pesticides, and allow waivers for data gaps for microbials as compared to conventional chemical pesticides. Additionally, there is increased public concern in some of these countries for plant pesticide products in their food supplies.

As can be seen by the above, and in other chapters of this book, *Bt*'s and other Biopesticides have been shown to be inherently safer than most synthetic chemical pesticides, due to the specific mode of action, narrow host range specificity and rapid degradation in the environment. Although these qualities in the past have reduced product efficacy, recent advances of improved product formulations and application techniques have improved efficacy dramatically.

Most countries have recognised the above and have reduced, or are in the process of reducing, the barriers for registration and introduction of these products into the mainstream of agricultural practices. Because of the reduction in the barriers to these products, use has increased in the last 10 years significantly. While still a small percentage of the total pesticide usage globally, *Bt* is becoming increasingly more important for use in agricultural, forestry and public health settings.

ANNEX 1 *(matrix 1 to 9)*:

OECD GUIDELINES FOR REGISTRATION REQUIREMENTS

Information Requirements for Registration of Microbial Pesticides excluding GMO (OECD on Microbial Steering Committee (Wendy Sexsmith, PMRA, June 1999)

CODES used :
R = information is required, the item must be addressed in submission package, i.e. Tier 1
CR = information is conditionally required, the item must only be addressed if condition is met, i.e. Tier 1
Tier 2 = information is only required if triggered by certain findings in Tier 1
Step I (in Matrix 7-ai for EU) = information is required, may be sufficient basis to waive requirements in Step II
i.e. Step I+II = Tier 1
Step I and Step II (in Matrix 7-ep for EU) = information is required or conditionally required, i.e. Step I+II = Tier 1
Step I (in Matrix 8 and 9 for EU) = basis for waiving requirements in Step II; therefore Step II = Tier 1

W:\ASRA-PERSON\LYNN\MICROB\OECD\oecd-drs.wpd   (May 6, 1999)

Matrix 1. Identity, Composition and Biological Properties		EU	Canada	USA	different
Applicant	ai	R	R	R	
	ep	R	R	R	
Producer, manufacturer	ai	R	R	R	
	ep	R	R	R	
International regulatory status	ai		R		
	ep		R		
Names, species description, accession no. of sample in culture collection; for end-use product, trade name and any developmental code names used in dossier	ai	R	R	R	
	ep	R	R	R	
Test procedures and criteria used to characterize strain or serotype of micro-organism	ai	R	R	R	
Content of micro-organism in appropriate terms, eg. # active units per unit volume; development phase of micro-organism (eg. spore)* in end-use product	ai	R	R	R	
	ep	R	R	R, not *	
Identity and content of impurities, additives, contaminating micro-organisms, metabolites; function of additives	ai	R	R	R	
	ep	R	R	R	
Analytical profile of batches	ai	R	R	CR	X
Formulation type (eg.emulsifiable concentrate)	ep	R	R	R	
Biological properties of micro-organism					
Historical uses, natural occurrence, origin of isolate, method of isolation		R	R	R	
Identification test for the micro-organism		CR	CR	R	
Development stages of micro-organism, including virulence and survival time of resting stages; parasitism, competitors, vectors		R	R		
Physiological properties; effect of environmental conditions (eg. pH, temperature, humidity, light) on infectivity, dispersal, colonization		R	R	R	
Relationships to known plant/animal/human pathogens		R	R	R	
Genetic stability and factors affecting it		R			X
Information on production of toxins (nature, identity, stability)		R	R	R	
Resistance/sensitivity to antibiotics / anti-microbial agents		R	R	R	
Available information on development of resistance		R		CR	
Target organism(s), mode of action, infective dose, transmissibility		R	R	R	
Any available information on host specificity range and the occurrence of species closely related to the target pest		R	R	R	

Matrix 2. Function, Application and Handling		EU	Canada	USA	different
Function and field of use categories eg. fungicide for forestry	ai	R	R	R	
	ep	R	R	R	
Details of intended use: target pest, crops/products to be protected or treated	ai	R	R	R	
	ep	R	R	R	
Application rate	ep	R	R	R	
Method of application (equipment, diluent type and volume)	ep	R	R	R	
Number and timing of applications, duration of protection	ep	R	R	R	
Necessary waiting periods for re-entry, or harvest, or handling, or planting succeeding crop, or applying chemicals	ep	R	R	R	
Type of packaging	ep	R	R	R	
Specifications for packaging and measures of its suitability	ep	R		R	
Procedures to clean equipment and protective clothing (EU requires measures of their effectiveness)	ep	R		CR	
Procedures for destruction or decontamination of product and packaging	ai	R		CR	
	ep	R	R	CR	
Recommendations re: handling, storage, transport or fire	ai	R	CR	R	
	ep	R	R	R	
Measures for environmental safety, in case of an accident (containment, decontamination, to render organisms uninfective or non-infected)	ai	R	R		
	ep	R	R		

Matrix 3. Physical-Chemical Properties		EU	Canada	US	different
Colour and odour	ai			R	
	ep	R		R	
Physical state	ai		R	R	
	ep	R	R	R	
Density	ai		R	R	
	ep		R	R	
pH, acidity, alkalinity	ai			R	
	ep	R		R	
UV/Visible absorption	ai			R	
	ep			R	
Boiling point	ai			R	
	ep			R	
Viscosity, surface tension	ai		R	R	
	ep	R	R	R	
Explosivity, corrosive character, oxidizing properties	ai		R	R	
	ep	R	R	R	
Flashpoint, flammability, spontaneous ignition	ep	R			
Technical characteristics as appropriate: wettability, persistent foaming, suspensibility, emulsifiability,	ai		R	R	
flowability, particle size distribution	ep	R	R	R	
Adherence to target plants or seeds	ep	R	R		
Compatibility with products in recommended tank mixes or with chemical pesticides routinely used on the crop	ep	R		CR	
Storage stability, shelf life, maintenance of purity and	ai		R	R	
potency	ep	R	R	R	

Matrix 4. Manufacturing, Quality Control and Analytical Methods		EU	Canada	USA	different
Production process, describing techniques used to ensure a uniform product and quality control methods to check for standardization; product specifications or quality criteria	ai	R	R	R	
	ep	R	R	R	
Method to establish the identity and purity of seed stock from which batchs are produced and results obtained, including information on variability	ai	R	R	R	
Method to prevent loss of virulence of seed stock	ai	R	R	R	
Method to determine impurities or toxic metabolites	ai	CR	CR	CR	
Method to measure content (purity) of micro-organism	ai	R	R	R	
	ep	R	R	R	
Method to identify and quantify microbial contaminants	ai	R	R	R	
	ep	R	R	R	
Method for establishing storage and shelf-life stability	ai	R	R	R	
	ep	R	R	R	
Method to detect the micro-organism in the environment	ai		CR	CR	?
Method to distinguish a mutant micro-organism from the wild parent strain	ai	CR	CR	R	
Methods to determine and quantify residues of viable or non-viable micro-organism and metabolites on food, feed, animal tissue, in soil, water or air, where relevant.	ai	CR	CR	CR	
	ep	CR	CR	CR	

Matrix 5. Residues of micro-organism and microbial toxin		EU	Canada	USA	different
Estimated persistence & multiplication of micro-organisms and toxins, in/on treated products, food and feed, based on biological information in matrix 1.	ai	R	CR	CR	
	ep	CR	CR	CR	
Non-viable residues and viable residues on food/feed stuffs, potable water, or fish that might be affected by treatment	ai/	CR	CR	CR	
	ep	CR			
Summary and evaluation of residue behaviour	ai	CR	CR	CR	
Proposed maximum residue level and preharvest/withholding period		CR	CR	CR	
Residues in succeeding or rotational crops		CR	CR	CR	
Residues in animals that come in contact with treated crop material			CR	CR	
Residue storage stability and freezer stability		CR	CR	CR	
Metabolism		CR			

Matrix 6. Efficacy	EU	Canada	US	different
Preliminary range-finding tests	R	R	CR	X
Field tests	CR	R	CR	
Dose - efficacy relationships	R	R	CR	
Level, duration and consistency of protection/control	R	R	CR	X
Effects on yield and quality of treated plants	CR	R	CR	
Compatibility of micro-organism with chemical control measures	R?	CR	CR	
Effects on transformation processes (wine or breadmaking, brewing)	CR	CR	CR	
Phytotoxicity / phytopathogenicity to target plants and rotational crops	CR	R	CR	
Information on use in integrated pest management systems		CR	CR	
Development of resistance, appropriate mitigation strategy		CR	CR	

Matrix 7. Effects on Human Health		EU	Canada	US	different
Consider potential hazards of micro-organism, with knowledge (in matrix 1) of its ability to colonize, cause damage, and produce toxins and other relevant metabolites.		Step I		R	
Sensitization by inhalation and dermal exposure	ai	Step II			X
Skin sensitization	ep	Step II-CR			
Skin irritation	ep	Step II-CR	R	R	
Eye irritation	ep	Step II-CR		R	
Acute toxicity: percutaneous or dermal	ai	Step II-CR			
	ep	Step I- R	R	R	
Acute toxicity, pathogenicity and infectivity: oral	ai	Step II	R	R	
Acute toxicity: oral	ep	Step I - R	CR	R	
Acute toxicity, pathogenicity and infectivity: inhalation or intratracheal instillation	ai	Step II	R	R	
Acute toxicity: inhalation or intratracheal instillation	ep	Step I-CR	CR	R	
Acute toxicity, pathogenicity and infectivity: intraperitoneal or subcutaneous or intravenous, single dose	ai	Step II	R	R	
Genotoxicity testing	ai	Step II (3 *in vitro* + 1 test) Step III (*in vivo* germ)	CR-fungus/ actinomycete	CR	X
Tissue / cell culture study, for viruses	ai	Step II	R	R	
Short-term toxicity, pathogenicity, infectivity by most appropriate route of exposure	ai	Step II		Tier 2	X
Longterm toxicity, pathogenicity, infectivity, kinetic studies	ai	Step III		Tier 3	
Medical surveillance on manufacturing plant personnel, sensitization / allergenicity reports, published clinical reports	ai	Step I	R	R	
Data on exposure of operators and bystanders	ep	Step III			
Available toxicological data on formulants (MSDS)	ep	Step III			
Supplementary studies with tankmix partners	ep	Step III			
Proposed first aid measures and medical treatment	ai	Step II	R on label		
Summary: mammalian toxicology, pathogenicity, infectivity, and overall evaluation	ai	Step I	R	R	
	ep	Step I - R	R	R	

Matrix 8. Effects on Non-Target Organisms (studies are usually based on the active substance, unless effects of the end-use product cannot be predicted from data available for the micro-organism)		EU	Canada	US	different
Justification that a valid assessment of impact on each non-target organism can be based on information provided in sections 2, 3, 5 and 7, together with information on the formulation and use pattern of the pest control product(s).		Step I	similar basis for waiver of Tier 1	similar basis for waiver of Tier 1	
Effects on non-target plants	algae	Step II			X
	other plants	Step II	Tier 1	Tier 1	
Effects on birds	oral	Step II	Tier1:1 sp	Tier1:2 sp	
	inhalation		Tier 1	Tier 1	X
Effects on fish	species 1	Step II	Tier 1	Tier 1	
	species 2		Tier 1	Tier 1	X
Effects on freshwater invertebrates	species 1	Step II	Tier 1	Tier 1	
	species 2		Tier 1	Tier 1	X
Estuarine and marine animal			Tier 1	Tier 1	
Effects on bees		Step II	Tier 1	Tier 1	
Effects on arthropods other than bees (beneficial insects)		Step II	Tier 1	Tier 1	
Effects on non-arthropod invertebrates (incl. earthworms)		Step II	Tier 1	CR	
Effects on micro-organisms			Tier 1		
Effects on wild mammals			Tier 1	Tier 1	
Long term studies		Step III	Tier 3	Tier 3	
Field studies		Step III	Tier 4	Tier 4	

Matrix 9. Fate and Behaviour in the Environment (studies are based on the active substance, unless effects of the end-use product cannot be predicted from data available for the micro-organism)	EU	Canada	US	different
Estimated fate of micro-organism, given information on its origin, properties, survival and residual metabolites, in section 2	Step I		CR	
Spread, mobility, persistence and multiplication in soil/water/air	Step II	Tier 2 or Tier 3	Tier 2	X
Summary and evaluation of environmental impact	R	R	R	

Chapter 6.1

Bacillus thuringiensis application in agriculture

Amos Navon
Department of Entomology, Agricultural Research Organization, The Volcani Center, Bet Dagan 50250, Israel

Key words: *Bacillus thuringiensis* (*Bt*) products, agricultural biopesticides, insect control, application strategies, interactions

Abstract: *Bacillus thuringiensis* (*Bt*) has been the leading microbe in agriculture since the 1960s. During the last 40 years, substantial knowledge and experience of uses of *Bt* against lepidopteran and coleopteran insects in the field have been accumulated and aspects covered include with regard to: natural and genetically modified products, larval age, insect feeding behaviour, environmental constraints, safety, timing of application, formulation, application technologies. The knowledge of interactions of *Bt* with other entomopathogenic microbes, natural enemies and natural and selective insecticides, is useful in selecting compatibilities and to promote synergistic effects between this microbe and other means of pest control. The new choice of *Bt* products, uses of the microbe against 1st instar defoliators, and the combinations of *Bt* with parasitoids and predators have been effectively introduced into pest control strategies. Novel *Bt* products against wider insect host range, new formulations and application technologies which will prolong residual activities of the microbe can increase the use of *Bt* in insect pest management strategies

1.   INTRODUCTION

Bacillus thuringiensis (*Bt*) has been the leading biopesticide against lepidopterous pests in agriculture since its discovery in *Bt* subsp. *kurstaki* strain HD-1 [15]. This strain, does not produce β-exotoxins and therefore is free of its undesired effects [53]. It controls a wide range of lepidopteran insect species [18]. New *Bt* strains controlling coleopteran insects, predominantly the Colorado potato beetle, *Leptinotarsa*

355

J.-F. Charles et al. (eds.),
Entomopathogenic Bacteria: From Laboratory to Field Application, 355–369.
© 2000 *Kluwer Academic Publishers. Printed in the Netherlands.*

decemlineata [29, 31] extended the insect host range. The costs of commercial *Bt* products have been higher than those of chemical insecticides for use on major crops such as cotton [21]. Lower efficacy of *Bt* versus chemical insecticides, the limited *Bt* persistence in row crops and the narrow insect spectrum and the incomplete plant cover due to leaf growth have further limited the number of crops on which *Bt* is used for plant protection. Nevertheless, growing public concern for environmental safety and health, and the problem of insect resistance to chemical insecticides has led to further use of the microbe products, mostly on high-value vegetable and fruit crops. The intensive use of *Bt* for pest control, induced resistance, best known against the diamond backmoth *Plutella xylostella* [55]. At the research level, studies with conventional *Bt* in the 1990s were characterised by: 1) identification of new *cry* genes 2) development of new genetically engineered *Bt* products and conventional formulations and 3) widening knowledge of *Bt* interactions with entomophagous, microbial and chemical control agents which could provide effective tools for integrated insect pest management (IPM) programs. To improve integration of *Bt* in pest control, more knowledge of application strategies and rational uses is needed. This chapter on *Bt* application in agriculture will evaluate the feasibility of strategies of using *Bt* products, and their integration in IPM programs at the start of the new millennium.

2.   CONSIDERATIONS OF *BT* USES IN THE FIELD

2.1   Bioassays

Screening *Bt* strains has been an essential line of work in almost any laboratory engaged in developing pest management programs with the microbe. *Bt* standardisation by means of potency bioassays to determine the insecticidal power of the commercial spore-crystal products has started in the 1960s [7]. Within the last 40 years, *Bt* bioassays have been used world-wide to select promising strains held by domestic collections, academic institutions, biocontrol companies and international collections in the US and Institut Pasteur, France [42] [33]. Some *Bt* strains selected by the screening programs became useful commercial products for pest control (Table 1). Primarily, the standardisation of *Bt* was limited to products based on subsp. *kurstaki*. However, within the last decades, new subspecies of the microbe active against insect of other orders have been discovered, and genetically modified *Bt* products were developed. These changes pointed on the need to re-standardise the official bioassays. New

Table 1a. Natural *Bt* products registered for use in agricultural*

Bt subspecies	Company	Product	Target insects	Crop
kurstaki HD-1	Abbott Laboratories Chicago IL, US	Biobit®, Dipel®, Foray®	Lepidoptera	Field and vegetable crops, greenhouse, orchard fruits & nuts, ornamentals, forestry, stored products
kurstaki HD-1	Thermo Trilogy Corp. Columbia MD, US	Javelin®, Steward®, Thuricide®, Vault®	Lepidoptera	
kurstaki	Abbott	Bactospeine®, Futura®	Lepidoptera	
kurstaki	Thermo Trilogy	Able®, Costar®	Lepidoptera	
kurstaki	Abbott	Florbac®, Xentari®	Lepidoptera armyworms	Row crops
tenebrionis	Abbott	Novodor®	Colorado potato beetle, Elm bark beetle	Potato, tomato, eggplant- CPB Ornamentals, shade trees - EBB
tenebrionis	Thermo Trilogy	Trident®	Coleoptera	Potato, tomato, eggplant
kurstaki	BioDalia, Dalia, Israel	Bio-Ti®	Lepidoptera	Avocado, tomato, vinyard, pine forests
kurstaki	Rimi, Tel Aviv, Israel	Bitayon® (granular Feeding baits)	*Btrachedra amydraula*	Datepalms
galleriae	Tuticorin Alkali Chemicals & Fertilizers ltd. India	Spicturin®	Lepidoptera	Cruciferous crop plants
YB-1520	Huazhong Agric. University, China	Mainfeng pesticide	Lepidoptera	Row crops, fruits trees
	Scient. & Technol. Develop. China	*Bt* 8010 Rijin	Lepidoptera	Row crops, rice. Maize, fruit trees, forests, ornamentals
CT-43 (non motile strain)	Huazhong Agric. University, China	Shuangdu	Lepidoptera, Coleoptera, Diptera	Row crops, garden plants, forests

*based on [4, 54]

Table 1b. Genetically modified *Bt* products registered for use in agricultural*

Bt subspecies	Company	Product	Target insects	Crop
aizawai recipient, *kurstaki* donor	Thermo Trilogy	Agree®, Design® (transconjugant)	Lepidoptera (Resistant *P. xylostella*)	Row crops
kurstaki recipient, *aizawai* donor	Ecogen Inc. Langhorne PA, US	Condor®, Cutlass® (transconjugant)	Lepidoptera	Row crops
kurstaki	Ecogen	CryMax®, Leptinox®	Lepidoptera	Vegetables, horticultural, ornamental
kurstaki	Ecogen	Leptinox® (recombinant)	Lepidoptera, armyworms	Turf, hay, row crops, sweet corn
kurstaki recipient	Ecogen	Raven® (recombinant)	Lepidoptera Coleoptera	Row crops Potato, tomato, eggplant - CPB
δ-endotoxin encapsulated in *Pseudomona s fluorescens*	Mycogen Corp. San Diego, CA, USA	MVP MATTCH MTRACK (CellCap®)	Lepidoptera Lepidoptera Coleoptera	Row crops - armyworms Potato, tomato, eggplant - Beetles

*based on [4, 54]

bioassay procedures have been standardised with regards to diet [46]. However, the possibility of using new international standardisation of *Bt* products, discussed in one of the recent Society of Invertebrate Pathology meetings [45] was not accepted, mainly because: 1) as mentioned in [13], insect strains and qualities differed among laboratories, and therefore mortalities could not be easily compared, 2) each of the *Bt* manufacturers has developed its own bioassay procedures for labelling the *Bt* products, and changing the protocols would not be feasible, for commercial reasons (see also Chapter 5.1).

On the other hand, field bioassays of *Bt* [44] are needed to determine efficacy of selected microbes or commercial products.

2.2   *Bt* products

Bt subsp. *kurstaki* products have been used continuously since the 1970s. The crystals in the products caused intoxication in the larvae, and the spores potentiated [38] and synergised [14] the crystal toxin. The HD-1 strain of this subsp. was produced first by Abbott Laboratories (Chicago, IL) followed by many other agrochemical and fermentation companies which added new *Bt* products to the biopesticide market (Table 1). Two developments in *Bt* production have occurred since the 1980s [57]: 1) the natural and recombinant *Bt* products are being produced in the US by four companies only, and 2) developing countries have selected domestic *Bt* strains for pest control in their own countries to save costs of pest control.

Products based on *Bt* subsp. *aizawai* were used to control armyworm species such as *Spodoptera* sp. and other insects that were not susceptible to *Bt* subsp. *kurstaki* strains. The vast knowledge accumulated on molecular biology of *Bt* [9] led to developing of useful techniques in genetic manipulations with which the *Bt cry* genes became a useful tool for widening the insect host range and increasing the content of toxic proteins in the microbial cell. Two principles of genetic manipulations have been used: 1) Transconjugation: conjugation and transduction processes have been used to transfer recombinant plasmid from a donor *Bt* strain to another strain used as a recipient. 2) Recombinant *Bt*: in this technology, series of genetic manipulations including plasmid curing, conjugation, transformation and site-specific recombination have been used. The genetic manipulations of *cry* genes showed that certain combinations of Cry proteins exhibited synergistic intoxication of insect pests and increased the production of Cry proteins [4]. These improvements in the efficacy of *Bt* products against Lepidoptera including armyworms can result in a better cost effectiveness of the microbe.

2.3   Larval age

One of the most important economic considerations in pest management with *Bt* is the use of the microbe against young larvae, preferably neonates. This is because it has been shown in laboratory and field bioassays that third instar lepidopteran [46] and coleopteran [19] larvae are less susceptible to the *Bt* products. These differences in larval susceptibility to *Bt* are probably aspects of a more general phenomenon among lepidopteran and coleopteran pests. Therefore, the costs of controlling 1st instar larvae can be a fraction of that of controlling older larvae. Furthermore, initial nibbling of *Bt* spray on the plant by neonate larvae is followed by cessation of feeding and gut paralysis within minutes [24], with negligible damage to plants, whereas mature larvae feeding on *Bt* treated plants may cause some damage to the crop that will reduce the quality of the agricultural product. 2) Also, mature larvae may recover from the *Bt* intoxication [6], and even complete the developmental cycle. This situation can be enhanced by incomplete spray coverage, by rapid degradation of the crystal toxin by UV radiation, and by rain or overhead irrigation wash-off.

2.4   Insect feeding behaviour

The larval feeding habits markedly affect the efficacy of pest management with the microbe, as they determine the availability of the crystal toxin to the insect [43]. Defoliators are effectively controlled with *Bt*, but the control of bollworms, borers and feeders on hypogeal plant parts has been marginal because the amounts of *Bt* available to them are commonly below lethal doses. Knowledge of feeding behaviour is a fundamental information requirement in the development of new formulations and optimisation of biopesticide uses. *Bt* dose-transfer modelling based on monitoring the effects on larval feeding behaviour and locomotion [25], has identified factors that need to be improved to ensure better control of lepidopteran larvae. The use of a phagostimulant mixture such as COAX in a sprayable starch encapsulation [37], or of a yeast extract in a dustable granular formulation as a feeding stimulant [48] are promising new tactics for manipulating the feeding behaviour of bollworm and borer larvae. The principles behind the use of using this formulation, in addition to increasing residual toxic activity are: 1) the larvae are attracted/arrested to selectively feeding on the *Bt* product and therefore will feed less on the plant, 2) for the same *Bt* dose per acre, the amounts of spore-crystal materials available to the larvae are several times higher than in a leaf meal with the *Bt* spray. With these approaches the insect control efficacies of the new *Bt* formulations can exceed those obtained with *Bt* sprays.

2.5   Timing of application

The effectiveness of *Bt* application depends strongly on its timing. The following guidelines for timing of *Bt* application can be useful: 1) Application early in the season, before high field populations of parasitoids and predators have been reduced by chemical insecticides. It has been shown in the US that *Bt* use against heliothine infestation early in the cotton season, followed by traditional insecticides [26], effectively reduced the *Heliothis virescens* populations. 2) Timing the *Bt* application according to monitoring of egg hatching: the most effective control of the Colorado potato beetle in potato was achieved up to 4 days after 30%

of the eggs had hatched, whereas premature or delayed application resulted in significantly more defoliation and larval recovery from the *Bt* intoxication [23]. This timing can be determined by pest monitoring with the aid of traps based on sex-pheromones or other insect attractants; a significant increase in trap catches may serve as an indication of the onset of egg laying. 3) *Bt* application after sunset instead of in the morning can increase the persistence of the *Bt* product in warm countries where activity of the microbe persist for only 2-3 days [22].

2.6   Formulations and residual activity

Low persistence of the spore-crystal product on the plant is one of the main problems in *Bt* application: low persistence (48 h) of *Bt* products on cotton was observed with *Bt* subsp. *kurstaki* products [22]. Leaf growth indirectly cause a reduction in *Bt* persistence as it causes natural dilution of the product on the plant, mostly in broad leaf crops such as cotton a leaf vegetables. Low residual activity of *Bt* subsp. tenebrionis products used against the Colorado potato beetle on potato in the field was limited to 48 h [20].

Residual activity is determined by the effectiveness of the formulation. A comprehensive information has been published recently on the use of additive in *Bt* formulations [8]. The additives include: wetters, stickers, sunscreens synergists and phagostimulants. It is widely accepted that UV inactivation of the crystal toxin is the major cause of the rapid loss of *Bt* activity. But, in practice, the commercial spray formulations do not include sunscreens. Several chromophores have been selected to shield *Bt* preparations against inactivation by sunlight [17, 12]. However, environmental safety and palatability to the insects are prerequisites for using these dyes and pigments in commercial products. In a different approach to photoprotection of *Bt*, a melanin-producing mutant of the microbe increased the UV resistance and insecticidal activity [49].

A promising approach to improving residual activity is exemplified by the use of starch-encapsulated *Bt* in the control of *Ostrinia nubilalis* in corn [16], or embedding *Bt* in a dusting formulation based on wheat flour for the control of vegetable and field lepidopteran pests [48]. These protections of the crystal included phagostimulants which arrested larvae to feed on the granule and ingest lethal amount of the microbe. Rainfasting was increased by *Bt* encapsulation with biopolymers to reduce the microbe wash-off [50]. Selecting environmentally safe and cost effective formulations to increase the residual activity of commercial *Bt* products is one of the major needs for widening the use of the microbe.

2.7   Safety

Extensive toxicity studies have showed that *Bt* isolates devoid of β-exotoxin are not toxic or pathogenic to mammals [36]. Nevertheless, in work with spore-crystal products, precautions should be taken to avoid contact with open wounds and to protect the eyes from infection via spray splashing or dusting [52]. In the western world, *Bt* strains which produce β-exotoxin cannot be registered for use in agriculture because of their toxicity to mammals [53].

2.8   Application equipment

Unlike the practice in forestry where most control programs are conducted by aerial sprays (see Chapter 6.2), *Bt* in agriculture is applied mostly with the ground sprayers used for the application of chemical insecticides. Since the effect of the crystal toxin in the larva is through feeding, high volumes of aqueous spray per unit area are required for adequate coverage of the plant canopy with the microbe. This approach is very costly because substantial amounts of the product are lost by dripping off the plants. In addition, the use of high volume of aqueous spray can be impractical in areas where water is not available. In recent years efforts have been made to reduce the spray volume and to achieve better control of the droplets: avocado orchards [59] were effectively air sprayed with *Bt* products from a helicopter to control *Boarmia* (*Ascotis*) *selenaria* (M. Wysoki, ARO, Israel, unpublished data), and aerial spraying of *Bt* in ULV application in the control of *Anarsia lineata* in almond orchards also seemed promising [51]. The use of an air-assisted sleeve boom (Degania Sprayers, Degania Bet, Israel) on row crops increased penetration of the spray and plant coverage, and reduced drift. Also, horizontal spraying from nozzles located within the rows in parallel to the plant canopy markedly improved the delivery of the *Bt* product to the lower surfaces of cotton leaves and to the flower buds and bolls (A. Navon, unpublished data).

3.   COMBINATIONS OF *BT* WITH OTHER MEANS OF PEST MANAGEMENT

The use of *Bt* in IPM program offers the advantage of replacing undesired chemical insecticides by safe bioinsecticides which could be applied with other biological means, provided that *Bt* is synergistic or compatible with them. The rational of this strategy is that biological means such as entomopathogenic microbes and nematodes, natural enemies, natural insecticides in combination with *Bt* can improve pest control, mostly when *Bt* efficacies are suboptimal. In this subchapter, laboratory experiments and field uses of *Bt* in combination with other pest control agents are described, and compatibilities and antagonistic interactions in this strategy, are discussed.

3.1   Entomopathogenic microbes

The combination of *Bt* with baculoviruses has mainly shown an additive effect [43]. In principle, larval feeding cessation caused by *Bt* may reduce the amounts of virus needed to kill the larva. Therefore, this combination may not be justified in terms of cost-effectiveness. The combination of *Bt* with fungi has not been evaluated, in any detail; however, in view of the high efficacy of commercial insecticidal fungi against sucking insects, combinations of *Bt* and fungi in suitable formulations may provide a useful strategy in complex pest situations where moths and white flies / aphids infest the same crop.

The combination of *Bt* and insecticidal nematodes seemed not suitable for the control of resistant *P. xylostella* [3]; the two biopesticides applied together provided little advantage over the use of either one of them alone.

3.2   Natural enemies

Egg parasitoids, such as *Trichogramma platneri* against *B. senenaria* in avocado [59] and *T. caoeciae* for the control of *O. nubilalis* [27], have been found highly compatible with *Bt*, as the egg is not a target stage for the microbe. The release of *Trichogramma* sp. in a *Bt*-based IPM program in tomato contributed to the improved profitability of this program, compared with the use of chemical insecticides [56]. The combination of *Bt* with natural enemies of larval pests necessitates careful screening of compatibilities; the level of host intoxication has a direct impact on the parasitoid performance: for example, longer exposures of the pest to lethal *Bt* levels resulted in lower survival of *Myiopharus doryphorae*, the tachinid parasitoid of the Colorado potato beetle [34], and of the braconid *Microplitis croceipes* in *Helicoverpa armigera* [5]. Such incompatibility would result in undesired loss of a parasitoid progeny in the host larvae and, therefore reduced parasitoid populations. Therefore, sequential use of *Bt* and parasitoids instead of joint application, is the recommended strategy;. it is more useful when the parasitoids infest 3rd instar and older larvae, which have survived *Bt* application against neonate larvae. The parasitoid *Cotesia marginiventris* emerged more successfully from *Heliothis virescens* when exposure of the pest to *Bt* was delayed for 48 h following parasitisation [1]. In *Bt*-sensitive *P. xylostella* the microbe had a negative effect on the host parasitoid *Cotesia plutellae*, but this competition between *Bt* and the parasitoid improved the performance of *C. plutellae* when *P. xylostella* was highly resistant to *Bt* [10]. In contrast to the adverse effects of *Bt* observed on immature stages of larval parasitoids, *Bt* products in aqueous mixtures consumed by adult parasitoids of target [5] or a non-target pest [59] led to increased parasitoid longevity. This probably resulted of the nutritious fermentation residues in the commercial microbe.

The control of the Colorado potato beetle in the field showed synergy between *Bt* and an egg-larval predator *Prillus bioculatus* (Pentatomidae) of the pest [11]. *Bt* had no detrimental effects on natural predators of the heliothine populations in cotton [60].

3.3   Plant allelochemicals and natural insecticides

Limonoids isolated from citrus which had antifeeding effects seem to be antagonistic to *Bt* in its activity against the Colorado potato beetle [41]. Also, cotton condensed tannins were found antagonistic to *Bt*, probably because they deterred the larvae and/or inactivated the toxin [47]. Neem has antifeeding effects [58] which stop larval feeding, therefore, it may impair the effect of *Bt* in some insects. On the other hand, plant phenols [35] and caffeine [39] increased the toxicity of *Bt*. The screening of phytochemical effects at tritrophic levels is useful for the following purposes: 1) to evaluate compatibilities of *Bt* with allelochemicals used in crop breeding programs for developing natural resistance to insects, 2) to select plant allelochemicals that synergise with *Bt* products and that could be used in the microbe formulations for pest control; and 3) to develop rotational uses of *Bt* and insecticides of botanical origin, to overcome insect resistance to the microbe.

3.4   Chemical insecticides

Mixtures of chemical insecticides and *Bt* have been evaluated mostly to potentiate the microbe. Several constraints on using such mixtures can be listed : 1) most of the chemical insecticides were not compatible with biological control based on egg and larval parasitoids and predators, whereas *Bt* did not affect several beneficial insects [2, 28, 32, 40, 60]; 2) mixing of *Bt* with the chemical instead of using them sequentially does not avoid the continuous selection pressure for resistance to the microbe, 3) some chemical insecticides have antifeeding effects on the pest and, therefore, the in combination with *Bt* could be antagonistic, as probably, the case with the pyrethroid esfenvalerate tested with *Bt* against *P. xylostella* [30]. In such cases, the chemical may reduce the ingestion of the microbe and so reduce its efficacy; 4) *Bt* products reduce larval feeding and thereby minimise the efficacy of chemical insecticides which affect the insects *per os*.

4. FUTURE PROSPECTS

The future use of chemical insecticides seems to further decline and more restriction on their registration which will result in a smaller market of these products. Also, insect resistance to the chemicals will continue. This will increase the use of *Bt*, but competition from selective natural and synthetic chemical will not be avoided. Effects of *Bt* transgenic crops on the market of *Bt* products cannot be accurately predicted at this stage. Increasing the acreage of *Bt* crops such as row crops may reduce the use of *Bt* sprays. On the other hand, niche markets for the microbe will increase in cash crops free of *Bt* genes predominately in bioorganic agriculture. It is anticipated that the following essential developments and knowledge will improve microbial pest management with *Bt* products: 1) novel formulations that will increase the residual activity of the microbe in the field; 2) improved application technologies developed in consideration of the environmental constraints and mode of action of *Bt*; 3) larger choice of IPM strategies including *Bt* with selective chemicals, natural enemies and other microbial agents; 4) biotechnological efforts to increase *Bt* activities and insect host range; 5) cost-effective levels of *Bt* compatible with those of chemical control agents.

Unlike the situation in forestry (see Chapter 6.2), the adoption of *Bt* application programs in agriculture has been slow, because of the constraints discussed in this subchapter. A helpful route for rapid integration of the microbe into pest control programs would be through transferring the knowledge gained from successful IPM in bioorganic farming to conventional agriculture.

REFERENCES

[1] Atwood, DW, Young SY III & Kring TJ (1997) Development of Cotesia (Hymenoptera: Braconidae) in tobacco budworm (Lepidoptera: Noctuidae) larvae treated with *Bacillus thuringien* and thiodicarb. J. Econ. Entomol. 90, 751-756

[2] Atwood DW, Young SY, III & Kring TJ (1997) Impact of Bt and thiodicarb alone and in combination on tobacco budworm, mortality and emergence of the parasitoid *Microplitis croceipes*. vol 2 pp. 1305-1310. Proc. Beltwide Cotton Conf., National Cotton Council, New Orleans, USA

[3] Bauer ME, Kaya HK, Tabashnik BE & Chilcutt CF (1998) Suppression of Diamondback moth (Lepidoptera:Plutellidae) with an entomopathogenic nematode (Rhabditida: Steinernematidae) and *Bacillus thuringiensis*. J. Econ. Entomol. 91, 1089-1095

[4] Baum JA, Timothy BJ & Carlton, BC (1999) *Bacillus* thuringiensis, p. 189-209. *In* Hall FR & Menn JJ (eds.), Biopesticide use and delivery, Humana Press, NJ. USA

[5] Blumberg D, Navon A, Keren S, Goldenberg S & Ferkovich SM (1997) Interactions among *Helicoverpa armigera* (Lepidoptera: Noctuidae), its larval endoparasitoid *Microplitis croceipes* (Hymenoptera: Braconidae), and *Bacillus thuringiensis*. J. Econ. Entomol. 90, 1181-1186

[6] Bryant JE (1994) Application strategies for *Bacillus thuringiensis*. Agric. Ecosys. Environ. 49, 65-75

[7] Burges HD (1967) The standardization of products based on *Bacillus thuringiensis*, p. 306-308. Proceedings of the International Colloquium on Insect Pathology and Microbial Control, Wageningen, The Netherlands

[8] Burges HD & Jones KA (1999) Formulation of bacteria, viruses and protozoa to control insects, p. 34-127. *In* Burges HD (ed.), Formulation of microbial biopesticides, Kluwer Acedemic Publisher, Dordrecht, The Netherlands

[9] Cannon RJC (1996) *Bacillus thuringiensis* use in agriculture: A molecular perspective. Biol. Rev. 71, 561-636

[10] Chilcutt CF & Tabashnik BE (1997) Host-mediated competition between the pathogen *Bacillus thuringiensis* and the parasitoid *Cotesia plutella* of the diamondback moth (Lepidoptera: Plutellidae). Environ. Entomol. 26, 38-45

[11] Cloutier C & Jean C (1998) Synergism between natural enemies and biopesticides: a test case using stinkbug *Perillus bioculatus* (Hemiptera: Pentatomidae) and *Bacillus thuringiensis tenebrionis* against Colorado potato beetle (Coleoptera: Chrysomelidae). J. Econ. Entomol. 91, 1096-1108

[12] Cohen E, Rozen, H, Joseph T & Margulis L (1991) Photoprotection of *Bacillus thuringiensis* var. *kurstaki* from ultra-violet irradiation. J. Invertebr. Pathol. 57, 343-351

[13] Dent DR (1993) The use of *Bacillus thuringiensis* as an insecticide, p. 19-44. *In* Jones DG (ed.), Exploitation of Microorganisms, Chapman & Hall, London. UK

[14] Dubois NR & Dean DH (1995) Synergism between Cry1A insecticidal crystal proteins and spores of *Bacillus thrinigiensis*, other bacterial spores, and vegetative cells against *Lymantria dispar* (Lepidoptera: Lymantriidae) larvae. Environ. Entomol. 24, 1741-1747

[15] Dulmage HD (1970) Insecticidal activity of HD-1, a new isolate *of Bacillus thuringiensis* var. *alesti*. J. Invertebr. Pathol. 15, 232-239

[16] Dunkle RL & Shasha BS (1988) Stasrch-encapsulated *Bacillus thuringiensis*: A new method for increasing environmental stability of entomopathogens. Environ. Entomol., 17, 120-126

[17] Dunkle RL & Shasha BS (1989) Response of starch encapsulatred *Bacillus thuringiensis* containing UV screens to sunlight. Environ. Entomol. 18, 1035-1041

[18] Entwistle PE, Cory, JS Bailey MJ & Higgs, S (eds.) (1993*) Bacillus thuringiensis*, an Environmental Biopesticide: Theory and Practice, 311 pp. John Wiley & Sons, Chichester, UK

[19] Ferro DH & Lyon SM (1991) Colorado potato beetle (Coleoptera: Chrysomelidae) larval mortality: Operative effects of *Bacillus thuringiensis* subsp. *san diego*. J. Econ. Entomol. 84, 806-809

[20] Ferro DH, Yuan QC, Slocombe A & Tutle A (1993) Residual activity of insecticides under field conditions for controlling the Colorado potato beetle (Coleoptera: Chrysomelidae). J. Econ. Entomol. 86, 511-516

[21] Forrester NW (1994) Use of *Bacillus thuringiensis* in integrated control, especially on cotton pests. Agric. Ecosys. Environ. 49, 77-83

[22] Fuxa J (1989) Fate of released entomopathogens with reference to risk assessement of genetically engineered microorganisms. Bull. Entomol. Soc. Am. 35, 12-24

[23] Ghidiu GM & Zehnder GW (1993) Timing of the initial spray application of *Bacillus thuringiensis* for control of the Colorado potato beetle (Coleoptera: Chrysomelidae) in potatoes. Biological Control 3, 348-352

[24] Gould F, Anderson, A, Landism D & Van Mellert H (1991) Feeding behavior and growth of *Heliothis virescens* larvae on on diets containing *Bacillus thuringiensis* formulations or endotoxins. Entomol. Exp. Appl. 58, 199-210

[25] Hall FR, Chapple AC, Taylor RAJ & Downer RA (1995) Modeling the dose acquisition process of *Bacillus thuringiensis*, p. 68-78. *In* Hall FR & Barry JW (eds.), Biorational Pest Control Agents Formulation and Delivery, ACS Symposium Series 595, London, UK

[26] Hand SS & Luttrell RG (1997) Strategies for foliar application of *Bacillus thuringiensis* in cotton Vol. 2 pp. 1151-1157. Proc. Beltwide Cotton Conf. National Cotton Council, New Orleans, USA

[27] Hassan SA (1983) Results of laboratory testing of a series of pesticides on egg parasites of the genus *Trichogramma* (Hymenoptera: trichogrammatidae) Nachrichtenbl. Dtsch Pflkanzenschutzdienst (Braunschw.) 35, 21-25

[28] Hassan E & Graham-Smith S (1995) Toxicity of endosulfan, esfenvalerate and *Bacillus thuringiensis* on adult of *Microplitis demolitor* Wilkinson and *Trichogrammatoidea bactrae* Nagaraja. Z. Pflanzenk. Pflanzensch. 102. 442-428

[29] Herrnstadt C, Soares GG, Wilcox ER & Edwards DL (1986) A new strain of *Bacillus thuringiensis* with activity against coleopteran insects. Bio/Technology, 4, 305-308

[30] Hoy CW & Hall FR (1993) Feeding behaviour of *Plutella xylostella* and *Leptinotarsa decemlineata* on leaves treated with *Bacillus thuringiensis* and esfenvalerate. Plant Sci. 38, 335-340

[31] Krieg A, Huger AM, Langenbruch GA & Schnetter, W (1983) *Bacillus thuringiensis* var. *tenbrionis*: ein neuer, glarven von Coleoptaran wirksamer pathotype. Z. angew. Entomol. 96, 500-508

[32] Kring TJ & Smith TB (1995) *Trichogramma pretiosum* efficacy in cotton under Bt-insecticide combinations. Vol 2, pp. 856-857. Proc. Beltwide Cotton Conf., National Cotton Council. San Antonio, TX, USA

[33] Lambert B, Peferoen M (1992) Insecticidal promise of *Bacillus thuringiensis*. Facts and mysteries about a successful biopesticide. BioScience 42, 112-121

[34] Lopez R & Ferro DN (1995) Larviposition response of *Myiopharus doryphorae* (Diptera: Tachinidae) to Colorado potato beetle (Coleoptera: Chrysomelidae) larvae treated with lethal and sublethal doses of *Bacillus thuringiensis* Berliner subsp. *tenebrionis*. J. Econ. Entomol. 88, 870-874

[35] Ludlum CT, Felton GW & Duffey SS (1991) Plant defenses: chlorogenic acid and polyphenol oxidase enhance toxicity of *Bacillus thuringiensis* subsp. *kurstaki* to *Heliothis zea*. J. Chem. Ecol. 17, 217-237

[36] McClintock JT, Schaffer CR & Sjoblad RD (1995) A comparative review of the mammalian toxicity of *Bacillus thuringiensis*-based pesticides. Pest. Sci. 45, 95-105

[37] McGuire MR & Shasha BS (1995) Starch encapsulation of microbial pesticide, p. 229-237. *In* Hall FR & Barry JW (eds.), Biorational Pest Control Agents Formulation and Delivery, ACS Symposium Series 595, London, UK

[38] Moar WJ, Trumble JT & Federici BA (1989) Comparative toxicity of spores and crystals from the NRD-12 and HD-1 strains of *Bacillus thuringiensis* subsp. *kurstaki* to neonate beet armyworm (Lepidoptera: Noctuidae). J. Econ. Entomol. 82, 1593-1603

[39] Morris ON, Trorrier M, McLaughlin NB & Converse V (1994) Interaction of caffeine and related compounds with *Bacillus thuringiensis* ssp. kurstaki in Bertha armyworm (Lepidoptera: Nuctuidae) J. Econ. Entomol. 87, 610-617

[40] Muckenfuss AE & Shepard BM (1994) Seasonal abundance and response of Diamondback moth, *Plutella xylostella* (L.) (Lepidoptera: Plutellidae), and natural enemies to esfenvalerate and *Bacillus thuringiensis* subsp. *kurstaki* Berliner in coastal South Carolina. J. Agric. Entomol. 11: 361-373

[41] Murray KD, Alford AR, Groden E, *et al.*, (1993) Interactive effects of an antifeedant used with *Bacillus thuringiensis* var. *san diego* delta-endotoxin on Colorado potato beetle (Coleoptera: Chrysomelidae). J. Econ. Entomol. 86, 1793-1801

[42] Nakamura LK & Dulmage HT (1988) *Bacillus thuringiensis* cultures available from the U.S. Department of Agriculture. U.S. Department of Agriculture, Technical Bulletin No 1738, 38

[43] Navon A (1993) Control of lepidopteran pests with *Bacillus thuringiensis* pp. 125-146. *In* Entwistle PF, Cory JS, Bailey MJ &Higgs S (eds.), *Bacillus thuringiensis*, an Environmental Biopesticide: Theory and Practice. John Wiley & Sons, New York

[44] Navon A (2000) Bioassays of *Bacillus thuringiensis*. *In* Bioassays of Entomopathogenic Microbes and Nematodes. CABI Publishing, UK (in press)

[45] Navon A & Gelernter W. (1996) Proposals for addressing standardization issues. p. 29. Abst. of SIP 29th Ann. Meeting & IIIrd Internat. Colloq. On *Bacillus thuringiensis*. Cordoba, Spain

[46] Navon A, Klein M & Braun S (1990) *Bacillus thuringiensis* potency bioassays against *Heliothis armigera*, *Earias insulana*, and *Spodoptera littoralis* larvae based on standardized diets. J. Invertebr. Pathol., 55, 387-393

[47] Navon A, Hare, JD & Federici, BA (1993) Interactions among *Heliothis virescens* larvae, cotton condensed tannin and the CryIA(c) endotoxin of *Bacillus thuringiensis*. J. Chem. Ecol. 19, 2485-2499

[48] Navon A, Keren S, Levski S, Grinstein A & Riven J. (1997) Granular feeding baits based on *Bacillus thuringiensis* products for the control of lepidopterous pests. Phytoparasitica 25 (suppl), 101S-110S

[49] Patel KR, Wyman JA, Patel KA & Burden, BJ (1996) A mutant of *Bacillus thuringiensis* producing a dark-brown pigment with increased UV resistance and insecticidal activity. J. Invertebr. Pathol. 67, 120-124

[50] Ramos LM, McGuire MR & Galan Wong LJ (1998) Utilization of several biopolymers for granular formulations of *Bacillus thuringiensis*. J. Econ. Entomol. 91, 1109-1113

[51] Roltsch WJ, Zalom FG, Barry JW, Kirfman GW & Edstrom JP (1994) Ultra-low volume aerial application of *Bacillus thuringiensis* variety *kurstaki* for the control of peach twig borer in almond trees. Appl. Eng. Agric. 11, 25-30

[52] Sample JR & Buettner H (1983) Ocular infection caused by a biological insecticide. J. Infec. Dis. 148, 614

[53] Sebesta K, Farkas J, Horska K & Vankova J (1981) Thuringiensin, the beta-exotoxin of *Bacillus thuringiensis*, p. 249-282. *In* Burges HD (ed.), Microbial control of pests and plant diseases 1970-1980, Academic Press, London, UK

[54] Shah PA & Goettel MS (eds.) (1999) Directory of microbial control products and services, p. 31, *In* Society of invertebrate Pathology, Gainesville, FL 32614-7050, USA

[55] Tabashnik BE, Cushing NL, Finson N & Johnson MW (1990) Field development of resistance to *Bacillus thuringiensis* in diamondback moth. Journal of Economic Entomology, 83, 1671-1676

[56] Trumble J & Alvarado-Rodriguez B (1993) Development of economic evaluation of an IPM program for fresh market tomato production in Mexico. Agric. Ecosys. Environ. 43, 267-284

[57] van Frankenhuyzen K (1993) The challenge of *Bacillus* thuringiensis, p. 1-35. *In* Entwistle PF, Cory JS, Bailey MJ &Higgs S (eds.), *Bacillus thuringiensis*, an Environmental Biopesticide: Theory and Practice, John Wiley & Sons, New York

[58] Walter JF (1999) Commercial exerience with neem products, p. 155-163. *In* Hall R & Menn JJ (ed.), Biopesticide Use and Delivery, Humana Press, NJ, USA

[59] Wysoki M (1989*) Bacillus thuringiensis* preparations as a means for the control of lepidopterous pests in Israel. Isr. J. Entomol. 23, 119-129

[60] Young SY, Kring TJ, Johnson DR & Klein CD (1997) *Bacillus thuringiensis* alone and in mixtures with chemical insecticides against heliothines and effects on predator densities in cotton. J. Entomol. Sci. 32, 183-191

Chapter 6.2

Application of *Bacillus thuringiensis* in forestry

Kees van Frankenhuyzen
*Canadian Forest Service, Great Lakes Forestry Centre, P.O. Box 490,
Sault Ste. Marie, Ontario, Canada P6A 5M7*

Key words: *Bacillus thuringiensis* (*Bt*), forestry, spruce budworm, *Bt*-efficacy model.

Abstract :   Current world-wide use of *Bt* subsp. *kurstaki* in forestry was preceded by
its development for control of the spruce budworm, *Choristoneura
fumiferana* (Lepidoptera: Tortricidae), a major defoliator of coniferous
forests in North America. Since its widespread operational acceptance in
the mid 1980s, *Bt* has been used on a cumulative total of about 8.5 million
ha in Canada and 3 million ha in the United States. In Europe, *Bt* has been
used for control of various defoliators on 1.7 million ha since 1990. In
most forest protection programs, undiluted high-potency products
containing 12.7-25.4 billion international units (BIU) per litre are applied
in volumes of 1.2-2.5 litres per ha. Further optimisation of aerial forestry
applications, currently in progress, uses a detailed process-oriented model
that simulates the efficacy of *Bt* sprays against the spruce budworm.

1.   INTRODUCTION

Bacillus thuringiensis (*Bt*) is presently the most successful microbial
insecticide, with world-wide application for protection of crops, forests,
and human health [6]. World-wide commercialisation of *B. thuringiensis*
subsp. *kurstaki* (*Bt*) was preceded by its development for control of the
spruce budworm, *Choristoneura fumiferana* (Lepidoptera: Tortricidae), a
major defoliator of coniferous forests in North America. Large-scale use
against the spruce budworm established for the first time that *Bt* was
indeed a viable and effective alternative to synthetic insecticides.

J.-F. Charles et al. (eds.),
Entomopathogenic Bacteria: From Laboratory to Field Application, 371–382.
© 2000 *Kluwer Academic Publishers. Printed in the Netherlands.*

The development of *Bt* for control of forest insects is reviewed in the first part of this chapter, with a detailed overview of current operational use. In the second part, the importance of understanding *Bt* dose acquisition processes for the improvement of field efficacy is illustrated, using the spruce budworm as a model for forest defoliators.

2.   FIELD DEVELOPMENT

2.1   Development of *Bt* as a viable alternative

The discovery of *Bt* and its early history as an insect control agent has been reviewed in detail elsewhere [19], and is only briefly summarised here. Although the first commercial product was already available in Europe in 1938, the focus of commercialisation shifted to North America in the early 1950s, with the production and registration of Thuricide® in 1957. The availability of commercial products initiated a period of intermittent field testing throughout the 1960s, in agriculture as well as in forestry. The results were generally inconsistent and *Bt* did not measure up against available synthetic insecticides. Two developments in the 1960s were of particular significance in improving efficacy and accelerating commercialisation. The first was the discovery of the *kurstaki* isolate HD-1, which was quickly adopted for commercial production. The second was the establishment of an International Unit (IU) to standardise the potency of commercial products.

Formulations based on HD-1 generally improved *Bt* field efficacy during the 1970s. Most of the field development during that decade took place in North America [13, 17]. Large-scale aerial spraying against defoliating forest Lepidoptera, in particular the spruce budworm, *Choristoneura fumiferana*, and the gypsy moth, *Lymantria dispar*, presented an opportunity for field testing on a scale that was commercially attractive and in markets that had high damage thresholds. Field tests with new formulations focused on developing appropriate aerial application prescriptions and fine-tuning formulation requirements. Although the effectiveness of *Bt* sprays improved, results remained inconsistent and treatment costs were much higher than with chemical insecticides. Cost effectiveness started to improve in the late 1970s, when producers achieved significant advances in both production and formulation technologies. By the end of the decade, *Bt* was generally considered an operational alternative for control of spruce budworm and gypsy moth.

Table 2. Operational use of *B. thuringiensis* for control of other defoliating forest insects in Canada between 1985 and 1999

	Number of BIU (x 10^3) applied against: [2,3]						Sprayed ha[1]
Year	JPBW	WSBW	EHL	GM	WMTM	TOTAL	(x 10^3)
1985	4,400	0	70	7	0	4,477	222.5
1986	10,488	0	162	6,488	0	17,139	703.7
1987	2,109	0	151	4,829	0	7,090	271.5
1988	0	55	826	827	0	1,708	56.9
1989	428	15	273	777	0	1,493	49.7
1990	0	0	1,347	4,074	0	5,422	180.7
1991	0	90	509	2,194	0	2,793	93.1
1992	0	1,067	29	4,000	0	5,095	116.5
1993	3	1,027	1,357	131	0	2,519	82.2
1994	644	630	531	103	0	1,265	40.8
1995	1,530	0	2,689	52	0	4,272	141.7
1996	769	0	4,244	18	0	5,031	167.5
1997	0	484	589	0	0	1,073	34.0
1998	0	636	288	0	5,150	6,074	131.0
1999	0	651	489	1,621	0	2,761	353.4
Total[4]	20,230	4,659	13,465	20,735	5,150	64,234	2,564.7

Sprayed hectares = number of hectares treated (in thousands) x number of applications
[2] BIU applied = number of hectares treated x BIU/ha x number of applications
[3] JPBW = Jackpine budworm, *Choristoneura pinus pinus*; WSBW = Western spruce budworm, *Choristoneura occidentalis*; EHL = Eastern hemlock looper, *Lambdina fiscellaria fiscellaria*; GM = gypsy moth, *Lymantria dispar*, including Asian gypsy moth eradication program in Vancouver, BC, 1992; WMTM = whitemarked tussock moth, *Orgyia leucostigma*
[4] Discrepancies between totals shown and row or column totals are due to rounding errors in the main body of the table
Source: Forestry Insecticide Database, Canadian Forest Service, Great Lakes Forestry Centre

Operational use in the early 1980s before cost and efficacy were competitive with those of synthetic chemicals stimulated further cost reductions [19]. Product potency continued to increase and new high-potency formulations were designed for undiluted (neat) application in ultra-low volumes (ULV). By the mid-1980s, the dosage rate of 30 billion (10^9) international units (BIU) per ha that was recommended for spruce budworm control was routinely applied in 2.4 litres of undiluted product. The use of such low spray volumes reduced application costs by increasing spray aircraft efficiencies, while the higher product potency increased the efficacy and reliability of control operations [17]. The application of undiluted high-potency products undoubtedly made the most important contribution to reducing the constraints of high cost and unreliable efficacy in forestry. Those improvements, together with a shift in political climate that favoured the use of biologicals, resulted in the widespread acceptance of *Bt* by the mid-1980s as a fully operational insecticide for control of spruce budworm, gypsy moth and other forest defoliators.

2.2   Current use

Operational use for control of the spruce budworm in Canada increased from <5% of the total area sprayed in the early 1980s to almost 100% 15 years later. This was due primarily to the political decision by various jurisdictions to curb the aerial application of synthetic insecticides in public forests. In most of Canada, *Bt* is now the only insecticide used for budworm control, apart from a recently registered biorational (tebufenozide) which has been used operationally on about 33,000 ha since 1995. *Bt* use declined sharply in the early 1990s as the result of a general collapse of spruce budworm populations in Eastern Canada, although its use in western provinces increased (Table 1, page 380). By 1999, ~158 x 10^{15} international units had been sprayed on a cumulative total of almost 6 million ha. Use against other defoliators amounted to ~64 x 10^{15} international units applied to a cumulative total of 2.5 million ha between 1985 and 1999 (Table 2).

Operational use of *Bt* followed a similar pattern in the United States (Table 3). In Maine, the percentage of the spruce budworm control program in which *Bt* was used rose rapidly from 2% in 1979 to 81% in 1985 when the outbreak collapsed, for treatment of a total of 460,000 ha. In Washington and Oregon, aerial spray programs to control Western spruce budworm, *C. occidentalis*, switched from primarily carbaryl in 1983 to 85-100% *Bt* after 1985, with use on a total of 547,000 ha between 1983 and 1998 [15]. The main *Bt* market in the United States, however, is for control of gypsy moth. Since 1980, about 2.4 million ha of deciduous forests in the eastern United States have been treated with *Bt* as part of Federal and State Co-operative Suppression Programs. Annual use increased from a low of 6% of the total area treated with *Bt* to a high of 82%, with diflubenzuron (Dimilin®) being the second most widely used insecticide on 1.7 million ha. Operational use for population suppression and foliage protection usually involves one application of undiluted high-potency products in dose rates ranging from 30 to 50 BIU/ha, from a wide range of aircraft types and spray equipment. Multiple applications of *Btk* were used to eradicate gypsy moth in Oregon (1985-1987) [4] and Utah (1998-1993), and for eradication of the Asian strain of the gypsy moth in 1992 in British Columbia (20,000 ha), Washington and Oregon

(200,000 ha), and again in 1994 in North Carolina (50,000 ha) [13]. A similar approach was used for eradication of the white-spotted tussock moth, *Orgyia thyellina*, in Auckland, New Zealand, in 1996 and 1997 (http://www.maf.govt.nz/MAFnet/evergreen/).

Improvements in the formulation and application of *Bt* resulting from more than a decade of operational experience, together with the introduction of differential Global Positioning Systems to aid operational navigation of spray aircraft [12], set the stage for a sharp increase in the use of *Bt* in Europe's forestry market during the 1990s (Table 4). *Bt* use in Europe had grown steadily between the early 1970s and mid-1980s for control of various defoliators. For example, the use of *Bt*-based products

Table 3. Operational use of *B. thuringiensis* for control of three forest defoliators in the US between 1980 and 1998 (number of treated hectares in thousands)

Year	*C. fumiferana*[1]		*C. occidentalis*[2]		*L. dispar*[3]		
	No. ha treated	% *Bt*	No. ha treated	% *Bt*	No. ha treated	% *Bt*	% Dimilin®
1980	480.0	15	0	0	32.5	21	0
1981	469.0	11	0	0	141.7	6	0
1982	329.1	10	80.7	0	294.2	9	10
1983	338.5	14	251.2	2	242.3	79	7
1984	267.2	33	0.3	100	207.3	42	49
1985	164.4	81	16.5	100	210.4	57	47
1986	0	0	5.3	0	238.5	37	61
1987	0	0	64.3	100	282.7	45	53
1988	0	0	273.2	92	303.4	36	63
1989	0	0	5.8	86	323.0	51	48
1990	0	0	29.0	100	614.9	56	43
1991	0	0	46.1	100	446.7	66	32
1992	0	0	88.7	84	388.9	68	30
1993	0	0	27.0	96	238.1	63	35
1994	0	0	0	0	262.9	57	41
1995	0	0	11.3	100	190.6	64	34
1996	0	0	8.1	100	140.6	63	32
1997	0	0	0	0	39.4	67	16
1998	0	0	6.5	100	61.7	82	<1
Total ha	1,742	2,048	521[5]	914	547[5]	4,660	2,435

Sources:
[1] Annual Pest Forum reports, Canadian Forest Service, Ottawa
[2] [15]; data shown are for Oregon and Washington only
[3] GM Digest, US Forest Service, Forest Health Protection, Morgantown, WV; data for European strain only
[4] Number of hectares treated (in thousands) with *B. thuringiensis* and synthetics

[5] Total number of hectares (in thousands) treated with one or more applications of *B. thuringiensis*

Table 4. Operational use of *B. thuringiensis* for control of defoliating forest insects[1] in Europe between 1990 and 1998[3]

| Year | No. hectares (in thousands) treated with: | | | | % of treated area |
	Bt	Dimilin®	Pyrethroids	Other[2]	sprayed with *Bt*
1990	81.5	578.5	25.2	210.1	9.5
1991	70.8	471.3	26.1	271.5	9.5
1992	85.9	410.7	23.8	195.4	13.5
1993	125.7	302.5	38.1	175.5	26.0
1994	300.0	788.2	98.3	149.6	24.3
1995	193.9	305.4	16.0	18.0	36.3
1996	204.3	295.0	10.0	21.8	38.5
1997	500.5	299.0	20.0	20.0	59.6
1998	197.7	226.8	26.0	0.0	43.8
Total ha	1,760.4	3,677.4	283.5	1,061.9	26.0

[1] including *Thaumetopoea pityocampa, Lymantria monacha, L. dispar, Dendrolimus* sp., *Bupalus piniaria, Panolis flammea, Tortix viridana* and *Operophtera brumata*
[2] including areas for which information on control agent was not available
[3] Source: J.M. Sanders, Abbott Laboratories, England

in Bulgaria rose from 2% of the total area treated in 1975 to 53% in 1984, for treatment of up to 40,000 ha annually [25]; treatment typically involved application of diluted products in high volumes. In the early 1990s, high-potency products and undiluted application in ultra-low volumes were introduced by the *Bt* manufacturers and quickly became standard practice [16]. The use of high-potency products at 20-50 BIU/ha for control of the pine processionary caterpillar, *Thaumetopoea pityocampa,* in Italy is reported in [1]. In Poland, high-potency products were applied at 50 BIU in 4.0 litres/ha (AU5000) on 148,000 ha in 1994 to control an outbreak of nun moth, *Lymantria monacha* [9]. The ULV application technology was introduced in Russia by Abbott Laboratories in 1996 in a co-operative program for control of the Siberian moth, *Dendrolimus superans sibericus,* on 120,000 ha using 38 BIU in 3 litres/ha [5]. *Bt* was used for the first time in Sweden in 1997, on about 4000 ha against pine looper, *Bupalus piniaria,* and again in 1998 on 1000 ha against nun moth [10]. Total use in Europe between 1990 and 1998 for control of defoliators is estimated to have involved treatment of almost 1.8 million ha of deciduous and coniferous forests (Table 4). At present, *Bt* is second to Dimilin® as the most widely used product, while the use of synthetic pyrethroids continues to decline.

3. THE BIOLOGICAL INTERFACE: REDUCING THE EFFICACY BOTTLENECK

3.1 Superseding the empirical approach

Much of the field development of *Bt* as a successful forest insect control product has been a highly empirical process. The choice of application parameters, such as dosage and volume application rates and droplet size spectra, was largely based on economic and technical considerations [17], rather than a sound knowledge of key processes that govern dose ingestion and expression [8]. However, years of research on the *Bt*-budworm interface is now providing a rational framework for further optimisation of *Bt* products and their application for spruce budworm control. A more detailed review of that research is presented below because it may be used as a template for improving *Bt* efficacy against other forest defoliators.

3.2 The *Bt*-budworm interface

Operational experience in Canada demonstrated that *Bt* efficacy is more dependent than that of fenitrothion on proper timing of spray application and favourable weather after the spray application [2]. This is not surprising, since *Bt* has to be ingested and has no other toxicity mechanisms. Its single mode of entry restricts the window opportunity for spray application, because larvae have to be feeding on elongating shoots in order to contact the dose. Proper timing of application in terms of larval development and bud phenology is further complicated by poor foliar persistence of spray deposits, which is often limited to a few days [21]. A third limitation is that ingestion of a sublethal dose causes temporary cessation of larval feeding.

Several lines of evidence suggest that feeding inhibition resulting from sublethal dose ingestion limits the acquisition of an efficacious dose by spruce budworm larvae. First of all, ingestion by spruce budworm larvae causes immediate feeding inhibition, which is permanent in lethally dosed larvae but temporary in larvae that have acquired a sublethal dose [18, 20]. Those larvae resume feeding after a recovery period which can last from one to several days, depending on the dose and ambient temperature [7, 20, 22]. During this time, spray deposits are degraded rapidly [21]; therefore, if larvae do not ingest a lethal dose within 1-2 days following spray application, the probability of them doing so is thought to decline rapidly. Field studies confirm that most dose acquisition occurs within one to two days following aerial application [8]. Secondly, comparison of instar-specific lethal dose requirements [24] with the dose theoretically contained in droplets of various sizes suggests that much of the active ingredient is deposited in droplets that are too small to contain a lethal dose [22]. Our calculations suggest that current application prescriptions, which seldom yield an average density of more than one droplet per needle, result in at best one LD_{50} per needle. This means that spruce budworm larvae must ingest several droplets for mortality to result, which renders acquisition of an efficacious dose subject to the adverse effects of feeding inhibition [11]. Since dose levels as low as one tenth of an LD_{50} can inhibit larval feeding for several hours [22], temporary cessation of larval feeding is likely to be a predominant phenomenon in treated spruce budworm populations.

All the above factors combine to determine the probability of feeding larvae ingesting an efficacious dose. Thus, field efficacy of *Bt* sprays is determined by multiple interacting processes that govern dose ingestion and expression after spray deposition in the larval microhabitat. These processes include weathering of spray deposits, larval development, instar-dependent feeding behaviour, feeding rates, and the insect's behavioural and physiological response to the ingested dose, all of which are influenced by weather conditions following application.

3.3   The spruce budworm-*Bt* efficacy model

Knowledge of the those interactions was recently integrated with current knowledge of budworm phenology, feeding behaviour and population dynamics into a *Bt*-efficacy simulation model [3]. The model accurately predicted foliage protection and population reduction in balsam fir stands in extensive aerial application trials [14]. Validation of the model suggests that our current understanding of the processes underlying *Bt* efficacy against spruce budworm is reasonably complete. It also supports the use of the model as a tool for optimising decision making in spruce budworm control programs, such as selection of optimal application rates and atomisation, and timing of application to achieve specified control objectives at various population densities [14]. Thus, the model can be used to generate testable hypotheses towards optimisation of *Bt* field efficacy, replacing the *ad hoc* nature of field development trials as used to date with an approach that is based on a detailed understanding of pest-pathogen interactions.

3.4   Implications for improving *Bt* application

Research on the spruce budworm-*Bt* biological interface has opened several options for improving the effectiveness of *Bt* in forestry applications. One way to maximise treatment success is to ensure that larvae ingest a lethal dose in the first one or two spray droplets, so that possible adverse effects of feeding inhibition on dose acquisition are minimised. This could be achieved by increasing product potency so that droplets in the physically optimum size range contain a lethal or near-lethal dose [22]. However, products with potencies exceeding 25.4 BIU/litre are not available and may not be practical because of physical constraints in terms of viscosity and sprayability of higher concentrates. A more efficient way to increase the dose in small droplets would be by using isolates or strains that have higher spruce budworm toxicity per IU. Our calculations indicate that a ten-fold increase in budworm toxicity should be sufficient to reduce the constraints imposed by feeding inhibition [22]. Finally, adverse effects of feeding inhibition on dose acquisition can be minimised by using larger spray droplets through coarser atomisation of the spray cloud. Dose acquisition by spruce budworm larvae in the field was enhanced by increasing the volume median diameter of the spray cloud from 80 to 160 µm, but we were unable to demonstrate a significant improvement in spray efficacy [23]. Model simulations show that droplet diameter is a critical determinant of efficacy [3], and that increasing product potency and increasing droplet density per needle have equivalent but diminishing-returns effects on efficacy [14]. Simulations need to be applied to determine the optimum trade-off between droplet density and dose per droplet (product potency and droplet diameter). A considerable field research effort will then be needed to compare the outcome of model simulations in the real budworm world. It is likely, however, that more reliable application of *Bt* for spruce budworm control will involve the use of higher potency products together with higher volume application rates in order to increase both the dose per droplet and the number of droplets per needle. If so, the registration of dosage application rates exceeding the current maximum of 30 BIU/ha might be required.

Table 1. Operational use of B. thuringiensis for control of Choristoneura fumiferana in Canada between 1980 and 1999

Year	Sprayed[1] ha	Number of BIU ($\times 10^3$)[2] applied in[2,3]									
		NFL	NS	NB	QUE	ONT	MA	SAS	ALB	BC	Total
1980	65.3	150	572	0	629	204	0	0	0	0	1,557
1981	51.9	38	647	0	243	109	7	0	0	0	1,046
1982	51.0	94	306	0	768	61	0	0	0	0	1,231
1983	58.2	0	412	309	490	55	0	0	0	0	1,267
1984	360.6	62	414	1,119	7,792	63	0	0	0	0	9,450
1985	675.7	103	1,491	2,430	14,832	587	0	0	0	0	19,444
1986	356.8	0	1,684	3,345	837	3,684	0	0	0	0	9,551
1987	404.8	0	932	2,703	6,081	1,536	10	0	0	0	11,264
1988	434.5	0	0	6,333	6,083	420	35	0	0	4	12,875
1989	432.5	0	0	3,282	6,822	917	149	0	0	0	11,171
1990	1061.8	0	0	4,372	19,091	1,948	143	0	300	0	25,856
1991	526.7	0	0	3,058	5,849	2,420	0	0	1,668	15	13,012
1992	261.1	480	0	2,543	170	0	0	232	1,755	34	5,215
1993	195.1	0	0	1,821	0	8	0	1,888	418	0	4,135
1994	24.7	0	0	0	0	0	0	630	673	0	1,303
1995	204.5	0	0	111	0	0	911	513	3,708	0	5,243
1996	213.5	0	0	0	0	0	416	0	5,070	0	5,486
1997	112.0	0	0	0	0	0	0	2,400	832	0	3,232
1998	201.0	0	0	0	0	0	0	5,618	360	0	5,486
1999	283.0	0	0	0	0	0	0	4,939	3,551	0	8,490
Total[4]	5,974	930	5,462	31,316	71,173	11,934	1,675	16,220	18,337	54	158,104

[1] Sprayed hectares = number of hectares treated (in thousands) x number of applications. [2] BIU applied = number of hectares treated x BIU/ha x number of applications. [3] NFL Newfoundland, NS Nova Scotia, NB New Brunswick, QUE Quebec, ONT Ontario, MA Manitoba, SAS Saskatchewan, ALB Alberta, BC British Columbia. [4] Discrepancies between totals shown and row or column totals are due to rounding errors in the main body of the table. Source: Forestry Insecticide Database, Canadian Forest Service, Great Lakes Forestry Centre

REFERENCES

[1] Battisti A, Longo S, Tiberi R & Triggiani O (1998) Results and perspectives in the use of *Bacillus thuringiensis* var. *kurstaki* and other pathogens against *Thaumetopoea pityocampa* in Italy (Lep., Thaumetopoeidae). Anz. Schädlingskde, Pflanzenschutz, Umweltschutz 71, 72-76

[2] Carter NE (1991) Efficacy of *Bacillus thuringiensis* in New Brunswick, 1988-1990, p. 113-116. *In* Preprints of the 72nd Annual Meeting, Woodlands Section, Canadian Pulp and Paper Association, Montreal

[3] Cooke BJ & Régnière J (1996) An object-oriented, proces-based stochastic simulation model of *Bacillus thuringiensis* efficacy against the spruce budworm, *Choristoneura fumiferana* (Lepidoptera: Tortricidae). Intern. J. Pest Management 42, 291-306

[4] Dreistadt SH & Dahlsten DL (1989) Gypsy moth eradication in Pacific coast states: history and evaluation. Bull. Entomol. Soc. Am. 35, 13-19

[5] Dubois NR, Baranchikov NY, Soldatov VV & Dean DH (1997) Susceptibility of the Siberian moth, *Dendrolimus superans sibericus*, to *Bacillus thuringiensis* and its toxins: laboratory and field studies. Abstract, 30th Annual meeting, Society for Invertebrate Pathology, Banff, Alberta

[6] Entwistle PE, Cory, JS Bailey MJ & Higgs S (eds.) (1993*) Bacillus thuringiensis*, an environmental biopesticide: theory and practice, John Wiley & Sons, Chichester, UK, 311pp.

[7] Fast PG & Régnière J (1984) Effect of exposure time to *Bacillus thuringiensis* on mortality and recovery of the spruce budworm (Lepidoptera: Tortricidae). Can. Entomol. 116, 123-130

[8] Fleming RA & van Frankenhuyzen K (1992) Forecasting the efficacy of operational *Bacillus thuringiensis* Berliner applications against spruce budworm, *Choristoneura fumiferana* Clemens. Can. Entomol. 124, 1101-1113

[9] Glowacka B (1996) The control of the nun moth, *Lymantria monacha* L., with the use of *Bacillus thuringiensis* in Poland. IOBC Bull. 19, 57-60

[10] Linedlöw A (1998) Insect damage in Swedish forests during 1997: an overview. Växtskyddsnotiser 62, 14-16

[11] Payne NJ & van Frankenhuyzen K (1995) Effect of spray droplet size and density on efficacy of *Bacillus thuringiensis* against the spruce budworm, *Choristoneura fumiferana*. Can. Entomol. 127, 15-23

[12] Picot JJC & Kristmanson DD (1997) Forestry pesticide aerial spraying. Spray droplet generation, dispersion and deposition. Environmental Science and Technology Library Vol. 12, Kluwer Academic Publishers, The Netherlands

[13] Reardon R, Dubois NR & McLane W (1994) *Bacillus thuringiensis* for managing gypsy moth: a review. FHM-NC-01-94. National Center of Forest Health Management, USDA Forest Service

[14] Régnière J & Cooke BJ (1998) Validation of a process-oriented model of *Bacillus thuringiensis* variety *kurstaki* efficacy against spruce budworm (Lepidoptera: Tortricidae). Environ. Entomol. 27, 801-811

[15] Sheehan KA (1996) Effects of insecticide treatments on subsequent defoliation by western spruce budworm in Oregon and Washington: 1982-1992. General technical report PNW-GTR-367, USDA Forest Service, Pacific Northwest Research Station, Portland, OR, 54 pp.

[16] Svestka M (1995) The use of biological preparations against leaf-eating pests in the forests of Czech Republic. p. 12-16. In: Biological and Integrated Forest Protection. H. Manilowski & G. Tsankov (eds.),. Forest Research Institute, Warsaw, Poland, 1994

[17] van Frankenhuyzen K (1990) Development and current status of *Bacillus thuringiensis* for control of defoliating forest insects. For. Chron. 66, 498-507

[18] van Frankenhuyzen K (1990) Effect of temperature and exposure time on toxicity of *Bacillus thuringiensis* Berliner spray deposits to spruce budworm, *Choristoneura fumiferana* Clemens (Lepidoptera: Tortricidae). Can. Entomol. 122, 69-75

[19] van Frankenhuyzen, K (1993) The challenge of *Bacillus thuringiensis.* p. 1-35. *In* Entwistle PE, Cory, JS Bailey MJ & Higgs S (eds.), *Bacillus thuringiensis*, an environmental biopesticide: theory and practice, John Wiley & Sons, Chichester, UK

[20] van Frankenhuyzen K & Nystrom CW (1987) Effect of temperature on mortality and recovery of spruce budworm exposed to *Bacillus thuringiensis* Berliner. Can. Entomol. 119, 941-945

[21] van Frankenhuyzen K & Nystrom CW (1989) Residual toxicity of a high-potency formulation of *Bacillus thuringiensis* to spruce budworm (Lepidoptera: Tortricidae). J. Econ. Entomol. 82, 868-872

[22] van Frankenhuyzen K & Payne NJ (1993) Theoretical optimization of *Bacillus thuringiensis* Berliner for the control of eastern spruce budworm, *Choristoneura fumiferana* Clem. (Lepidoptera: Tortricidae): estimates of lethal and sublethal dose requirements, product potency, and effective droplet sizes. Can. Ent. 125, 473-478

[23] van Frankenhuyzen K, Payne NJ, Cadogan L, Mickle B & Robinson A (1996) Effect of droplet size spectrum and application rate on field efficacy of *Bacillus thuringiensis*. Spray Efficacy Research Group report 1995/02/final, 38 pp.

[24] van Frankenhuyzen K, Gringorten L, Dedes J & Gauthier DG (1997) Susceptibility of different instars of the spruce budworm (Lepidoptera: Tortricidae) to *Bacillus thuringiensis* var. *kurstaki* estimated with a droplet-feeding method. J. Econ. Entomol. 90, 560-565

[25] Weiser J (1986) Impact of *Bacillus thuringiensis* on applied entomology in eastern Europe and in the Soviet Union. Mitteilungen aus der Biologischen Bundesanstalt für Land- und Forstwirtschaft (Berlin-Dahlem), Heft 223, 37-49

Chapter 6.3

Bacterial control of vector-mosquitoes and black flies

Norbert Becker

German Mosquito Control Association, Ludwigstrasse 99, 67165 Waldsee, Germany

Key words: mosquitoes, black flies, bacterial control

Abstract: For centuries man has attempted to control mosquitoes in order to protect himself from mosquito borne diseases and annoyance. The control of mosquitoes using microbial control agents like *Bacillus thuringiensis* subsp. *israelensis* (*Bti*) and *Bacillus sphaericus* (*Bsp*) offer man a means to serve all of his interests, protect himself from mosquitoes and protect nature reserves by taking into account the requirements of modern environmental protection. After careful screening, several hundred tons of *Bti* and *Bsp* are now used annually world-wide in mosquito control campaigns without evidence of any harmful impact on the environment. This chapter will deal with application histories from various parts of the world, with a specific example from Upper Rhine Valley, Germany. There, *Bti* and *Bsp* are applied by KABS (a voluntary mosquito control organisation) using in a combined, mosaic-like integrated biological control strategy. There are considerations of ecological impact as well as close collaboration with the environmental protection authorities. An efficient reduction of more than 90% of the mosquito population was achieved against dominant species like *Aedes vexans* and *Culex pipiens pipiens*. Approximately 300 river kilometres and 600 km^2 inundation areas are treated annually with various formulations of *Bti* (tablets, wettable powders, water dispersible granules, fluid concentrates, sand granules, ice granules and corn cobs). *Bti* is also extensively used for mosquito control in the USA when larvicides are required. Examples from China, Thailand and other tropical regions are also described. In these areas, *Bti* and *Bsp* are used to control important mosquito vectors including *Anopheles sinensis*, *Aedes albopictus*, *Aedes aegypti* and *Culex pipiens quinquefasciatus* resulting in a substantial reduction of cases of malaria, lymphatic filariasis and dengue. In the Onchocerciasis Control Programme (OCP) in West Africa, various insecticides are used against the larval stages of black flies (*Simulium damnosum s.l.*), the vector of *Onchocerca volvulus*, which is the agent of onchocerciasis. More than 18,000 km in the OCP area are treated regularly. Due to the resistance to chemical insecticides several hundred thousands

J.-F. Charles et al. (eds.),
Entomopathogenic Bacteria: From Laboratory to Field Application, 383–398.
© 2000 *Kluwer Academic Publishers. Printed in the Netherlands.*

litres of *Bti* are applied annually against black fly larvae. This has resulted
resulting in a rapid decrease of the prevalence of this infection among the
people protected by the OCP.

1.   INTRODUCTION

The discovery of mosquitocidal bacilli such as *Bacillus thuringiensis
israelensis* (*Bti*) in 1976 and potent strains of *Bacillus sphaericus* in
recent years inaugurated a new chapter in the control of mosquitoes and
blackflies. These bacilli produce protein toxins during sporulation which
are highly toxic to mosquito larvae and in the case of *Bti* also against
blackfly larvae. The special properties of these bacilli such as
environmental safety, relative ease of mass production, formulation and
application, the stability of proper formulations as well as the suitability
for integrated control programs based on community participation and
the relative low costs for development and registration were the reasons
for the fast development and utilisation of these bacilli in many mosquito
and blackfly control programs.

The exceptional specificity and thus the environmental safety of the
bacterial control agents were confirmed in numerous tests. Beside plants
and Mammals none of the tested taxa such as Cnidaria, Turbellaria,
Rotatoria, Mollusca, Annelids, Acari, Crustacea, Ephemeroptera,
Odonata, Heteroptera, Coleoptera, Trichoptera, Pisces and Amphibians
appeared to be affected when exposed in water containing large amounts
of bacterial preparations [5]. Even within the dipterans, the toxicity of
Bti is restricted to mosquitoes and to a few nematocerous families. In
addition to larval mosquitoes and blackflies, only those of the closely
related dixids are similarly sensitive to *Bti*. Larval psychodids,
chironomids, sciarids, and tipulids generally are far less sensitive than
mosquitoes or blackflies. In contrary to *Bti*, the toxins of *B. sphaericus*
are toxic to a much narrower range of insects. Certain mosquito species,
such as *Culex pipiens quinquefasciatus* and *Anopheles gambiae* are
highly susceptible whereas *Aedes aegypti* larvae are more than 100-fold
less susceptible. Blackfly larvae and other insects (except Psychodidae),
mammals, and other non-target organisms are not susceptible to *B.
sphaericus*.

The high potential of *B. sphaericus* as a bacterial control agent lies in
its spectrum of efficacy and its ability to recycle or to persist in the
larval breeding sites under certain conditions. Which means that a long-
term control can be achieved and time-span between re-treatments could

be extended and personnel costs reduced. These abilities open up the possibility of the successful and cost-effective control of *Culex* species, particularly of *C. quinquefasciatus* the most important vector of lymphatic filariasis and which breeds primarily in highly polluted waterbeds in urban areas.

The insecticidal effect of the *Bti* crystal emanates from 4 major toxin proteins which are referred to as Cry4Aa, Cry4Ba, Cry11Aa and the non specific Cyt1Aa, with molecular weight of 135, 128, 72 and 27 kDa respectively. These toxins act by synergism, none of them individually impose full larval toxicity. It is assumed that this synergistic effects also reduce the likelihood of resistance development (see Chapter 2.3). The parasporal crystal toxin of *B. sphaericus* is a binary toxin, consisting of two polypeptides, 51.4 kDa and 41.9 kDa. Both are required for mosquitocidal activity indicating a high degree of synergism too. While there is no clear evidence for, that *Bti* susceptible Dipteran larvae possesses specific receptors for binding of the *Bti* toxins, as it has been shown for most lepidoteran active Cry-toxins, it has been shown that the binary toxin of *B. sphaericus* have specific receptors in both *Culex* and *Anopheles* [16, 19]. On the contrary to *Bti*, to which no reports of appearance of field resistance is observed for *B. sphaericus* there are several cases ([17, 20]; see also Chapter 6.4).

A basic requirement for the successful use of bacterial control agents was the development of effective formulations suited to the biology and habitats of the target organisms (Table 1). *Bti* preparations can be obtained as wettable powders, fluid concentrates, granules, pellets, tablets, or briquettes. The development of suitable formulations based on *B. sphaericus* are also in good progress, as shown Table 1. A few hundred grams of powder or even less, a half of a litre to two litres of liquid concentrate or a few kilograms of granules per hectare, are usually enough to kill all mosquito larvae. In some situations a long-term effect can be achieved if higher amounts are used. In recent years, with the production of tablet, briquette or pellet formulations, progress has been made to obtain a long-term effect. Sustained-release floating granules are being developed.

New tablet formulations such as Culinex® tablets based on *Bti* material sterilised by γ-radiation to prevent contamination of drinking water spores can successfully be used for control of container breeding mosquitoes such as *A. aegypti* as the main vector of dengue. Tablets or briquettes based on *B. sphaericus* or *Bti* are very effective to control *Culex* sp. larvae close to human settlements.

Bacterial control agents have a considerable safety advantage over synthetic insecticides :

1. Neither the operator nor the occupants of treated sites become exposed to potentially dangerous chemicals. For this reason, such preparations are particularly well suited for use by volunteers.
2. Applications of bacterial control agents do not harm beneficial animals such as fish, crustaceans or predacious insects. After their application the predator can continue to feed upon newly hatching mosquito larvae.
3. They are biodegradable, no toxic residues remain after their use. Their environmental safety permits bacterial control agents to be accepted by both public official and the general public. Thus not only ecological but also economic advantages are achieved.

Table 1. Different *Bacillus thuringiensis israelensis* and *Bacillus sphaericus* products used in mosquito and blackfly* control programmes

Product	Formulation	Potency (ITU/mg)**
B. thuringiensis israelensis		
Aquabac[®1]	Primary Powder	7,000
Bactimos WP[®2]	Powder	5,000
Bactimos PP[®2]	Primary Powder	10,000
Bactimos G[®2]	Granules	200
Teknar HP-D[®3]	Fluid concentrate	1,200
Teknar TC[®3]	Technical Powder	>10,000
Teknar G[®3]	Granules	200
Vectobac 12AS[®2]	Fluid Concentrate	1,200
Vectobac TP[®2]	Technical Powder	5,000
Vectobac WDG[®2]	Water dispersible powder	3,000
Bactecide[®4]	Water dispersible powder	?
Culinex[®5]	Tablets	8,000
Biotouch[®6]	Fluid Concentrate	1,000
Bacillus sphaericus		
Spherimos[®2]	Fluid concentrate	120
Vectolex[®2]	Concentrate granules	50
Spherico[®7]	Fluid concentrate	1,700

* In blackfly control *Bti* only. ** The product potency is tittered against the international standard IPS82 on *Aedes aegypti* and for *B. sphaericus* with standard SPH88 on *Culex pipiens*.
[1]Becker Microbial Products, USA; [2]Abbott (recently Valent BioScience Corp., USA); [3]Thermo Trilogy, U.SA; [4]Biotech International Ltd, India,; [5]Culinex (Gmbh),Germany; [6]Zohar Dalia, Israel; [7]Geratec, Brazil

Only a few years after its discovery *Bti* was being used on a large-scale for mosquito control. More than 200 tonnes of *Bti* are now used annually world-wide. In Europe alone more than 50 tonnes of various *Bti* formulations are applied in control programmes in France, Germany, Hungary, Italy, Russia, Spain, Switzerland, Slovenia and Yugoslavia, without evidence of any harmful impact on the environment. In the USA *Bti* has become one of the leading mosquito larvicides. In parts of Hubei Province in China the application of *Bti* at two-week intervals has succeeded in reducing the number of malaria cases by more than 90 % [22].

2. MOSQUITOES

2.1 Mosquito control in Germany

In the Upper Rhine valley, mosquitoes can be a major public nuisance. The dominant species include the floodwater mosquitoes *Aedes vexans, A. sticticus* and *A. rossicus* [2].

C. pipiens is the main mosquito pest in houses in the Rhine Valley. Their breeding sites usually consist of rainwater containers near houses and of other temporary water bodies occurring in summer. In reaction to this nuisance, about 100 towns and villages on both sides of the Rhine River joined to form a voluntary mosquito control organisation (KABS). The territory of this organisation now covers approximately 300 river kilometres with about 600 km^2 of inundation area [3].

The main goal of this organisation is to reduce the abundance of mosquitoes to a tolerable level without damaging ecologically sensitive riverbank areas. This has been achieved through the widespread use of *Bti*.

On the basis of precise mapping of all mosquito breeding sites, a mosquito control strategy for each individual area is worked out together with the environmental protection authorities. By allowing the ecological conditions to direct the choice of the application technique used, any disturbance of wild-life and the natural flora can be minimised..

The application of *Bti* in a combined, mosaic-like pattern, operating in accordance with ecological considerations and under the supervision of biologists allows an effective as well as environmentally non-intrusive mosquito control operation in protected areas and nature reserves. There are usually about 300 well-trained people active in the KABS, supported by two helicopters. The annual budget amounts to approximately US $ 1.5 million. Depending on the number of breeding sites, there are two to eight people responsible for mosquito control in every community.

An integrated control program, applying different but compatible methods, reduces the development of resistance and permits more specific control. Wherever feasible the program encourages engineering methods for reducing the productivity of breeding sites. The protection and encouragement of all the natural predators of mosquito larvae is very important. As the predators are not affected by the microbial agents, they can continue to feed upon newly hatching mosquito larvae long after the breeding sites have been treated. This also has economical consequences, since in most cases no further treatments are then necessary. Microbiological methods for reducing the abundance of nuisance mosquitoes provide the main intervention measures used in routine control measures. Today the mosquitoes in more than 90% of the area are controlled by the KABS exclusively with *Bti*. *B. sphaericus* preparations are increasingly being used against *C. p. molestus*. Various available commercial *Bti* products were tested in the laboratory and in the field in order to determine the most effective dosage of each product for routine treatments. As a result of these tests, in routine treatments against first and second instar larvae or in shallow breeding sites, 250 g of *Bti* powder (activity: 5,000 International Toxic Units (ITUs) per mg) are mixed with 10 litres of screen-filtered pond water for each hectare treated. The mixture is applied by field workers with a high-pressure knap-sack sprayer. In deeper breeding sites or when third of early fourth instar larvae are present, 500 g of *Bti* wettable powder (WP) are used. This corresponds to 1.25×10^9 ITUs and 2.5×10^9 AA (*A. aegypti*) ITUs per hectare respectively. To achieve a sufficiently high mortality rate under field conditions, one litre of liquid *Bti* concentrate (Vectobac 12AS® or Teknar HP-D®) is mixed with 9 litres of screen-filtered pond water per hectare (1.2×10^9 ITUs per hectare).

When high water levels on the Rhine cause widespread inundation or when dense vegetation occurs, *Bti* ice, sand, or corn cob granules are applied by helicopters equipped with a granule-sprayer. Two natural compounds (water and *Bti*) are formulated to Icybac granules for cost-effective and environmentally friendly mosquito control. A special ice machine is capable of quick transforming the *Bti*-water suspension into ice granules. Ice granules are ready to use and easy to apply. Melting ice granules release *Bti* in the feeding zone of mosquito larvae, allowing a cost-effective control operation. Ice granules can be easily applied in densely vegetated areas by helicopters even when it is raining or at wind with moderate speeds. The special advantages of the new formulation are: melting on the water surface-release of the control agent in the feeding zone of mosquito larvae; no loss of material by friction; excellent penetration of the vegetation even when the leaves are wet; enhanced swath of application; saving money by enhanced efficacy and reduced consumption of active material per acre.

Between 1981 and 1999 about 170,000 hectares of mosquito breeding sites were successfully treated with about 55 tons of *Bti* wettable powder and 30,000 litres of liquid concentrate. The wettable powder was also used to produce approximately 1,000 tons of *Bti* granules. Interventions against *C. p. molestus* near houses are based on the provision of information to the general public on the biology of these mosquitoes and on strategies for their control. People are asked to destroy all unnecessary water bodies near their homes, to empty rain water containers at least once a week, or to cover them thoroughly. Fish can be used as predators in larger water bodies, or *Bti* and *B. sphaericus* tablets applied in rainwater containers, allowing efficient control for several weeks. Since 1992 several millions of fizzy tablet based on *Bti* and registered as Culinex® tablets have been successfully used against *C. pipiens*. The *Bti* material used for the production of the tablets is hygienically pure and free of bacteria or spores. Therefore, contamination of drinking water is avoided.

As a result of these combined measures, a monitoring system with CDC-light traps documented that mosquito abundance in the Upper Rhine Valley has been reduced each year by over 90%. Similar results against nuisance mosquitoes were achieved in other parts of the world for instance in many mosquito control programs in the United States of America or in the Magadino plain (in Southern Switzerland), where, floodwater mosquitoes such as *Aedes vexans* are successfully controlled by *Bti* (P. Lüthy, personal communication).

2.1.1 Monitoring the environmental impact

It has been essential to document the environmental impact of *Bti* application in order to provide a scientific basis for rebutting the arguments commonly brought against mosquito control by its opponents. Before large-scale application of the *Bti* method was undertaken, the most important members of various aquatic groups (Cnidaria to Amphibia) were screened in the laboratory and in small-scale field trials for their susceptibility to *Bti*. This work showed that in addition to mosquitoes and blackflies only a few species of midges were affected by *Bti*. For the most part these midges were much less susceptible to *Bti* than the target organisms [23].

2.1.2   Monitoring the direct impact of *Bti* treatments

The development of insects in treated and untreated water is continuously monitored using emergence traps (photo eclectors). The occurrence of insects in treated areas is assessed by regular light trap catches. All investigations have shown that the numbers of *Aedes* mosquitoes are drastically reduced but that all other insects continue to develop in the water and, as winged adults, provide a food resource for birds, amphibians and bats.

2.1.3   Monitoring the indirect impact of *Bti* treatments

The effect on the food-chain of a reduction in the number of mosquitoes was also studied. To this end, the food of birds (e.g. *Delichon urbica* and *Acrocephalus scirpaceus*), Amphibia (e.g. *Rana* spp., *Bufo* spp. and *Hyla arborea*) and the niche utilisation and feeding preferences of bats (e.g. *Myotis daubentoni*) were determined. All investigations have shown that *Aedes* mosquitoes form no part, or only a very minor part, of the food chain .

2.1.4   Resistance

Mosquito populations are checked at regular intervals for the development of resistance. No resistance has been detected after more than 10 years of treatment with *Bti* [4]. It could be shown that due to the more simple mode of action of *B. sphaericus* even in the field resistance against *B. sphaericus* may occur. To prevent resistance to *B. sphaericus* developing in *Culex*, *B. sphaericus* and *Bti* are used alternately in the control management plan for this species.

All the studies carried out to date have shown that the introduction of *Bti* and *B. sphaericus* has reduced the numbers of nuisance mosquitoes to a tolerable level, but that the diversity and beauty of the ecosystem as a whole has not been damaged.

2.2   Control of vector-mosquitoes

In Hubei Province (People's Republic of China) located along the Yangtze River in the subtropical monsoon region of China the high groundwater level and the widely distributed drainage system, provide not only optimal conditions for growing rice but also ideal breeding conditions for mosquitoes. The 69 mosquito species known to date include major malaria vector species, such as *Anopheles sinensis* and

Anopheles anthropophagus. In the recent past, more than 20 million people living on both sides of the Yangtze River are threatened by the malaria agent *Plasmodium vivax*. During 1985 about 82,000 cases of malaria were reported, which corresponds to an annual incidence of about 168 cases/100,000 people. The control of *A. sinensis* with insecticides has recently become increasingly difficult, due to developing resistance and ecological and toxicological risks. Other diseases transmitted by mosquitoes in Hubei include Japanese B encephalitis (transmitted by *C. tritaeniorhynchus*), as well as Bancroftian filariasis (transmitted by *C. p. quinquefasciatus*). Dengue is transmitted by *A. albopictus*, but only few cases having been reported.

For more than 30 years, chemical insecticides for adult and larval control have allowed relatively simple control of almost all vector species in Hubei. Some 40 tons of DDT are now used annually for residual applications and about 3 tons of deltamethrin for bednet impregnation. Since the development of resistance, the high costs, and the environmental risks of chemicals have caused these to become unacceptable. The "Institute of Parasitic Diseases" started very early to investigate microbial methods for the control of mosquito-borne diseases. Various local strains of *Bt* and *B. sphaericus* were tested in the laboratory against *C. p. quinquefasciatus, A. albopictus* and *A. sinensis*.

While the *Bt* preparations were effective against larvae of *A. albopictus* they were less active against larvae of *C. p. quinquefasciatus*. On the other hand, a relatively low dosage of *B. sphaericus* killed *Culex* larvae, whereas *Aedes* larvae are less sensitive. Relative high concentrations of both microbial control agents were necessary to obtain satisfactory control of larvae of *A. sinensis*.

In routine treatments, fluid formulations are applied using a high-pressure sprayer attached to a 600-litre tank pulled by a minitractor. The fluid formulation of either *Bt* (strain 187) or *B. sphaericus* (strain C3-41) are usually applied, to give a dosage of 3–5 ppm. According to experience in the field, the treatments were necessary every seven days during summer or every ten days during spring and autumn, depending on the temperature. In the last few years, about 10 tons of *Bt* 187 and 14 tons for *B. sphaericus* C3-41 have been produced each year in Hubei Province by using natural resources, which was enough to treat about 12.000 hectares of mosquito breeding sites. One litre of *Bt* 187 with a potency of 400 ITUs/mg costs about one US-dollar, and of *B. sphaericus* C3-41 with a potency of 270 ITUs/mg US $ 1.20.

The impact of the treatments was recorded by measuring both the density of adult mosquitoes and the incidence of malaria before and after the campaign. Both mosquito population density and malaria incidence were reduced by more than 90%.

In large field tests in tropical areas against *A. albimanus, A. rangeli, A. nigerrimus* and *A. sundaicus* excellent results were achieved when 1–2 kg of *Bti* wettable powder (potency: >5,000 ITUs/mg) were applied in weekly intervals.

2.2.1   Control of dengue vectors

Up until now in most dengue epidemic countries, beside environmental sanitation the organophosphate insecticide temephos has been the most commonly used chemical for mosquito control in water containers. However, many people do not like the rotten egg-like odour of this chemical in water used for consumption and the household. Furthermore, resistance against temephos could be problem.

For these reasons, there was the need to develop new approaches for controlling the vector of dengue haemorrhagic fever. One of the most promising alternatives to temephos are the Culinex®-*Bti* tablets, which has been used in Germany for many years with excellent results [1].

The *Bti* tablets are especially suited for the control of *A. aegypti* because

1. they are safe for humans and the environment. This is a considerable advantage over many traditional insecticides as neither the operator nor the house occupier is exposed to danger. For this reason, *Bti* tablets are particular suitable for programmes based on Community Participation and the use by volunteers, such as house occupiers.
2. application is simple. No precautionary measures are necessary. The calculation of the effective dosage is easy. One or two tablets are usually enough to treat a container with 50-100 litres of water.
3. *Bti* tablets are particular suitable components for integrated control programmes. Because of their selective effect they do not kill mosquito predators, such as fish or copepods which can be introduced in the containers inhabited by larvae of *A. aegypti*.
4. *Bti* tablets can be easily distributed. They can be sold on the market, since this formulation is accepted by the users.

The new fizzy tablet formulation was tested against *A. aegypti* in Jakarta, Indonesia and in Cucuta, Colombia. A long-term effect of about 30 days could be achieved when 1–2 tablets per container were applied. The higher the dosage the longer the long-term effect. One tablet per 50 litres provided control for about half a month. The tablets are not only effective but also well accepted by the users. The interest in using *Bti*-tablets in urban areas in the tropics is already growing.

In Brazil, significant outbreaks of both dengue and recently yellow fewer, transmitted by *Aedes aegypti*, have required intensive control efforts, mainly by chemicals. Recently, resistant *A. aegypti* populations were detected (Rio de Janeiro, Sao Paulo) and it is now recommended to treat those areas with resistance using *Bti* as granular formulations, *e.g.* corncobs, as these are more suitable to large scale field operations. The field rates defined are 1 gram of granules for each 50 litres of volume up to 5 grams for 250 litres. In volumes larger than 250 litres, the rates are 5 grams per square meter surface (P.T.R. Vilarinhos, personnal communication).

2.2.2 Control of lymphatic filariasis vectors

In recent large-scale field trials against *C. quinquefasciatus* (main vector of *Wucheria bancrofti*), in north Cameroon, Brazil, India, Sri Lanka and Tanzania under the auspices of WHO/TDR a remarkable impact of *B. sphaericus* use has been observed in reducing vector biting density by 80% through bi-monthly treatment of mosquito larval habitats; in addition there was a significant decline in the proportion of *Culex* carrying filarial infective larvae.

The design of the intervention strategy was based on:
1. Sufficient information on the climatic conditions such as occurrence of rainy and dry seasons which influence the mosquito densities and the efficacy of the treatment (e.g. the control agent can be flashed during heavy rainfalls).
2. Precise knowledge on the fluctuation of the mosquito population. Based on this knowledge the timing for the treatment has to be determined in order to reduce the mosquito population at that time, when the population is most sensible in its development to avoid the maximum or at the beginning of a low level of development to reduce the population to a very low level. Thus the prerequisites for the development of the following mosquito populations are drastically reduced.
3. The efficacy and long-term effect of the product in various breeding types. The long-term effect can vary from breeding type to breeding type. Therefore the sequence or re-treatments have to be adapted to the local situation e.g. shaded, stagnant water bodies have to be treated bi-monthly or less but drains with flowing water weekly or bi-weekly. The phases of transmission of the parasite have to also be considered in order to reduce the mosquito population especially when transmission occurs and to such a level that transmission is avoided. In most projects *B. sphaericus* fluid concentrate was used at very high

dosages of 5–10 g/m^2 in cess pits/pools in intervals of 3–6 month.
Drainage systems with slowly flowing water were treated with a dosage
of 2 g/m^2 at least once a month.

In these tests *B. sphaericus* has proven to be an effective and
selective mosquito control agent for use against *C. quinquefasciatus* in
integrated control programmes. In France, Spain and Germany *B.
sphaericus* is used by tons against *C. pipiens* with great success. However,
it seems that resistance to *B. sphaericus* is more likely than against *Bti*.
In 1994 signs of resistance against *B. sphaericus* were observed in
Southern France in an isolated *Culex* population which was treated with
B. sphaericus for 7 years with a total of 18 treatments ([20]; see
Chapter 6.5). Therefore, resistance management by rotation of
insecticides e.g. *Bti* and *B. sphaericus* is recommended.

3.   BLACKFLIES

3.1   The Onchocerciasis Control Programme

In the Onchocerciasis Control Programme (OCP) in West Africa, the
organophosphate temephos has been used exclusively since 1975 against
the larval stages of *Simulium damnosum s.l.*, the vector of *Onchocerca
volvulus*, the agent of onchocerciasis. During the wet season, a 20%
emulsion in a dose of 150 cc of formulation per m^3/sec and during the dry
season in a dose of 300 cc per m^3/sec was applied [13, 21].

The program now operates in 11 West African countries, protecting
2-3 million people living along a river network that totals 50,000 river
km. At the end of the 80's, about 18,000 km of this system was treated
regularly, requiring about 10,000 hours of flight time and about 800,000
litres of insecticides per year at a cost of some $US 29 millions [12]. The
program provides a model for successful international co-operation under
the guidance of the UN. Based on concerted efforts by four sponsoring
international agencies (UNDP, FAO, World Bank and WHO), an
important step toward improved socio-economic development in the
countries involved is going on [18].

By 1979, in some parts of the OCP, the first signs of resistance of *S.
damnosum s.l.* against temephos began to appear [7, 11]. Therefore, in
1981, experiments with *Bti* (Teknar$^®$) were conducted under the guidance
of the WHO in the areas most affected by this developing resistance.
After early difficulty and consequent improvement in the potency of the
Bti product used, satisfactory results were obtained. Large scale

application of *Bti* proved effective at application rates of 1,200 cc of formulation per m^3/sec.

In order to avoid additional increases in resistance, an "integrated" OCP effort was developed by about 1985. In those regions where no resistance had developed, temephos continues to be used. Where resistance had occurred, *Bti* was to be applied where the river discharge rate is less than 15–70 m^3/sec. Above this level, the strategy was to alternate the use of insecticides such as chlorphoxim, permethrin, carbosulfan and later phoxim, pyraclofos and etofenprox [9, 12].

As a result of this strategy, *Bti* (Teknar HP-D® and Vectobac 12AS®) is now used at lower dosage (500 to 720 cc per m^3/sec) in increasing amounts in weekly intervals at low discharges, mainly during the dry season. Whereas only 8,000 litres of *Bti* during 1981 were applied out of a total of 222,000 litres of insecticides (about 3.6%), about 750,000 litres from a total of 923,000 litres of insecticides were applied in 1988, resulting in 81% of the region being protected by *Bti* [6, 12]. Thus, thousands of kilometres of river in 11 countries in West Africa have come to be treated with more than 700,000 litres of *Bti* liquid concentrate every year, now reduced to 200,000 litres with the reduction of the larvicide coverage.

The substitution of temephos by *Bti* presents certain technical difficulties [8]. *Bti* sprays must be applied more carefully, covering the entire width of the river, rather than at a single point, as in the case of synthetic chemicals. Non-corrosive tanks and nozzles must be used. In addition, the presence of alga at a density of 1,500 to 3,000 cells per ml may require a doubling of the operational dose of *Bti*.

Since the inception of the OCP, few new cases of infection with *Onchocerca volvulus* have been recorded. Prevalence of this infection among the people protected by the OCP is rapidly decreasing. The nine million children born within the OCP area since operations began are free from onchocercal infection. By the year 2000, or even before, the number of children thus protected will have grown to 15 million. The very fertile low-lying valleys can now once again be used by the human population, permitting enhanced socio-economic development of the region.

3.2 Use of *Bti* against blackflies in temperate climates

In temperate parts of the world, where no blackflies transmit pathogens that affect human health, these insects may be exceedingly annoying [14]. Cattle are also severely affected. Temperate regions subject to these problems generally are ecologically sensitive, as in the case of mountain resorts.

Certain peculiar problems affect the use of *Bti* in mountain streams. The quantity of *Bti* to be applied varies with: discharge rate of the water, profile of the stream, turbidity, presence of pollutants, water temperature, pH, degree of vertical mixing, settling due to presence of pools and characteristics of the substrate. Although the ideal particle size seems to be about 35 μm, the rate of application varies several fold, depending on these listed characteristics [15]. Larger particles settle faster than smaller particles. This provides liquid formulations with a distinct advantage over powders, the small particles in a liquid formulation carry better, a crucial factor in the treatment of fast-moving streams. Some 5-30 ppm of *Bti* per minute generally provide satisfactory control of larval blackflies over a span of 50-250 m in moderate sized streams [10].

The non-target effects of *Bti* in these sites are minimal. Of the wide variety of organisms present, only filter-feeding chironomids are sensitive to *Bti*, but at very high rates of application. Such midges survive when the rate of application was 17 times greater than normal. *Bti* is the sole larvicide that can be used in such sensitive ecosystems.

4.   FUTURE PROSPECTS

Bacteriological control agents have been established as a commercially viable and promising alternative to conventional pesticides. Their exceptional qualities such as high efficacy and the outstanding environmental safety have made the extraordinary development of appropriate formulations an effective and economic control of mosquitoes and blackflies is now possible in the majority of cases. As a result, interest in these methods is increasing world-wide year by year. Hundreds of tonnes of *Bti* and *B. sphaericus* preparations are now being used annually, with success and with increasing interest. So far there have been no cases of negative effects on the various ecosystems. Microbiological methods offer an ecologically defensible compromise between the desire of man to protect himself from troublesome mosquitoes or blackflies and requirements of contemporary environmental protection policies not to damage highly sensitive ecosystems. *Bti* and *B. sphaericus* may also be promising agents in the battle against dangerous diseases such as malaria, filariasis, and Arbovirus diseases. Used in suitable formulations, these microbial agents are useful supplements to or replacements for broad-spectrum chemicals. Further improvements of these microbial preparations, particularly to extent their long-term effect and, for example, to enhance *Anopheles* control thereby, will accelerate this process still further.

ACKNOWLEDGEMENTS

I am especially indebted to the German Mosquito Control Association (KABS/GFS), the UNDP/World Bank/WHO Special Program for Research and Training in Tropical Diseases (TDR). The co-operation of the Scientific board of the KABS/GFS and the following persons is gratefully recognised: Dr. Paul Schädler, Professor Dr. Yoel Margalith, Professor Dr. Herbert W. Ludwig and Dr. Wolfgang Schnetter.

REFERENCES

[1] Becker N, Djakaria S, Kaiser A, Zulhasril O & Ludwig H (1991) Efficacy of a new tablet formulation of an asporogenous strain of *Bacillus thuringiensis israelensis* against larvae of *Aedes aegypt*i. Bull. Soc. Vector Ecol. 16, 176-182

[2] Becker N & Ludwig H (1981) Untersuchungen zur Faunistik und Ökologie der Stechmücken (Culicinae) und ihrer Pathogene im Oberrheingebiet. Mitt. dtsch. Ges. allg. angew. Ent. 2, 186-194

[3] Becker N & Ludwig M (1983) Mosquito Control in West Germany. Bull. Soc. Vector Ecol. 8, 85-93

[4] Becker N & Ludwig M (1993) Investigations on possible resistance in *Aedes vexans* field populations after a 10-year application of *Bacillus thuringiensis israelensis*. J. Am. Mosq. Contr. Assoc. 9, 221-224

[5] Becker N & Margalit J (1993) Use of *Bacillus thuringiensis israelensis* against mosquitoes and blackflies, p. 147-170. *In* Entwistle P, Cory J, Bailey MJ & Higgs S (ed.), *Bacillus thuringiensis*: an environmental biopesticide, John Wiley & Sons, Ltd

[6] Davidson EW (1990) Microbial control of vector insects, p. 199-212. *In* Baker RR Dunn PE (ed.), New directions in biological control: Alternatives for suppressing agricultural pests and diseases. Proceedings of a UCLA colloquium, AR Liss, New York

[7] Guillet P, Escaffre H, Quedrago M & Quillévéré D (1980) Mise en évidence d'une résistance au temephos dans le complexe *Simulium damnosum* (*S. sanctipauli* et *S. soubrense*) en Côte d'Ivoire. Cah. ORSTOM, sér. Ent. Med. Parasit. 18, 291-298

[8] Guillet P, Kurtak DC, Philippon B & Meyer R (1990) Use of *Bacillus thuringiensis israelensis* for Onchocerciasis Control in West Africa, p. 187-201. *In* de Barjac H & Sutherland DJ (ed.), Bacterial Control of Mosquitoes and Blackflies, Rutgers University Press

[9] Hougard J-M, Poudiougo P, Guillet P *et al.* (1993) Criteria for the selection of larvicides by the Onchocerciasis Control Programme in West Africa. Ann. Trop. Med. Parasitol. 87, 435-442

[10] Knutti HJ & Beck WR (1987) The control of black fly larvae with Teknar, p. 409-418. *In* W KKCMR (ed.), Black fly Ecology, population management, an annotated world list., University Park, Penn

[11] Kurtak D (1986) Insecticide resistance in the Onchocerciasis Control Program. Parasitology Today 2, 20-21

[12] Kurtak D, Back C, Chalifour A *et al.* (1989) Impact of *B.t.i.* on black-fly control in the Onchocerciasis Control Program in West Africa. Israel J. Entomol. 23, 21-38

[13] Lévêque C, Fairhurst CP, Abban K *et al.* (1988) Onchocerciasis control programme in West Africa: ten years monitoring of fish populations. Chemosphere 17, 421-440

[14] Molloy D (1990) Progress in the biological control of blackflies with *Bacillus thuringiensis* with emphasis on temperate climates, p. 161-186. *In* de Barjac H & Sutherland DJ (ed.), Bacterial Control of Mosquitoes and Blackflies, Rutgers University Press

[15] Molloy D, Wraight SP, Kaplan B, Gerardi J & Peterson P (1984) Laboratory evaluation of commercial formulations of *Bacillus thuringiensis* var *israelensis* against mosquito and black fly larvae. J. Agric. Entomol. 1, 161-168

[16] Nielsen-LeRoux C & Charles J-F (1992) Binding of *Bacillus sphaericus* binary toxin to a specific receptor on midgut brush-border membranes from mosquito larvae. Eur. J. Biochem. 210, 585-590

[17] Rao DR, Mani TR, Rajendran R, Joseph AS & Gajanana A (1995) Development of high level resistance to *Bacillus sphaericus* in a field population of *Culex quinquefasciatus* from Kochi, India. J. Amer. Mosq. Control. Assoc. 11, 1-5

[18] Samba (1994) The onchocerciasis Control Programme in West Africa: an example of effective public health management. World Health Organization, Public Health in Action 1, 107 pp.

[19] Silva-Filha M-H, Nielsen-LeRoux C & Charles J-F (1997) Binding kenetics of *Bacillus sphaericus* binary toxin to midgut brush border membranes of *Anopheles* and *Culex* spp. mosquito larvae. Eur. J. Biochem. 247, 754-761

[20] Sinègre G, Babinot M, Quermel JM & Gaven B (1994) First field occurence of *Culex pipiens* resistnce to *Bacillus sphaericus* in southern France. *In* VIII European Meeting of Society for Vector Ecology, 5-8 September 1994, Faculty of Biologia, University of Barcelona, Spain p. 17

[21] Walsch JF, Davies JB & Cliff B (1981) World Health Organization Onchocerciasis Control Programme in the Volta River Bassin, p. 85-103. *In* Laird M (ed.), Blackflies, the future for biological control methods in integrated control, Academic Press, Canada

[22] Xu B, N. Becker, X. Xianqi and H. Ludwig. 1992. (1992) Microbial control of malaria vectors in Hubei Province, P.R.-China. Bull. Soc. Vector Ecol. 17, 140-149

[23] Yiallouros M, Storch V & Becker N (1999) Impact of *Bacillus thuringiensis* var *israelensis* on larvae of *Chironomus thumim thummi* and *Psectrocladius psilopterus* (Diptera: Chrironomidae). J. Invertebr. Pathol. 74, 39-47

Chapter 6.4

Resistance management for agricultural pests

Richard T. Roush

Department of Applied and Molecular Ecology, Waite Campus, PMB1, University of Adelaide, Glen Osmond, South Australia, 5064, Australia

Key words: *Bacillus thuringiensis*, resistance, diamondback moth, *Plutella xylostella*, *Leptinotarsa decemlineata*

Abstract: Resistance to *Bacillus thuringiensis* sprays is already a widespread problem in one major agricultural insect pest, the diamondback moth, *Plutella xylostella*. There is no reason to think that resistance won't occur in other pests and to other entomopathogenic bacteria and their products. The primary means for slowing resistance to *Bt* sprays and similar products is to avoid their excessive use, especially by using the sprays in concert with naturally occurring predators and parasites. High doses of sprays, and mixtures of bacterial toxins with or without mixtures of chemical insecticides, are more likely to be harmful than helpful to delaying resistance. Especially given the difficulties of accurately distinguishing between resistant and susceptible insects for *Bt* resistance, meaningful resistance monitoring is likely to be difficult.

1.   INTRODUCTION

Resistance has evolved in so many species of insects, probably more than 500, and to so many insecticides [10] that most entomologists accept that resistance is inevitable for nearly every toxicant. Thus, few entomologists were surprised that resistance evolved to *Bacillus thuringiensis* (*Bt*). Resistance in the Indianmeal moth, *Plodia interpunctella* [24], may be have been a bit unusual because it was a pest of stored grain, and indeed there may be some evidence that the Indianmeal moth was predisposed to resistance, as discussed below. On the other hand, given the status of the diamondback moth (*Plutella*

399

J.-F. Charles et al. (eds.),
Entomopathogenic Bacteria: From Laboratory to Field Application, 399–417.

xylostella) as the key pest of cabbage and other cole crops and its history of resistance to other insecticides, it should have been no surprise that would be the first pest to evolve resistance to *Bt* in the field.

There is probably nothing unique about the diamondback moth except that it was the first pest against which *Bt* sprays were regularly and intensively applied. Resistance to *Bt* sprays based primarily on the Cry1A toxins of *Bt kurstaki* strains has become a serious concern for control of the diamondback moth in many tropical and subtropical regions [16, 26, 42, 44, 47], with many farmers now forced to change to alternate insecticides, including to sprays based on *Bt aizawai* (relying largely on Cry1C). Resistance is now appearing to these products in at least Hawaii and South Carolina in the USA [4, 21, Chapter 3.4 of this volume]. Laboratory experiments have also selected resistance to *Bt* toxins in several other insect species (Chapter 3.4).

The potential for resistance to transgenic crops using insecticidal δ-endotoxin genes from *Bt* has been a increasing international concern, particularly among opponents of genetic engineering. As a consequence, resistance management strategies for *Bt*-transgenic crops are now required at least in the United States (by the US EPA) and Australia (by the National Registration Authority), even though there is as yet no evidence that resistance is evolving to these crops in the field. The diamondback moth has provided ample evidence that resistance can and does evolve to *Bt* sprays. Ironically, strains of diamondback moth that evolved resistance in the field in Florida to *Bt* sprays prosper on *Bt* transgenic plants [23], a reversal of the usual concern that *Bt* crops will weaken the effectiveness of *Bt* sprays. However, there are to my knowledge no resistance management programs required or implemented for *Bt* sprays, even though several have been proposed by researchers. The challenge now faced by those who develop and sell other products that use *Bt* is to match the proactive approach taken for *Bt* transgenic crops. Otherwise, more *Bt* products are likely to suffer the fate of *Bt* sprays on the diamondback moth.

Unfortunately, it seems likely that the most effective strategy for managing most *Bacillus* and other entomopathogenic bacteria (and sprays made from them) is simply to avoid their overuse. Extravagant claims have been made for the potential of high doses and toxin mixtures, but these are not supported by conclusive experiments or theory that adequately address management of dose and residue decay and should probably be relegated to the status of myths. This is in contrast to *Bt* transgenic crops, where the application of sophisticated high dose and multiple gene (pyramiding) approaches may delay resistance enormously because of the relatively greater control of exposure that is possible with transgenic plants [31, 32, 33, 34]. On the other hand, the low persistence that is generally characteristic of *Bt* sprays is an advantage for resistance management, and one that should be exploited rather than casually discarded by unnecessarily increasing the persistence of formulations or introducing the genes into other micro-organisms that can persist in the environment. Before discussing these issues, it is useful to provide some background on the evolutionary process that produces resistance as an economic problem.

2. FACTORS THAT INFLUENCE SELECTION

The main factors that drive selection for resistance can be listed as initial resistance allele frequency and fitnesses of resistant and susceptible genotypes. Selection rate depends essentially just on the fitnesses, the factors over which we can have some influence in resistance management. Fitnesses can be partitioned into several components for ease of discussion: dominance of resistance (e.g., the relative survival of heterozygotes, which implicitly incorporates the influence of dose as described below), the proportion of the population exposed, and the costs of fitness in the absence of exposure to the toxins [29, 31, 32, 33, 34].

Most mathematical models about resistance assume that resistance is due to a single gene. Although there is some evidence that more than one gene is involved in some cases of *Bt* resistance (Chapter 3.4), a close inspection of the data even in these cases suggests that the majority of resistance is due to a single major gene [52]. As summarised by Van Rie and Ferré (see Chapter 3.4), there may be a single major gene that confers resistance to a limited group of *Bt* toxins (Cry1A and Cry1F) through reduced binding at a target site in the insect midgut [50]. Thus, single gene models probably form a good approximation for resistance to *Bt*. Perhaps more generally, no evidence has come to this reviewer's attention to suggest that resistance management recommendations would be any different if resistance was assumed to be under complex control. A resistance manager still has to reduce the fitness advantages to resistant genotypes, whether they are under the control of one or many loci.

Absolute numbers of pests exposed each generation may also play a major role at least at the local level because this may increase the likelihood that resistance genes can be present at any given time, as has been proposed for herbicide resistance [19]. Perhaps the most readily available indicator of the potential for the evolution of resistance in any given species or environmental circumstance is its past history of resistance.

2.1   Initial allele frequencies

Short of releasing susceptible insects (an idea that has been proposed and even tested for chemical pesticides, but is generally too expensive to be practical), we have no real control over initial resistance allele frequencies. Resistance will evolve more quickly if the frequency of resistance alleles is higher rather than lower, but generally not directly proportionally. That is, a 10 fold lower initial resistance allele frequency generally does not mean that resistance will evolve 10 fold slower. To the contrary, resistance may take less than two fold longer to occur [31, 32, 33].

The frequency of resistance alleles prior to the first use of a pesticide is rarely known [37], but recent work suggests that in at least some species, the initial frequency of *Bt* resistance may be on the order of 10^{-3} to 10^{-4} (Chapter 3.4). Gould and colleagues [13] estimated a frequency of about 10^{-3} from field collections of a cotton bollworm, *Heliothis virescens* (Chapter 3.4), although their data seem inconsistent with the predictions given in their paper for single gene inheritance of the resistance studied. Evidence for single gene inheritance is clearer from an earlier selection experiment [13] in which resistance must have had a frequency of about 10^{-3} in the founding population ([14], Chapter 3.4). However, this is probably the only case of a successful selection for single gene resistance in perhaps 10 similar attempts in *H. virescens* (some of which are cited in Chapter 3.4, but others have gone unpublished), so one might suppose that a more general approximation for the frequency of resistance for this species is 10^{-4}.

Further, recent studies in European corn borer (*Ostrinia nubilalis*) [2], diamondback moth [1], and the cotton bollworm *Helicoverpa armigera* [1] failed to detect resistance in samples of more than 1,000 alleles, suggesting a frequency of less than 10^{-3} [41]. On the other hand, resistance in susceptible colonies of the Indian meal moth appeared to be greater than 10^{-2} (Roush & McGaughey, unpublished data), consistent with the assumption that the Indianmeal moth has been naturally exposed to *Bt* in grain dust (Chapter 3.4).

Nonetheless, even a frequency of 10^{-4} is much higher than expected for chemical insecticides, which are now generally suspected to be 10^{-6} or less [11]. If anything, the relatively easy isolation of resistance from the field in species such as *H. virescens* should increase the level of care applied to managing resistance to *Bt* and other naturally occurring insecticides.

2.2 Dominance of resistance

For any gene locus with one resistance allele, there would be three genotypes: SS susceptible homozygotes, RS heterozygotes, and RR resistant homozygotes. Resistance is often described as either dominant or recessive, which simply describes the closer resemblance of the heterozygotes to either the resistant or susceptible parent, respectively. Dominance depends on the intrinsic characteristics of the resistance mechanism and the dose applied; a higher dose that kills heterozygotes will make resistance more recessive. So long as some fraction of the population escapes exposure, resistance that is recessive will be selected more slowly than resistance that is dominant [36, 37]. This has been the theoretical basis of the "high dose" or "high kill" strategy [29, 31, 45], which is further discussed below. However, it also illustrates the ambiguity of the term "selection pressure". In principle, a dose so high that it kills all heterozygotes (and maybe even resistant homozygotes) can slow resistance, i.e., high dose can give only low selection pressure. It is more precise to discuss the effects of treatment practices on the survival of the various genotypes (especially the heterozygotes) and the fraction of the population that escapes exposure.

2.3 Escapes: refuges

In studies of field populations, one of the most consistent influences on the rate at which evolution evolves is the proportion of the population that escapes pesticide exposure each generation in untreated "refuges" where the insects can live without selection [35, 36, 46]. The benefits of refuges have been specifically demonstrated for *Bt* [22]. The refuge may include non-crop plants or crop plants that are not treated. Refuges may also include plant parts in which insects are protected from exposure (say a corn borer inside a corn stalk), but it is critically important to realise that not all susceptible insects which survive treatment are necessarily in a refuge. In many cases, the insects that survive failed to receive a lethal exposure, which can still result in selection through reduced longevity or fecundity, rather than were totally unexposed. In recognition of this, refuges that are untreated with *Bt* are a major part of resistance management strategies for *Bt*-transgenic crops [12, 31, 32, 33, 34].

2.4   Fitness costs of resistance

Resistant strains of arthropods studied in the laboratory often show reproductive disadvantages in developmental time, fecundity, and fertility compared to susceptible strains. In the field, the frequencies of resistant individuals usually decline over time in the absence of pesticide use. It is a principle from population genetics theory that resistance alleles must have fitness costs, but often these costs seem too low or too inconsistent to be reliable for resistance management. Further, although proven fitness costs are an effect of specific resistance mechanisms, it is often difficult to associate fitness disadvantages specifically with resistance [35, 36, 37].

Resistant and susceptible strains may differ in fitness attributes independently of resistance, such from inbreeding during selection for resistance in the laboratory. Declines in the frequencies of resistant individuals in the field can be from dilution by inward migration of susceptible individuals [36]. In any case, as will be discussed below with respect to the high dose strategy (see Chapter 3.1), the key factor is the costs of resistance to the heterozygotes, which are the most common carriers of resistance. Until the very last stages of selection, resistant homozygotes are too rare to cause much of a disadvantage to the resistance alleles in the population [32, 37]. Even when resistance does decline in the field, it can often be rapidly re-selected, suggesting that susceptibility is probably never fully restored, but only to the level that resistance is not easily detected by the monitoring techniques [37].

Resistance to *Bt* is no exception to these general rules. Several resistant strains of *Bt* resistant diamondback moths [15, 44, 48] and other species (Chapter 3.4) have shown fitness disadvantages in the lab, but not always [49, 52]. Only one experiment seems to have explicitly tried to measure costs of resistance to *Bt* resistant heterozygotes (as opposed to the resistance homozygotes that would dominate selected strains), in this case by establishing a segregating population of RR, RS, and SS individuals. Resistance did not decline over time, suggesting that there were no apparent costs to resistance [52].

Fitness costs do not obviously assist in making choices between various resistance management tactics [29, 32], but empirical field trials may help one to judge the number of uses of a pesticide per year that will not increase the long term average frequency of resistance. In principle, if "back selection" due to fitness costs at least compensates for "forward selection" during the use of the pesticide, resistance will not increase. In practice, experiments to determine this level of selection are likely to be difficult.

3. MYTHS ABOUT MANAGEMENT OF RESISTANCE

3.1 High doses

Contrary to popular myth, there is no general advantage to applying high doses of pesticides, and indeed high doses may even inhibit the biological control component of integrated management programs, if not by direct mortality, by starving natural enemies when their food supply has been destroyed. The use of higher doses is especially inappropriate once resistance has been found in the field, because the strategy depends critically on a low initial gene frequency (see below). The high dose strategy also assumes consistent and uniform high doses that kill greater than about 95% of the heterozygous (RS) insects, an assumption usually not met by sprays [31, 45]. Neither theory nor experiments support the use of high rates in the field as a resistance management tactic [29,45]. To the contrary, in field trials with *Bt* sprays against diamondback moth, higher doses caused higher resistance [27]. On the other hand, due to the more uniform control of dose that is possible with transgenic plants, the high dose strategy shows promise for crops expressing *Bt* genes [31].

The high dose strategy is based on the assumption that the expected frequencies of the SS, RS and RR genotypes are a simple binomial probability function, where p represents the frequency of the resistance allele, and q the frequency of the susceptible allele, the frequencies of RR, RS and SS are: p^2, $2pq$, and q^2, respectively. While resistance is still rare (as when a pesticide is first introduced), the most common carriers of a resistance allele should be the heterozygotes. For example, if the frequency of resistance is 10^{-3} (which is only modestly rare), the frequency of heterozygotes will be approximately 2×10^{-3}, whereas the frequency of resistant homozygotes will be 10^{-6}, about 2,000 fold less common. The heterozygotes are the most common and therefore the most important resistance genotype. Control the heterozygotes and you can largely control resistance. The resistant homozygotes will be so greatly outnumbered by susceptible homozygotes immigrating from refuges that there will be essentially no matings between resistant homozygotes [29, 31, 32, 33, 45].

As a practical matter, these assumptions can be rarely met for pesticides that must be sprayed or drenched on a crop or pest breeding site. Most application methods fail to provide uniform coverage of the crop (thereby allowing some minimally treated heterozygotes to survive) and sprays generally expose a range of life stages, some of which will be less susceptible to the toxicant [6, 53].

The high dose strategy also fails to delay resistance significantly once the resistance allele frequency exceeds 10^{-2} [31] unless the refuge size is very large (20% or greater). Thus, the use of higher doses is especially inappropriate once resistance has been found in the field [45], which is generally difficult to do before the resistance allele frequency exceeds 1% (as discussed in chapters of section 5). The level of expression of *Bt* toxins found in transgenic cultivars is often high enough to constitute a high dose for some pests when they feed as recently hatched larvae [31], but this may be one of the few places where the high dose strategy can be successfully applied.

3.2   Toxin mixtures and mixtures with insecticides

Contrary to another popular myth, it is not necessarily true that mixtures of pesticides or toxins will delay resistance. Mixtures of insecticides have generally failed to delay resistance in the field compared to the rotational use or sequential introduction of the same insecticides [18, 43], and theoretical models showed that mixtures will significantly delay resistance only when several conditions are met [29, 33, 34]. To be most effective, mixtures require low initial frequencies of the resistance genes, refuges (as with the high dose strategy, such that resistant genotypes are rare and can be diluted), high mortality from each of the insecticides when used alone, a high spatial correlation of residues (not just equal decay rates), and a lack of cross-resistance between the toxins. In sum, the key is that almost all individuals resistant and exposed to one pesticide must be killed by the other for mixtures to be highly effective [29].

These conditions, especially a high spatial correlation of residues, are probably rarely met for sprays. For example, experiments with mixtures of *Bt* serotypes, applied at doses that did not provide high levels of control when used individually, failed to delay resistance in Indianmeal moth [25]. Not only is this experiment a good model for the field (where control with *Bt* sprays probably rarely exceeds 90%), the results are just as would be predicted from the arguments outlined above.

There are a number of laboratory experiments that have supported the use of pesticide and toxin mixtures, but it has to be noted that they do not realistically account for incomplete pesticide coverage and residue as would occur in the field for agricultural pests. Field experiments should be seen as much more reliable, and they are not encouraging [18]. As with the high dose strategy, mixtures of toxins (or pyramids as they are called in plant breeding) are much more promising for transgenic plants because of their greater uniformity of "coverage" through internal plant expression [32, 33, 34].

As an example of the problems that can result from mixtures of toxins, *Bt* sprays often include a mixture of specific toxins that appear to attack different binding sites within the insect gut. The resistance of the diamondback moth to *Bt* has resulted from the use of *B. thuringiensis* subspecies *kurstaki* (*Btk*) [44, Chapter 3.4), which produces Cry1A and Cry2 toxins [20]. Resistance appears to be due to reduced binding of the Cry1A (and Cry1F) toxins to the insect's mid-gut membrane, with relatively little cross resistance to toxins from other families, especially Cry1C [50, Chapter 3.4]. Subsequent to widespread resistance to *Btk, B. thuringiensis* subspecies *aizawai* (*Bta*) was marketed. *Bta* produces Cry1A and Cry1C and Cry1D proteins [20], where Cry1C is apparently the toxin with sufficient activity to control resistant larvae (see Chapter 6.2).

In at least some diamondback moth populations, resistance to *Bt* appears to decline in the absence of continuing sprays [15, 44, 48]. However, given that *Bta* includes Cry1A toxins, it seems likely that use of *Bta* would maintain enough exposure to Cry1A toxins to retard the decline in resistance to *Btk*. To investigate this, *Btk*-resistant diamondback moth larvae from Florida were divided into 4 treatment groups and selected with (1) *Btk*, (2) *Bta*, (3) purified Cry1C toxin, or (4) left unselected. When tested with 100 µg/ml Cry1Ab in leaf dip assays after four generations of selection, the unselected and Cry1C selected colonies showed 58-70% mortality, but the *Btk* and *Bta* colonies both showed only 4% mortality. Thus, there was enough Cry1A toxin in the *Bta* product to maintain resistance for Cry1A-resistance gene(s) [51]. In the field, this would eliminate the possibility of even occasional use of *Btk* products relying on Cry1A.

Toxin specific spray products have been developed commercially by transforming particular *Bt* genes into *Pseudomonas* [9], which for resistance management seems to be the desirable alternative to *Bt* strains with a mixture of toxins. In general, products with shared toxins, especially for those to which resistance is already widespread in the targeted pests, should be avoided.

4.   PROMISING TACTICS FOR RESISTANCE MANAGEMENT FOR BACTERIA AND SPRAYS

Resistance management plans have been characterised as proactive and reactive. Proactive plans anticipate the potential for resistance and attempt to delay it before resistance is ever detected. Reactive plans respond to a resistance crisis that has already occurred in the field, often with tactics such as pesticide mixtures and higher rates that desperately aim only to control the resistant pests. Because the frequency of resistance is already high, these tactics have little likelihood of actually slowing the evolution of resistance, as was discussed above. The focus of resistance management should be proactive, that is, resistance management should ideally be begun upon introduction of the product.

It is commonly argued that resistance managers rarely recommend more than to reduce the use of a particular pesticide, that is, to simply apply good integrated pest management (IPM). Reducing the overall number and area of applications through good IPM is critically important, but there are also other key resistance management tactics that are not obvious from general principles of IPM. For example, as discussed below (see Chapter 4.7), rotational use of pesticides across generations is superior to mosaics or shorter term rotations of pesticide use. This is true even if the same amount of each pesticide applied.

Because many pests (such as the cotton bollworm *Helicoverpa zea*) affect several crops (e.g., cotton, corn, tomatoes, soybeans), integrated management of pests and their resistance often requires a unified cropping system approach rather than focusing on specific crops. The tactics to be considered here include: reduced numbers of applications and areas treated, low doses, low persistence formulations, integration with other control tactics, pesticide rotations, pesticide mixtures, and selective vs. broad spectrum products.

4.1   Reduce number and area of applications

First and foremost, the most important tactic to slow resistance is to reduce proportion of the population exposed; reduce frequency of sprays and especially the proportion (or area) of the population treated. For any pest that has shown serious problems with resistance in the past, any strategy that relies solely on pesticide treatments, even with entomopathogenic bacteria, is unlikely to be durable. Integration of applications with other control tactics (see Chapter 6.2) is essential. So long as such bacteria are not used so intensively as to eliminate prey for natural enemies, the selectivity of agents like *Bt* sprays is ideal for integration with other biocontrols.

How does one reduce the use of sprays? Perhaps the single most important feature of any resistance management effort (but still largely unheralded) is a sampling scheme and system of treatment thresholds that targets sprays only when and where applications are truly needed. It is arguable that the most important contribution ever to resistance management has been the development of practical sampling schemes such as "leaf presence/absence" techniques and insect pheromone traps.

Further, the best way to leave a refuge is through "spot treatment". There are many examples in which spot treatment is often sufficient to control pests without treating the entire population. For example, early season infestations of Colorado potato beetles are often concentrated along the edges of fields. Spot treatments only where the pest densities have exceeded an economic threshold (where potential damage exceeds the costs of control) means that the rest of the crop contributes to the refuge [39]. In such cases, spraying the entire field will not generate economic returns in the short term and will only worsen resistance.

4.2   Avoid excessive dose rates

As argued above, high doses are unlikely to delay resistance for sprays and other such applications. Further, they may excessively reduce prey for natural enemies, thereby causing the enemies to starve or disperse. This would reduce a major advantage of the selectivity of biopesticides, their integration with natural enemies to reduce the overall number of sprays needed.

4.3   Avoid excessive persistence of residues

The persistence of a pesticide is a two-edged sword. Users want enough persistence to control the pest, but excessive persistence can continue to select for resistance long after the pest has been suppressed below damaging numbers. A persistent pesticide can have the effect of regular prophylactic treatments: rapid resistance. Thus, the advice to avoid persistent pesticides or formulations has been among the oldest and most widely accepted in resistance management [3], and is well documented by experiment [7] and theory [29].

Most biopesticides probably will have short persistence. Although it is now technically feasible to express *Bt* toxins in aquatic micro-organisms for control of mosquitoes and other medically important pests [9], it must be questioned as to whether such delivery systems will be counterproductive by accelerating resistance.

4.4   Integrate with other control tactics

There are many ways to integrate resistance management with other tactics, and I'll discuss only a few of the lesser known but very effective tactics here. For example, it is a common practice now in Australian cotton production to intentionally disturb the soil surface in order to trap cotton bollworm (*Helicoverpa* spp.) pupae in the ground and prevent the successful emergence of adults. This means that insects selected for resistance are killed, which enhances the dilution effect of refuges.

The diamondback moth is another key pest which has a more general weakness; it has a narrow host range, and can feed only on crucifers. In other species with such a limited host range, the local elimination of the host via crop rotation has proven to be a very effective pest and resistance management tactic, such as for the Colorado potato beetle [40]. A mandatory crucifer-free period has been undertaken for diamondback moth control in regions of at least two countries, Mexico (A.M. Shelton, personal communication) and Australia. After the introduction of a summer break in crucifer production in southern Queensland during 1990, frequencies of resistance to the chemical insecticide permethrin appeared to drop sharply (S. Heisswolf, personal communication).

A mandatory break in crucifer production will never be easy to implement. Neither was crop rotation for the Colorado potato beetle, but growers who found a way to rotate their crops have had fewer insecticide resistance problems [40] and are able to rely on insecticides in those cases when they still need them. An important part of the long term and sustainable future of production for some crops may be the establishment of grower cooperatives in neighbouring growing regions to produce crucifers (and complementary vegetable crops) in alternate times of the year to defeat pests such as the diamondback moth while stabilising crop markets. Regions which do this will remain highly profitable and retain access to relatively cheap and effective insecticides, whereas regions that cannot practice host-free periods will probably eventually suffer an economic decline.

4.5   Be cautious of mixtures of toxins and with chemicals

All other things being equal, selective products are probably less likely than broad spectrum insecticides to select for resistance, for two ecological reasons. First, a pesticide that is so broad spectrum that it eliminates the natural enemies or competitors of a pest will probably be sprayed more often, simply because the natural enemies are less able to suppress the pest. Second, when there are multiple pest targets for the same pesticide, applications against either pest will select for resistance both pests simultaneously. For a hypothetical example, consider a pesticide that is effective against the Colorado potato beetle and the European corn borer in potatoes. Applications against corn borers could select for resistance in the Colorado potato beetle even when it was at such low densities that control was not required. Resistance would be selected in potato beetles even when there was no economic benefit to their control.

Bt and other products using entomopathogenic bacteria are being introduced into a system where there will continue to be chemical insecticides. Some people will be tempted to mix chemical insecticides with bacterial products. However, this temptation should be avoided because of the increased potential of the mixture to reduce the abundance of beneficial species. A better way to use both the chemical and bacterial insecticides is in rotations.

4.6   Design careful rotations of pesticides

Rotating the use of pesticides over an entire area in a so-called "window strategy" based on the calendar has proven to be a very effective resistance management tactic both in terms of adoption and efficacy [8, 29]. The use of different compounds at roughly the same time in neighbouring fields constitutes a mosaic of treatment patterns and should be avoided. Mosaics are simply the worst way to deploy a set of pesticides [29]. The problem is that you have simultaneous selection with several pesticides, resulting in much lower pesticide durability [30].

With respect to rotations, rotations are always best across generations. Rotations within a generation are essentially the same as a mosaic: part of the population gets treated with each of the two pesticides, selecting for resistance to both simultaneously. Consider a case where selection for resistance is so strong that you get resistance in a single generation. If you have two pesticides and do this with the first pesticide, you at least have the second pesticide to fall back on for a total of two generations of control. If instead, you split the population in half, depending on what assumptions you make about the dominance of resistance, roughly half of the population will be resistant to each of the pesticides in the next generation, which is essentially a failure after one generation. Make a more elaborate model, or do the experiment [30], and one finds that rotation across generations is roughly twice as good as field-to-field mosaics or rotation within a generation when two or three pesticides are available. When there are overlapping generations, one should try to go on a cycle at least longer than the mean generation time.

Not all pesticides will have similar efficacy. For example, *Bt* products may well have lower efficacy than synthetic insecticides, and may therefore be more appropriate to use early in the cropping cycle when the crop can often withstand higher densities of larvae, and preservation of predators and parasites is often most beneficial. Thus, one can develop a window strategy for all products, placing each in the system at times that take advantage of their particular seasonal strengths, if any. Each new pesticide might be allotted a 1-3 month period depending on local conditions (e.g., duration of the cropping season, which months are hottest and therefore have the most insect generations). Rotational systems over a period of months or even years have proved to be extremely successful for cotton in Australia and Zimbabwe [8, 29].

The advantages are not specifically dependent on fitness costs to resistance. Such fitness costs improve the durability of a pesticide and the effectiveness of any resistance management program [32], but the genetic conditions that favor rotations over other tactics on the basis of fitness costs alone appear to be rare [5, 29].

5.   RESISTANCE MONITORING

Resistance monitoring can be very important for determining if a resistance management strategy is failing and whether improvements are required. Unfortunately, resistance monitoring efforts in the past have often failed to go beyond documenting failures and have rarely predicted failures before they occurred; monitoring will be useless if there is not also some follow-up management.

The most cost-effective systems for resistance monitoring rely on a diagnostic dose or a few diagnostic doses [38]. Such systems require testing of a range of methods to find those that are easily used even in remote locations [28] and which discriminate as much as possible between resistant and susceptible individuals. However, a key problem for routine monitoring efforts are the large sample sizes required. Hundreds or thousands of individuals must be tested to detect resistance at a 0.1-1% frequency at any given location, especially where bioassays are difficult and do not neatly distinguish between susceptible and resistant genotypes (Roush and Miller 1986). Even with detection at a 0.1% frequency, control failures in the field could occur quite quickly. Given the difficulties of bioassays with *Bt* and other biopesticides, and the limited knowledge of their modes of action, it seems most prudent to invest in highly proactive resistant management plans designed to delay resistance before it is ever detected.

6.   IMPLEMENTATION

Identifying an appropriate resistance management strategy is relatively easy compared to gaining adoption of the strategy, and indeed, this has been the most limiting factor to the success of resistance management efforts. Ultimately, any resistance management strategy will be most effective if it has the support of the private sector. The pesticide industry has established the Fungicide, Insecticide, and Herbicide Resistance Action Committees (FRAC, IRAC, and HRAC, respectively) to assist in this process.

However, contrary to popular myth, it is the pesticide users and not the pesticide manufacturers who have the most to lose when resistance evolves in those cases where resistance is such a problem that we must really be concerned with managing it. If the pesticide does not more than pay for its purchase cost in terms of improved price for the crop, the grower would be foolish to use it. On the other hand, this cost far exceeds the actual profit to a company. Thus, on a per unit basis, the cost of a lost pesticide to a grower, especially when there are few if any alternative pesticides, must considerably exceed the revenue lost to a pesticide company if the pesticide fails. In the case of the Colorado potato beetle, for example, resistance to all of the existing pesticides can easily cost more than $ 750 per hectare [39] in the northeastern US and Canada; a recently registered insecticide that provides effectively complete control is sold for less than $ 200 per hectare, of which perhaps less than half is profit to the company. Because government scientists in some measure represent the interests of the general public and growers, the public sector thus has a particular obligation to address resistance issues.

Both the pesticide companies and pesticide users have strong economic incentives to manage resistance, yet resistance is still generally poorly managed. Ultimately, a stronger partnership is required between the public and private sectors to assure that the promise of resistance management is fulfilled.

7.   CONCLUSIONS

This is an exciting time for resistance management. Several new bacterial pesticides or control agents are now available or may be developed. The danger is that people will become complacent when offered new pesticides, and assume that more will be offered in the future if resistance becomes a problem. If not fully integrated with other control tactics that are more difficult to implement, there is a real possibility that the new biological tools described in this book will be squandered and will fail to live up to their full potential to provide a sustainable system of pest management.

Although resistance management is often perceived as a complex problem, the list of potential tactics is generally so short that choosing which ones would be useful is not difficult. Most of these tactics are also complementary and are most effective when adopted before selection commences. Thus, even though resistance management can always be improved with additional data, one should also aim to adopt a resistance management plan at the first introduction of the product.

Pesticide resistance management would perhaps more appropriately be called "susceptibility management" because it is susceptibility that one aims to preserve. Entomopathogenic bacteria have revealed new types of susceptibility within populations; it is now up to us to manage them.

REFERENCES

[1] Ahmad, M (1999) Initial frequencies of alleles for resistance to *Bacillus thuringiensis* toxins in field populations of *Plutella xylostella* and *Helicoverpa armigera*. PhD thesis, University of Adelaide, 209 pp.
[2] Andow DA, Alstad DN, Pang YH, Bolin PC & Hutchinson WD (1998) Using an F2 screen to search for resistance alleles to *Bacillus thuringiensis* toxin in European corn borer (Lepidoptera: Crambidae) J. Econ. Entomol 91, 579-584
[3] Brown AWA (1967) Insecticide resistance-genetic implications and applications. World Rev. Pest Control 6, 104-114
[4] Cao J, Tang JD, Strizhov N, Shelton AM & Earle ED (1999) Transgenic broccoli with high levels of *Bacillus thuringiensis* Cry1C protein control diamondback moth larvae resistant to Cry1A or Cry1C. Molecular Breeding 5, 131-141
[5] Curtis CF (1987) Genetic aspects of selection for resistance, p. 151-161. *In* Ford MG, Holloman DW, Khambay BPS & Sawicki RM (eds.), Combating resistance to xenobiotics, Ellis Horwood, Chichester, England

[6] Daly J, Fisk JH & Forrester NW (1988) Selective mortality in field trials between strains of *Heliothis armigera* (Lepidoptera: Noctuidae) resistant and susceptible to pyrethroids: functional dominance of resistance and age class. J. Econ. Entomol. 81, 1000-1007

[7] Denholm IA, Farnham W, O'Dell K & Sawicki RM (1983) Factors affecting resistance to insecticides in house-flies, *Musca domestica* L. (Diptera: Muscidae). I. Long-term control with bioresmethrin of flies with strong pyrethroid- resistance potential. Bull. Entomol. Res. 73, 481-489

[8] Forrester NW, Cahill M, Bird LJ & Layland JK (1993) Management of pyrethroid and endosulfan resistance in *Helicoverpa armigera* (Lepidoptera: Noctuidae) in Australia. Bull. Entomol. Res., Suppl 1

[9] Gelernter W & Schwab GE (1993). Transgenic bacteria, viruses, algae and other microorganisms as *Bacillus thuringiensis* toxin delivery systems, p. 89-104. *In* Entwistle PF, Corey JS, Bailey MJ & Higgs S (eds.), *Bacillus thuringiensis,* an environmental pesticide: theory and practice. John Wiley, New York

[10] Georghiou GP & Lagunes-Tejeda A (1991) The occurrence of resistance to pesticides in arthropods. FAO, Rome, 318 pp.

[11] Goss PJE & McKenzie JA (1996) Selection, refugia, and migration: Simulation of evolution of resistance of dieldrin resistance in *Lucilia cuprina* (Diptera: Calliphoridae). J. Econ. Entomol 89, 288-301

[12] Gould F (1998) Sustainability of transgenic insecticidal cultivars: integrating pest genetics and ecology. Annu. Rev.Entomol. 43, 701-726

[13] Gould, F, Anderson A, Jones A *et al.* (1997) Initial frequency of alleles for resistance to *Bacillus thuringiensis* toxins in field populations of *Heliothis virescens.* Proc. Natl. Acad. Sci. USA. 94, 3519-3523

[14] Gould, F, Anderson A, Reynolds A, Bumgarner L & Moar WJ (1995) Selection and genetic analysis of a *Heliothis virescens* (Lepidoptera: Noctuidae) strain with high levels of resistance to *Bacillus thuringiensis* toxins. J. Econ. Entomol. 88, 1545-1559

[15] Groeters FR, Tabashnik BE, Finson N & Johnson MW (1994) Fitness costs of resistance to *Bacillus thuringiensis* in the diamondback moth. Evolution 48, 197-201

[16] Hama H, Suzuki K & Tanaka H (1992) Inheritance and stability of resistance to *Bacillus thuringiensis* formulations in the diamondback moth, *Plutella xylostella* (Linnaeus) (Lepidoptera: Yponomeutidae). Appl. Entomol. Zool. 27, 355-362

[17] Heckel DG, Gahan LC, Gould F & Anderson A (1997) Identification of a linkage group with a major effect on resistance to *Bacillus thuringiensis* Cry1Ac endotoxin in the tobacco budworm (Lepidoptera: Noctuidae) J. Econ. Entomol 90, 75-86

[18] Immaraju JA, Morse JG & Hobza RF (1990) Field evaluation of insecticide rotation and mixtures as strategies for citrus thrips (Thysanoptera: Thripidae) resistance management in California. J. Econ. Entomol. 83, 306-314

[19] Jasienuik M, Brule-Babel AL & Morrison IN (1996) The evolution and genetics of resistance in weeds. Weed Science 44, 176-193

[20] Koziel MG, Carozzi NB, Currier TC, Warren GW & Evola SV (1993) The insecticidal crystal protein of *Bacillus thuringiensis*: past, present, and future uses. Biotechnol. Genet. Eng. Rev. 11, 171-228

[21] Liu YB & Tabashnik BE (1997) Inheritance of resistance to the *Bacillus thuringiensis* toxin Cry1C in the diamondback moth. Appl. Environ. Microbiol. 63, 2218-2223

[22] Liu Y-B. & Tabashnik BE. (1997) Experimental evidence that refuges delay insect adaptation to *Bacillus thuringiensis*. Proc. R. Soc. Lond. B 264, 605-610

[23] Metz TD, Roush RT, Tang JD, Shelton AM & Earle ED (1995) Transgenic broccoli expressing a *Bacillus thuringiensis* insecticidal crystal protein: implications for pest resistance management strategies. Molecular Breeding 1, 309-317

[24] McGaughey WH (1985) Insect resistance to the biological insecticide *Bacillus thuringiensis*. Science 229, 193-194

[25] McGaughey WH & Johnson DE (1987) Toxicity of different serotypes and toxins of *Bacillus thuringiensis* to resistant and susceptible Indianmeal moth (Lepidoptera: Pyralidae). J. Econ. Entomol. 80, 1122-1126

[26] Perez CJ & Shelton AM (1997) Resistance of *Plutella xylostella* to *Bacillus thuringiensis* Berliner in Central America. J. Econ. Entomol. 90, 87-93

[27] Perez CJ, Shelton AM & Roush, RT (1997) Managing diamondback moth (Lepidoptera: Plutellidae) resistance to foliar applications of *Bacillus thuringiensis*: Testing strategies in field cages. J. Econ. Entomol. 90, 1462-1470

[28] Perez CJ, Tang JD & Shelton, AM (1997) Comparison of leaf-dip and diet bioassays for monitoring *Bacillus thuringiensis* resistance in field populations of diamondback moth (Lepidoptera: Plutellidae). J. Econ. Entomol. 90: 94-101

[29] Roush RT (1989) Designing resistance management programs: How can you choose? Pesticide Science 26, 423-441

[30] Roush RT (1993) Occurrence, genetics and management of insecticide resistance. Parasitology Today 9, 174-179

[31] Roush RT (1994) Managing pests and their resistance to *Bacillus thuringiensis*: Can transgenic crops be better than sprays? Biocontrol Science & Technol. 4, 501-516

[32] Roush RT (1997) Managing resistance to transgenic crops, p. 271-294. *In* Carozzi N & Koziel M (eds.), Advances in insect control: the role of transgenic plants. Taylor and Francis, London

[33] Roush RT (1997) *Bt*-transgenic crops: Just another pretty insecticide or a chance for a new start in resistance management? Pesticide Science 51, 328-334

[34] Roush RT (1998) Two toxin strategies for management of insecticidal transgenic crops: Can pyramiding succeed where pesticide mixtures have not? Phil. Trans. Roy. Soc. London, Series B, Biol. Sci., 353,1777-1786

[35] Roush RT & Croft BA (1986) Experimental population genetics and ecological studies of pesticide resistance in insects and mites, pp 257-270. *In:* Pesticide Resistance: Strategies and Tactics for Management, (National Research Council, Ed). National Academy Press, Washington DC, USA

[36] Roush RT & Daly JC (1990) The role of population genetics in resistance research and management, p 97-152. *In* Roush RT & BE Tabashnik BE (eds.), Pesticide resistance in arthropods, Chapman and Hall, New York

[37] Roush RT & McKenzie JA (1987) Ecological genetics of insecticide and acaricide resistance. Ann. Rev. Entomol. 32, 361-380

[38] Roush RT & Miller GL (1986) Considerations for design of insecticide resistance monitoring programs. J. Econ. Entomol. 79, 293-298

[39] Roush RT & Tingey WM (1992) Evolution and management of resistance in the Colorado potato beetle, *Leptinotarsa decemlineata*, p 61-74. *In* Denholm I, Devonshire AL & Holloman DW (eds.), Resistance '91: achievements and developments in combating pesticide resistance, Elsevier Applied Science, Essex, England

[40] Roush RT, Hoy CW, Ferro DN & Tingey WM (1990) Insecticide resistance in the Colorado potato beetle (Coleoptera: Chrysomelidae): Influence of crop rotation and insecticide use. J. Econ. Entomol. 83, 315-319

[41] Schneider, JC (1999) Confidence interval for Bayesian estimates of resistance allele frequencies J. Econ. Entomol. 92, 755

[42] Shelton AM, Robertson JL, Tang JD *et al.* (1993) Resistance of diamondback moth (Lepidoptera: Plutellidae) to *Bacillus thuringiensis* subspecies in the field. J. Econ. Entomol. 86, 697-705

[43] Tabashnik BE (1989) Managing resistance with multiple pesticide tactics: theory, evidence, and recommendations. J. Econ. Entomol. 82, 1263-69

[44] Tabashnik BE (1994) Evolution of resistance to *Bacillus thuringiensis*. Annu. Rev. Entomol. 39, 47-79

[45] Tabashnik BE & Croft BA (1982) Managing pesticide resistance in crop-arthropod complexes: interactions between biological and operational factors. Environmental Entomology 11, 1137-1144

[46] Tabashnik BE & Croft BA (1985) Evolution of pesticide resistance in apple pests and their natural enemies. Entomophaga 30, 37-49

[47] Tabashnik BE, Cushing NL, Finson N & Johnson MW (1990) Field development of resistance to *Bacillus thuringiensis* in diamondback moth (Lepidoptera: Plutellidae) J. Econ. Entomol. 83, 1671-1676

[48] Tabashnik BE, Finson N, Groeters FR, Moar WJ, Johnson MW, Luo K, Adang MJ. 1994. Reversal of resistance to *Bacillus thuringiensis* in *Plutella xylostella*. Proc. Natl. Acad. Sci. USA 91, 4120-4124

[49] Tabashnik BE, Finson N, Johnson MW, Heckel DG. 1995. Prolonged selection affects stability of resistance to *Bacillus thuringiensis* in diamondback moth (Lepidoptera: Plutellidae). J. Econ. Entomol. 88, 219-224

[50] Tabashnik BE, Liu YB, Finson N, Masson L, Heckel DG (1997) One gene in diamondback moth confers resistance to four *Bacillus thuringiensis* toxins. Proc. Natl. Acad. Sci. USA 94, 1640-644

[51] Tang JD, Shelton AM, Moar WJ & Roush RT (1995) Consequences of shared toxins in strains of *Bacillus thuringiensis* for resistance in diamondback moth. Pesticide Resistance Management Newsletter. 7, 5-7 (also on World Wide Web at http://www.msstate.edu/Entomology/v7n1/s95rpm.html#art03)

[52] Tang JD, Gilboa S, Roush RT & Shelton AM (1997) Inheritance, stability, and fitness of resistance to *Bacillus thuringiensis* in a field colony of *Plutella xylostella* (L.) (Lepidoptera: Plutellidae) from Florida. J. Econ. Entomol. 90, 732-741

[53] Zehnder G. W & Gelernter WD (1989) Activity of the M-ONE formulation of a new strain of *Bacillus thuringiensis* against Colorado potato beetle (Coleoptera: Chrysomelidae): relationship between susceptibility and insect life stage. J. Econ. Entomol. 82, 756-761

Chapter 6.5

Management of resistance to bacterial vector control

Lêda Regis[1] & Christina Nielsen-LeRoux[2]
[1]Centro de Pesquisas Aggeu Magalhães-FIOCRUZ, Av. Moraes Rêgo s/n 50670-420
Recife PE, Brazil, and [2]Institut Pasteur, 28 Rue du Dr. Roux,75724 Paris, France

Key Words: *Bacillus sphaericus*, *Bacillus thuringiensis israelensis*, Culicidae,
              Mosquito susceptibility, Resistance management

Abstract: The vast majority of entomopathogenic bacteria used in vector control
              belongs to the species *Bacillus thuringiensis israelensis* (*Bti*) to which
              no field resistance has been reported. Thus consideration of the
              management of resistance to bacterial larvicides essentially concerns *B.
              sphaericus*. Resistance to this bacterium has been recorded in mosquito
              populations both in the laboratory and under field conditions, after
              periods of continuous exposure to *B. sphaericus* strains 2362, 1593 or
              C3-41. *B. sphaericus* has several advantages over other agent for
              controlling mosquitoes such as *Culex* and *Anopheles* species. It may be
              applied rationally as a part of an integrated control program. Case
              histories, the reasons for the development of *B. sphaericus* resistance and
              application strategies for the continued success of this environmentally
              safe larvicide are discussed.

1.   INTRODUCTION

The mechanism of action of insecticidal *Bacillus* endotoxins involves a number of steps. This was considered to be an advantage over synthetic insecticides, because the multiple levels of interaction should prevent the development of insect resistance. However it has been shown that resistance may develop to some *B. thuringiensis* and *B. sphaericus* strains in laboratory and field conditions. For mosquito and black fly control, resistance under field conditions has only been reported for *B. sphaericus*, whereas no resistance to *Bacillus thuringiensis israelensis* (*Bti*) has

J.-F. Charles et al. (eds.),
Entomopathogenic Bacteria: From Laboratory to Field Application, 419–438.
© 2000 Kluwer Academic Publishers. Printed in the Netherlands.

occurred, possible due to its multitoxin crystals. Since 1982, *Bti* formulations have been continually increasing in use world-wide, as a useful and effective larvicide in areas in which mosquitoes and black flies have become resistant to chemicals or in which environmental concerns led to demands for selective, environmentally safe control measures.

This is the case, for example, in some large control programs such as the German Mosquito Control Association (KABS) (Germany), OCP (Onchocerciasis Control Programme, Africa) and PCS (Simulium Control Program, Brazil) [3, 15, 24], in which no resistance to *Bti* has been detected in the respective target insect species, despite more than ten years of intensive exposure to its toxins (see chapter 6.2)

B. sphaericus was introduced into large-scale programs some years later. Various formulations have been used and this bacterium has been found to be very effective with long-term residual action against *Culex* sp. larvae breeding in polluted water [2, 16, 20, 35, 36, 48, 58] in which breeding sites *B. sphaericus* has advantage over *Bti*. *Culex pipiens quinquefasciatus*, the most common mosquito in tropical urban environments, is the principal vector of Bancroftian filariasis. *B. sphaericus* controls this species and has also proved useful for controlling *Anopheles* sp., the vectors of malaria [19], the most prevalent vector-borne disease in humans. Resistance to *B. sphaericus* was first reported in 1994, when resistance in mosquito populations intensively exposed to the bacterium was observed at about the same time in France, India and Brazil [33, 43, 45]. Resistance to *B. sphaericus* in field populations of mosquitoes is now well documented. Understanding the circumstances facilitating resistance development and its reversal in field populations will help us to design strategies to keep this important weapon effective in the fight against disease vectors.

This chapter will deal principally with case histories of field resistance to *B. sphaericus,* with what we know about the mechanisms, genetics and stability of resistance to *B. sphaericus* and suggestions for its management. It will also discuss the advantages of *Bti* and some other environmentally safe control measures.

2.   CASE HISTORIES OF *B. SPHAERICUS* RESISTANCE IN MOSQUITO POPULATIONS

B. sphaericus has been used to control mosquito populations in various countries (see Table 1, page 439). The main case histories of resistance development are discussed below.

2.1   Brazil

A field trial testing the efficacy of *B. sphaericus* strain 2362 efficacy against *C. pipiens quinquefasciatus* was carried out in a 1.2 km^2 area in the Coque district of Recife, where more than 3,000 breeding sites had been treated 37 times over a 2-year period [35, 36]. The local *Culex* population was 10.6 times more resistant to *B. sphaericus*, than untreated populations [43]. This low-level resistance gradually declined and population susceptibility was fully restored 16 months after the cessation of treatment [44]. In contrast, no change in *C. pipiens quinquefaciatus* susceptibility was recorded after 13 treatments *with B. sphaericus* 2362 over on 18-month period, in a 5.7 km^2 area at Olinda city, about 20 km from Coque (L. Regis *et al.*, unpublished data).

2.2   France

Since 1986, *B. sphaericus* has been used by the Entente Interdépartementalee de Démoustication (EID) to control *Culex pipiens pipiens* larvae in urban breeding sites in 210 towns and villages on the French Mediterranean Coast [46]. In 1994, a larval population living in sanitary voids was found to be highly resistant to *B. sphaericus* (> 20,000 times) [45] . These breeding sites were treated with large doses of the bacterium for 7 years, favouring the persistence of toxic activity possibly due to bacterial recycling. The following year, high-level resistance (> 1,000 times) to *B. sphaericus* was found in the *Culex* larvae from 50% of 16 breeding sites examined. However, no resistance was recorded for mosquito larvae from an oxidation lagoon after 3 years of exposure to *B. sphaericus* sprayed at short intervals (G. Sinègre, personal communication).

2.3   India

A relatively high level of resistance was detected in a field population of *C. quinquefasciatus* exposed for a 2 year-period to 35 rounds of spraying with *B. sphaericus* strain 1593, in an 8 km^2 area in Kochi, India. After the 2 year-period a moderate level of resistance was observed: *Culex* larvae from the treated area were 146 times more resistant than those from unsprayed locations. If the resistant strain was subjected to moderate selection pressure in the laboratory, resistance rapidly increased and, by generation 18, the LC$_{50}$ was 6,223 times higher than that for susceptible strains [33].

2.4   China

The Changping Village (8 km^2 area) in Guangdong Province was treated with *B. sphaericus* strain C3-41 three times a month over an 8-year period to control *C. p. quinquefasciatus*. Resistance (22,000 times increase) was finally detected after 7 to 8 years of treatment. *B. sphaericus* treatments were then stopped and replaced by *B. thuringiensis israelensis* treatment for six months after which the susceptibility to *B. sphaericus* toxin of the *C. p. quinquefasciatus* population was partially restored [59, 61].

B. sphaericus resistance was also reported in Tunisia after a short treatment period (G. Sinègre & A. Bouattour, personal communications). This clearly demonstrates that the development of resistance to *B. sphaericus* in field populations of mosquitoes is a serious threat to the continued success of this environmentally benign larvicide. Evaluation of the factors affecting resistance development is necessary if we are to design appropriate resistance management strategies.

3.   MECHANISMS AND GENETICS OF RESISTANCE IN TERMS OF STABILITY AND REVERSIBILITY

The mode of action of an active compound and the mechanisms of resistance to it are obviously strongly correlated. So, the main steps in the interactions between the bacteria, their principal toxic factor and the larvae, may be targets for changes affecting the insect susceptibility. It is well known that the interaction between the *B. thuringiensis* (*Bt*) insecticidal crystal proteins (ICPs) and larval midgut cells frequently determines the host specificity of ICPs [5, 14, 55]. This has been clearly shown for many lepidopteran, some coleopteran and fewer dipteran species. It has also been shown that the absence of functional toxin receptors may cause resistance (see Chapter 3.4) and that the solubilisation and proteolysis steps may also be involved in resistance [10, 29, 54] as may any alteration in post-binding events such as membrane integration of toxins, pore formation and intracellular events. The *Bti* crystal toxins also target midgut cells [6, 34] but little is known about specific toxin/receptor interactions for *Bti* toxins active against Dipterans [9] . However, Cry1C from *Bt aizawai*, which has weak activity against *Aedes aegypti,* was recently reported to have specific midgut receptors in the larvae of this species [1].

The development of resistance to individual *Bti* toxins has proved that resistance can develop, but the strength of that resistance is inversely correlated to the number of functional toxin genes expressed in *Bti* during selection. The Cyt toxin has a particular large effect [56]. In contrast, the mode of action of *B. sphaericus* and of its principal virulence factor, the crystal Bin toxin, has been shown to involve specific receptors present on the midgut membranes of susceptible *C. p. pipiens* and *Anopheles gambiae* larvae [26, 42] (see also Chapter 3.5).

The activity of *B. sphaericus* is due mainly to the presence of the Bin toxin. Its components, BinA and BinB, seem to interact with the target cells like a single-site insecticide [7, 26, 42]. It is therefore not surprising that resistance is selected in certain situations. As discussed in Chapter 3.5, resistance to *B. sphaericus* is related to the absence of a functional receptor in some cases, whereas the mechanism of resistance is unknown in others [27, 28, 43] (Table 2).

The aim of studying the genetics and mechanisms of resistance is to improve our understanding of this phenomenon and if possible, to prevent its development in field conditions. Predictions about resistance development are often based on the results of research conducted in the laboratory, put into multifactorial models dealing with real field conditions [8].

Table 2. C. pipiens populations resistant to *B. sphaericus*

Country *Culex* sp. and colony	Field (F) or Lab (L) selected	Resistance level[1]	Receptor binding[2]	Resistance dominance	Ref.
USA *Cpq*[3] Geo	L	100,000	No	Recessive	[25]
USA *Cpq* L-SEL	L	37	Yes (unpublished data)	Recessive	[37]
France *Cpp*[4] SPHAE	F	20, 000	Yes	Recessive	[45] [28]
Brazil *Cpq* Coque	F	10	Yes	Recessive	[43]
India *Cpq* Kochi	F	150	NA	NA	[33]
China *Cpq* RFCq	F	20,000	NA	NA	[59]

[1]The level of resistance is calculated as the ratio of the LC50 values of the resistant

population to that of susceptible laboratory colonies. The level is that observed in the field or the final level after selection in the laboratory. [2]The toxin could bind to the receptor with unchanged affinity. [3]*Culex quinquefasciatus.* [4]*Culex pipiens.*
NA: information not available.

All studies with *B. sphaericus*-resistant *Culex* colonies have shown that the gene for resistance is recessive. One major somatic gene is probably involved for the laboratory-selected population (Geo), whereas the major gene involved in the field-selected resistant colony (SPHAE), is sex-linked [28]. This is consistent with the finding that two different mechanisms may be responsible for resistance to *B. sphaericus* in *C. pipiens* (Table 2*).

The fact that resistance is determined by a recessive allele of a single major gene has implications for the susceptibility of the population and for the reversibility of resistance. All F1 progenies of crosses between homozygous resistant (rr) and homozygous susceptible (SS) individuals should not differ significantly in susceptibility from homozygous susceptible individuals. Thus, under field conditions, the migration of susceptible homozygous *Culex* adults into *B. sphaericus*-treated zones would restore susceptibility in 100% of the progeny of matings with homozygous resistant, and 75% of the progeny of matings with heterozygous (Sr) individuals. This does not mean that the mosquitoes in the treated area will rapidly become susceptible, but the migration of susceptible adults from untreated areas should dilute the frequency of the resistance alleles in the population and influence the stability and level of resistance.

Other factors important in the development and stability of resistance are the selection pressure, the initial frequency of the resistance allele and its cost to fitness. Intensive selection experiments have shown that very different levels of stable resistance can occur, depending on the size of the population subjected to selection, and the kind of mutation (mechanisms of resistance involved [27, 38].

Very few ecological studies have been carried out into the natural variation of susceptibility to *B. sphaericus* and *Bti.* Georghiou *et al.* (1991) [11] found up to 20% variation in susceptibility to *B. sphaericus* amongst wild *C. p. quinquefasciatus* populations from California, whereas the variation in susceptibility to *Bti* was much smaller. The geographic variation in susceptibility to *B. thuringiensis* of *Heliothis virescens* (Lepidoptera) and *H. zea* populations has been found to be of the order of 4 to 16 times in the southern states of the USA This may indicate that the frequency of genes for resistance to insect pathogens is higher than the normal mutation rate (one in a million). Results from single-pair crossing studies with *Bt*-resistant *Plutella xylostella* (Lepidoptera) showed that the frequency of the resistance gene in a susceptible population was about 1,000 times higher than expected [13]. Similar results were reported by Tang *et al.* (1997) [49] for the same species in Florida, with a high estimated natural frequency of the resistance allele.

Some cost to fitness of *B. sphaericus* resistance in mildly resistant laboratory-selected colonies [37] has been reported. No evidence for such a phenomenon has been found for the highly resistant colonies from California (Geo) (M.C. Wirth & C. Nielsen-LeRoux, unpublished data) or for the highly resistant field-selected colony SPHAE (C. Nielsen-LeRoux & C. Chevillon, unpublished data).

4.   FACTORS INFLUENCING THE RATE OF DEVELOPMENT OF RESISTANCE IN THE FIELD

How do the factors described above relate to the observed resistance development against *B. sphaericus* in the field? The remarkable differences in the levels of resistance found in field populations exposed to *B. sphaericus* in India, Brazil, China and France, from 10 fold in Brazil up to 22,000 fold in China, seem to be due to different levels of selection pressure and/or frequencies of the resistance gene in the mosquito population before treatment. The data from Brazil and India are comparable because the size of the area, and many other circumstances facilitating resistance development in the treated areas were similar. This includes important features determining selection pressure such as the rate and frequency of *B. sphaericus* application and the duration of the treatment period (2 years). Despite these similarities, the exposed *C. quinquefasciatus* population in India was a 146 times more resistant than untreated mosquitoes whereas the resistance of the mosquito population in Brazil increased only by a factor of 10. Although efforts were made to treat every potential and actual breeding site, the existence of unsprayed refuges cannot be excluded for either area. Mosquito migration from surrounding unsprayed areas probably is thought to have occurred in Kochi, India [33], whereas very little mosquito immigration can have occurred in Coque due to a barrier zone isolating the target control zone [36]. We therefore think that the difference is more likely to be due to a low frequency of resistance genes in the *C. p. quinquefasciatus* population in Coque (Recife city, Brazil). Indeed selection for resistance in the laboratory is currently underway in Recife and no significant resistance is observed until generation 18 (C.F. Oliveira & L. Regis, unpublished data). However, the difference may also be due to there being different mutations resulting in various levels of resistance.

In France and China, very high levels of resistance, (> 10,000 and 22,000 times normal respectively), were found in mosquito populations intensively exposed to *B. sphaericus* for more than 7 years [61]. High rates of *B. sphaericus* application at every breeding site at short intervals (every 10 days) for a long time, may have led to the continuous exposure of successive generations of mosquitoes to *B. sphaericus* and the Bin toxin, thereby accounting for the high level of resistance of the mosquito population in China. Few treatments were carried out in the sanitary voids (indoor closed breeding sites) of Allende, France, but selection pressure was high due to recycling of the bacteria, as shown by the long periods of persistence of larvicidal activity and confirmed by bacterial analysis. This resulted in continuous selection pressure responsible for the high level of resistance reported in the local, isolated mosquito population [45].

The creation of refuges exposure to *B. sphaericus*, by leaving some breeding sites within the area untreated, could help to delay the development of resistance. However, this would be a problem in terms of reducing mosquito population size, because vector mosquito species are *r*-strategists (very high birth rates and short generation time) and therefore their populations recover quickly even after severe reduction [40]. The importance of treating every potential breeding site for reducing population density is well known. However, observations from Brazil and China suggest that providing temporal refuges (i.e. interrupting treatment for limited periods) can lead to the abolition of resistance. In Brazil, the population susceptibility was fully restored 16 months after the treatment stopped [44] and in China the resistance level was drastically reduced after only six months without *B. sphaericus* spraying [61]. However in Coque, Brazil, stopping of *B. sphaericus* spraying did not result in the absence of this bacterium from breeding sites. Surprisingly, the density of the adult mosquito population remained low for as much as 10 months after the cessation of treatment. The long lasting persistence of *B. sphaericus* spores was confirmed in some habitats by bacterial analysis. It seems therefore that spore recycling was responsible for both continued population control and the slowness of the decrease in resistance. Then laboratory investigations indicating a recessive character of *B. sphaericus* resistance and the importance of selection pressure correlates with the low stability of resistance in the field and the variation of level of resistance in various regions.

5. CROSS-RESISTANCE AND TOXIN RECEPTOR INTERACTION

The greater likelihood of resistance to *B. sphaericus* developing under particular conditions requires further investigation into the differences in toxicity between natural *B. sphaericus* strains. Control programs are subject to practical considerations and it would therefore be of interest to determine whether the resistance developed to certain strains confers cross-resistance to other strains of the same species or to different species of toxin-producing organisms.

All *B. sphaericus*-resistant *Culex* populations have been selected by strain 2362, 1593 or C3-41. All these strains have the same serotype and identical genes encoding the Bin toxin. However, small differences exist in the amino acid sequence of *B. sphaericus* Bin toxins ([4, 31, 60]; see also Chapter 2.3) that may be important to the structure/function of the toxin/receptor complex and therefore to larvicidal activity and specificity. Table 3 summarises results of various studies on cross-resistance.

Cross-resistance to other *B. sphaericus* strains has been reported in three different studies :

a) In one study, the two mildly resistant colonies from California, both selected in the laboratory using *B. sphaericus* 2362, showed significant levels of cross-resistance to *B. sphaericus* strains 1593 and 2297 (Table 3).

b) In another study [60], four *B. sphaericus* strains were tested against the SPHAE colony which is highly resistant to *B. sphaericus* 2362. Cross-resistance to the *B. sphaericus* IAB-881, IAB-872 and BS-197 strains but not to LP1-G was found to occur.

c) A third study involved four *B. sphaericus* strains belonging to different serotypes: the highly toxic IAB-59, IAB-872 and IAB-881 strains and the weakly toxic strain ISPC-5, the toxic element of which has not been identified (Nielsen-LeRoux *et al.*, in preparation). The amino acid sequence of the Bin toxin proteins from IAB-59 (type 1 Bin toxin) is different from that of 2362/1593 (type 2 Bin toxin) and 2297 (type 3 Bin toxin) [4, 17]. These 4 *B. sphaericus* strains were bioassayed against four different *B. sphaericus*-resistant *Culex* populations: 3 highly-resistant colonies, Kochi, SPHAE and GEO, and one mildly resistant colony, L-SEL. Strong cross-resistance to strains IAB-881 and IAB-872 was found in all the mosquito colonies. The weakly toxic ISPC-5 strain was only tested on the Kochi colony, which showed no cross-resistance (D.A. Rao, personal communication [32]). A low level of cross-resistance to the IAB-59 strain was found in all three highly resistant colonies.

Table 3. Level of resistance and cross-resistance to various *Bacillus sphaericus* (*Bsp*) strains and to *B. thuringiensis israelensis* Bti in various *Bsp*-resistant *Culex pipiens* sp. colonies

Resistant colony	*Bsp* 2362*	*Bsp* C3-41	*Bsp* 2297	*Bsp* IAB-881 IAB-872	*Bsp* IAB-59	*Bsp* LP1-G	Bti
				Bacillus species and strains			
France	≤20,000	Yes	Yes[h]	Yes	No[a]	No[b]	No[c]
USA (Geo)	≤100,00	NT	Yes[h]	Yes	No[a]	No[b]	No[d]
USA (L-SEL)	37 (4.5)	NT	Yes[e]	Yes	Yes	**	No[e]
India (Kochi)	≤1,700	NT	Yes[g]	Yes	No[a]	**	No[f]

*Level of resistance to the *B. sphaericus* standard SPH-88 (strain 2362), based on LC_{50} relatively to susceptible reference colonies. NT: Not tested.
[a]The level of cross-resistance varied from 11 to 43 compared to susceptile reference colonies; [b]No significant difference in LC50 values between resistant and susceptible colonies (Nielsen-Le-Roux *et al.* unpublished, Yuan *et al.* unpublished); [c]Nielsen-LeRoux *et al.* [28]; [d]Nielsen-LeRoux *et al*, 1995;[27] [e]Rodcharoen & Mulla [37]; [f]Rao *et al.* [33]; [g]Poncet *et al.* [30]; [h]Thiéry *et al.* [52].

The membrane receptor-toxin interaction is crucial for the activity of the crystal toxin. It is therefore important to investigate whether Bin toxins from different *B. sphaericus* strains use the same receptor binding site, to estimate the risk of cross-resistance at this level. Preliminary results from *in vitro* binding competition studies have shown that Bin toxins from LP1-G, C3-41, 1593 and IAB-59 strains use the same binding site (Nielsen-LeRoux *et al.*, unpublished), which suggests that there is a risk of developing cross-resistance to these toxins. However, the results of the *in vivo* bioassays provide little evidence of cross-resistance (low-level cross-resistance to LP1-G and IAB-59 strains). Recent findings show that the Bin toxins from these strains are not toxic to *B. sphaericus*-resistant larvae (Nielsen-LeRoux *et al.*, unpublished data). These data also suggest that there are other toxic compounds in IAB-59 and LP1-G strains.

These reports indicate that at least two *B. sphaericus* strains, LP1-G and IAB-59, are potential alternatives to strains 2362 and 1593 and could be used in future *B. sphaericus* formulations to manage resistance depending of their production and field performance.

All the reported *B. sphaericus*-resistant *Culex* colonies have been tested for susceptibility to *Bti*. All are affected by *Bti*, and some are even more susceptible to its toxins than are the *B. sphaericus*-sensible control colonies. In fact, the *B. sphaericus*-resistant colonies from Brazil, India, and China were 3 to 7 times more susceptible to *Bti* than controls [33, 43, 61] whereas the laboratory-selected resistant colonies Geo and

L-SEL, did not differ significantly from the controls in susceptibility to *Bti* [27, 37] as was also the case for SPHAE [28].

6. STRATEGY FOR THE MANAGEMENT OF RESISTANCE TO *B. SPHAERICUS*

A resistance management strategy should be set up before beginning a control program using *B. sphaericus*. Such a strategy should be based on 1) Monitoring the susceptibility of mosquito population before and during treatment by comparison with a laboratory colony, or with populations from untreated areas; 2) Promoting the discontinuity of selection pressure by using another or several other control agents.

As resistance to *B. sphaericus* is a recent phenomenon, little information is available to determine the frequency of susceptibility assessment required to detect resistance. In Coque, Brazil, the resistance was detected early in its development, after 2 years of exposure to *B. sphaericus* but within a similar period, mosquitoes in India became 146 times more resistant, demonstrating that the speed of resistance development depends on various local conditions, as discussed above. Monitoring mosquito susceptibility before exposure to *B. sphaericus* and every 4 or 6 months thereafter, should make it possible to detect any change in susceptibility during the treatment period.

Several aspects are important when measuring mosquito population susceptibility:

6.1 Monitoring

a) At each time, large numbers of egg rafts or first instar larvae should be collected from various breeding sites, widely distributed within the area, to ensure a broad genetic background. Larvae from each collection should be reared separately in the laboratory and tested individually in bioassays.

b) Bioassays should follow the standard World Health Organization protocol (see Chapter 5.1) It is very important to use the same *B. sphaericus* product for all bioassays. It is preferable to prepare stock suspensions of toxins to be stored frozen at -20°C.

c) It is recommended that the susceptibility of a laboratory colony and/or a field population collected in an untreated area should be determined simultaneously to compare their susceptibility to that of the larvae from the treated area. Comparison of LC_{50} values and slopes of the dose/response regression lines should make it possible to detect any change in susceptibility if the regression lines are parallel. Precise estimation of the resistance level requires a homozygous susceptible colony.

6.2   Choice of control method

Most resistance management strategies seek to reduce or abolish selection pressure. This can be done by dealing with the rate of application and the presence of the toxins in time and/ or space and by using different insecticides with different modes of action. The characteristics of mosquito population, as stated above, suggest that it is best to interrupt *B. sphaericus* treatment for periods of time and to use other larvicides. However, due to recycling, the interruption of *B. sphaericus* treatment does not necessarily result in the breeding sites being free from the spores and toxins of this bacterium. In some isolated, shaded, shallow larval breeding sites, the recycling of bacteria may result in long-term effect. If the bacteria (spores/crystals) are present in the larval feeding zone this may lead to a certain selection pressure, which is however unlikely to be strong if the entire mosquito population of the area is considered.

6.2.1   Chemicals

B. sphaericus is suitable for use with many synthetic and biological mosquito control agents. This is very important for the design of integrated control programs. The choice of control measures should be based on factors/agents active against different target cells in the insect. It has been demonstrated that known mechanism of resistance to conventional insecticides do not confer cross-resistance to bacterial toxins and *vice versa*. None of the common metabolic mechanisms (dehydrochlorinases, multifunction oxidases, non-specific esterases and glutathione-S-transferases), or those involving a reduction of sensitivity at the site of action have an effect on *Bt* or *B. sphaericus* activity [12]. It has been shown in the *B. sphaericus*-resistant colony, SPHAE, which also carries genes conferring resistance to organophosphorous compounds, that *B. sphaericus* resistance does not depend on the presence of these other resistance genes [28]. However, the use of synthetic larvicides for mosquito control in delicate ecosystems raises environmental concerns especially in coastal and estuarine urban areas. Although resistance management should be integrated and based on several measures, we will consider mostly those control agents that respect the environment.

Slow-release formulations of methoprene, a Juvenil Hormone analogue, are commercially available for use in mosquito control programmes and it may be possible to use them in rotation with *B. sphaericus.* Methoprene poses only minor environmental risks [18] However, it has adverse effects on arthropods such as shrimps and crabs and should therefore be used with caution in estuarine areas.

6.2.2   Biological

Despite some controversy the efficacy and operational viability of certain predators and parasites such as larvivorous fishes, *Toxorhynchites* sp., *Romanomermis* sp., for the control of mosquito populations have been demonstrated (see [2] for review). Many other organisms, including parasitic microsporidia, entomopathogenic fungi, viruses and natural plant extract like Neem, have potential as control agents for mosquitoes, but further investigations are required to optimise their efficacy at the operational level.

The main advantage of using predators, parasites and other pathogens to control mosquito larvae in selected habitats with the conditions required for the establishment and recycling of a mosquito population, is that such sites could function as spatial refuges from exposure to *B. sphaericus* without increasing the density of the mosquito population.

6.2.3   Future alternatives

Other attempts to overcome and to reduce the risk of developing resistance to bacterial toxins have involved combining toxins with various modes of action. *Bti* toxins are major candidates because of the number and also because of absence of cross-resistance to *B. sphaericus.* *B. sphaericus* strain 2297 has been tested as a recombinant in which *cry* genes from *Bti* have been introduced into the chromosome by homologous recombination *in vivo* [30] . The gene encoding the cytolytic protein from the *Bt* serovar *medellin* has been expressed in *B. sphaericus* strain 2297: it partially restored toxicity against two *B. sphaericus*-resistant *Culex* colonies probably due o the ability of Cyt toxin to create pores and then facilitating the entrance of Bin toxins too [57]. Susceptibility was also partially restored, in resistant colonies, with *B. sphaericus* recombinant, producing Cry11A from *Bti* or Cry11Ba from *Bt* serovar *jegathesan* [41]. The Cyt1A of *Bti* also has an effect in decreasing resistance to *B. sphaericus* or to the Cry toxins of *Bti* in laboratory-selected colonies ([57] and M.C. Wirth, personal communication). Several investigations have also been conducted in an attempt to introduce and to overproduce the entomopathogenic toxins in organisms living in the environment and serving as natural larval food, such as Cyanobacteria [50].

6.3   Application strategy

Unlike classical biological control agents, larvicides based on entomopathogenic bacteria can cause 100% larval mortality within 24 to 48 hours which is necessary for to reduce significantly the size of the mosquito population. *B. sphaericus* and *Bti* are amongst the most important commercially available biopesticides active against mosquito larvae. The first steps in the modes of action of these bacteria are similar, but their toxins have different target molecules and the time taken to have a lethal effect, *Bti* killing mosquito larvae more rapidly than *B. sphaericus* [53]. *B. sphaericus* persists for longer in water polluted with organic matter [25, 47] such as is typical for *C. p. quinquefasciatus* urban breeding sites.

This suggests that in mixtures of the two, the shorter lethal time of *Bti* may result in its having the stronger toxic effect, probably even masking that of *B. sphaericus*. Indeed, some experiments have shown that powders in which these two bacteria are mixed in various ratios are not more toxic to the tested mosquito species than the individual bacteria ([23] and I. Thiéry, personal communication).

The use of *Bti* in rotation, replacing *B. sphaericus* for short periods would be a better option for managing resistance in programs based on *B. sphaericus* spraying, as has been done in China [61]. However, the frequency of treatments must be increased to overcome the lack of persistence of *Bti* in breeding sites. Alternatively, the mosaic model could be used. This is illustrated by the new "stable zone strategy" proposed by Lenormand & Raymond [22], in which the various insecticides are applied to an area smaller than a certain critical size, determined by mosquito migration distance. In any situation, the dose applicated is important. Theoretically, the ultra high dose strategy is a way to prevent resistance development in insect populations. If the dose delivered to the target is sufficiently high to kill heterozygous the onset of resistance can be delayed. However in practice, it is difficult to maintain a high-dose schedule in control programs for environmental and economic reasons [39].

7.   CONCLUSIONS AND PERSPECTIVES

The vast majority of entomopathogenic bacteria used in vector control belong to *Bti*, to which no field resistance has been reported. Thus consideration of the management of resistance to bacterial larvicides essentially concerns *B. sphaericus*. Resistance to *B. sphaericus* has been recorded in mosquito populations both in laboratory, after subjecting several generations to intense selection pressure, and in field conditions, after periods of continuous exposure to *B. sphaericus* strains 2362, 1593 and C3-41. However, *B. sphaericus* has several advantages over other control measures against mosquitoes as *C. p. quinquefasciatus* and *Anopheles* sp., and should therefore be considered to be an efficient control agent, if applied rationally as part of an integrated control program.

It is important to monitor the susceptibility of the mosquito population before and during treatment. Spatial and temporal refuges from exposure to *B. sphaericus* are very important for the prevention of resistance. Other measures include multiplying control methods by integrating other available measures such as the physical protection of breeding sites, certain growth regulators and other environmentally safe agents. For example, classical biological control could be used continuously in certain habitats and *Bti* could be used as a temporary alternative in certain conditions.

Other mosquitocidal *B. thuringiensis* strains, *B. thuringiensis* serovar *medellin* and *jegathesan*([51] and Chapter 2.4), and eventually at least two *B. sphaericus* strains (IAB-59 and LP1-G) active against mosquito populations resistant to *B. sphaericus* (strains 2362, 1593) are possible alternatives to the commercially available strains, but further research into their potential value is required. New toxic elements and investigations into the ability of these strains to reduce or to delay the development of resistance to *B. sphaericus* strain 2362/1593/C3-41 are required. In addition, the use of recombinant *B. sphaericus* expressing toxins from other mosquitocidal bacteria should be considered. There is a risk attached to introducing *B. sphaericus* crystal toxin genes alone into natural mosquito larval foods such as Cyanobacteria, because this would expose the larvae to continuous selection pressure and then to resistance. Increasing our understanding of the mode of action of the *B. sphaericus* Bin toxin will help to determine other mechanisms of resistance. And the identification of mutations that may be responsible for resistance in the gene encoding the toxin receptor may enable us to predict and possibly to reduce the risk of the development of resistance.

ACKNOWLEDGEMENT

Thanks to Christine Chevillon (University Montpellier II, France), and Maria Helena Silva-Filha (CPqAM/FIOCRUZ, Brazil), for critical reading of the manuscript.

REFERENCES

[1] Abdul-Rauf M & Ellar DJ (1999) Toxicity and receptor Binding Properties of *Bacilllus thuringiensis* Cry1CToxin active agaisnt both Lepidoptera and Diptera. J. Invertbr. Pathol 73, 52-58

[2] Barbazan P, Baldet T, Darriet F *et al.* (1997) Control of *Culex quinquefasciatus* (Diptera: Culicidae) with *Bacillus sphaericus* in Maroua, Cameroon. J. Am. Mosq. Control Assoc. 13, 263-269

[3] Becker N & Ludwig M (1993) Investigations on possible resistance in *Aedes vexans* field populations after a 10-year application of *Bacillus thuringiensis israelensis*. J. Am. Mosq.Control Assoc. 9, 221-224

[4] Berry C, Jackson-Yap J, Oei C & Hindley J (1989) Nucleotide sequence of 2 toxin genes from *Bacillus sphaericus* Iab59 - Sequence comparisons between 5 highly toxinogenic strains. Nucleic Acids Res. 17, 7516

[5] Bravo A, Koen H, Jansens S & Peferoen M (1992) Immunocytochemical analysis of specific binding of *Bacillus thuringiensis* insecticidal crystal proteins to lepidopteran and coleopteran midgut membranes. J. Invertebr. Pathol. 60, 247-253

[6] Charles J-F (1987) Ultrastructural midgut events in Culicidae larvae fed with *Bacillus sphaericus* 2297 spore/crystal complex. Ann. Inst. Pasteur/Microbiol. 138, 471-484

[7] Charles J-F, Silva-Filha M-H, Nielsen-LeRoux C, Humphreys M-J & Berry C (1997) Binding of 51-and 42 kDa individual components from *Bacillus sphaericus* crystal toxin to mosquito larval midgut membranes from *Culex* and *Anopheles* sp. (Diptera: Culicidae). FEMS Microbiol. lett. 156, 153-159

[8] Denholm I & Rowland MW (1992) Tactics for managing pesticide resistance in arthropods : theory and practice. Ann. Rev. Entomol. 37, 91-112

[9] Feldman A, Dullemans A & Waalwijk C (1995) Binding of the CryIVD toxin of *Bacillus thuringiensis* subsp. *israelensis* to larval dipteran midgut proteins. Appl. Environm. Microbiol. 61, 2601-2605

[10] Ferré J, Real MD, Van Rie J, Jansens S & Peferoen M (1991) Resistance to the *Bacillus thuringiensis* bioinsecticide in a field population of *Plutella xylostella* is due to a change in a midgut membrane receptor. Proc. Natl. Acad. Sci. USA 88, 5119-5123

[11] Georghiou G, Wirth M, Ferrari J & Tran H (1991) Baseline susceptibility and analysis of variability toward biopesticide in Califonia populations of *Culex quinquefasciatus.* Annual Report, University of Califonia Riverside 25-27

[12] Georghiou GP (1994) Principles of insecticide resistance management. Phytoprotection, 75, 51-59

[13] Gould F, Anderson A, Jones A *et al.* (1997) Initial frequency of alleles for resistance to *Bacillus thuringiensis* toxins in field populations of *Heliothis virescens*. Proc. Natcl. Acad. Sci. USA 94, 3519-3523

[14] Hofmann C, Lüthy P, Hütter R & Pliska V (1988) Binding of the delta endotoxin from *Bacillus thuringiensis* to brush-border membrane vesicles of the cabbage butterfly (*Pieris brassicae*). Eur. J. Biochem. 173, 85-91

[15] Hougard J-M & Back C (1992) Perspectives on the bacterial control of vectors in the tropics. Parasitology Today 8, 364-366

[16] Hougard JM, Mbentengam R, Lochouarn L *et al.* (1993) Control of *Culex quinquefasciatus* by *Bacillus sphaericus*: results of a pilot campaign in a large urban area in equatorial africa. Bull WHO 71, 367-375

[17] Humphreys MJ & Berry C (1998) Variants of the *Bacillus sphaericus* binary toxin : implications for differential toxicity strains. J. Invertbr. Pathol. 71, 184-85

[18] Ishii T (1985) Field trials of Altozid 10F against mosquitoes in Japan, p. 143-163. *In* Laird M & Miles (ed.), J Integrated Mosquito Control Methodologies, Vol 2. London Academic Press Inc. Ltd

[19] Karch S, Asidi N, Manzambi M & Salaun JJ (1992) Efficacy of *Bacillus sphaericus* against the malaria vector *Anopheles gambiae* and other mosquitoes in swamps and rice fields in Zaire. J. Am. Mosq. Control Assoc. 8, 376-380

[20] Kumar A, Sharma VP, Thavaselvam D *et al.* (1996) Control of *Culex quinquefasciatus* with *Bacillus sphaericus* in Vasoco City, Goa. J. Am. Mosq. Control. Assoc. 12, 409-413

[21] Laird M & Miles J (1985) *In* Integrated Mosquito Control Methodologies, Academic Press Inc. Ltd, London

[22] Lenormand T & Raymond M (1998) Resistance management : the stable zone strategy. Proc. R. Soc. Lond. B 265, 1985-1990

[23] Majora Brazao e Silva C (1997) Avaliaçao dos efeitos interrativos decorrentes do uso combinado de biomassas das epécies entomotogênicas: *Bacillus thuringiensis* serovar *israelensis* IPS82 e *Bacillus sphaericus* 2362. Fundaçao Oswaldo Cruz

[24] Mardini L (1998) Programa Estadual de controle biologico *Simulium* sp (Diptera: Simulidae) No Rio Grando do Sul, Brasil. Proc. VI SICONBIOL, Rio de Janeiro, Brasil 41-43

[25] Nicolas L, Dossou-Yovo J & Hougard J-M (1987) Persistence and recycling of *Bacillus sphaericus* 2362 spores in *Culex quinquefasciatus* breeding sites in West Africa. Appl. Microbiol. Biotechnol. 25, 341-345

[26] Nielsen-LeRoux C & Charles J-F (1992) Binding of *Bacillus sphaericus* binary toxin to a specific receptor on midgut brush-border membranes from mosquito larvae. Eur. J. Biochem. 210, 585-590

[27] Nielsen-LeRoux C, Charles J-F, Thiéry I & Georghiou GP (1995) Resistance in a laboratory population of *Culex quinquefasciatus* (Diptera: Culicidae) to *Bacillus sphaericus* binary toxin is due to a change in the receptor on midgut brush-border membranes. Eur. J. Biochem. 228, 206-210

[28] Nielsen-LeRoux C, Pasquier F, Charles J-F *et al.* (1997) Resistance to *Bacillus sphericus* involves different mechanisms in *Cuelx pipiens* (Diptera : Culicidae) larvae. J. Med. Entomol. 34, 321-327

[29] Oppert B, Kramer KJ, Johnson DE, Macintosh SC & McGaughey WH (1994) Altered protoxin activation by midgut enzymes from a *Bacillus thuringiensis* strain of *Plodia interpunctella*. Biochem. Biophys. Res. Comm. 198, 940-947

[30] Poncet S, Bernard C, Dervyn E *et al.* (1997) Improvement of *Bacillus sphaericus* toxicity against Diperan larvae by integration, via homologous recombination, of the Cry11A toxin gene from *Bacillus thuringiensis* subsp. *israelensis*. Appl. Environ. Microbiol. 63, 4413-4420

[31] Priest FG, Ebdrup L, Zahner V & Carter P (1997) Distribustion and characterization of mosquitocidal toxin genes in some strains of *Bacillus sphaericus*. Appl. Environ. Microbiol. 63, 1195-1198

[32] Rao DR (1996) Management of resistance of *Culex quinquefasciatus* to *Bacillus sphaericus*. Annual Report : Centre for Research in Medical Entomology, ICMR, Madurai, India 18

[33] Rao DR, Mani TR, Rajendran R, Joseph AS & Gajanana A (1995) Development of high level resistance to *Bacillus sphaericus* in a field population of *Culex quinquefasciatus* from Kochi, India. J. Amer. Mosq. Control. Assoc. 11, 1-5

[34] Ravoahangimalala O, Charles J-F & Schoeller-Raccaud J (1993) Immunological localization of *Bacillus thuringiensis* serovar *israelensis* toxins in midgut cells of intoxicated *Anopheles gambiae* larvae (Diptera: *Culicidae*). Res. Microbiol. 144, 271-278

[35] Regis L, Furtado AF, Oliveira C & *et al.* (1996) Integrated Control of the Filariasis Vector with Community participation, in an Urban Area of Recife. Cadernos Saúde Pública 12, 473-482

[36] Regis L, Silva-Filha MHNL, de Oliveira CMF *et al.* (1995) Integrated control measures against *Culex quinquefasciatus*, the vector of filariasis in Recife. Mem. Inst. Oswaldo Cruz 90, 115-119

[37] Rodcharoen J & Mulla MS (1994) Resistance development in *Culex quinquefasciatus* (Diptera : Culicidae) to the microbial agent *Bacillus sphaericus*. J. Econ. Entomol. 87, 1133-1140

[38] Rodcharoen J & Mulla MS (1996) Cross-resisatnce to *Bacillus sphaericus* strains in *Culex quinquefasciatus*. J. Am. Mos. Control. Assoc. 12, 247-250

[39] Roush RT (1993) Occurence, genetics and management of insecticide resistance. Parasitology Today 9, 174-179

[40] Schofield C (1991) Vector population responses to control intervention. Ann. Soc. Belg. Méd. Trop 71, 201-217

[41] Servant P, Rosso M-L, Hamon S *et al.* (1999) Production of Cry11A and Cry11Ba toxins in *Bacillus sphaericus* confers toxicity towards *Aedes aegypti* and resistant *Culex* populations. Appl. Environ. Microbiol. 65, 3021-3026

[42] Silva-Filha M-H, Nielsen-LeRoux C & Charles J-F (1997) Binding kenetics of *Bacillus sphaericus* binary toxin to midgut brush border membranes of *Anopheles* and *Culex* spp. mosquito larvae. Eur. J. Biochem. 247, 754-761

[43] Silva-Filha M-H, Regis L, Nielsen-leRoux C & Charles J-F (1995) Low level resistance to *Bacillus sphericus* in a field-treated population of *Culex quinquefasciatus* (Diptera: Culicidae). J. Econ. Entomol. 88, 525-30

[44] Silva-Filha MH & Regis L (1997) Reversal of low-level resistance to *Bacillus sphaericus* in a field populatioin of the Southern House Mosquito (Diptera: Culicidae) from Urban Area of Recife, Brazil. J. Econ. Entomol. 90, 299-303

[45] Sinègre G, Babinot M, Quermel JM & Gaven B (1994) First field occurence of *Culex pipiens* resistnce to *Bacillus sphaericus* in southern France. In VIII European Meeting of Society for Vector Ecology, 5-8 September 1994, Faculty of Biologia, University of Barcelona, Spain

[46] Sinègre G, Babinot M, Vigo G & Jullien J-L (1993) *Bacillus sphaericus* et démoustication urbaine. Bilan de cinq années d'utilisation expérimentale de la spécialité Spherimos® dans le sud de la France. Entente Interdépartementale pour la Démoustication du Littoral Méditerranéen. Document E.I.D.L.M. N°62, 21 pp

[47] Skovmand O & Bauduin S (1997) Efficacy of granular formulation of *Bacillus sphaericus* agaisnt *Culex quiqnquefasciatus* and *Anopheles gambiae* in West African countries. J. Vector Ecol. 22, 43-51

[48] Sundararaj R & Reuben R (1991) Evaluation of a microgel droplet formulation of *Bacillus sphaericus* 1593 M (Biocide-S) for control of mosquito larvae in rice fields in Southern India. J. Am. Mosq. Control Assoc. 7, 556-559

[49] Tang JD, Gilboa S, Roush RT & Shelton A (1997) Inheritance, stability, and lack-of-fitness cost of field-selected resistance to *Bacillus thuringiensis* in diamondback Moth (Lepidoptera : Plutellidae) from Florida. J. Econ. Entomol. 90, 732-741

[50] Thanabalu T, Hindley J, Brenner S, Oei C & Berry C (1992) Expression of the mosquitocidal toxins of *Bacillus sphaericus* and *Bacillus thuringiensis* subsp. *israelensis* by recombinant *Caulobacter crescentus*, a vehicule for biological control of aquatic insect larvae. Appl. Environ. Microbiol. 58, 905-910

[51] Thiéry I, Fouque F, Gaven B & Lagnau C (1999) Residual activity of *Bacillus thuringiensis* serovar *medelin* and *jegathesan* on *Culex pipiens* and *Aedes aegypti* larvae. J. Am. Mosq.Control Assoc. 15, 371-379

[52] Thiéry I, Hamo S, Delécluse A & Orduz S (1998) The introduction into *Bacillus sphaericus* of *Bacillus thuringiensis* subsp. *medelin* cyt1Ab1 gene results in higher susceptibility of resistant mosquito larva populations to *B. sphaericus*. Appl. Environ. Microbiol. 64, 3910-3916

[53] Thiéry I, Hamon S, Cosmao Dumanoir V & de Barjac H (1992) Vertebrate safety of *Clostridium bifermentans* serovar *malaysia*, a new larvicidal agent for vector control. J. Econ. Entomol. 85, 1618-1623

[54] Van Rie J, Jansen S, Höfte H, Degheele D & Van Mellaert TH (1990) Receptors on the brush border membrane of the insect midgut as determinants of the specificity of *Bacillus thuringiensis* delta-endotoxins. Appl. Environ. Microbiol. 56, 1378-1385

[55] Van Rie J, Mc Gaughey W, Johnson D, Barnette D & Van Mellaert TH (1989) Mechanism of resistance to the microbial insecticide *Bacillus thuringiensis*. Science 247, 72-74

[56] Wirth M & Georghiou G (1997) Cross resistance among CryIV toxins of *Bacillus thuringiensis* subsp. *israelensis* in *Culex quinquefasciatus* (Diptera: Culicidae). J. Econ. Entomol. 90, 1471-1477

[57] Wirth M, Georghiou GP & Federici B (1997) CytA enables cryIV endotoxins of *Bacillus thuringiensis* to overcome high levels of CryVI resistance in the mosquito in *Culex quinquefasciatus*. Proc.Natl.Acad.Sci. USA 94, 10536-10540

[58] Yadav R, Sharma V & Upadhyay A (1997) Field trial of *Bacillus sphaericus* strain B-101 (serotype H5a,5b) against Filariasis and Japanese encephalitis vectors in India. J. Am. Mosq. Control Assoc. 13, 158-163

[59] Yuan Z, Cai Q, Zhang Y & Liu E (1998) High-level resistance to *Bacillus sphaericus* C3-41 in field collected *Culex quinquefasciatus*. Abstract of the VIIth Intern. Coll. Invert. Pathol. Microb. Control, Saporo, Japan, August 1998, 40

[60] Yuan Z, Nielsen-LeRoux C, Pasteur N, Charles JF & Frutos R (1998) Detection of the binary toxin genes of several *Bacillus sphaericus* strains and their toxicities against susceptible and resistant *Culex pipiens*. Acta Entomol. Sinica 41, 337-342

[61] Yuan Z, Zhang Y, Cai Q & Liu Ey (2000) High-level field resistance to *Bacillus sphaericus* C3-41 in *Culex quinquefasciatus* from southern China. Biocontrol. Sci. & Technol. 10 *(in press)*

Table 1. Examples of large-scale use of Bacillus sphaericus against mosquito populations and the development of resistance.

Target species	B. sphaericus strain	Treatment frequency	Treatment period	Treated area	Locality /country	Resistance RR[1] (LC$_{50}$)	Ref.
A. stephensi	B-101	Weekly	9 months	2 km^2	Panaji city/ India	–	[20]
C. quinquefasciatus	2362	Monthly / fortnightly	26 months	1.2 km^2	Recife city/ Brazil	10	[43]
C. quinquefasciatus	1593	Fortnightly	2 years	8 km^2	Kochi/ India	146	[33]
C. quinquefasciatus	2362	Monthly	18 months	5 km^2	Olinda city/ Brazil	No	[36]
C. pipiens	2362	Every 21 days to 6 months	7 years	210 villages	Mediterranean Coast/ France	>20,000^2	[46]
C. quinquefasciatus	C3-41	3 times / month	8 years	8 km^2	Dongguan city/ China	22,000	[61]
C. quinquefasciatus	2362	Every 3 months	4 years	200 ha	Yaoundé/ Cameroon	–	[16]
C. quinquefasciatus	2362	Twice / year	2 years	2,000 ha	Maroua/ Cameroon	–	[2]

[1](RR) The level of resistance is calculated as the ratio of the LC$_{50}$ values of the resistant population to that of susceptible laboratory colonies; (–) not reported. [2]at isolated breeding sites.

Chapter 7.1

Biotechnological improvement of *Bacillus thuringiensis* for agricultural control of insect pests: benefits and ecological implications

Vincent Sanchis
Unité de Biochimie Microbienne, Institut Pasteur, 28 rue du Dr. Roux, 75724 Paris Cedex 15, France and Station de Recherches de Lutte Biologique, INRA, La Minière, 78285 Guyancourt Cedex, France.

Key words: *Bacillus thuringiensis*, biopesticides, δ–endotoxin, asporogenic strains, site-specific recombination, transgenic plants, resistance management

Abstract: *Bacillus thuringiensis* (*Bt*) strains and toxins are highly diverse. Our understanding of their regulation and the development of efficient host-vector systems has made possible to overcome a number of the problems associated with *Bt*-based insect control measures. Recombinant DNA technology has been used to develop new *Bt* strains for more effective pest control in various crops. *Bt* insecticidal toxin genes have also been introduced into bacteria that colonise plants and inserted directly into plants to make them resistant to specific insect pests. This article presents an overview of the principal approaches used to improve *Bt* and describes the achievements of biotechnology and the prospects for future improvement.

1.   INTRODUCTION

In the last few decades the more widespread use of agricultural fertilisers and pesticides has been a major factor making it possible to increase the world's food supply, essentially by increasing agricultural yields. As a result, not only has the control of pests and diseases become one of the most costly aspects of plant production, but pesticides have become indispensable tools in the maintenance of high agricultural productivity [41]. For example, in 1997, the world-wide market for

J.-F. Charles et al. (eds.),
Entomopathogenic Bacteria: From Laboratory to Field Application, 441–459.
© 2000 *Kluwer Academic Publishers. Printed in the Netherlands.*

insecticides alone was estimated at approximately 8 billion US dollars per annum. At the same time, factors such as increasing public awareness about the fragility of the environment, better understanding of the chemical hazards, higher development costs for new chemicals and the increasing problem of insecticide resistance, have led us to search for safer alternative control methods. Therefore, much attention has been given to the use of microbial control agents that cause disease in insects and to the development of transgenic crop varieties expressing genes for resistance to insects. Today, biopesticides account for 2% of the pesticide market and 95% of all microbial pesticide products world-wide are based on the bacterial agent *Bacillus thuringiensis* (*Bt*). *Bt* is a spore forming bacterium that produces highly specific insecticidal proteins, called δ-endotoxins or Cry proteins, during the stationary phase, at the same time as sporulation. Commercial *Bt* products generally consist of a mixture of spores and crystals produced in large fermenters at a cost competitive with that of chemical insecticides and are applied as foliar sprays, much like synthetic insecticides. The genes encoding δ–endotoxins have also become the principal source of insecticidal genes for introduction into plants to produce insect-resistant transgenic plants and the large-scale use of insect-resistant crops expressing δ–endotoxin genes began in 1996 and 1997.

Biopesticides containing *Bt* are environmentally friendly and effective in a variety of situations. However, their performance is often considered to be poorer than that of chemicals in terms of reliability, spectrum of activity, speed of action and cost effectiveness. Indeed, *Bt* products are not as potent or persistent in the field as the reference chemical products: *Bt* products act slowly, have a narrow activity spectrum (minimising the size of their potential market) and are not stable in the environment after spraying because they are rapidly inactivated by exposure to sunlight [43] or other environmental factors. Consequently, the duration of pest control is often too short and their use on many crops is not cost effective because too many applications are required. Therefore, the driving force for their adoption by growers has been the continuous provision of more powerful and efficient products and the ability of these products to control insect pests resistant to other insecticides, rather than their low ecotoxicity and other ecological advantages. Today, *Bt* products are still essentially restricted to niche markets, favoured by environmental pressures, such as forestry, organic farming and glasshouse fresh vegetable production, in which the application of conventional chemical agents is restricted and to the development of insect-resistant varieties of the world's major crops (cotton, corn, rice soybean, potatoes), which are prone to insect attack.

This review presents the various biotechnological approaches that have been used to improve *Bt* products for the control of agricultural insect pests and have contributed to the successful use of this biological control agent. I will also discuss some of the potential problems that could arise from the inadequate use or deployment in the field of some of these genetically engineered products. Finally, possible solutions to these problems will be described and future prospects for research to extend the potential and preserve the future of this biocontrol agent will be discussed.

2.   IMPROVEMENT OF *BT* STRAINS

2.1   Construction of new *Bt* strains by conjugation

The first step towards improving *Bt* strains involves the isolation of new strains with higher insecticidal activity against targeted insect pests and the cloning of *cry* genes encoding new insecticidal crystal proteins. Such genes have been isolated from various *Bt* strains isolated from several sources (soil, insects, plants etc.) from a number of geographic areas. More than 200 *cry* genes have now been cloned and sequenced. Bioassays of purified crystal proteins on various insect pests have shown that each *cry* gene product is toxic to only a few insects species. The characterisation of the specific activities of individual *Bt* strains and cloned *cry* genes led to the development of genetically modified strains with optimised activity against a given insect pest or a broadened toxicity spectrum. Conjugation was initially used to manipulate *Bt* strains genetically because no transformation procedure was available. Early studies had shown that many *cry* genes were present on large extrachromosomal plasmids, some of which were conjugative (see chapter 2.5, and [18]). Conjugation has been used by Ecogen to construct *Bt* strains carrying new combinations of *cry* genes ; for example, strain EG2348, the active ingredient of the Condor® bioinsecticide product (Table 1) contains a combination of *cry* genes encoding crystal proteins particularly active against specific lepidopteran pests of soybean crops. Another strain, EG2424, the active ingredient of Foil® (Table 1) contains two plasmids, one carrying a *cry* gene whose product is active against a lepidopteran, *Ostrinia nubilalis* (European corn borer), a major pest of corn in the US and Europe, the second encoding a crystal protein with activity against a coleopteran, *Leptinotarsa decemlineata* (Colorado potato beetle), a pest of potatoes.

Table 1. Examples of commercially available genetically engineered microbial insecticides based on *Bacillus thuringiensis*

Products	Companies	Target insects	Crops
Condor®	Ecogen		Soybean, Forestry
Cutlass®	Ecogen		Vegetables crops
CryMax®	Ecogen	Lepidopteran larvae	Corn
CryStar®	Ecogen		Fruit trees
MVP®	Mycogen		Grapes
Mattch®	Mycogen		Cotton
Agree®	Ecogen	Lepidopteran larvae	Corn
Foil®	Ecogen	Lepidopteran and	Vegetables
Raven®	Ecogen	Coleopteran larvae	Potatoes
M-Trak®	Mycogen	Coleopteran larvae	Potatoes Tomatoes

This approach was successful for improving the insecticidal properties of *Bt* and its development as a biopesticide. However, it was only applicable to *cry* genes carried by conjugative plasmids and could not be used to associate genes on plasmids from the same incompatibility group. Moreover, the conjugative approach could not be used to over express a given gene or to exploit interesting genes expressed only weakly in *Bt*.

2.2 Expression of *Bt cry* genes in recombinant heterologous microbial hosts

A second strategy for improving the exploitation of *Bt* or increasing its entomopathogenic potential involves diversifying or improving the administration of the toxin, by using molecular biological methods. This approach has been used by a number of laboratories trying to express genes encoding Cry proteins in diverse organisms occupying the same ecological environment as the targeted insects; *cry* genes have thus been introduced into *Azospirillum* sp. [53], *Pseudomonas cepacia* [50] and *P. fluorescens* [37], three bacteria that colonise the roots and leaves of many plants, and *Rhizobium leguminosarum* [49], a symbiotic nitrogen-fixing bacterium responsible for root nodule formation in many legume species. This approach, which involves synthesising the toxin in a bacterium other than *Bt*, has been used by Mycogen to produce two commercial products : MVP® for controlling lepidopterans and M-Trak® for controlling coleopterans (Table 1). The *Bt* toxins are produced in *P. fluorescens* and the bacteria are then killed by means of a physical chemical process and the toxins remain enclosed in the cells as crystalline inclusions. This process significantly increases the efficacy of the Cry proteins, increasing their persistence in the environment by protecting them against degradation and inactivation by UV irradiation. In addition, the registration procedure for these new products is relatively straightforward because the bacteria have been killed.

Using a different approach, Crop Genetics International (CGI) has transferred a *cry* gene into an endophytic bacterium, *Clavibacter xyli* var. *cynodontis*. This bacterium colonises the vascular system of various plants including maize. The gene introduced into this bacterium encodes a protein toxic to the larvae of *O. nubilalis* [27]. This insect is particularly difficult to control due to the way it attacks plants. The young caterpillars burrow into the apical bud and then penetrate into the interior of the stem, creating a network of holes in the soft tissue. Thus, the insect rapidly finds shelter from classical insecticides and the damage it causes is not immediately apparent. CGI wanted to produce maize plants colonised by an endophytic bacterium synthesising a *Bt* toxin, thereby protecting the plants against *O. nubilalis* infestations. The efficacy of such a process depends on the recombinant bacterium producing sufficient toxin and on the development of a method for inoculating seeds with the endophytic bacterium (by encasing the seeds or dusting them with powder). Encouraging results have been obtained and a product called InCide®, developed using this technology, is about to be released commercially.

2.3 Transformation and genetic manipulation of *Bt* strains

Another important step in the production of new *Bt* strains was the development of a transformation system for *Bt*. This has opened up new possibilities for improving the entomopathogenic potential of *Bt*, making it possible to fully exploit the high natural diversity of *Bt* strains and toxins and to modify strains of *Bt*, changing their characteristics. In 1989, several laboratories reported that vegetative *Bt* cells were readily transformed with plasmid DNA by electroporation [8, 29]. In 1991, *Bt* strains were shown to possess several restriction-modification systems and some strains were shown to restrict methylated DNA [32]. Therefore, a much higher efficiency of transformation was achieved with plasmid DNA prepared from a *dam- dcm- Escherichia coli* strain and most *Bt* strains can now be transformed with frequencies of 10^2 to 10^5 transformants / μg of plasmid DNA. The construction of genetically engineered *Bt* strains also required the development of an effective host/vector system with vector plasmids able to replicate and persist in a stable manner in this bacterium. Plasmids carried by *Bt* were cloned, the regions required for their replication and stability identified and shuttle vectors (*E. coli/Bt*) were constructed using the replication regions of these *Bt* resident plasmids [5, 6]. This has made it possible to reintroduce various cloned *cry* genes back into *Bt* and to construct genetically engineered *Bt* strains containing new combinations of toxin genes. The genetic manipulation of *Bt* strains is also required for the disruption or elimination of deleterious genes and for introducing new genes, not only into the vector plasmids, but also directly into the bacterial chromosome or plasmids naturally present in *Bt*. Heat-sensitive vectors capable of integration have therefore been constructed [30]. Such vectors have been used to integrate, by homologous recombination, a *cry* gene directly into the *Bt* chromosome [25] and into a resident *Bt* plasmid [9, 30]. These thermosensitive plasmids have also been used to disrupt genes involved in sporulation, such as *spo0A* and *sigK* (see below), to produce asporogenic strains of *Bt* [28]. The first generation of plasmid vectors used for the construction of *Bt* recombinant strains generally contained, in addition to the origin of replication functional in *Bt,* a second origin of replication functional in Gram-negative bacteria such as *E. coli* and antibiotic resistance genes for selection. These DNA sequences were undesirable when seeking regulatory approval for the release of these recombinant strains and their use as commercial biopesticides.

For application in the field, recombinant strains of *Bt* devoid of DNA from other sources have been constructed [7, 47]. These strains, from which antibiotic resistance genes and other non-*Bt* DNA sequences were selectively eliminated, were obtained *in vivo* using a new generation of site-specific recombination vectors based on the specific resolution site of the *Bt* class II transposons, Tn*4430* and Tn*5401* [7, 33, 46]. These vectors contain only the DNA sequences necessary for selection and replication in *E. coli* at the earliest stages of assembly. The sequences that are undesirable in the final *Bt* recombinant strains are flanked with two internal resolution sites (IRSs) in the same orientation. In an appropriate host background (*Bt* strains containing Tn*4430* or Tn*5401*), site-specific recombination between the duplicate IRSs, catalysed by the recombinase of Tn*4430* or Tn*5401*, eliminates the resistance marker genes and other non-*Bt* DNA after introduction of the vector into the bacterium and selection of the transformants [7, 47]. The first live recombinant *Bt* product to be produced using this technology was developed and sold by Ecogen under the trade name Raven®. The active ingredient of Raven® is a *Bt* strain that produces two types of Cry3 proteins active against beetles (Cry3A and Cry3B) and a protein active

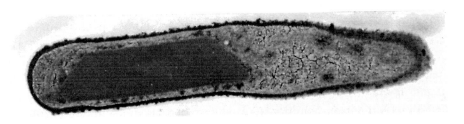

Figure 23. Electron micrograph of an asporogenic *spo0A* mutant of *Bt* strain 407 producing large amounts of Cry3A crystal protein. The toxins accumulate to form large crystal inclusions which remain encapsulated within the ghost cells

against caterpillars, Cry1Ac (Table 1). The development of an efficient DNA transfer system for *Bt* also made it possible to analyse the expression of the cloned crystal protein genes in their natural host. The principal result of this work was the demonstration that the *cry3A* gene was activated at the onset of sporulation, independently of the factors involved in the initiation of sporulation [3]. This clearly implies that it should be possible to produce the Cry3A toxin in a sporulation-deficient genetic background. This was demonstrated when the *cry3A* gene was expressed in a Spo0A mutant of *Bt*, resulting in the production of large amounts of Cry3A toxin (Figure 1) [28].

The scope for genetic manipulation of *Bt* was also increased by the demonstration that *in vivo* recombination occurs in this bacterium. The gene encoding the *cry3A* coleopteran-specific toxin was inserted into a resident plasmid of a *Bt* strain toxic to lepidopteran insects, using a thermosensitive vector eliminated after recombination. The resulting strain had insecticidal activity against both lepidopterans and coleopterans [30]. The total amount of crystal protein produced by the recombinant strain was almost twice that of the native strain. This is presumably because the expression systems of the *cry* genes are different, so they do not compete for rate-limiting elements of gene expression such as specific sigma factors [4].

2.4   Construction of non-sporulating *Bt* strains

Sporogenic *Bt* strains have been used for more than 40 years and have a good safety record for vertebrates and other non-target organisms. However, in a few cases, *Bt* has been shown to be responsible for infections in man [45]. A case of severe war wounds infected by *Bt* serotype H34 has been described and experimental evidence that this strain is pathogenic in a mouse model of cutaneous infection presented [23]. *Bt* spores are also known to cause infection in the mulberry silkworm, *Bombyx mori*, and extensive use of *Bt* formulations on vegetables in silkworm-rearing areas or countries like India or Japan frequently results in the accidental infection of silkworm larvae [35]. A recent study also reported adverse effects of a *Bt* formulation on a beneficial organism commonly produced and used for the control of lepidopteran pests [36]. It was therefore of interest to develop asporogenic *Bt* recombinant strains capable of producing large amounts of Cry proteins.

A sporulation-dependent lepidopteran active *cry1C* gene has been expressed in a sporulation-deficient background by fusing the coding sequence of the *cry1C* gene to the non-sporulation dependent *cry3A* promoter. This resulted in large amounts of Cry1C accumulating in a sporulation-deficient Spo0A mutant of *Bt* [46]. Another report showed that as much Cry1Aa protein was produced from a wild-type type sporulation-dependent *cry1Aa* gene introduced by electroporation into a *Bt* mutant blocked at late sporulation, as was produced in the Spo$^+$ strain [9]. In both cases, the toxins accumulated in the mother cell compartment forming crystal inclusions that remained encapsulated within the cell wall. This should make it possible to develop *Bt* products that are safer and more environmentally friendly than the current commercially available sporogenic strains, by producing non-living recombinant *Bt* insecticides. For example, we constructed a non-sporulating derivative of a wild-type *Bt kurstaki* strain by disrupting the chromosomal *sigK* gene, which encodes the σ^{28} late sporulation-specific sigma factor. The σ^{K-} strain produced large amounts of crystal proteins that remained encapsulated in the cells, which did not lyse. The encapsulation of the crystals in *Bt* ghost cells protected them from deactivation by ultraviolet radiation (Sanchis *et al.*, unpublished results) as is the case for *Bt* toxins encapsulated in *P. fluorescens*. Furthermore, the SigK$^-$ strain does not produce viable spores and is therefore unable to compete with wild-type *Bt* strains in the environment, thereby minimising any unforeseen environmental effects arising from the dissemination of large numbers of viable spores. In greenhouse and field trials, this new recombinant strain was as effective for controlling a cabbage pest complex as some currently available commercial *Bt* products (Sanchis *et al.*, unpublished results).

Table 2. Examples of species in which *Bt* transgenic plants have been produced and field tested.

Species (common name)						
Alfalfa	Cabbage	Cotton	Eggplant	Poplar	Rice	Tobacco
Apple	Coffee	Cranberry	Grape	Potato	Spruce	Tomato
Broccoli	Corn	Crysanthemum	Peanut	Rapeseed (Canola)	Sweetgum	Walnut

3.   EXPRESSION OF *CRY* GENES IN PLANTS

In 1987, the biotechnology company Plant Genetic Systems was the first to transform tobacco with a *cry* gene from *Bt,* and this gene was expressed [54]. Tobacco plants transformed with the region of the *cry1* gene encoding the toxic fraction of the molecule were significantly protected against insect attack. The transformation technique used was based on transfer of the Ti plasmid from *Agrobacterium.* This method can be used to introduce foreign DNA into plants other than tobacco. However, its application is restricted to a limited number of plant species, most of them dicotyledons. The development of new methods of transformation, such as electroporation or particle bombardment has made it possible to transfer DNA into most plants, including tomato [15] and monocotyledons such as maize [26]. Technically, all plants species can be rendered toxic to insects by introduction of a *cry* gene from *Bt* and some 20 species have now been transformed by various methods (Table 2).

However, despite the use of strong promoters, the first transformed plants produced too little toxin protein (less than 0.001% of the leaf soluble proteins) for effective agronomic use. The genes of *Bt,* unlike those of plants, have a large proportion of A and T bases (66%). This is not consistent with optimum codon usage in plants and the resulting transcripts may be unstable. This characteristic has been shown to be largely responsible for the problems encountered in attempts to express *cry* genes in plants [40]. Partial or complete resynthesis of *Bt* toxin genes, reducing the A + T content to 51%, to optimise the codon usage, for expression in plants, resulted in much higher expression levels, with the toxin accounting for 0.02 to 0.5 % of leaf soluble protein and more than 100 ng of toxin produced per mg of total protein [40]. This strategy has been successful in various plants: cotton, rice and maize have been transformed with modified *cry1* genes and potato has been transformed with a modified *cry3A* gene [1, 16, 26, 39]. Another strategy, involving the expression of unmodified *cry* genes in plant chloroplasts has also been shown to be very effective in terms of the

Table 3. Genetically engineered *Bt* plants approved for sale as of October 1997.

Crops	Target insects	Trade name	Company
Potato	*Leptinotarsa decemlineata* (Colorado potato beetle)	New Leaf	Monsanto
Cotton	bollworms and budworms	Bollgard	Monsanto
Corn	*Ostrinia nubilalis* (European corn borer)	YieldGard	Monsanto
		KnockOut	Novartis
		Nature Guard	Mycogen
		Bt-Xtra	Dekalb
		StarLink	AgrEvo

amount of toxin produced [34]. It is possible to have many copies of the *cry* gene into the chloroplasts and the translation and transcription apparatus of these organelles is typically prokaryotic. Toxin production may therefore account for 3 to 5% of the leaf soluble protein. Plants transformed by either of these techniques produce significant amounts of toxin, providing excellent protection against insects. A first generation of plants (cotton, maize and potato), expressing genes encoding insecticidal *Bt* toxins, went on the market in the US in 1996 (Table 3).

Second generation transgenic plants are now being developed. In particular, efforts are being made to produce transgenic plants expressing at least two *cry* genes encoding toxins recognising different receptors.

4.   ECOLOGICAL RISKS ASSOCIATED WITH THE USE OF TRANSGENIC *BT* CROPS

4.1   The development of resistant insect populations

As the large-scale use of crops genetically engineered to produce *Bt* toxins becomes a reality world-wide (10 million acres of *Bt* crops were planted world-wide in 1997 and the global acreage is thought to have doubled in 1998), the possible development of resistance to *Bt* toxins in insect pests becomes a critical issue. There are already known cases of resistance to Cry proteins used as sprays [19, 51]. In transgenic plants, *Bt* insecticidal proteins will be produced continually and may be broken down less quickly than those applied as sprays. They may thus be more persistent. This is likely to create strong selection pressure in insect pests and will probably result in insects rapidly building up resistance to *Bt*. Therefore, the first ecological risk associated with the use of transgenic plants involves the development of *Bt* resistant insect populations. This would result in both sprays and transgenic plants becoming ineffective as insect control agents. A number of strategies have been suggested for decreasing the rate at which insects adapt to the *Bt* toxins produced in transgenic plants [44]. They include the engineering of plants to produce *Bt* toxins only in plant tissues prone to insect attack; the use of rotations, in which transgenic plants are alternated with non-transgenic plants, or mosaics, in which mixtures of transgenic and non-transgenic plants are grown together; the creation of refuges, in which a portion of a field is planted with non-transgenic plants and the production of plants expressing several *Bt* toxin genes or other genes from various sources such as proteinases inhibitors. These strategies are essentially based on the notion that the genes conferring resistance to *Bt* toxins are not dominant. Therefore, if enough susceptible insects are allowed to survive at each generation, they will reproduce, possibly mating with resistant insects, and their offspring will consist mostly of susceptible heterozygotes. This should greatly decrease the risk of homozygous resistant individuals becoming predominant in the population. However, these resistance management strategies are still only hypothetical and their reliability must be validated both in the laboratory and in the field.

4.2 Effects of *Bt* plants on non-target insects and soil organisms

A second concern associated with *Bt*-transgenic crops is that Cry toxins are produced in an active form in plants whereas in bacteria they are produced as inactive protoxins that must be dissolved and activated in the insect's gut to become toxic. Very little is known about the effects of activated *Bt* toxins on non-target insect species and too little is known to exclude the possibility that the host range of the transgenic organism is not significantly altered or broadened by genetic manipulation. Furthermore, transgenic plants producing *Bt* toxins may persist in the soil for a long period of time [2, 38] and preliminary results suggest that the toxins bind to clay minerals in the soil, which protects them from degradation [21]. Thus toxins accumulate in the soil in an active form and this may affect soil invertebrates not normally in contact with *Bt* toxins. The environmental impact of this is difficult to assess [17]. Moreover, the remote possibility of horizontal gene transfer to other bacterial organisms must also be reconsidered given the greater persistence of the DNA in the environment [17, 31]. Thus, constructs containing antibiotic resistance markers (or ubiquitous bacterial origins of replication for genetically engineered microbes) are not desirable and efforts should be made to develop methods to eliminate these DNA sequences, which are useful only in the primary selection procedure.

4.3   Dissemination of the transgenic characters

Another ecological risk associated with the use of transgenic plants is that of the dissemination of the *Bt* toxin genes into cultivated varieties of the same crop or related wild species that are interfertile to some extent with the transgenic variety. It is unclear whether such gene flow is likely to be frequent and whether it could have adverse effects in a given agricultural or ecological context. The probability of gene transfer, particularly via the pollen, is not negligible, especially for transgenic characters that are advantageous, such as those conferring greater tolerance to insect pests [52]. This highlights our lack of knowledge of the long-term ecological impact of releasing transgenic plants and again points out the need to restrict transgenic constructs to minimal insertions of foreign genes and to develop methods for removing the antibiotic selection markers, after selection of the genetically modified plants, when they are no longer needed. Transgenes could be biologically contained by insertion into the chloroplast genome which is, in most crops, maternally inherited. This would prevent genes incorporated into the chloroplast from being transferred to related plant species via pollen [34]. Chloroplast transformation has only been achieved in tobacco, but the transfer of this technique to other agronomically important plant species should result in the development of new varieties producing very large amounts of insecticidal toxins in which the absence of gene flow via pollen can be guaranteed. However, this approach is only useful for leaf eating insects and cannot be used to control stem and fruit borers or root and tuber-damaging subterranean insects.

5.   FUTURE CHALLENGES AND PROSPECTS

5.1   Managing insect resistance to transgenic *Bt* plants

The use of *Bt* genes in plants is one of the most effective and economical ways to deploy insecticidal toxins and control major insect pests that are difficult to reach with sprays because of their ecological behaviour (e.g. sap sucking insects, borers and root-feeding caterpillars). Moreover, only insects feeding on the plant would be exposed to the insecticide and both beneficial and non-target arthropod species sensitive to the toxin would be preserved. Nevertheless, an assessment of the risk of the targeted pests developing resistance to the *Bt* toxins produced in the plants is critical for the long-term use and effectiveness of *Bt*.

Information about the resistance risk is required for rational decisions about *Bt* gene incorporation into a crop (especially in the case of perennial plants like trees) before the plant is transformed. This key strategic information can be obtained by analysing the crop and the associated pest and parasitoid ecosystem in relation to the primary determinants of the evolution of resistance (e.g. selection intensity, size of refuges or presence of other host plants, pest mortality due to natural enemies, population dynamics, mating behaviour, gene flow). The second step, if the decision to use transgenic plants has been taken, is to include plans for preventing and managing the development of resistance. Such plans are generally based on two strategies that may well work if correctly implemented: a high dose of *Bt* and refuges. Theoretically, the delivery by the plant of a dose of *Bt* toxin high enough to kill virtually all targeted pests would allow only rare highly resistant pests to survive. The use of refuges is designed to prevent these resistant insects from mating with other resistant insects; it relies on these being refuges where susceptible insects can escape exposure to the toxin and are likely to mate with the resistant insects that survive on the transgenic plant, thus diluting the resistance trait in the population [20, 44]. The setting up of a network to collect insects from representative distribution areas and compare their susceptibility in the field with a baseline established before the use of transgenic crops has also been proposed. However, it is difficult to implement an effective monitoring program for the early detection of resistance in the field. Farmers would probably be the first to warn about resistance building up after experiencing the failure of transgenic plants to control the insect pests. Given the diversity of situations in which these plants are likely to be used, it would be better to set up case by case research programs to check whether susceptible individuals remain on the site, and whether the surviving resistant insects are in contact and mate randomly with the susceptible ones thereby delaying the appearance of a resistant homozygous population.

Another problem is the size of the refuges. Estimates on how large such refuges need to be differ greatly. Until recently, the refuges required for *Bt* cotton were only 4% unsprayed or 20% sprayed with other insecticides and no refuges of specific size were required for *Bt* corn varieties. However, the Canadian government has announced that companies selling *Bt* crops must implement resistance management plans ensuring large refuges of a minimum of 20% unsprayed non-Bt corn refuge on each farm planting *Bt*-corn. This measure concerns all *Bt* crops issued after December 3, 1998, but the Canadian government is also amending existing authorisations for *Bt* corn such that the new resistance management plan must be implemented. The Environmental Protection

Agency (EPA) in the USA has also conditionally approved new *Bt* corn subject to the establishment of larger refuges; refuges of 20 to 30% unsprayed (or 40% sprayed refuge) are now required by EPA for the new varieties of Novartis and AgrEvo's *Bt* corn. However, farmers may be uneasy at allowing a fraction of the pest population to survive and it is then important to determine whether requesting farmers to plant 20 to 40% of their corn acreage with non-*Bt* corn would be acceptable and economically viable for them.

5.2   Searching for new insecticidal proteins to engineer insect resistance in plants

Several important insect pests, including *Diabrotica* species (corn rootworms) or *Anthonomus grandis* (cotton boll weevil) are not susceptible to currently available *Bt* Cry toxins. Screening programs, using various screening methods, including PCR and immunocytochemical techniques, could (as they have in the past) lead to the discovery of new Cry proteins toxic to some of these insects. However, although the discovery of new *cry* genes encoding toxins with new insecticidal specificities or that bind to different receptors in the insect gut would be of great value, it may not be the most appropriate answer for future resistance management strategies because broad-spectrum resistance to *Bt* toxins has already been observed in some cases [19]. Therefore, many research projects and screening programs are currently underway to identify new insecticidal proteins with different modes of action. Screening *Bt* strains for the production of insecticidal proteins at various physiological stages (other than sporulation, when Cry proteins are produced) has shown that *Bt* produces various insecticidal proteins during vegetative growth, one of which, Vip3A, is highly toxic to lepidopteran pests such as *Agrotis ipsilon* (black cutworm) and *Spodoptera exigua* (beet armyworm). The corresponding genes have been cloned [13]. Similarly, a random screening of supernatants from *Streptomyces* cultures showed that this bacterium secretes a cholesterol oxidase that is active against *A. grandis* (boll weevil) [42]. These proteins are highly toxic with LC50s in the range of 50 to 200 ng of protein per square centimetre or millilitre of diet and will probably become the second generation of insecticidal proteins for engineering insect resistance in plants [14]; they could be used in combination with or instead of Cry toxins.

Other genes encoding proteins with potential insecticidal properties such as proteinase inhibitors, lectins and chitinases have also been isolated from various sources (plants, insects and microorganisms). Artificial diet bioassays have shown that these proteins have a deleterious effect *in vivo* on various insect pests including aphids [11]. The genes involved have been expressed in various plants (e.g.; tobacco, potato, cotton) and in some cases this has reduced the damage caused by insect pests [12, 24, 48]. However, these proteins are effective insecticides only if produced in large amounts in the plant. It is too early to assess the long-term value of the expression of these genes in plants, but their expression in association with other resistance genes with much more potent insecticidal activity, such as the *Bt cry* and *vip* genes or the *Streptomyces* cholesterol oxidase gene is being explored. The use of these genes may delay the appearance of resistance or synergy may be found.

Another important novel approach involves engineering plants to produce new non-protein compounds that kill or repel insects. Indeed, the biosynthesis of secondary metabolites in plants is part of their natural defence against various organisms including phytophagous insects. However, the synthesis of these molecules usually involves a number of enzymatic steps. A number of genes are therefore involved, presenting a difficulty for genetic engineering [22]. However, as almost all plants have secondary metabolites and some metabolic pathways are common to many plants, genetic engineering strategies aimed at the limited modification of existing pathways to enable a plant to produce metabolites useful for crop protection (such as toxins, semiochemicals and feeding suppressants) are worthy of exploration [10]. To date, the feasibility and long-term value of this strategy has not been assessed.

6. SUMMARY AND CONCLUSIONS

Increasing global agricultural production and preserving the environment are major challenges facing our society in the Twenty First Century. Much depends on our capacity to implement effective and environmentally friendly measures to protect crops from insect damage. *Bt* and its toxins have great potential to become major tools in integrated pest management programs, and their use is likely to increase in coming years. The development of biotechnological products based on *Bt* to improve the efficacy of *Bt* has followed various approaches: *Bt* strains with new combinations of *cry* genes have been obtained using the efficient conjugation system this bacterium possesses. *Bt* toxin genes have been introduced into epiphytic and endophytic bacteria that colonise plant tissues and occupy the same ecological niches as the targeted insects. The production of insecticidal proteins in *Bt* has also been increased by combining various *cry* genes with different promoters into the same strain. The persistence of toxins in the field has been increased by encapsulation within recombinant *Pseudomonas fluorescens* cells (CellCap® encapsulation process developed by Mycogen Corp.) or asporogenic *Bt* strains; this protects the toxins against UV degradation and has the advantage that the transgenic microorganisms released into the environment are non-viable. Finally, insect-resistant transgenic crops and trees have been developed by the expression of *Bt* toxin genes in plants, by transformation. This method is very efficient for delivery and is ideal for controlling borers and root-feeding insects, providing the plant with a very high level of insect resistance. In the future, continuing improvement of first generation products and research into new sources of resistance is essential to ensure the long-term control of insect pests. Chimeric toxins could be produced so as to increase toxin activity or direct resistance towards a particular type of insect. Bacterial strains could also be modified so as to enlarge their spectrum of activity or to control resistant insects. The search for new insecticidal toxins, in *Bt* or other microorganisms, may also provide new weapons for the fight against insect damage.

There is currently no evidence of greater hazards resulting from the use of these genetically engineered organisms. They should however be used with care and it would be advisable to establish the scientific bases for the management of transgenic plants and to make an assessment in the field of the resistance management strategies currently adopted (as they are all still theoretical) so that invasions of resistant insects do not rapidly render them obsolete. The monitoring of interactions of transgenic organisms with the indigenous microflora and plants must also be carefully considered and managed to avoid the possible deleterious effects of interfering with the ecological balance and to ensure full public acceptance of this technology.

REFERENCES

[1] Adang MJ, Brody MS, Cardineau G *et al.* (1993) The reconstruction and expression of a *Bacillus thuringiensis cry*IIIA gene in protoplasts and potato plants. Plant Mol. Biol. 21, 1131-1145

[2] Addison JA (1993) Persistence and nontarget effects of *Bacillus thuringiensis* in soil: a review. Can. J. Forest. Res. 23, 2329-2342

[3] Agaisse H & Lereclus D (1994) Expression in *Bacillus subtilis* of the *Bacillus thuringiensis cryIIIA* toxin gene is not dependent on a sporulation-specific sigma factor and is increased in a *spo0A* mutant. J. Bact. 176, 4734-4741

[4] Agaisse H & Lereclus D (1995) How does *Bacillus thuringiensis* produce so much insecticidal crystal protein ? J. Bact. 177, 6027-6032

[5] Arantes O & Lereclus D (1991) Construction of cloning vectors for *Bacillus thuringiensis*. Gene 108, 115-119

[6] Baum JA & Gilbert MP (1991) Characterization and comparative sequence analysis of replication origins from three large *Bacillus thuringiensis* plasmids. J. Bact. 173, 5280-5289

[7] Baum JA, Kakefuda M & Gawron-Burke C (1996) Engineering *Bacillus thuringiensis* bioinsecticides with an indigenous site-specific recombination system. Appl. Environ. Microbiol. 62, 4367-4373

[8] Bone EJ & Ellar DJ (1989) Transformation of *Bacillus thuringiensis* by electroporation. FEMS Microbiol. Lett. 58, 171-178

[9] Bravo A, Agaisse H, Salamitou S *et al.* (1996) Analysis of *cryIAa* expression in *sigE* and *sigK* mutants of *Bacillus thuringiensis*. Mol. Gen. Genet. 250, 734-741

[10] Chilton S. (1997) Genetic engineering of plant secondary metabolites for insect protection, p. 237-269. *In* Carozzi N & Koziel M (ed.), Advances in insect control : the role of transgenic plants, Taylors & Francis Ltd.

[11] Czapla TH (1997) Plant lectins as insect control proteins in transgenic plants, p. 123-138. *In* Carozzi, N. & Koziel, M. (ed.), Advances in insect control : the role of transgenic plants, Taylors & Francis Ltd.

[12] Down RE, Gatehouse AMR, Hamilton WDO *et al.* (1996) Snowdrop lectin inhibits development and decreases fecundity of the glasshouse potato aphid (*Aulacorthum solani*) when administred *in vitro* and via transgenic plants in laboratory and glasshouse trials. J. Insect Physiol. 42, 1035-1045

[13] Estruch JJ, Warren GW, Mullins MA *et al.* (1996) Vip3A, a novel *Bacillus thuringiensis* vegetative insecticidal protein with a wide spectrum of activities against lepidopteran insects. Proc. Natl. Acad. Sci. USA 93, 5389-5394

[14] Estruch JJ, Carozzi NB, Desai N *et al.* (1997) Transgenic plants: an emerging approach to pest control. Nat. Biotechnol. 15, 137-141

[15] Fischhoff DA, Browdish KS, Perlak FJ *et al.* (1987) Insect-tolerant transgenic tomato plants. Bio/Technology 5, 807-813

[16] Fujimoto H, Itoh K, Yamamoto M *et al.* (1993) Insect-resistant rice generated by introduction of a modified δ–endotoxin gene of *Bacillus thuringiensis*. Bio/Technology 11, 1151-1155

[17] Fuxa JR (1989) Fate of released entomopathogens with reference to risk assessment of genetically engineered microorganisms. Bulletin of the ESA, 12-24

[18] González JMJ, Brown BJ & Carlton BC (1982) Transfer of *Bacillus thuringiensis* plasmids coding for delta-endotoxin among strains of *B. thuringiensis* and *B. cereus*. Proc. Natl. Acad. Sci. USA 79, 6951-6955

[19] Gould F, Martinez-Ramirez A, Anderson, A *et al.* (1989) Broad-spectrum resistance to *Bacillus thuringiensis* toxins in *Heliothis virescens* Proc. Natl. Acad. Sci. USA 89, 7986-7990

[20] Gould F (1994) Potential problems with high-dose strategies for pesticidal engineered crops. Biocontrol Sci. Technol. 4, 451-461

[21] Grecchio C, Stotzky G (1998) Insecticidal activity and biodegradation of the toxin from *Bacillus thuringiensis* subsp. *kurstaki* bound to humic acids from soil. Soil Biol. Biochem. 30, 463-470

[22] Hallahan DL, Pickett JA, Wadhams LJ *et al.* (1992) Potential secondary metabolites in genetic engineering of crops for resistance, p. 215-248. *In* Gatehouse AMR, Hilder VA & Boulter D (ed.), Plant genetic manipulation for crop protection, CAB International

[23] Hernandez E, Ramisse F, Ducoureau JP *et al.* (1998) *Bacillus thuringiensis* subsp. *konkukian* (serotype H34) superinfection : case report and experimental evidence of pathogenicity in immunosuppressed mice. J. Clin. Microbiol. 36, 2138-2139

[24] Hilder VA, Gatehouse AMR, Sheerman SE *et al.* (1987) A novel mechanism of insect resistance engineered into tobacco. Nature 333, 160-163

[25] Kalman S, Kiehne KL, Cooper N *et al.* (1995) Enhanced production of insecticidal proteins in *Bacillus thuringiensis* strains carrying an additional crystal protein gene in their chromosomes. Appl. Environ. Microbiol. 61, 3063-3068

[26] Koziel MG, Beland GL, Bowman C *et al.* (1993) Field performance of elite transgenic maize plants expressing an insecticidal protein derived from *Bacillus thuringiensis*. Bio/Technology 11, 194-200

[27] Lampel JS, Canter GL, Dimock MB *et al.* (1994) Integrative cloning, expression, and stability of the *cry1A(c)* gene from *Bacillus thuringiensis* subsp. *kurstaki* in a recombinant strain of *Clavibacter xyli* subsp. *cynodontis*. Appl. Environ. Microbiol. 60, 501-508

[28] Lereclus D, Agaisse H, Gominet M *et al.* (1995) Overproduction of encapsulated insecticidal crystal proteins in a *Bacillus thuringiensis* spo0A mutant. Bio/Technology 13, 67-71

[29] Lereclus D, Arantes O, Chaufaux J *et al.* (1989) Transformation and expression of a cloned δ−endotoxin gene in *Bacillus thuringiensis*. FEMS Microbiol. Lett. 60, 211-218

[30] Lereclus D, Vallade M, Chaufaux J *et al.* (1992) Expansion of insecticidal host range of *Bacillus thuringiensis* by in vivo genetic recombination. Bio/Technology 10, 418-421

[31] Lorenz MG & Wackernagel W (1996) Mechanism and consequences of horizontal gene transfer in natural bacterial populations, p. 45-57. *In* Tomiuk J, Wöhrmann K & Sentker A (ed.), Transgenic organisms: biological and social implications, Birkhauser Verlag

[32] Macaluso A & Mettus AM (1991) Efficient transformation of *Bacillus thuringiensis* requires nonmethylated plasmid DNA. J. Bact. 173, 1353-1356

[33] Mahillon J & Lereclus D (1988) Structural and functional analysis of Tn*4430*: identification of an integrase-like protein involved in the co-integrate-resolution process. EMBO J. 7, 1515-1526

[34] McBride KE, Svab Z, Schaaf DJ *et al.* (1995) Amplification of a chimeric *Bacillus* gene in chloroplasts leads to an extraordinary level of an insecticidal protein in tobacco. Bio/Technology 13, 362-365

[35] Mohan KS, Asokan R & Gopalakrishnan C (1997) Development and field performance of a sporeless mutant of *Bacillus thuringiensis* subsp. *kurstaki*. J. Plant Biochem. Biotechnol. 6, 105-109

[36] Nascimiento ML, Capalbo DF, Moraes GJ *et al.* (1998) Effect of a formulation of *Bacillus thuringiensis* Berliner var. *kurstaki* on *Podisus nigrispinus* Dallas (Heteroptera: Pentatomidae: Asopinae). J. Invertebr. Pathol. 72, 178-180

[37] Obukowicz MG, Perlak FJ, Kuzano-Kretzmer K *et al.* (1986) Integration of the delta endotoxin gene of *Bacillus thuringiensis* into the chromosome of root-colonizing strains of pseudomonads using Tn5 Gene 45, 327-331

[38] Palm CJ, Schaller DL, Donegan KK *et al.* (1996) Persistence in soil of transgenic plant produced *Bacillus thuringiensis* var. *Kurstaki* δ−endotoxin. Can. J. Microbiol. 45, 1258-1262

[39] Perlak FJ, Deaton RW, Armstrong TA *et al.* (1990) Insect resistant cotton plants. Bio/Technology 8, 939-943

[40] Perlak FJ, Fuchs RL, Dean DA *et al.* (1991) Modification of the coding sequence enhances plant expression of insect control protein genes. Proc. Natl. Acad. Sci. USA 88, 3324-3328

[41] Pimentel D (1991) CRC Handbook of Pest Management in Agriculture. Vol. 1, 2nd edn, CRC Press

[42] Purcell JP, Greenplate JT, Jennings MG *et al.* (1996) Cholesterol oxidase : a potent insecticidal protein active against boll weevil larvae. Biochem. Biophys. Res. Commun. 196, 1406-1413

[43] Pusztai M, Fast M, Gringorten L *et al.* (1991) The mechanism of sunlight-mediated inactivation of *Bacillus thuringiensis* crystals. Biochem. J. 273, 43-47

[44] Roush RT (1989) Designing Resistance Management Programs : How can you choose? Pesticide Science 26, 423-441

[45] Samples JR & Buettner H (1983). Corneal ulcer caused by a biological insecticide (*Bacillus thuringiensis*). Am. J. Ophthalmol. 95, 258-260

[46] Sanchis V, Agaisse H, Chaufaux J *et al.* (1996) Construction of new insecticidal *Bacillus thuringiensis* recombinant strains by using the sporulation non-dependent expression system of *crylIIA* and a site specific recombination vector. J. Biotechnol. 48, 81-96

[47] Sanchis V, Agaisse H, Chaufaux J *et al.* (1997) A recombinase-mediated system for elimination of antibiotic resistance gene markers from genetically engineered *Bacillus thuringiensis* strains. Appl. Environ. Microbiol. 63, 779-784

[48] Shade RE, Schroeder HE, Pueyo JJ *et al.* (1994) Transgenic pea seeds expressing the alpha-amylase inhibitor of the common bean are resistant to bruchid beetles. Bio/Technology 12, 793-796

[49] Skot L, Harrison SP, Nath A *et al.* (1990) Expression of insecticidal activity in *Rhizobium* containing the δ–endotoxin gene cloned from *Bacillus thuringiensis* subsp. *tenebrionis*. Plant & Soil 127, 285-295

[50] Stock CA, McLoughlin TJ, Klein JA *et al.* (1990) Expression of a *Bacillus thuringiensis* crystal protein gene in *Pseudomonas cepecia*. Can. J. Microbiol. 36, 879-884

[51] Tabashnik BE (1994) Evolution of resistance to *Bacillus thuringiensis*. Annu. Rev. Entomol. 39, 47-79

[52] Thuriaux P (1996) Les flux de gènes, p. 99-110. *In* Kahn A (ed.), Les plantes transgéniques en agriculture, John Libbey Eurotext.

[53] Udayasuriyan V, Nakamura A, Masaki H *et al.* (1995) Transfer of an insecticidal protein gene of *Bacillus thuringiensis* into plant-colonizing *Azospirillum*. World J. Microbiol. Biotechnol. 11, 163-167

[54] Vaeck M, Reynaerts A, Höfte H *et al.* (1987) Transgenic plants protected from insect attack. Nature 327, 33-37

Chapter 7.2

Genetic engineering of bacterial insecticides for improved efficacy against medically important Diptera

Brian A. Federici, Hyun-Woo Park, Dennis K. Bideshi & Baoxue Ge
Department of Entomology and Interdépartementale Graduate Program in Genetics, University of California-Riverside, Riverside, California 92521 U.S.A.

Key words: *Bacillus thuringiensis*, *Bacillus sphaericus*, promoters, Cry proteins synthesis, mRNA stability, stem-loop structures, mosquitoes, blackflies, Lepidoptera, Coleoptera, resistance avoidance

Abstract: Bacterial insecticides have been in use for control of agricultural pests and vector and nuisance mosquitoes and blackflies for more than two decades. Nevertheless, these insecticides constitute less than 2% of the market world-wide due primarily to their low to moderate efficacy in comparison to chemical insecticides. Recombinant DNA techniques have made it possible to improve the efficacy of bacterial insecticides from 2 to 10-fold by markedly increasing the synthesis of insecticidal proteins, and by enabling new combinations of insecticidal proteins from different bacteria to be produced within single strains. This chapter reviews the use of promoters, 3' and 5' enhancer elements, and chaperone-like proteins in conjunction with shuttle expression vectors to improve the efficacy of bacterial insecticides, with an emphasis on those used in mosquito and blackfly control. The prospects for additional improvements in efficacy and extending the use of this technology to other bacterial species are also discussed.

1.   INTRODUCTION

Malaria, filariasis, dengue and the viral encephalitis remain the most important diseases of humans, with an estimated 2 billion people world-wide living in areas where these are endemic. The etiologic agents of these diseases are transmitted by mosquitoes, and disease control efforts therefore have relied heavily on broad spectrum chemical insecticides to

461

J.-F. Charles et al. (eds.),
Entomopathogenic Bacteria: From Laboratory to Field Application, 461–484.
© 2000 *Kluwer Academic Publishers. Printed in the Netherlands.*

reduce mosquito populations. However, insecticide use is being phased out in many countries due to resistance in target populations. Furthermore, many governments restrict use of these chemicals because of concerns about their effects on non-target organisms, especially on humans and other vertebrates, through contamination of food and water supplies.

Alternatives to broad spectrum chemical insecticides exist or are under development for disease control and include vaccines, new more specific chemical insecticides, bacterial insecticides, and transgenic mosquitoes refractive to pathogen transmission. However, significant problems exist with all of these. In most cases, with the exception of some viral diseases, vaccines have not proven effective. New chemical insecticides are expensive, still impact many non-target arthropod populations, and are prone to resistance. Bacterial insecticides are also expensive, and though almost ideally specific for mosquitoes and related biting nematoceran flies, resistance remains a potential problem. And while transgenic mosquitoes hold promise, a range of technical problems will prevent these from being operationally effective for reducing malaria and filariasis prevalence for at least twenty years, probably longer. In the interim, improved recombinant bacterial insecticides developed using knowledge of insecticidal bacterial proteins and their synthesis hold excellent promise for providing cost-effective control of mosquito and blackfly populations.

Over the past decade, the insecticidal proteins of two bacteria, *Bacillus thuringiensis* subsp. *israelensis* (*Bt*) and *Bacillus sphaericus*, have received considerable study with the aims being to define the genetic and biochemical bases of their high toxicity to mosquitoes, and to understand their different modes of action [11, 20, 39, 53]. During the same period, studies of the molecular genetics of *Bt* Cry proteins have identified a variety of regulatory elements responsible for the high levels of endotoxin synthesis characteristic of most *Bt* strains [4, 9]. And to complement these studies, shuttle expression vectors have been developed for studying *Bt* proteins alone and in combination [6, 40, 47, 58]. The knowledge and genetic tools resulting from these studies have made it possible to develop improved recombinant bacteria that are more potent than wild-type strains and have a broader target spectrum. In Chapter 7.1 the techniques and bacteria developed for control of agricultural pests are reviewed. The present chapter focuses on properties of mosquitocidal bacteria and their endotoxins, and shows how these have been genetically manipulated to markedly improve efficacy. Though the focus is on mosquitoes, in most cases the proteins and strains discussed are also active against the larvae of related nematoceran flies including blackflies, craneflies, and chironomid midges [33], and thus have utility well beyond mosquito control. The chapter begins with a summary of the properties of the most important mosquitocidal bacteria, and then shows how regulatory elements have been used to enhance the synthesis of their endotoxins. The properties and efficacy of recently constructed *Bti/B. sphaericus* recombinants are then discussed. The chapter closes with examples of how additional further genetic manipulation will likely result in strains that are even more effective than those currently available.

2.   PROPERTIES OF MOSQUITOCIDAL BACTERIA

2.1   *Bacillus thuringiensis* subsp. *israelensis*

Until the mid-1970's, *B. thuringiensis* was only thought to produce endotoxins active against lepidopterous insects. Then in 1976 a new subspecies, *Bt* subsp. *israelensis* (H14) was isolated from a mosquito breeding site in Israel that proved highly toxic to larvae of mosquitoes and blackflies [31], with an LC_{50} ranging from 10-13 ng/ml against the fourth instar of many mosquito species. The parasporal body of this subspecies differs substantially from the classic *Bt* bipyramidal crystal toxic to lepidopteran larvae (Figure 1). It contains four major proteins, Cyt1A (27.3 kDa), Cry4A (128 kDa), Cry4B (134 kDa) and Cry11A (72 kDa) [23, 34] in three different inclusion types assembled into a spherical parasporal body held together by lamellar envelope [37, 38]. Studies of the deduced amino acid sequences of *Bti*'s Cry proteins have shown that they are related to other Cry endotoxins [23, 34]. However, Cyt1A differs markedly from these in its amino acid sequence and toxicology. It is highly cytolytic to a range of vertebrate and invertebrate cells *in vitro*, having an affinity for unsaturated fatty acids in the lipid portion of cell membranes [60]. While it may act by forming transmembrane pores, other evidence suggests it has a detergent-like mode of action, perturbing the membrane by binding to specific fatty acids [18, 19].

Though Cyt1A's mode of action remains to be resolved, it is nevertheless an extremely important protein. Numerous studies have revealed that *Bti*'s high toxicity is due to synergistic interactions among its Cry proteins [22, 48], and especially between Cyt1A and the Cry proteins [22, 37, 67, 69, 71]. Even more importantly, recent studies have either shown or provided strong evidence that Cyt1A can delay the development of resistance to Cry proteins in mosquitoes [30], and can overcome resistance to these if it develops [66]. For example, resistance

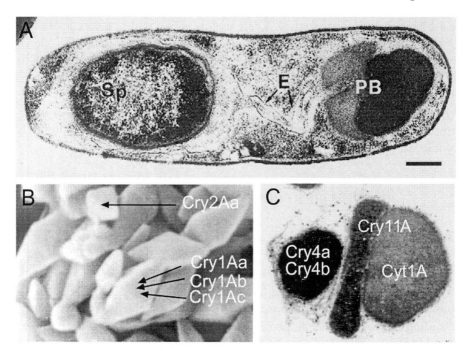

Figure 1. Sporulating cell of *Bacillus thuringiensis* and typical parasporal bodies. A.
Sporulating cell of *B. thuringiensis* subsp. *israelensis*. B and C. Respectively, parasporal
bodies of *Bt* subsp. *kurstaki* and *Bt* subsp. *israelensis* showing toxin individual
inclusions and their toxin composition. Sp, Spore. E, Exosporium. PB, Parasporal body.

levels of greater than 900-fold to Cry11A in laboratory populations of
Culex quinquefasciatus were suppressed completely when Cry11A was
combined with Cyt1A in a 3:1 ratio [66]. In addition, more recently it
has been shown that Cyt1A can overcome very high levels of resistance
to the binary toxin of *B. sphaericus* 2362 [64], and can extend the target
spectrum of this species to *Aedes aegypti* [67].

The high efficacy that *Bti* showed in laboratory and field trials during
the early 1980's led rapidly to its development as a commercial bacterial
larvicide for control of mosquito and blackfly larvae [13, 33, 44;
Chapter 6.2]. Four commercial products, Vectobac® (Abbott
Laboratories), Teknar® (Thermo-Trilogy, Inc.), Bactimos® (AgrEvo) and
Acrobe® (Becker Microbial Products) are used in many countries for the
control of vector and nuisance mosquitoes and blackflies. Teknar® and
Vectobac® proved to be particularly important for the World Health
Organization's Onchocerciasis Control Programme in West Africa, where
they have been used to control the blackfly vectors of *Onchocerca*

volvulus, which causes River Blindness in humans, for almost two decades [33; Chapter 6.2].

Despite its intensive use in numerous mosquito and blackfly ecosystems, and the development of resistance under intensive selection in the laboratory, resistance to *Bti* has not been reported in the field [12]. Laboratory studies suggest that this lack of resistance is due primarily to presence of Cyt1A in the parasporal body [30, 66]. Cyt1A's capacity to synergism endotoxin proteins, including the *B. sphaericus* binary toxin [67], and delay resistance, are important properties for the improvement of dipteran larvicides.

2.2   *Bacillus sphaericus*

Many mosquitocidal strains of *B. sphaericus* have been isolated over the past thirty years, and the most toxic of these, including strains 1593 and especially 2362, belong to flagellar serotype 5a5b [11, 20; Chapter 2.3]. The principal toxin in these strains is a binary toxin composed of two proteins, a 51.4 binding domain and 41.9 toxin domain that co-crystallise into a single small parasporal body. Strain 2362 has an LC_{50} of 18 ng/ml against the fourth instar of *Culex* mosquitoes [11]. A very similar binary toxin occurs in strain 2297 and forms a much larger, but not as toxic, parasporal body [14, 15, 20]. After ingestion by mosquito larvae, the 51.4 and 41.9 kDa proteins are cleaved by proteinases yielding peptides of 43 and 39 kDa, respectively, that form the active toxin [11, 20]. These associate, bind to a receptor on the midgut microvilli, and cause lysis of midgut cells after internalisation [24].

In addition to the binary toxin, many strains of *B. sphaericus* produce other mosquitocidal toxins during vegetative growth that are referred to as Mtx toxins [Chapter 2.3]. Two of these have been well studied, Mtx1 (100 kDa) and Mtx2 (30.8 kDa), but these are not as toxic as the binary toxin.

The target spectrum of *B. sphaericus* is more limited than that of *Bti*, being restricted to mosquitoes, and its highest activity is against *Culex* and *Anopheles* species. Some species of *Aedes* that have been tested, such as *A. aegypti*, are not very sensitive to *B. sphaericus*. Moreover, even though strain 2362 was isolated from a blackfly adult (*Simulium damnosum*), *B. sphaericus* strains have little or no activity against nematoceran flies other than mosquitoes. Nevertheless, *B. sphaericus* does appear to have better initial and residual activity than *Bti* against mosquitoes in polluted waters. As a result, a commercial formulation, VectoLex® (Abbott Laboratories), based on strain 2362, is marketed in many countries, especially for control of *Culex* larvae in polluted waters.

A disadvantage of *B. sphaericus* strains is that the binary toxin is in essence a single toxin. Laboratory studies have shown that it is much more likely to result in resistance than *Bti* [52]. In fact, resistance to *B. sphaericus* has already been reported in field populations of *Culex* mosquitoes in France, Brazil, and India [51, 56, 57].

2.3 Other mosquitocidal bacteria

The discovery and successful use of *Bti* and *B. sphaericus* strains in mosquito and blackfly control programmes stimulated a world-wide search for more potent isolates of these and other bacteria. Many isolates of *Bt* and *B. sphaericus*, and even other bacterial species such as *Clostridium bifermentans* have been discovered with mosquitocidal properties. Most of these produce toxins related to those already known [see Chapter 2.3]. One of the more interesting *Bt* isolates is the PG-14 isolate of *Bt* subsp. *morrisoni* discovered in the Philippines. This isolate is as toxic as *Bti* and produces the same complement of endotoxin proteins (Cyt1A, Cry4A, Cry4B, and Cry11A), plus an additional 144 kDa Cry1 protein toxic to lepidopterans [38]. The latter protein comprises about 40% of the parasporal body, and by itself is not toxic to mosquitoes. Because PG-14 is as toxic to mosquitoes as *Bti*, these results imply that Cyt1A and/or the other Cry proteins potentiate the 144 kDa protein, making it mosquitocidal.

Another interesting isolate is *Bt* subsp. *jegathesan* from Malaysia [54]. This isolate produces a complex of seven Cry and Cyt proteins, several of which are related to those of *Bti*, but have different toxicological properties. One of these is Cry11B, a protein of 80 kDa that is approximately 10-fold more toxic to mosquitoes than the related Cry11A protein that occurs in *Bti* [25]. The discovery of proteins such as Cry11B demonstrates the value of searching for new insecticidal isolates, even if the new strains prove not to be as effective as *Bti* and *B. sphaericus*.

3. FACTORS FOR ENHANCING ENDOTOXIN SYNTHESIS

Until recently, most attempts to achieve significant improvements in the potency of *Bti* and *B. sphaericus* strains through genetic engineering met with limited or only moderate success. However, these and other studies of *Bt* toxins identified several factors such as helper proteins and mRNA stabilising sequences that can be used to enhance endotoxin synthesis. In addition, shuttle expression vectors have been developed for introducing toxin-encoding genes into in *B. thuringiensis* and *B. sphaericus* [6, 40, 47, 58]. Some enhancing factors where identified in strains of *Bt* used against agricultural pests whereas others originated in *Bti*. It has recently been possible to construct a markedly improved recombinant bacterium for mosquito control by combining several of these factors with the *B. sphaericus* binary toxin and engineering these into *Bti* using the *Escherichia coli-B. thuringiensis* shuttle vector pHT3101 [16]. Before discussing the construction and properties of this bacterium (section 4), the various genetic elements and genes used to enhance synthesis are discussed along with examples of their use to increase Cry and Cyt synthesis above levels that occur in wild type isolates. To avoid duplicating material covered in Chapter 7.1, only genetic elements not discussed there or alternative uses of these are discussed here. In most of the studies summarised below, the *E. coli-B. thuringiensis* shuttle vector pHT3101 [40] was used to evaluate the properties of these elements in *B. thuringiensis*.

3.1 The 20 kDa chaperone-like protein of the *cry11A* operon

Early attempts to express the *cyt1A* gene in *Escherichia coli* were unsuccessful until it was noticed that inclusion of a large region flanking the 5' end of the gene resulted in moderate levels of Cyt1A synthesis. Subsequent studies of this region showed that the critical region encoded a 20 kDa protein that occurred as the third open reading frame (ORF) of the *cry11A* operon [1, 26, 62]. The precise function of this protein has not been determined, but antibody to it co-precipitates Cyt1A from cell homogenates, showing that the 20 kDa protein binds to Cyt1A. Thus, the 20 kDa protein is referred to as a helper protein, and may act as a molecular chaperone. Yields of Cyt1A in *E. coli* and *B. thuringiensis* are much lower when the gene for the 20 kDa protein is not included in expression vectors, and cell mortality is high. Based on Cyt1A's affinity for the lipid portion of membranes, it appears that the 20 kDa protein binds to Cyt1A during its synthesis, helping it fold while at the same time preventing Cyt1A from interacting with the inner lipid-containing membrane and associated replication complex.

The increased yields of Cyt1A obtained when the gene coding for the 20 kDa protein was included in *cyt1A* constructs suggested that even higher yields of this and other proteins might be obtained by expressing

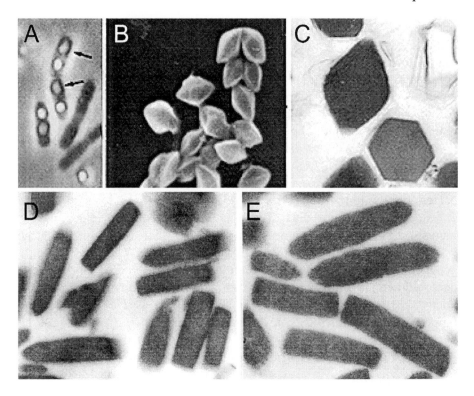

Figure 2. Enhanced synthesis of Cyt1A and Cry11A using the 20 kDa chaperone-like
protein encoded by the *cry11A* operon. A. Sporulating *Bt* cells containing Cyt1A
crystals (arrows). B and C. Respectively, scanning and electron micrographs of Cyt1A
crystals produced with the aid of the 20 kDa protein. D and E. Respectively, Electron
micrographs of Cry11A crystals produced without and with the aid of the 20 kDa
protein.

endotoxin genes and the 20 kDa protein gene under separate promoters.
To test this possibility, constructs were made in which the expression of
the 20 kDa protein gene was placed under the control of the strong *Bt*I
and *Bt*II sporulation-dependent promoters of *cry1Ac*. This construct was
then cloned into the *E. coli-B. thuringiensis* shuttle vector pHT3101 or
other vectors and used to produce different proteins including Cyt1A
[70], Cry11A [72], Cry2A [29], Cry4A [73], and truncated Cry1C [50].
In all cases, yields were enhanced substantially, with net increases ranging
from 1.7-fold for Cry11A to greater than 10-fold for Cyt1A (Figure 2).

Once plasmids based on pHT3101 were constructed for producing the
Cyt1A and Cry endotoxins, they could be used to introduce these proteins
into *Bt* strains in which they are lacking. For example, the Cyt1A protein

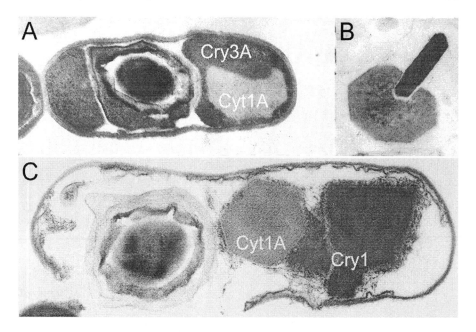

Figure 3. Recombinant *B. thuringiensis* strains that produce Cyt1A. A. *Bt* subsp.
morrisoni (strain tenebrionis) with Cry3A and Cyt1A crystals. B. A Cry3A crystal
embedded in a Cyt1A crystal, resulting from synthesis of Cry3A during vegetative
growth and Cyt1A during sporulation. C. Cyt1A crystal along with a crystal of the Cry1
complex in a fully sporulated cell of *Bt* subsp. *aizawai*.

has been introduced into *Bt* subsp. *morrisoni* (strain tenebrionis) and *Bt*
subsp. *aizawai* to test the compatibility of this cytolytic toxin with Cry
proteins other than Cry4 and Cry11 proteins, and determine whether it
might improve the efficacy of these strains (Figure 3). While there is no
evidence that Cyt1A is active against lepidopterous insects or can
potentiate Cry1 toxins, it is toxic to the cottonwood leaf beetle,
Chrysomela scripta [27], and may be capable of enhancing the toxicity
of Cry3A and/or delaying resistance to this toxin.

3.2   The 29 kDa scaffolding protein of the *cry2A* operon

The 29 kDa protein is encoded as *orf2* of the *cry2A* operon [63], and
is also referred to as a helper protein. Like the 20 kDa protein, its role in
endotoxin synthesis is not known precisely, but evidence suggests that it
is primarily a scaffolding protein that establishes a lattice within *Bt* cells

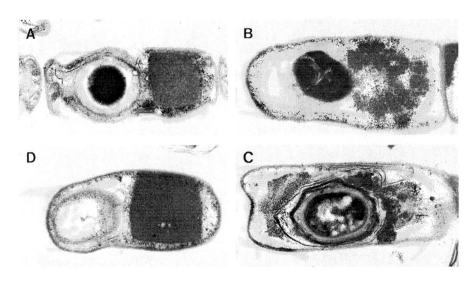

Figure 4. Effect of the *cry2A* operon-encoded 29 kDa protein on Cry2A crystal formation.
A. Cry2A crystal produced by the *cry2A* operon. B and C. Aggregates of Cry2A produced
in the absence of the 29-kDa protein. D. Larger Cry2A crystals resulting from
concomitant synthesis of Cry2A and the 29 and 20 kDa proteins.

to facilitate the formation of Cry2A crystals [29]. The 29 kDa protein contains a 15-amino acid sequence repeat almost perfectly repeated eleven times in tandem [63]. Modelling the tandem repeats shows that they can potentially form a lattice-like structure similar to the ice nucleating protein of *Pseudomonas fluorescens* [29]. Cry2A is capable of forming small crystalline aggregates in the absence of the 29 kDa protein (Figure 4), but it does not form a typical cuboidal Cry2A crystal [29]. In the HD-1 isolate of *B. thuringiensis*, from which the *cry2A* operon was cloned, the Cry2A crystal is embedded at the apex of the short axis of bipyramidal crystal composed of Cry1Aa, Cry1Ab, and Cry1Ac [43]. How this specific association of the Cry2A and Cry1A proteins is organised is not known. However, an important function of the 29 kDa protein may be to bind the Cry2A crystal to the Cry1A bipyramidal crystal to increase the chance all four proteins are eaten by the same insect.

Other evidence suggests that the 29 kDa protein may also act as a molecular chaperone increasing net Cry2A synthesis [21, 29]. In constructs used to synthesise Cry2A, yields were 25% higher in those that contained the 29 kDa protein [29].

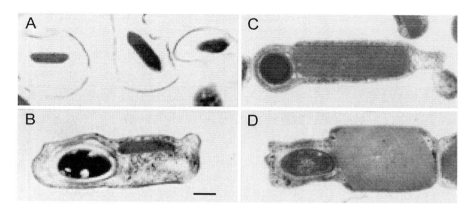

Figure 5. Use of *cytlA* promoters and the 5' STAB-SD mRNA stabilising sequence to enhance Cry3A synthesis. A. Cry3A crystals in sporulated wild-type *Bt* subsp. *morrisoni* (strain tenebrionis) cells. B. Cry3A crystal produced using *cytlA* promoters without STAB-SD. C and D. Sporulated cells showing longitudinal and transverse sections through large Cry3A crystals produced using *cytlA* promoters and STAB-SD.

3.3   5' and 3' mRNA stabilising sequences

Although the 20 kDa and 29 kDa helper proteins increased Cry endotoxins yields by 25 to 70%, these increases were not sufficient to justify their use in commercial strains. Other tactics under development for enhancing endotoxin yields included increasing gene copy number using higher copy number plasmids or γ-irradiation mutagenesis. The latter strategy was used to create a strain of *Bt* subsp. *morrisoni* (strain tenebrionis) that produced 5-fold the amount of Cry3A obtained with the parental strain [2]. The improved efficacy was high enough to warrant commercial development of this strain, and it now serves as the active ingredient of Novodor® (Abbott Laboratories), a product used to control the Colorado potato beetle, *Leptinotarsa decemlineata*.

Other studies of Cry3A synthesis in the wild type isolate showed that the expression of *cry3A* was primarily under the control of a promoter active during the vegetative growth and the stationary phase [3]. The yield of Cry3A under this promoter was moderate, but expression of this gene in asporogenous mutants increased Cry3A yields significantly, again by as much as 5-fold [41, 42]. Most of this increase was attributed to a Shine-Dalgarno sequence referred to as STAB-SD located just downstream from the 5' end of the major *cry3A* transcript (T-129) that stabilised the transcript-ribosome complex [5, 35]. Even higher levels of Cry3A, in the range of 10 to 12-fold that obtained in the parental strain, were obtained by expressing the *cry3A* gene including the STAB-SD sequence under the

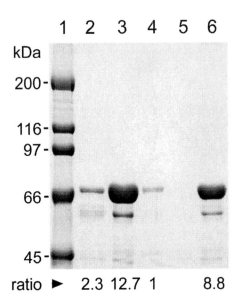

Figure 6. Effect of using *cyt1A* promoters in combination with the STAB-SD mRNA
stabilising sequence to increase Cry3A yield. Lane 4 shows the amount of Cry3A
obtained with the wild type strain whereas lanes 2 and 3 show, respectively, Cry3A
amounts obtained using *cyt1A* promoters alone or in combination with STAB-SD. Lane 6
shows the amount of Cry3A obtained with the gamma-irradiated strain used in
Novodor®. Each lane contain solids from an equal amount of media. SDS-polyacrylamide
gel stained with Coomassie blue.

control of the strong *cyt1A* sporulation-dependent promoters (Figures 5
& 6). The yield of Cry3A obtained with a construct that used *cyt1A*
promoters to drive expression but which lacked the STAB-SD sequence
was only 2-fold that of the parental strain, showing that the STAB-SD
sequence was responsible for most of the increase in Cry3A [45].

To determine whether the *cyt1A* promoters in combination with the
5' STAB-SD sequence could be used to increase the yield of other Cry
proteins, Cry2A and Cry11A were synthesised using this expression
system. These proteins were selected because like Cry3A, they are both
in the 70 kDa mass range and produce relatively small crystals in wild
type isolates compared to the typical large bipyramidal crystals produced
by Cry1 proteins. After combining either the *cry2A* or *cry11A* gene with
cyt1A promoters and the STAB-SD sequence, the constructs were cloned
separately into pHT3101 and expressed in the acrystalliferous 4Q7 strain
of *B. thuringiensis* subsp. *israelensis* [46]. In comparison to

constructs that lacked the STAB-SD sequence and used the normal promoters, the increases obtained using the *cytlA* promoter/STAB-SD combination were 4-fold greater for Cry2A, but only 1.3-fold greater for CrylIA. These results show that the protein being synthesised has a significant effect on the increase in yield that can be obtained with the *cytlA* promoters and STAB-SD sequence.

Less study has been devoted to the stem-loop structures that occur at the 3' end of *cry* and *cyt* genes. However, these also increase endotoxin yield by stabilising transcripts through inhibiting their rapid degradation by exonucleases [32, 68]. Variation in the size and stability of stem-loop structures has an effect on transcript half-life, and therefore the amount of endotoxin protein synthesised. By using more stable stem-loop structures, it is possible to increase endotoxin yield. For example, the yield of CryllA was increased by approximately 30 % by replacing the normal *cryllA* stem-loop structure with the more stable stem-loop that occurs at the 3' terminus of the *cytlA* gene [28; Figure 7].

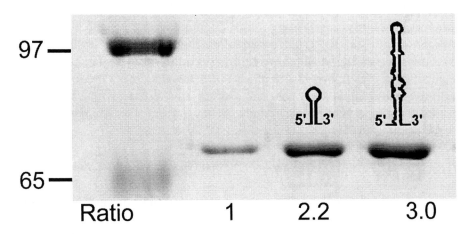

Figure 7. Effect of two different stem-loop structures on CryllA synthesis. Lane 2, no stem-loop structure. Lanes 3 and 4, respectively, the *cryllA* (ΔG value–17.2 kcal) and *cytlA* (ΔG value of –27.6) stem-loop structures. Yield ratios are indicated at the bottom of the gel (SDS-PAGE).

4.   IMPROVEMENT OF MOSQUITOCIDAL BACTERIA

The existence of highly mosquitocidal strains of *Bti* and *B. sphaericus*, each with a unique set of toxins, as well as other mosquitocidal bacteria suggested that it might be possible to construct improved recombinant bacteria that combined the best properties of these. For example, a basic principle of resistance management is that it is more difficult to develop resistance to a multiplicity of toxins than to a single toxin. In addition, multiple toxins with different modes of action have the potential to be less prone to resistance than toxins with the same mode of action. Studies of *Bti* have validated the multiplicity and different modes of action principles for mosquitocidal bacteria [30, 65, 66], as has the development of resistance in the field to *B. sphaericus* for the single versus multiple toxin principle [51, 56, 57]. Thus, there were two obvious possibilities for making improved recombinant mosquitocidal bacteria, introduce *Bti* or related mosquitocidal endotoxin genes into the best *B. sphaericus* strains, and introduce *B. sphaericus* toxin genes into *Bti*. Both of these approaches have been used over the past decade to construct a variety of *Bt* and *B. sphaericus* recombinants that produce different combinations of *Bt* (mostly *Bti*) and *B. sphaericus* proteins. The next two sections review the properties of these recombinants. The bioassay methods used in these studies varied considerably from one study to another. In some cases toxicity is reported in terms of media dilutions or spore counts, whereas in others values were based on protein dry weight or the dry weight of total media solids. In addition, the species and instar used in bioassays were often different from one study to another. Thus, to avoid cumbersome comparisons among of results of these studies, improvements in toxicity or lack thereof are discussed relative to the internal controls used in each study. In most cases, the comparisons are made at the LC_{50} level.

4.1   *Bt* subsp. *israelensis* endotoxins in *B. sphaericus*

Most recombinants made to date have introduced the Cry or Cyt proteins of *Bti* and related mosquitocidal subspecies into *B. sphaericus*, with *B. sphaericus* 2297 being the typical host. In general, production of *Bti* or other *Bt* toxins in recombinant *B. sphaericus* strains made these considerably more toxic to mosquito species insensitive to *B. sphaericus*, such as *A. aegypti*, or to species normally sensitive to *B. sphaericus*, but which had developed resistance to the *B. sphaericus* binary toxin, such as *C. pipiens*. Yet most *B. sphaericus* recombinants producing Cry or Cyt toxins were either only equal in toxicity to parental strains (i.e., *Bti* or *B. sphaericus*) or only slightly more toxic, or were unstable.

In one of the first sets of *B. sphaericus/Bti* recombinants, a *Bti* DNA fragment encoding the Cry11A and Cyt1A genes was cloned into pPL603E and introduced into *B. sphaericus* 2362 by protoplast transformation [7]. One recombinant produced Cyt1A, Cry11A, and the *B. sphaericus* binary toxin, and was 10-fold more toxic to *A. aegypti* than parental *B. sphaericus* 2362, but was not nearly as toxic to this species as *Bti*. Initially, this recombinant appeared to be stable, but it was eventually found to be unstable [8].

In two other early recombinants, a plasmid containing *cry4B* was transformed into *B. sphaericus* strains 1593 and 2297 by protoplast transformation. Parental *B. sphaericus* 1593 and 2297 strains had low toxicity to *A. aegypti*. However, production of Cry4B in the transformants increased toxicity 100-fold to this species [61], making *B. sphaericus* transformants as toxic to *A. aegypti* as *Bti*. Against *Anopheles dirus* and *C. quinquefasciatus*, the Cry4B *B. sphaericus* transformants were similar in toxicity to the parental strains, being slightly more or less toxic depending on the recombinant strain and mosquito species tested.

In a related study, the *cry4B* or *cry11A* genes of *Bti* were transferred into *B. sphaericus* 2297 by electroporation using the shuttle vector pMK3 [47]. The parental *B. sphaericus* 2297 strain was non-toxic to *A. aegypti*, whereas the *B. sphaericus* Cry4B and *B. sphaericus* Cry11A 2297 transformants were both moderately toxic to this species, but not as toxic as *Bti*. In this study, it was found that the Cry4B transformant was approximately 10-fold more toxic to *A. aegypti* than the Cry11A transformant, and the authors suggested that the higher toxicity of the former was due to synergism between the Cry4B and the *B. sphaericus* binary toxin [47].

A more recent attempt to improve *B. sphaericus* used the transposon Tn*917* to insert the major *Bti* toxin genes or fragments thereof into the chromosome of *B. sphaericus* 2362 [8]. A series of recombinants were obtained which produced one or more of the *Bti* proteins in *B. sphaericus* 2362 along with the *B. sphaericus* binary toxin. As in previous studies, though not as toxic as *Bti*, many of the *B. sphaericus* 2362 recombinants obtained in this study were as much as 10-fold more toxic to *A. aegypti* than the parental *B. sphaericus* 2362 strain. However, against *C. quinquefasciatus* and *Anopheles gambiae*, the recombinant toxicity was only in the range of parental *B. sphaericus* 2362 or *Bti*. In another study, integrative plasmids were used to introduce the *cry11A* gene into *B. sphaericus* 2362, resulting in recombinants that produced both the

B. sphaericus binary toxin and Cry11A [49]. These recombinants were much more toxic to *A. aegypti* than parental *B. sphaericus* 2297, and were similar in toxicity to the parental strain against *C. quinquefasciatus*. However, one of the Cry11A *B. sphaericus* 2297 recombinants {2297(::pHT5601)} had toxicity to *Anopheles stephensi* comparable to *Bti*. In addition, the recombinant 2297(::*cry11A*) partially suppressed resistance to *B. sphaericus* 2297 in a strain of *C. quinquefasciatus* from India resistant to *B. sphaericus* 1593. Similar results were obtained when either the *cry11A* or *cry11B* (from *Bt* subsp. *jegathesan*), or both of these genes were inserted into the chromosome of *B. sphaericus* 2297 using integrative plasmids [55]. The production of Cry11A and/or Cry11B along with the *B. sphaericus* 2297 binary toxin increased the toxicity of this strain against *A. aegypti*, depending on the specific recombinant, from 5 to 11-fold. Against *C. pipiens*, most recombinants were similar in toxicity to parental *B. sphaericus* 2297, although one (2297*pro::cry11Ba*) was about twice as toxic. Recombinants producing Cry11A and/or Cry11B were able to partially suppress resistance to *B. sphaericus* 2297 in different populations of *C. pipiens* [55].

A different type of *B. sphaericus*/*Bt* recombinant was constructed by using the shuttle vector pMK3 to insert the *cyt1Ab1* gene from *Bt* subsp. *medellin* into *B. sphaericus* 2297 [59]. The production of the *B. sphaericus* 2297 binary toxin together with Cyt1Ab did not improve toxicity to *A. aegypti* or *C. pipiens*. However, the recombinant strain was able to restore the sensitivity of *B. sphaericus*-resistant populations of *C. pipiens* and *C. quinquefasciatus* by 10 to 20-fold.

Because they contained broad spectrum mosquitocidal Cry proteins, the *B. sphaericus* recombinants described above were typically considerably more toxic to *A. aegypti* than parental *B. sphaericus* strains. However, none of these recombinants was better than *Bti* against this species, and only a few were more toxic to *Culex* and *Anopheles* species than the parental *B. sphaericus* strains. Nevertheless, these studies were very valuable because they resulted in techniques for constructing recombinants, and showed that the various proteins of *Bti* and other *Bt*'s could be produced in substantial quantities in different strains of *B. sphaericus*. Moreover, they showed that producing *Bt* Cry and Cyt proteins in *B. sphaericus* extended its target spectrum to *A. aegypti*, and partially suppressed *B. sphaericus*-resistance in *Culex* species. Though not tested under field conditions, based on laboratory studies with *Bti*, it is probable that *B. sphaericus* strains containing Cry and/or Cyt toxins, even if not as effective as *Bti*, would be less prone to resistance, and therefore useful in polluted waters for *Culex* control.

4.2   *B. sphaericus* binary toxin in *Bt* subsp. *israelensis*

The strategy of improving mosquitocidal bacteria by producing the *B. sphaericus* binary toxin in *Bti* has been used much less frequently, primarily because *Bti* already has a broad host spectrum and is more effective against most mosquito species than *B. sphaericus*. In addition, *B. sphaericus* is more effective and persists better than *Bti* in polluted waters, so improving the host spectrum and toxicity of *B. sphaericus* with *Bti* proteins had the potential for producing an excellent recombinant strain. It is not clear at this point which host is the best for optimising toxin production and achieving broad spectrum mosquito control. It may be that *Bti* is the best for some endotoxin combinations, targets, and ecological situations and *B. sphaericus* for others. But producing the *B. sphaericus* binary toxin in *Bti* along with its normal toxin complement, especially using some of the enhancing elements (Section 3), show that very effective recombinants that use *Bt* as a host cell can be constructed.

In the first study where *B. sphaericus* toxins were produced in *Bti*, the binary toxin of *B. sphaericus* 1593 was cloned into the shuttle vector pBU4 yielding pGSP10 [17]. This plasmid was then transformed into the 4Q2-72 strain of *Bti*, a strain that only contains the large plasmid encoding the Cyt and Cry endotoxins typically found in this subspecies. Analysis of the recombinant *Bti* strain {4Q2-72(pGSP10)} showed that it produced the standard *Bti* toxins in normal amounts along with the 51.4 and 41.9 polypeptides of the *B. sphaericus* binary toxin. When tested against *A. aegypti*, *C. pipiens* and *A. stephensi*, the toxicity of the recombinant was no better than either the parental *Bti* or *B. sphaericus* strain.

In the above study, *B. sphaericus* promoters were used to express the *B. sphaericus* binary toxin in *Bti*, and none of the enhancing elements identified after this study was published were present in the plasmid used to produce the *B. sphaericus* binary toxin in *Bti*. Electron microscopy indicated that only small crystal of the binary toxin was produced in the *Bti* transformants [17]. This could account for the lack of improved toxicity. Therefore, more recently a *Bti* recombinant producing a large amount of the *B. sphaericus* binary toxin was developed [16]. To construct this strain, the *B. sphaericus* binary toxin was cloned from *B. sphaericus* 2362 using the polymerase chain reaction. *B. sphaericus* 2362 was selected as the source of the binary toxin genes because this pair is more toxic than the binary toxin that occurs in strain 2297 [11, 14, 20, 36]. To take advantage of the factors that can enhance toxin synthesis, expression of the *B. sphaericus* 2362 binary toxin was placed under the control of *cyt1A* promoters, and a STAB-SD sequence was added to the construct upstream from the coding region. This construct was then cloned into the shuttle vector pHT3101 and transformed into both acrystalliferous and crystalliferous (IPS-82) strains of *Bti*. In the transformed acrystalliferous strain, a large crystal of the *B. sphaericus* binary toxin, approximately the same size as the spore, was produced (Figure 8). A similar crystal was observed along with a typical *Bti* parasporal body in the *Bti* IPS-82/*B. sphaericus* recombinant (Figure 8). Analysis of the *Bti* IPS-82/*B. sphaericus* recombinant by gel electrophoresis showed that the typical *Bti* and *B. sphaericus* binary toxins were produced, though the levels of the *Bti* proteins appeared to be lower than those characteristic of the parental *Bti* (Figure 8). With respect to toxicity, the *Bti* recombinant that produced the *B. sphaericus* 2362 binary toxin was 10-fold more toxic to *C. quinquefasciatus* than either parental strain. Moreover, though not unexpected, this recombinant completely suppressed *B. sphaericus* resistance in a laboratory population of *C. quinquefasciatus* resistant to *B. sphaericus* 2362.

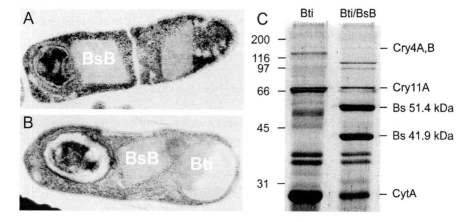

Figure 8. Recombinant strains of *Bt* subsp. *israelensis* that produce the *B. sphaericus* 2362 binary toxin. A. Acrystalliferous strain transformed with a plasmid that produces the *B. sphaericus* 2362 binary toxin using cyt1A promoters and the STAB-SD mRNA stabilising sequence. B. IPS-82 transformed with the same plasmid. A large crystal of the *B. sphaericus* binary toxin and a typical *Bti* crystal are obvious in the sporulated cell. C. SDS-PAGE analysis of IPS-82 (Lane *Bti*) and IPS-82 producing the *Bti* proteins and the *B. sphaericus* binary toxin (Lane *Bti*/*B. sphaericus*B).

5.   SUMMARY AND CONCLUSIONS

Over the past decade, a wide range of genes encoding mosquitocidal proteins along with expression vectors, helper proteins and enhancing factors have been identified and developed for constructing improved recombinant mosquitocidal bacteria. Most of the recombinant strains constructed to date are strains of *B. sphaericus* that produce Cry and/or Cyt proteins of *B. thuringiensis* subsp. *israelensis* and related *Bt* subspecies. These recombinant *B. sphaericus* strains have improved mosquitocidal properties in that they extend the target spectrum of this bacterium to species of *Aedes* mosquitoes, and can suppress resistance to *B. sphaericus* strains in *Culex* populations. In addition, markedly improved strains of *Bti* have been developed recently that produce high levels of the *B. sphaericus* 2362 binary toxin in addition to the standard *Bti* proteins. These recombinant strains of *Bti* and *B. sphaericus* should prove more effective in field situations because of their higher toxicity and multiplicity of toxins. Moreover, they should be less prone to the development of resistance in target populations owing to the multiplicity of toxins, especially the recombinants that include Cyt proteins. Further improvements in efficacy can be expected in the future by optimising insecticidal protein combinations and ratios, and by expressing these in alternate hosts such as asporogenic *B. sphaericus* and *Bt* strains and non-spore-forming bacteria. And methods are now available for removing non-*Bt* and *B. sphaericus* gene sequences [10]. The availability of these improved bacterial insecticides should result in more cost-effective control of vector and nuisance biting flies. Lastly, the use of these strains in operational programmes could yield data and models useful for managing resistance to insecticidal proteins in bacterial insecticides and insect-resistant crops used for pest control in agriculture.

REFERENCES

[1] Adams LF, Visick JE, & Whiteley HR (1989) A 20-kilodalton protein is required for efficient production of the *Bacillus thuringiensis* subsp. *israelensis* 27-kilodalton crystal protein in *Escherichia coli*. J. Bacteriol. 171, 521-530

[2] Adams LF, Mathewes S, O'Hara P, Petersen A & Gurtler H (1994) Elucidation of the mechanism of CryIIIA overproduction in a mutagenized strain of *Bacillus thuringiensis* var. *tenebrionis*. Mol. Microbiol. 13, 97-107

[3] Agaisse H & Lereclus D (1994) Structural and functional analysis of the promoter region involved in full expression of the *cryIIIA* toxin gene of *Bacillus thuringiensis*. Mol. Microbiol. 13, 97-107

[4] Agaisse H & Lereclus D (1995) How does *Bacillus thuringiensis* produce so much insecticidal crystal protein? J. Bacteriol. 177, 6027-6032

[5] Agaisse H & Lereclus D (1996) STAB-SD: a Shine-Dalgarno sequence in the 5' untranslated region is a determinant of mRNA stability. Mol. Microbiol. 20, 633-643

[6] Arantes O & Lereclus D (1991) Construction of cloning vectors for *Bacillus thuringiensis*. Gene 108, 115-119

[7] Bar E, Leiman-Hurwitz J, Rahamim E, Keynan A & Sandler N (1991) Cloning and expression of *Bacillus thuringiensis israelensis* δ-endotoxin DNA in *B. sphaericus*. J. Invertebr. Pathol. 57, 149-158

[8] Bar E, Sandler N, Makayoto M & Keynan A (1998) Expression of chromosomally inserted *Bacillus thuringiensis israelensis* toxin genes in *Bacillus sphaericus*. J. Invertebr. Pathol. 72, 206-213

[9] Baum JA & Malvar T (1995) Regulation of insecticidal crystal protein production in *Bacillus thuringiensis*. Mol. Microbiol. 18, 1-12

[10] Baum JA, Kakefuda M & Gawron-Burke C (1996) Engineering *Bacillus thuringiensis* bioinsecticides with an indigenous site-specific recombination system. Appl. Environ. Microbiol. 62, 4367-4373

[11] Baumann P, Clark MA, Baumann L & Broadwell AH (1991) *Bacillus sphaericus* as a mosquito pathogen: properties of the organism and its toxins. Microbiol. Rev. 55, 425-436

[12] Becker N & M Ludwig (1993) Investigations on possible resistance in *Aedes vexans* after a 10-year application of *Bacillus thuringiensis israelensis*. J. Amer. Mosq. Control Assoc. 9, 221-224

[13] Becker N & Margalit J (1993) Use of *Bacillus thuringiensis israelensis* against mosquitoes and blackflies, p. 147-170. *In* Entwistle PF, Cory JS, Bailey MJ & Higgs (ed.), *Bacillus thuringiensis*, An environmental biopesticide: theory and practice, John Wiley & Sons

[14] Berry C, Hindley J, Ehrhardt AF, Grounds T, de Souza I & Davidson EW (1993) Genetic determinants of host ranges of *Bacillus sphaericus* mosquito larvicidal toxins. J Bacteriol. 175, 510-518

[15] Berry C, Jackson-Yap J, Oei C & Hindley J (1989) Nucleotide sequence of two toxin genes from *Bacillus sphaericus* IAB59: sequence comparisons between five highly toxinogenic strains. Nucleic Acids Res. 17, 7516

[16] Bideshi DK, Park HW, Wirth MC, Walton WE & Federici BA (2000) Markedly improved recombinant bacterial insecticide for controlling mosquito vectors of human disease. Submitted

[17] Bourgouin C, Delécluse A, de La Torre F & Szulmajster J (1990) Transfer of the toxin protein genes of *Bacillus sphaericus* into *Bacillus thuringiensis* subsp. *israelensis* and their expression. Appl. Environ. Microbiol. 56, 340-344

[18] Butko, P, Huang, FM, Pusztai-Carey M & Surewicz WK (1996) Membrane permeabilization induced by cytolytic delta-endotoxin CytA from *Bacillus thuringiensis* var. *israelensis*. Biochemistry 35, 11355-11360

[19] Butko, P, Huang, FM, Pusztai-Carey M & Surewicz WK (1997) Interaction of the delta-endotoxin CytA from *Bacillus thuringiensis* var. *israelensis* with lipid membranes. Biochemistry 36, 12862-12868

[20] Charles J-F, Nielsen-LeRoux C & Delécluse A (1996) *Bacillus sphaericus* toxins: molecular biology and mode of action. Annu. Rev. Entomol. 41, 451-472.

[21] Crickmore N & Ellar DJ (1992) Involvement of a possible chaperonin in the efficient expression of a cloned CryIIA δ-endotoxin gene in *Bacillus thuringiensis*. Mol. Microbiol. 6, 1533-1537

[22] Crickmore N, Bone EJ, Williams JA & Ellar DJ (1995) Contribution of the individual components of the δ-endotoxin crystal to the mosquitocidal activity of *Bacillus thuringiensis* subsp. *israelensis*. FEMS Microbiol. Lett. 131, 249-254

[23] Crickmore N, Zeigler DR, Feitelson J *et al.* (1998) Revision of the nomenclature for the *Bacillus thuringiensis* pesticidal crystal proteins. Microbiol. Mol. Biol. Rev. 62, 807-813

[24] Davidson EW (1988) Binding of the *Bacillus sphaericus* (Eubacteriales: Bacillaceae) toxin to midgut cells of mosquito (Diptera: Culicidae) larvae: relationship to host range. J. Med. Entomol. 25, 151-157

[25] Delécluse A, Rosso M-L, & Ragni A (1995) Cloning and expression of a novel toxin gene from *Bacillus thuringiensis* subsp. *jegathesan* encoding a highly mosquitocidal protein. Appl. Environ. Microbiol. 61, 4230-4235

[26] Dervyn E, Poncet S, Klier A & Rapoport G (1995) Transcriptional regulation of the *cryIVD* gene operon from *Bacillus thuringiensis* subsp. *israelensis*. J. Bacteriol. 177, 2283-2291

[27] Federici BA & Bauer LS (1998) Cyt1Aa protein of *Bacillus thuringiensis* is toxic to the cottonwood leaf beetle, *Chrysomela scripta*, and suppresses high levels of resistance to Cry3A. Appl. Environ. Microbiol. 64, 4368-4371

[28] Ge B (1999) Molecular characterization of mosquitocidal protein synthesis and crystallization in *Bacillus thuringiensis*. Ph.D. Thesis, University of California, Riverside

[29] Ge B, Bideshi DK, Moar WJ & Federici BA. (1998) Differential effects of helper proteins encoded by the *cry2A* and *cry11A* operons on the formation of Cry2A crystals in *Bacillus thuringiensis*. FEMS Microbiol. Lett. 165, 35-41

[30] Georghiou GP & Wirth MC (1977) Influence of exposure to single versus multiple toxins of *Bacillus thuringiensis* subsp. *israelensis* on development of resistance in the mosquito *Culex quinquefasciatus* (Diptera: Culicidae). Appl. Environ. Microbiol. 63, 1095-1101

[31] Goldberg LJ & Margalit J (1977) A bacterial spore demonstrating rapid larvicidal activity against *Anopheles sergentii*, *Uranotaenia unguiculata*, *Culex univitattus*, *Aedes aegypti*, and *Culex pipiens*. Mosq. News 37, 355-358

[32] Glatron MF & Rapoport G (1972) Biosynthesis of the parasporal inclusion of *Bacillus thuringiensis*: half-life of its corresponding messenger RNA. Biochimie 54, 1291-1301

[33] Guillet P, Kurtak DC, Philippon B & Meyer R (1990) Use of *Bacillus thuringiensis israelensis* for onchocerciasis control in West Africa, p. 187-201. *In* de Barjac H & Sutherland DJ (ed.), Bacterial control of mosquitoes and black flies, Rutgers University Press

[34] Höfte H & Whiteley HR (1989) Insecticidal crystal proteins of *Bacillus thuringiensis*. Microbiol. Rev. 53, 242-255

[35] Hue KK, Cohen SD & Bechhofer DH (1995) A polypurine sequence that acts as a 5' mRNA stabilizer in *Bacillus subtilis*. J. Bacteriol. 177, 3465-3471

[36] Humphreys MJ & Berry C (1998) Variants of the *Bacillus sphaericus* binary toxins: implications for differential toxicity of strains. J. Invertebr. Pathol. 71, 184-185

[37] Ibarra, JE & Federici BA (1986) Isolation of a relatively nontoxic 65-kilodalton protein inclusion from the parasporal body of *Bacillus thuringiensis* subsp. *israelensis*. J. Bacteriol. 165, 527-533

[38] Ibarra JE & Federici BA (1986) Parasporal bodies of *Bacillus thuringiensis* subsp. *morrisoni* (PG-14) and *Bacillus thuringiensis* subsp. *israelensis* are similar in protein composition and toxicity. FEMS Microbiol. Lett. 34, 79-84

[39] Knowles BH & Dow JAT (1993) The crystal δ-endotoxins of *Bacillus thuringiensis*: Models for their mechanism of action on the insect gut. BioEssays 15, 469-476

[40] Lereclus D, Arantes O, Chaufaux J & Lecadet MM (1989) Transformation and expression of a cloned δ-endotoxin gene in *Bacillus thuringiensis*. FEMS Microbiol. Lett. 60, 211-218

[41] Lereclus D, Agaisse H, Gominet M & Chaufaux J (1995) Overproduction of encapsulated insecticidal crystal proteins in a *Bacillus thuringiensis spo0A* mutant. Bio/Technology 13, 67-71

[42] Malvar T & Baum JA (1994) Tn*5401* disruption of the *spo0F* gene, identified by direct chromosomal sequencing, results in CryIIIA overproduction in *Bacillus thuringiensis*. J. Bacteriol. 176, 4750-4753

[43] Moar WJ, Trumble JT & Federici BA (1989) Comparative toxicity of spores and crystals from the NDR-12 and HD-1 strains of *Bacillus thuringiensis* subsp. *kurstaki* to neonate beet armyworm (Lepidoptera: Noctuidae). J. Econ. Entomol. 82, 1593-1603

[44] Mulla MS (1990) Activity, field efficacy, and use of *Bacillus thuringiensis israelensis* against mosquitoes, p. 134-160. *In* de Barjac H & Sutherland DJ (ed.), Bacterial control of mosquitoes and black flies, Rutgers University Press

[45] Park HW, Ge B, Bauer LS & Federici BA (1998) Optimization of Cry3A yields in *Bacillus thuringiensis* by use of sporulation-dependent promoters in combination with the STAB-SD mRNA sequence. Appl. Environ. Microbiol. 64, 3932-3938

[46] Park HW, Bideshi DK, Johnson JJ & Federici BA (1999) Differential enhancement of Cry2A versus Cry11A yields in *Bacillus thuringiensis* by use of the *cry3A* STAB mRNA sequence. FEMS Microbiol. Lett. 181, 319-327

[47] Poncet S, Delécluse A, Anello D, Klier A & Rapoport G (1994) Transfer and expression of the CryIVB and CryIVD genes of *Bacillus thuringiensis* subsp. *israelensis* in *Bacillus sphaericus* 2297. FEMS Microbiol. Lett. 117, 91-96

[48] Poncet SA, Delécluse A, Klier A & Rapoport G (1995) Evaluation of synergistic interactions among CryIVA, CryIVB, and CryIVD toxic components of *Bacillus thuringiensis* subsp. *israelensis* crystals. J. Invertebr. Pathol. 66, 131-135

[49] Poncet S, Bernard C, Dervyn E, Cayley J, Klier A & Rapoport G (1997) Improvement of *Bacillus sphaericus* toxicity against dipteran larvae by integration, via homologous recombination, of the Cry11A toxin gene from *Bacillus thuringiensis* subsp. *israelensis*. Appl. Environ. Microbiol. 63, 4413-4420

[50] Rang C, Bes M, Lullien-Pellerin V, Wu D, Federici BA & Frutos R (1996) Influence of the 20-kDa protein from *Bacillus thuringiensis* ssp. *israelensis* on the rate of production of truncated Cry1C proteins. FEMS Microbiol. Lett. 141, 261-264

[51] Rao DR, Mani TR, Rajendran R, Joseph ASJ, Gajanana A & Reuben R (1995) Development of a high level of resistance to *Bacillus sphaericus* in a field population of *Culex quinquefasciatus* from Kochi, India. J. Am. Mosq. Control Assoc. 11, 1-5

[52] Rodcharoen J & Mulla MS (1994) Resistance development in *Culex quinquefasciatus* (Diptera: Culicidae) to *Bacillus sphaericus*. J. Econ. Entomol. 87, 1133-1140

[53] Schnepf E, Crickmore N, Van Rie J *et al.* (1998) *Bacillus thuringiensis* and its pesticidal crystal proteins. Microbiol. Mol. Biol. Rev. 62, 775-806

[54] Seleena P, Lee HL & Lecadet M-M (1995) A new serovar of *Bacillus thuringiensis* possessing 28a28c flagellar antigenic structure: *Bacillus thuringiensis* serovar *jegathesan*. J. Am. Mosq. Control Assoc. 11, 471-473

[55] Servant P, Rosso M-L, Hamon S, Poncet S, Delécluse A & Rapoport G (1999) Production of Cry11A and Cry11Ba toxins in *Bacillus sphaericus* confers toxicity towards *Aedes aegypti* and resistant *Culex* populations. Appl. Environ. Microbiol. 65, 3021-3026

[56] Silva-Filha MH, Regis L, Nielsen-LeRoux C & Charles JF (1995) Low-level resistance to *Bacillus sphaericus* in a field-treated population of *Culex quinquefasciatus* (Diptera: Culicidae). J. Econ. Entomol. 88, 525-530

[57] Sinègre G, Babinot M, Quermal JM & Gaven B (1994) First field occurrence of *Culex pipiens* resistance to *Bacillus sphaericus* in southern France. VII European Meeting, Society for Vector Ecology, Barcelona, Spain

[58] Sullivan MA, Yasbin RE & Young FE (1984) New shuttle vectors for *Bacillus subtilis* and *Escherichia coli* which allow rapid detection of inserted fragments. Gene 29, 21-26.

[59] Thiéry I, Hamon S, Delécluse A & Orduz S (1998) The introduction into *Bacillus sphaericus* of the *Bacillus thuringiensis* subsp. *medellin cyt1Ab1* gene results in higher susceptibility of resistant mosquito larva populations to *B. sphaericus*. Appl. Environ. Microbiol. 64, 3910-3916

[60] Thomas WE & Ellar DJ (1983) Mechanism of action of *Bacillus thuringiensis* var. *israelensis* insecticidal δ-endotoxin. FEBS Lett. 154, 362-368

[61] Trisrisook M, Pantuwatana S, Bhumiratana A & Panbangred W (1990) Molecular cloning of the 130-kilodalton mosquitocidal δ-endotoxin gene of *Bacillus thuringiensis* subsp. *israelensis* in *Bacillus sphaericus*. Appl. Environ. Microbiol 56, 1710-1716

[62] Visick JE & Whiteley HR (1991) Effect of a 20-kilodalton protein from *Bacillus thuringiensis* subsp. *isralensis* on production of the CytA protein by *Escherichia coli*. J. Bacteriol. 173, 1748-1756

[63] Widner WR & Whiteley HR (1989) Two highly related insecticidal crystal proteins of *Bacillus thuringiensis* subsp. *kurstaki* possess different host range specificities. J. Bacteriol. 171, 965-974

[64] Wirth MC, Walton WE & Federici BA (2000) Cyt1A from *Bacillus thuringiensis* restores toxicity of *Bacillus sphaericus* against *Culex quinquefasciatus* (Diptera: Culicidae). J. Med. Entomol. In press

[65] Wirth MC, Delécluse A, Federici BA & Walton WE (1998) Variable cross-resistance to Cry11B from *Bacillus thuringiensis* subsp. *jegathesan* in *Culex quinquefasciatus* (Diptera: Culicidae) resistant to single or multiple toxins of *Bacillus thuringiensis* subsp. *israelensis*. Appl. Environ. Microbiol. 64, 4174-4179.

[66] Wirth MC, Georghiou GP & Federici BA (1997) CytA enables CryIV endotoxins of *Bacillus thuringiensis* to overcome high levels of CryIV resistance in the mosquito *Culex quinquefasciatus*. Proc. Natl. Acad. Sci. USA 94, 10536-10540.

[67] Wirth MC, Walton WE & Federici BA (2000) Cyt1A from *Bacillus thuringiensis* synergizes activity of *Bacillus sphaericus* against *Aedes aegypti*. Appl. Environ. Microbiol. In press

[68] Wong HC & Chang S (1986) Identification of a positive retroregulator that stabilizes mRNAs in bacteria. Proc. Natl. Acad. Sci. USA 83, 3222-3237

[69] Wu D & Chang FN (1985) Synergism in mosquitocidal activity of 26 and 65 kDa proteins from *Bacillus thuringiensis* subsp. *israelensis* crystal. FEBS Lett. 190, 232-236

[70] Wu D & Federici BA (1993) A 20-kilodalton protein preserves cell viability and promotes CytA crystal formation during sporulation in *Bacillus thuringiensis*. J. Bacteriol. 175, 5276-5280

[71] Wu D, Johnson JJ & Federici BA (1994) Synergism of mosquitocidal toxicity between CytA and CryIVD proteins using inclusions produced from cloned genes of *Bacillus thuringiensis*. Mol. Microbiol. 13, 965-972

[72] Wu D & Federici BA (1995) Improved production of the insecticidal CryIVD protein in *Bacillus thuringiensis* using *cryIA(c)* promoters to express the gene for an associated 20-kDa protein. Appl. Microbiol. Biotechnol. 42, 697-702

[73] Yoshisue HK, Yoshida K, Sen H *et al.* (1992) Effects of *Bacillus thuringiensis* var. *israelensis* 20-kDa protein on production of the *Bti* 130-kDa crystal protein in *Escherichia coli*. Biosci. Biotechnol. Biochem. 56, 1429-1433

Chapter 7.3

Bacillus thuringiensis : risk assessment

André Klier

Unité de Biochimie Microbienne, URA 1300 CNRS, Institut Pasteur, 25 rue du Dr Roux, 75724 Paris Cedex 15, France

Key words: *Bacillus thuringiensis*, biopesticides, δ-endotoxin, risk/benefit ratio, safety, resistance, transgenic expression, virulence

Abstract: Use of biological control agents will dramatically increase in the next decades as farmers move towards environmentally safe agricultural practices, in response to the people request. *Bacillus thuringiensis* is one of these agents and the derivative products are estimated to rise at least 20 % per year. However, there are some species with an increased activity against specific insects and a broader host range. In addition, recombinant DNA technology allows now to reach new derivatives or to introduce the genetic determinants into new hosts, including plants and other microbes. It is obvious that technology combined with the diversity of the *Bt* species will increase the scope for the application of *Bt*. The benefits to agriculture and for the environment are considerable, but the possibility of adverse environmental impact for the fauna and/or the flora due to the large scale application of the new *Bt* derivative products needs to be considered and evaluated. Moreover, little is known about the ecology of *Bt* and the role of spores in the environment.

1. INTRODUCTION

Pests affecting primary production (agriculture, forestry, ...) and those which are vectors of tropical diseases (malaria, onchocerciasis, ...) are mostly controlled using chemical pesticides. The use of alternative biological control agents will probably increase over the next decade, as farmers move towards environmentally safe agricultural practices, in response to public opinion. *Bacillus thuringiensis* is the most widely used

J.-F. Charles et al. (eds.),
Entomopathogenic Bacteria: From Laboratory to Field Application, 485–504.
© 2000 *Kluwer Academic Publishers. Printed in the Netherlands.*

microorganism for insect pest control. However, the world-wide sales of products derived from *B. thuringiensis* are only 2% of the global insecticide market although it is predicted to rise by at least 20% per year [38]. There are limitations to the commercial application of these products, particularly their instability in the field and the narrow host spectrum of the active larvicidal ingredient. Overcoming these problems is the key to the greater use of this microorganism in the future. Significant efforts are being made to isolate new strains with higher potency against insect pest species and wider host ranges. Recombinant DNA technology supplies the tools to create new strains of *B. thuringiensis* and makes it possible to integrate *B. thuringiensis* toxin genes into alternative delivery system, such as other bacteria, virus and plants. Moreover, using genetic tools, it is now possible to combine different natural or engineered toxin genes into a single host. It is likely that the discovery of new isolates and the application of the recombinant DNA technology will increase the scope for the use of *B. thuringiensis*.

The benefits to agriculture and for the environment appear considerable, but the possibility of adverse environmental impact on the fauna (including humans) and/or the flora, due to the large scale application of these new *B. thuringiensis* derivatives needs to be considered and carefully evaluated. Although there is a well-documented history of successful application of *B. thuringiensis* which is safe for the environment and for human [48]. Various issues should be re-examined in particular its use with regard to the identification of new *B. thuringiensis* metabolites [16], to the new taxonomic relationships within the *B. cereus* group [29] and to the isolation of *B. thuringiensis* strains in the few clinical cases associated with *B. thuringiensis* [15, 31].

2.   TAXONOMY OF *BACILLUS THURINGIENSIS* AND ITS OCCURRENCE IN THE ENVIRONMENT

To evaluate the risk of releasing microorganism into the environment, it is necessary to understand how it interacts with its surroundings and the biota. Although *B. thuringiensis* is widely used to control insect larvae, there has been limited research on its role in the environment. Moreover, recent reports identified *B. thuringiensis* as a member of the *B. cereus* group, which also contains the pathogens *B. mycoides* and *B. anthracis* prompting a re-examination of *B. thuringiensis* as a potential opportunistic human pathogen [29]. It is therefore important to be able to distinguish *B. thuringiensis* from other *Bacilli*.

The life cycle of *B. thuringiensis* consists of two distinct phases : vegetative cell division and sporulation, when the production of the δ-endotoxin and the synthesis of the crystal take place. *B. thuringiensis* persists in soil predominantly as spores, with only limited multiplication of vegetative cells [2, 58], but the respective roles of spores and vegetative cells in persistence and possible spread of *B. thuringiensis* in the environment are not known.

2.1   Taxonomy

There is considerable debate and controversy as to whether the three species – *B. anthracis*, *B. mycoides* and *B. thuringiensis* – should be considered as varieties of *B. cereus* or classified as separate species. *B. thuringiensis* can only be distinguished from *B. cereus*, by the production during the sporulation phase of one or more inclusion bodies (δ-endotoxin), toxic for insect larvae of the orders Coleoptera, Diptera and Lepidoptera. Phenotypic differentiation of *B. thuringiensis* and *B. cereus* is not possible by on the basis of morphology or use of organic compounds, fatty acid content analysis, sugar utilisation, multilocus enzyme electrophoresis, enterotoxin production, or serological and phage-typing procedures [29]. Likewise, they cannot be differentiated genotypically by DNA homology testing, ribotyping, 16S rRNA sequencing, analysis of the 16S - 23S internal transcribed sequence, PCR analysis of genes encoding *B. cereus*-like toxic products, on pulsed-field electrophoresis [29]. According to all these characters and properties, *B. thuringiensis* is indistinguishable from *B. cereus*.

However, there are some reports of differences between the 16S rRNA sequences of a limited number of *B. cereus* and *B. thuringiensis* strains [23]. Beattie *et al.* [6] were also able to discriminate among members of the *B. cereus* group by Fourier transform infrared spectroscopy. RAPD fingerprinting can also distinguish *B. cereus* from *B. thuringiensis* [10]. Very recently, it was shown that specific primers designed to amplify part of the *B. cereus gyrB* gene do not amplify a similar DNA fragment from the *B. thuringiensis* genome [66]. It is not clear whether these slight variations are significant or are due to divergence within a single species. However, the transfer of encoding δ–endotoxin plasmids from *B. thuringiensis* to *B. cereus* makes the recipient *B. cereus* indistinguishable from *B. thuringiensis* [25] ; this is consistent with the two microorganisms being a single taxonomic group.

Several papers report the great variability of the genome size of *B. cereus* and to a lesser extent of *B. thuringiensis*. For example, *B. cereus* chromosome shows a 2.6-fold variation from 2.4 Mb to 6.3 Mb

[13]. Interestingly, accurate comparison of chromosomal maps suggests that all the chromosomes have a similar organisation in the half part near the replication origin, but are more divergent in the half near the terminus [12]. Similar variability in genome size has been observed in other bacterial species including *Clostridium* sp. and *Streptomyces* sp. This variability makes it difficult to compare different isolates belonging to the same group and raises questions about the species definition. The advantages in terms of coding capacity due to the extra DNA may be relevant to the virulence of different isolates of the *B. cereus* group. Nevertheless, these data on the variability of the genome size increase the taxonomic ambiguity within this group.

B. cereus is generally thought to be an opportunistic pathogen. There is substantial divergence in the virulence factor production between isolates. This applies whether or not *B. anthracis* and *B. thuringiensis* are considered to be pathogens of mammals and insect larvae, respectively. The genetic determinants encoding the specific toxic factors from *B. anthracis* and *B. thuringiensis* are plasmid borne. Possibly some of the plasmids can insert into the chromosome, leading to the diverse chromosome sizes [12].

The major taxonomic issue is whether *B. thuringiensis* and *B. cereus* are the same species. Phenotypic arguments favour a common identity whereas some genomic analysis reveals differences. The variability of the genome size and the coding capacity of both groups does not allow a straightforward conclusion. *B. cereus* may be a parental (but not separate) species and consequently, *B. thuringiensis* and *B. anthracis* be considered as derivatives and/or related species which have become specialised by acquisition of plasmids encoding specific toxic factors. Consequently, the *B. cereus* group could be defined as an aggregate of quasi-species.

2.2   Virulence factors

Virulence factors are described in more detail in chapter 1.3 of this book. A short overview is given in this review. The larvicidal activities of *B. thuringiensis* are due to proteinaceous crystals produced during sporulation. The crystals are composed of one or several polypeptides named δ-endotoxins and the genes (*cry*) encoding these polypeptides belong to a unique multigenic family [58]. The products formulated from *B. thuringiensis* are composed of a mixture of crystals and spores. The spores play an important synergistic role in the toxicity to some insect species [56].

However, *B. thuringiensis*, like other bacteria, produces during vegetative growth and early sporulation phase an assortment of antibiotics, enzymes, metabolites and toxins which are biologically active and may have effects both on target insect larvae (explaining the role of the spores) and on non-target organisms. β-exotoxin is associated with certain *B. thuringiensis* subspecies. It is a heat-stable derivative nucleotide, which is composed of adenine (or uracil), glucose and allaric acid. The β-exotoxin inhibits the RNA polymerase of all eukaryotic species and consequently has a non-specific effect on all such organisms [8]. *B. thuringiensis* products containing β-exotoxin are used for the control of houseflies in some countries, but environmental and health agencies currently prohibit the use of β-exotoxin-producing *B. thuringiensis* strains. This nucleotide analogue has never been found in *B. cereus* isolates. As the genetic determinants responsible for the biosynthesis of β-exotoxin are plasmid borne [43], the plasmid involved is therefore not resident in *B. cereus*.

However, *B. thuringiensis* produces a series of compounds which are also produced by the opportunistic pathogen *B. cereus* :

The α-exotoxin is a proteinaceous exotoxin, produced by some strains of *B. thuringiensis* and *B. cereus* [36]. The toxin is thermolabile, sensitive to proteinase degradation and has a haemolytic activity. It has been reported that injection of high doses of α-exotoxin are lethal for mice.

Several bacteria belonging to the Gram-positive group produce thiol-dependent haemolysins [55]. *B. thuringiensis* produces a 47 kDa thuringiolysin, inhibited by low cholesterol concentrations. Several reports describe similar properties for the thuringiolysin and the *B. cereus* cereolysin. The relationships between SH-dependent haemolysin and the α-exotoxin remain to be determined and it has not been excluded that the two proteins are indeed the same.

A second haemolysin of 29 kDa, named haemolysin II, is secreted by both *B. thuringiensis* and *B. cereus* [55]. The haemolytic activity of this polypeptide is not inhibited by cholesterol. The polypeptides purified from *B. thuringiensis* and *B. cereus* have the same molecular properties, but only 30 % of the *B. cereus* isolates are producers of haemolysin II, whereas all *B. thuringiensis* isolates are positive [11].

Several phospholipase C activities are produced by *B. thuringiensis* isolates [64]. One is specific for phosphatidyl inositol (PI-PLC) ; another hydrolyses phosphatidyl choline (PC-PLC) and a sphingomyelinase has been detected is some isolates. Phospholipase C lyses the membrane of eukaryotic cells. The phospholipases C are produced during the transition phase of bacterial growth and the corresponding structural genes are controlled by a pleiotropic regulator PlcR [41].

Using a non-quantitative commercial *B. cereus* enterotoxin immunoassay, Damgaard [14] reported that vegetative cells grown from spores of commercial *B. thuringiensis* products excreted a diarrhoeal enterotoxin. More recently, the same group used a Vero cell assay to show that some *B. thuringiensis* strains isolated from food were enterotoxigenic [16]. In the same paper, it was estimated that the amount of enterotoxin produced by the *B. thuringiensis* strains is at least 10-fold lower than that produced by the known pathogenic *B. cereus* strains. Shinagawa [59], using an immunological enterotoxin assay, concluded that about half *B. thuringiensis* isolates are positive, at variable levels. Finally, Tayaboli & Seligny [63] observed extensive damage on cultivated insect cells caused by vegetative cells grown from commercial formulated *B. thuringiensis* products. An extensive genomic analysis demonstrated that some *B. thuringiensis* species harbour genetic determinants responsible for enterotoxin synthesis [12, 13]. So, it seems that strains used commercially for their larvicidal properties can produce enterotoxin. However, it is not clear that the amounts produced are sufficient to induce outbreaks of gastro-enteritis in humans.

Recently, it was reported [19] that *B. thuringiensis* strains during the vegetative growth produced a new class of pesticidal proteins, the Vip. These proteins cause gut paralysis and complete lysis of gut epithelium cells of susceptible insect larvae. The genes encoding Vip3 are present in a number of *Bacillus* isolates but it is not clear whether the genes are present in *B. cereus* isolates.

In conclusion, the δ-endotoxins (Cry proteins) are the main toxic factors for insect larvae and allow further germination of *B. thuringiensis* spores present in dead or weakened larvae. *B. thuringiensis*, like *B. cereus*, produces non-specific virulence factors which may contribute to the death of the larvae by septicaemia or other unknown specific action. However, the overall and specific contribution of there additional factors remains to be determined and demonstrated, since for numerous insect larvae, spores are not required and transgenic plants (harbouring only *cry* genes) act efficiently against insect larvae.

2.3   Natural occurrence of *B. thuringiensis*

Members of the *B. cereus* group can be found in most ecological niches. Hansen *et al.* [27] reviewed the distribution of *B. thuringiensis* in the environment. Strains have been isolated world-wide from all habitats tested, including soil, dead insect larvae, stored-product dusts and plant leaf surfaces. However, little is known about the natural transmission and behaviour of *B. thuringiensis* in the environment.

Numerous *B. thuringiensis* species have been isolated from dead or dying insect larvae and most have activity against the insect species from which they were isolated [24]. However, isolates with no activity towards the source host have also been found. Frequently, these microorganisms have a relatively narrow range in the insect order and can proliferate within the bodies of these host larvae. When the larva dies, the carcass usually contains relatively large amounts of spores and crystals that may be released in the environment. Some limited cases [2] of recycling of natural *B. thuringiensis* in cadavers has been reported when competitive microorganisms were at a low density. Outbreaks of *B. thuringiensis* in susceptible insect populations are very infrequent. The outbreaks reported correspond to situations where the insect density is high, providing better opportunity for establishing the disease within the insect population [46].

In the past, soils have been the main source for isolation of *B. thuringiensis* strains. Several reports state that *B. thuringiensis* isolates represent between 0.5 and 0.005 % of all *Bacillus* species isolated from soil samples in the United States. Martin [47] and Ishii & Ohba [33] found up to 5×10^4 *B. thuringiensis* spores per gram of soil and Hansen *et al.* [28] counted about 9×10^2 spores containing the *cryIA* gene per gram of soil. So, *B. thuringiensis* spores are relatively abundant in all kinds of soils. However, the frequency of serotypes found in soil varies substantially and no association between particular isolates and types of soil has been reported. The question of the fate of the spores has been addressed by several authors [50]. It was demonstrated that vegetative growth can occur when nutrients are available [2]. However, it was not demonstrated that such germination could lead to an epizootic of *B. thuringiensis*.

B. thuringiensis has been found in the phylloplane. The phylloplane is an heterologous environment, in which proliferate a complex array of microorganisms including bacteria. Numerous *B. thuringiensis* subspecies have been recovered from coniferous and deciduous trees and vegetables, as well as other plants [60], with a density up to 10^2 *B. thuringiensis* spores/cm^2. Many of these subspecies have a broad diversity of specific activities to Coleopteran and Lepidopteran insects, but some have no activity against the insects tested. This environment may act as a reservoir of *B. thuringiensis* subspecies, which are mainly active against insect larvae feeding on the plant foliage.

B. thuringiensis seems also to be common in the stored product environment and in grain dusts. The *B. thuringiensis* index in this type of environment ranges from 0.3 to 0.7, which is very high [51]. The well-known serovar *kurstaki* was first discovered in grain stores [37]. Two other serotypes are predominant: *kenyae* and *aizawai*. Stored products constitute a protected and closed environment and some epizootics have been described [50]. However, in most cases, quantification of the reported epizootic event is incomplete.

In conclusion, *B. thuringiensis* seems to be a ubiquitous microorganism, widespread in various habitats. However, *B. thuringiensis* is present essentially as spores and, even if a germination can occur in favourable conditions, epizootics are rarely observed or documented.

3.   RISK ASSESSMENT

Risk is an estimate of the probability and severity of harm or a prediction of the likelihood of something going wrong in a certain set of circumstances. Risk assessment is the process of obtaining qualitative and quantitative data on the risk levels. In the case of *B. thuringiensis*, two cases have to be considered; either the use and the spread in the environment of crude microbial preparations or the use of alternative delivery systems including plants expressing *B. thuringiensis* toxins. In both cases, the effects on the fauna, including humans and flora must be carefully evaluated.

3.1   Effect of *B. thuringiensis* on organisms

3.1.1   Mammals

Bacterial pest control agents can in principle cause harmful effects *via* toxicity, *via* inflammation or *via* the combination of these effects. The presence of bacteria in tissues does not necessarily mean infection. The term colonisation is used to describe multiplication on the surface or within an organism without inducing tissue damage. Persistence is the recovery of part of the innoculum after a time interval. Persistence is not necessarily equated with infection: an infection is when the bacteria proliferates inducing tissue damage. Evidence of multiplication includes a significant increase in the bacteria count, recovery of vegetative stages when spores were injected and failure of the innoculum to clear. It cannot be determined solely on the basis of lesions since injection of foreign material may elicit an inflammatory process.

Various criteria and upper limits have been proposed to classify a bacterium as a toxic agent. For example, a microorganism is considered as toxic if an oral dose $\leq 10^6$ cfu per mouse causes mortality or clinical and pathological changes. According to these criteria,   studies of *B.*

thuringiensis pesticides demonstrate that the tested isolates from the products are neither toxic to mice nor pathogenic in laboratory assays [48]. Toxicity studies submitted to US Environmental Protection Agency (EPA) have failed to show any significant adverse effects on body weight gain or other clinical observations or at necropsy. Infectivity/pathogenicity studies showed that the intact rodent system responds as expected to eliminate *B. thuringiensis* gradually from the body after oral, pulmonary or intravenous challenge. Clearance of *B. thuringiensis* is not instantaneous, is similar to the rate for other non-pathogenic *Bacilli*.

Additional studies, involving intraperitoneal exposure, have been done in the course of registration procedures. This exposure is considered a highly challenging route of exposure, and evaluates the ability of a bacterium to cause infection or produce toxic metabolites in the peritoneal cavity. Human and animal exposure by this route is very unlikely to occur during the normal application of *B. thuringiensis*. Most of the studies demonstrated that *B. thuringiensis* persists for a variable length of time in mice, and in various tissues including heart blood, but that it is cleared with time [48]. It has also been reported [31] that lethal effect may be observed following intraperitoneally injection of doses $\geq 10^8$ cfu of certain isolates into mice. Lower doses (10^7 cfu/mouse) were not toxic. The basis for this toxicity observed at high doses is not understood. The finding has not been considered as evidence for a hazard associated with *B. thuringiensis* products since the route of administration is not relevant to human and animal exposure conditions and the amounts injected into the mice were extremely high.

3.1.2 Other animals and living organisms

There have been numerous and well documented studies conducted on animals and the fauna in general as part of the registration process for these products [48]. No significant adverse effects have been observed on birds, aquatic vertebrates, invertebrates and non-target insects including honey bees [48]. However, it was recently reported [44] that pollen from *B. thuringiensis* transgenic corn can significantly harm monarch larvae. This report has to be confirmed, since it is the first reported deleterious effect of *B. thuringiensis* δ-endotoxin on a non-target insect. Moreover, the data have to be quantified and the unexpected toxic effect remains to be understood.

3.1.3   Exposure and effects of *B. thuringiensis* on humans

For aeons, humans have been exposed to *B. thuringiensis* in their natural habitats, particularly from soil, water and the phylloplane and very few adverse effects to these natural *B. thuringiensis* have been documented.

The manufacture and field application of *B. thuringiensis* products, particularly spraying programmes in populated areas can result in aerosols and dermal exposure of workers and human populations. Several experimental ingestions by humans of high *B. thuringiensis* doses have been conducted. The effects were not severe and included nausea and few diarrhoeal symptoms [20]. Over a period of 40 years of production, there are also no reports of workers manufacturing *B. thuringiensis* products aerosol having been seriously affected. Exposure of workers to *B. thuringiensis* in spraying conditions has been studied. During the spray programmes, some workers experienced chapped lips, dry skin, eye irritation and normal drip and stuffiness, but no serious health problems resulted. Some studies report variable titres of specific antibodies against the spores and/or the crystals, but without any clinical syndromes.

B. thuringiensis has been applied in the environment since 1933, but it was not successfully commercialised until the late 1950's, when deep tank aerobic liquid formulation was used to produce spore and crystal preparations. Major applications of *B. thuringiensis* have been used in North America for the control of over 40 pest species. Some of these large scale applications were followed by public health surveillance programmes and in general, no or very limited harmful effects were reported among residents of the sprayed areas. *B. thuringiensis israelensis* has also been applied in large quantities for disease vector control. More than five millions of litres of *B. thuringiensis israelensis* were sprayed from 1982 to 1997, to control blackfly larvae (*Simulium damnosum*) in West Africa. At the peak of the campaign, about 50 000 km of rivers were successfully treated over an area of one million km^2. To assess the environmental impact of such treatments, a network of sampling stations was established and no significant effect on the fauna and flora was recorded. In other application programmes, *B. thuringiensis israelensis* was released on a large scale in mosquito-infected areas in Southern Switzerland [45] and along the Rhine River in Germany [7]. No adverse effect was reported.

In some countries, especially Asia, *B. thuringiensis israelensis* has been added to domestic containers of drinking water for mosquito control and no serious problem has been documented, even after more than 10 years of use.

To summarise, the extensive use of *B. thuringiensis* powder in the environment can be considered as safe for human and the risk/benefit ratio is largely in favour of the use of this biopesticide.

3.1.4 Clinical case reports

Formulated *B. thuringiensis* products have been increasingly used over the last two decades, but *B. thuringiensis* has been isolated from only a very few cases of human bacterial infections.

In 1983, a farm worker developed a corneal ulcer in one eye which was accidentally splashed with a commercial *B. thuringiensis kurstaki* product. A significant number of *B. thuringiensis* cells was isolated from the affected eye [57]. Treatment with antibiotics and corticosteroid led to a resolution of the corneal ulcer in two weeks. It is generally believed that the corneal ulcer was due primarily to *B. thuringiensis* infection since similar infections caused by *B. cereus* are common. However, the possibility that *B. thuringiensis* may have been a non-pathogenic contaminant of the ulcer was not considered.

In 1995, during an investigation of an outbreak of gastro-enteritis in a chronic case institution, bacteria were isolated from four individuals and identified as *B. thuringiensis*. The isolates showed cytotoxic effects characteristic of *B. cereus* [34]. This study included a systematic analysis of whether the *B. thuringiensis* species in commercial preparations could produce a diarrhoeal enterotoxin [14]. For most isolates the assay was positive. However, the amount of enterotoxin detected by this assay varied by a factor of more than 100 among the isolates and the *B. thuringiensis* strains produced much less than *B. cereus* strains. In another study [16], strains of *B. thuringiensis* isolated from various food products were demonstrated to express cytotoxicities to Vero cells as an indicator of enterotoxin activity. However, there is no report of a human infection from food products due to *B. thuringiensis*.

Human infections are rare and apart from gastrointestinal tract infection, these are only two clinical reports of *B. thuringiensis* infection. In 1997, Damgaard *et al.* [15] isolated *B. thuringiensis* from burn wounds of two patients. None of the isolates showed any toxicity to Vero cells. It was postulated that hot water used on these patients was contaminated with *B. thuringiensis*, which acted thus as an opportunistic pathogen in the hospital burn units.

The second case was described by Hernandez *et al.* [31]. A *B. thuringiensis* strain, *konkukian*, was isolated from soft tissue necrosis following severe war wounds caused by a land mine explosion. The strain was able to induce cutaneous inflammatory lesions after application of a suspension of 10^7 cfu/mouse. Interestingly, the lesions healed spontaneously after 2 days in non-immunosuppressed mice, whereas similar lesions became more severe in immunosuppressed animals. The same research group conducted further work with the bacteria [30]. Mice infected intranasally by a 10^8 spores in a suspension died within 8 hours from a clinical toxic shock syndrome. This quantity is very much higher than the doses recommended by environmental and health agencies. A derivative mutant of this strain without the crystalline inclusion had the same properties indicating that δ-endotoxin is not involved in the pathogenicity. Lower dose inoculates induced only local inflammatory reactions with bacterial persistence for 10 days. Instillation with the supernatant of a stationary phase culture containing 10^8 cells per ml of *B. thuringiensis konkukian* led to 80 % lethality and the toxic effect was strongly inhibited by cholesterol. So, it is postulated that thiol-dependent haemolysin is responsible for the lethal action. This is surprising as there was no evidence of spore germination or increase in cell number ; no vegetative cells were detected. This is not compatible with synthesis of thuringiolysin by germinated spores in the pulmonary tract. It was also reported that commercial *B. thuringiensis* H3a,3b and H14 showed similar properties to the *B. konkukian* strain whereas another strain, H12, did not cause the same symptoms.

Interestingly, Ramisse *et al.* (personal communication) demonstrated that a *B. thuringiensis* subspecies in which the *plcR* gene is disrupted lost its capacity to kill mice in a similar assay. Presumably the pleiotropic regulator affects the expression of some unknown virulence factors responsible for this lethal effect.

In another study in Oregon (USA), Green and co-workers reported the isolation of *B. thuringiensis* from the body fluids of 55 patients with different infectious diseases [26]. In 52 of them, it was considered as a contaminant. For the three other cases with pre-existing medical problems, no firm conclusion was established about the causal relationships between infection and *B. thuringiensis*.

3.2   *B. thuringiensis cry* genes and transgenic organisms

Research and development have made it possible to clone and express the *cry* genes encoding the δ-endotoxins in other *B. thuringiensis* strains or in alternative delivery systems, including plants. The primary rationale for using these possibilities was to overcome some limitations of the formulations based on natural *B. thuringiensis*, such as poor persistence, narrow host range and cost. However, the intensive use of these possibilities, especially in North America, have caused concern among some groups and individuals and there is an increasing awareness, especially in Europe, of the need to assess the possible consequence of the controlled release of Genetically Modified Organisms (GMO) into the environment.

The best-adapted host for *cry* genes is *B. thuringiensis* itself and a natural conjugative-like system has been extensively used to transfer *cry*-encoding plasmids from one strain to another. Some of the transconjugant *B. thuringiensis* strains have been registered and commercialised, at least in the United States [25]. Several tools, such as shuttle plasmid vectors [3], integrational vectors [42] and plasmids using site-specific recombination systems [5], are now available to reintroduce natural or modified *cry* genes. Moreover, homologous recombination procedures have been developed to disrupt some resident genes [17] and consequently, proteinase negative [18] or asporogenous derivatives [40] have been constructed. It is likely that similar procedures will be used to eliminate undesirable genetic traits, for example those determining non-specific potential virulence factors.

Alternative microbial delivery systems have been developed, but few have been registered [21]. The rationale was to use endophylic or epiphylic hosts to increase the persistence of Cry proteins in the field. For example a microorganism that can survive and multiply in the feeding zone of the larvae could be used as a host and so continuously produce significant amounts of crystal proteins. lepidopteran active toxins have been produced in the endophylic bacterium *Clavibacter xyli* [39] and in the plant colonisers *Pseudomonas fluorescens* [52]. Similarly, dipteran active toxins have been produced in blue-green algae [61] and in some protozoans. However, the synthesis of Cry proteins is often much lower than that in *B. thuringiensis*. Moreover, the release of genetically modified microorganisms, for which no possibility of control exists, is very rigorously considered by regulatory agencies.

Several *cry* genes encoding toxins with different insect specificities have been introduced into plants, first tobacco [4] and now various other major crops [53]. Since 1996, farmers in the United States have planted millions of acres of corn, cotton and potatoes that were genetically modified to produce *B. thuringiensis* toxins. As the toxins are produced continuously in all or in some plant tissues, there is no need to apply other insecticides. Moreover, the plant delivery system allows control of other insects such as sucking and boring pests and nematodes.

The consequences of GMO release and the risk/benefit balance for the environment must be carefully evaluated [27, 35]. It is obviously too easy to build frightening scenarios resulting for the presence of an unseen GMO and much of the opposition to the release of these organisms has been the result of misinformation supplied to the public. Nevertheless, some rules should be followed in the construction of GMOs for release. For example, it is not acceptable to most of the regulatory agencies that unknown DNA be inserted into the chromosome. Similarly, it is strongly recommended to avoid the use of antibiotic resistance genes as markers, as they could spread into the environment or transfer to other species. Moreover, it seems reasonable to ensure that all the modifications or mutations artificially made in the host genome be tested and evaluated. Finally, it is recommended that the expression of the heterologous genes be controlled in the transgenic organisms, in order to target expression to particular some tissues.

The impact of released GMOs or other detectable changes in the environment needs to be predicted and evaluated in experimental field assays. The impact of gene flows to wild species and the development of natural insect resistance to *B. thuringiensis* toxins are the main concerns. There is evidence that genes can move between and function in different microbial species [35]. There is also circumstantial evidence that trans kingdom transfers have occurred, particularly DNA sequence similarities between fungal and bacterial β-lactam antibiotic genes [54]. For microorganisms, it is generally recommended to minimise the possibility of gene transfer by using non-mobilisable or integratable plasmids. For plants, some data have been published on the possibility of cross-hybridisation among members of the family Brassicaceae. So, there is a potential risk of transferring a heterologous gene, for example a *B. thuringiensis cry* gene, from *B. napus* into wild-type species [62]. This event would be a disaster for the future use of *B. thuringiensis*-based insecticides. It has been proposed to avoid the possibility of cross-hybridisation to wild species by using, for example, sterilised plants.

Continuous expression of *B. thuringiensis cry* genes in alternative delivery systems will increase the selection pressure on the insect larvae. Several cases [58] of laboratory and field resistances have been reported in the literature, and it is generally accepted that the increase in the selection pressure will induce outbreaks of resistance. A number of general strategies, based either on population genetics and insect ecology or on combinations of toxin genes have been proposed as a means to decrease the rate at which resistance develops. It has been demonstrated [22] that combinations of individual toxins of *B. thuringiensis israelensis* delay the emergence of resistance in *Culex* insects. It was also demonstrated that some toxins show synergy with others and this can be used to overcome resistance [65]. For plants, it is now recommended to combine biochemically unrelated toxins in *B. thuringiensis* plants. Moreover, the companies involved in the *B. thuringiensis*-plant business support resistance-management programmes [49]. In terms of strategies applied to crop protection, several general procedures have been proposed, which are often too complicated and impractical for farmers. Moreover, the recent discovery of insect resistant to *B. thuringiensis* Dipel® ES due to a dominant autosomal gene led to a limitation of the efficacy of the high dose/refuge strategy [32]. So, the challenge for the future is to implement acceptable strategies that provide efficient short-term insect larvae control while delaying (or better avoiding emergence) of insect resistance to *B. thuringiensis* toxins, so that the long-term environmental benefits are preserved.

4.   CONCLUSIONS

Ultimately the public wants to know whether or not it is safe to continue the release of natural *B. thuringiensis* or derivative products to control insect pests and vectors. The commercial use of *B. thuringiensis* on agricultural and forest crops dates back nearly 40 years, when it was applied in France, and for control of tropical diseases, without severe adverse effect for the population, the fauna or the flora. Laboratory safety assays have been carried out over the years and give outstanding results confirming the specificity of the products and their safety for non-target organisms, including humans. There are, however, a few case reports of the occurrence of *B. thuringiensis* in patients with various infectious diseases. However, no study has clearly demonstrated a genuine risk to human health due to the use of *B. thuringiensis*. In view of the close relationships between *B. thuringiensis* and *B. cereus*, these diseases may be due to non-specific virulence factors produced at a high level by these isolates during the vegetative or early sporulation phase.

The long-term use of *B. thuringiensis* in the environment appears to be safe. Nevertheless, the worst scenario, a *B. thuringiensis* pathogenic for humans, has to be considered. Politicians often refer to the precautionary principle, which is to consider all the worst possibilities. In the case of *B. thuringiensis*, no significant doubt exists about the safety of the δ-endotoxins but there are concerns about the non-specific virulence factors. It is possible to use formulated *B. thuringiensis* products devoid of viable spores. These can be produced by γ-irradiation, as is recommended in Germany or Japan [7], or by construction of mutants unable to complete the sporulation process (see chapter 7.1 and [9]). In the absence of viable spores, no germination occurs and so non-specific virulence factors are not produced. Moreover, the release of γ-irradiated products does not lead to an increase in the number of spores in the environment. Another possibility would be to genetically manipulate *B. thuringiensis* strains of commercial interest to avoid the synthesis of these non-specific virulence factors. This can be done by disrupting the structural genes encoding these factors or better the regulatory genes encoding pleiotropic effectors, such as PlcR [1]. By either of these approaches, the risk associated with the intensive release of *B. thuringiensis* would be minimised and the worst scenarios become implausible.

Use of *B. thuringiensis*-based GMOs must be also governed by the same safety rules. It is obvious that the recombinant microorganisms or plants must be constructed using good laboratory practice. The frequency of transfer of heterologous genes must be reduced and special attention has to be given to the emergence of resistance. The success or failure of transgenic plants, from the commercial point of view, will depend on the attitude of the public and consumers. The attitude is highly reserved in Europe, as compared to the United States. It is one of the roles of scientists to demonstrate the safety of these new products and the resulting benefits for the environment, and the absence of risk.

Finally, because of their exceptionally good safety, *B. thuringiensis*-based products can play an increasingly major role in the field of environment. However, these products and their methods of use have to be made secure in order to eliminate any potential negative trait.

REFERENCES

[1] Agaisse H, Gominet M, Okstad OE, Kolsto AB & Lereclus D (1999) PlcR is a pleiotropic regulator of extracellular virulence factor gene expression in *Bacillus thuringiensis*. Mol. Microbiol. 32, 1043-4053

[2] Akiba Y (1986) Microbial ecology of *Bacillus thuringiensis* VI. Germination of *Bacillus thuringiensis* spores in the soil. Appl. Entomol. Zool. 21, 76-80

[3] Arantes O & Lereclus D (1991) Construction of cloning vector for *Bacillus thuringiensis*. Gene 108, 115-119

[4] Barston KA, Whiteley HR & Yang NS (1987) *Bacillus thuringiensis* δ-endotoxin expressed in transgenic *Nicotiana tobacum* provides resistance to lepidopteran insects. Plant Physiol. 85, 1103-1109

[5] Baum JA, Kafekuda M & Gawron-Burke C (1996) Engineering *Bacillus thuringiensis* bioinsecticides with an indigeneous site specific recombination systems. Appl. Environ. Microbiol. 62, 4367-4373

[6] Beattie SH, Holt C, Hirst D & Williams AG (1998) Discrimination among *Bacillus cereus*, *B. mycoides* and *B. thuringiensis* and some other species of the genus *Bacillus* by Fourier transform infrared spectroscopy. FEMS Microbiol. Lett. 164, 201-206

[7] Becker N & Margalit J (1993) Use of *Bacillus thuringiensis israelensis* against mosquitoes and blackflies, p. 147-170. *In* Entwistle PF, Cory JS, Bailey MJ & Higgs S (ed.), *Bacillus thuringiensis*, an environmental biopesticide: theory and practice, Wiley and Sons

[8] Beebee T, Korner A & Bond RPM (1972) Differential inhibition of mammalian ribonucleic acid polymerases by an exotoxin from *Bacillus thuringiensis*. Biochem. J. 227, 619-625

[9] Bravo A, Agaisse H, Salamitou S & Lereclus D (1996) Analysis of *crylAa* expression in *sigE* and *sigK* mutants of *Bacillus thuringiensis*. Mol. Gen. Genet. 250, 734-741

[10] Brousseau R, Saint-Onge A, Préfontaine G *et al.* (1992) Arbitrary primer polymerase chain reaction, a powerful method to identify *Bacillus thuringiensis* serovars and strains. Appl. Environ. Microbiol. 59, 114-119

[11] Budarina ZI, Sinev MA, Maycrov A *et al.* (1994) Hemolysin II is more characteristic of *Bacillus thuringiensis* than *Bacillus cereus*. Arch. Microbiol. 161, 252-257

[12] Carlson CR & Kolsto AB (1994) A small (2.4 Mb) *Bacillus cereus* chromosome corresponds to a conserved region of a larger (5.3 Mb) *Bacillus cereus* chromosome. Mol. Microbiol. 13, 161-169

[13] Carlson C R, Caugant DA & Kolsto AB (1994) Genotypic diversity among *Bacillus cereus* and *Bacillus thuringiensis* strains. Appl. Environ. Microbiol. 60, 1719-1725

[14] Damgaard PH (1995) Diarrhoreal enterotoxin production by strains of *Bacillus thuringiensis* isolated from commercial *Bacillus thuringiensis* based insecticides. FEMS Immunol. Med. Microbiol. 12, 245-250

[15] Damgaard PH, Granum PE, Bresciani, J *et al.* (1997) Characterization of *Bacillus thuringiensis* isolated from infections in burn wounds. FEMS Immunol. Med. Microbiol. 18, 47-53

[16] Damgaard PH, Larsen HD, Hansen BM *et al.* (1996) Enterotoxin producing strains of *Bacillus thuringiensis* isolated from food. Lett. Appl. Microbiol. 23, 146-150

[17] Delécluse A, Bourgouin C, Klier, A & Rapoport G (1991) Deletion by *in vivo* recombination shows that the 28 kDa cytolytic polypeptide from *Bacillus thuringiensis israelensis* is not essential for mosquitocidal activity. J. Bacteriol. 173, 3374-3381

[18] Donovan WP, Tan Y & Slaney AC (1997) Cloning of the *nprA* gene for neutral protease A of *Bacillus thuringiensis* and effect of *in vivo* deletion of *nprA* on insecticidal crystal protein. Appl. Environ. Microbiol. 63, 2311-2317

[19] Estruch J.J, Waren GW, Mullins MA *et al.* (1996) Vip3A a novel *Bacillus thuringiensis* vegetative insecticidal protein with a wide spectrum of activities against lepidopteran insects. Proc. Natl. Acad. Sci. USA 93, 5389-5394

[20] Fisher R & Rosner L (1959) Toxicity of the microbial insecticide, Thuricide. J. Agr. Food Chem. 7, 686-688

[21] Gawron-Burke C & Baum JA (1991) Genetic manipulation of *Bacillus thuringiensis* insecticidal crystal proteins genes in bacteria, p. 237-263. *In* Setlow JK (ed.), Genetic Engineering : principles and methods, Plenum Press

[22] Georghiou GP & Wirth MC (1997) Influence of exposure to single versus multiple toxins of *Bacillus thuringiensis* subps. *israelensis* on development of resistance in the mosquito *Culex quinquefasciatus* (Diptera : Culicidae). Appl. Environ. Microbiol. 63, 1095-1101

[23] te Giffel MC, Beumer RR, Klijn N *et al.* (1997) Discrimination between *Bacillus cereus* and *Bacillus thuringiensis* using specific DNA probes based on variable regions of 16S rRNA. FEMS Microbiol. Lett. 146, 47-51

[24] Goldberg LJ & Margalit J (1977) A bacterial spore demonstrating rapid larvicidal activity against *Anopheles sergentii, Uranotaenia unguiculata, Culex univitatus, Aedes aegypti* and *Culex pipiens.* Mosq. News 37, 355-358

[25] Gonzalez JM Jr, Brown BJ & Carlton BC (1982) Transfer of *Bacillus thuringiensis* plasmids coding for δ-endotoxin among strains of *B. thuringiensis* and *B. cereus.* Proc. Natl. Acad. Sci. USA 79, 6951-6955

[26] Green M, Heumann M, Sokolow R *et al.* (1995) Public health implications of the microbial pesticide *Bacillus thuringiensis*: an epidemiological study, Oregon 1985-1986. Am. J. Public Health 80, 848-852

[27] Hansen BM, Damgaard PH, Eilenberg J & Pedersen JC (1996) *Bacillus thuringiensis*, ecology and environmental effects of its use for microbial pest control, Ministry of Environment and Energy. Environmental project n° 316. Danish Environmental Production Agency, Copenhagen, Denmark.

[28] Hansen BM, Damgaard PH, Eilenberg, J & Pedersen JC (1998) Molecular and phenotypic characterization of *Bacillus thuringiensis* isolated from leaves and insects. J. Invertebr. Pathol. 71, 106-114

[29] Hendriksen NB. & Hansen BM (1998) Phylogenetic relation of *Bacillus thuringiensis*: implications for risk associated to its use as a microbiological pest control agent. IOBC Bulletin 2, 5-8

[30] Hernandez E, Ramisse F, Cruel T *et al.* (1999) *Bacillus thuringiensis* serotype H34 isolated from human and insecticidal strains serotypes 3a3b and H14 can lead to death of immuno competent mice after pulmonary infection. FEMS Immunol. Med. Microbiol. 24, 1-5

[31] Hernandez E, Ramisse F, Ducoureau JP *et al.* (1998) *Bacillus thuringiensis* subsp. *konkukian* (serotype H34) superinfection. Case report and experimental evidence of pathogenicity in immuno suppressed mice. J. Clin. Microbiol. 36, 2138-2139

[32] Huang F, Burchmann LL, Higgins RA & Mc Gaughey WH (1999) Inheritance of resistance to *Bacillus thuringiensis* toxin (Dipel ES) in the European Corn Borer. Science 284, 965-970

[33] Ishii T & Ohba M (1993) Characterization of mosquito specific strains coisolated from a soil population. Syst. Appl. Microbiol. 16, 494-499

[34] Jackson SG, Goodbrand RB, Ahmed R & Kasatiya S (1995) *Bacillus cereus* and *Bacillus thuringiensis* isolated in a gastroenteritic outbreak investigation. Lett. Appl. Microbiol. 21, 103-105

[35] Klier A (1992) Release of genetically modified microorganisms in natural environments : scientific and ethical problems, p. 183-190. *In* Gauthier ML (ed.), Gene tranfers and environment, Springer-Verlag

[36] Krieg A (1971) Concerning alpha exotoxin produced by *Bacillus thuringiensis* and *Bacillus cereus.* J. Invertebr. Pathol. 17, 134-135

[37] Kurstak E (1962) Données sur l'épizootie bactérielle naturelle provoquée par un *Bacillus* de type *Bacillus thuringiensis* sur *Ephestia kuhniella.* Entomophaga Mémoire hors série 2, 245-247.

[38] Lambert B & Peferoen M (1992) Insecticidal promise of *Bacillus thuringiensis*. Facts and mysteries about a successful biopesticide. Bioscience 42, 112-122

[39] Lampel JS, Canter GL, Dimock MB. *et al.* (1994) Integrative cloning, expression and stability of the CryIA(c) gene from *Bacillus thuringiensis* subsps. *kurstaki* in a recombinant strain of *Clavibacter xyli* subsps. *cynodontis.* Appl. Environ. Microbiol. 60, 501-508

[40] Lereclus D, Agaisse H, Gominet M. & Chaufaux J (1995) Overproduction of encapsulated insecticidal crystal protein in *Bacillus thuringiensis* SpoOA mutant. Bio/Technology 13, 67-71

[41] Lereclus D, Agaisse H, Gominet M *et al.* (1996) Identification of a gene that positively regulates transcription of the phosphatidyl inositol-specific phospholipase C gene at the onset of the stationary phase. J. Bacteriol. 178, 2749-2756

[42] Lereclus D, Vallade M, Chaufaux J *et al.* (1992) Expansion of the insectidical host range of *Bacillus thuringiensis* by *in vivo* genetic recombination. Bio/Technology, 10, 418-421

[43] Levinson BL, Kasyak KJ, Chiu SS *et al.* (1990) Identification of β-exotoxin production plasmids encoding β-exotoxin and a new exotoxin in *Bacillus thuringiensis* by using HPLC. J. Bacteriol. 172, 3177-3179

[44] Losey JE, Rayor LS & Carter ME (1999) Transgenic pollen harm monarch larvae.Nature 399 : 945.

[45] Luthy P (1989) Large scale use of *Bacillus thuringiensis* H14 in a mosquito infected area in a southern region of Switzerland. Proceedings and Abstracts Society for Invertebrate Pathology XXIInd Annual Meeting, p. 82, University of Maryland USA

[46] Margalit J & Dean D (1985) The story of *Bacillus thuringiensis* var. *israelensis*. J. Am. Mosq. Control Assoc. 1, 1-7

[47] Martin PAW & Travers RS (1989) Worldwide abundance and distribution of *Bacillus thuringiensis* isolates. Appl. Environ. Microbiol. 55, 2437-2442

[48] Mc Clintock JT, Schaffer CD & Sjoblod RD (1995) A comparative review of the mammalian toxicity of *Bacillus thuringiensis* based pesticides. Pest. Sciences 45, 95-105

[49] Mc Gaughey WH, Gould F & Gelernter W (1998) *Bt* resistance management. Nature Biotechnol. 16, 144-146

[50] Meadows MP (1995) *Bacillus thuringiensis* in the environment : ecology and risk assessment. p. 193-200. *In* Entwistle PF, Cory JS, Bailey MJ & Higgs S (ed.), *Bacillus thuringiensis*, an environmental biopesticide: theory and practice, Wiley and Sons

[51] Meadows MP, Ellar DJ, Butt J *et al.* (1992) Distribution, frequency and diversity of *Bacillus thuringiensis* in an animal feed mill. Appl. Environ. Microbiol. 58, 1344-1350

[52] Obukowitz MG, Perlak FT, Kusamo-Kretzner K *et al.* (1986) Integration of the δ-endotoxin gene of *Bacillus thuringiensis* into the chromosome of root-colonizing strains of *Pseudomonas* using Tn5. Gene 45, 327-331

[53] Peferoen M (1997) Progress and prospects for field use of *Bt* gene in crops. Trends Biotechnol. **15**, 173-177

[54] Penalva MA, Moya A, Dopazo J *et al.* (1990) Sequence of isopenicilline N synthase genes suggest horizontal transfer genes from prokaryotes to eukaryotes. Proc. R. Soc. Lond. 241, 164-168

[55] Pendelton I R, Bernheimer AW & Grushoff, P (1973) Purification and partial characterization of hemolysins from *Bacillus thuringiensis*. J. Invertebr. Pathol. 21, 131-135

[56] Salamitou S, Marchal M & Lereclus D (1996) *Bacillus thuringiensis* : un pathogène facultatif. Ann. Institut Pasteur/Actualités 7, 285-296

[57] Samples JR & Buettner H (1983) Corneal ulcer caused by a biologic insecticide (*Bacillus thuringiensis*). Am. J. Ophtalmol. 95, 258-260

[58] Schnepf E, Crickmore NB, Van Rie J *et al.* (1998) *Bacillus thuringiensis* and its pesticidal crystal proteins. Microbiol. Mol. Biol. Rev. 62, 775-806

[59] Shinagawa K (1990) Purification and characterization of *Bacillus cereus* enterotoxin and its application to diagnosis, p. 181-193. *In* Pohland AE, Dowell VR Jr and Richard JL (ed.), Microbial Toxins in Food and Feeds - Cellular and Molecular modes of action, Plenum Press

[60] Smith RA & Couche GA (1991) The phylloplane is a source of *Bacillus thuringiensis* variants. Appl. Environ. Microbiol. 57, 311-315

[61] Stevens SE Jr, Murphy R C, Lamoreaux WJ & Coons LB (1994) A genetically engineered mosquitocidal cyanobacterium. J. Appl. Phycol. 6, 187-197

[62] Stewart CN, All JN, Raymer PL & Ramachandran S (1997) Increased fitness of transgenic rapeseed under insect selection pressure. Mol. Ecol. 6, 773-779

[63] Tayaboli AF & Seligny VL (1997) Cell integrity markers for *in vitro* evaluation of cytotoxic response to bacteria containing commercial insecticides. Ecol. Environ. Safety 37, 152-162

[64] Tiball R W (1993) Bacterial phospholipases C. Microbiol. Rev. 54, 347-366

[65] Wirth MC, Georghiou GP & Federici BA (1997) CytA enables CryIV endotoxins of *Bacillus thuringiensis* to overcome high levels of CryIV resistance in the mosquito *Culex quinquefasciatus*. Proc. Natl. Acad. Sci. USA 94, 10536-10540

[66] Yamada S, Ohashi E, Agata N & Ventakeswaran K (1999) Cloning and nucleotide sequence analysis of *gyrB* of *B. cereus*, *B. thuringiensis*, *B. mycoides* and *B. anthracis* and their application to the detection of *B. cereus* in rice. Appl. Environ. Microbiol. 65, 1483-1490

Index

1

16S - 23S internal transcribed sequence
487
16S rRNA 4, 6, 7-8, 10, 13, 14-17, 487

2

23S rRNA 7

A

A + T content 449
AA ITU (International Toxic Unit on
    Aedes aegypti) 388
α-amylase 243, 245
Abbott Laboratories 286, 294, 319, 357,
    359, 375, 386, 464-465, 471
abiotic factor 33
Able® 357
Acari 42, 384
Acinetobacter calcoaceticus 261
Acrobe® 464
Acroneuria lycorias 258
acrystalliferous strain 472, 478
active larvicidal ingredient 486
ADP-ribosyl transferase 112
Aedes 102-104, 109, 324, 390, 465, 479
    aegypti 16, 25, 32, 46, 91, 103-104,
        107-108, 111-112, 123-125, 172-
        173, 175-176, 238, 240-241, 243,

249, 281, 384, 385, 386, 392-393,
    422, 464-465, 474-477
    Bora-Bora strain 281, 284, 294
    albopictus 391
    atropalpus 103, 111
    intrudens 103
    nigromaculis 103
    rossicus 387
    sticticus 387
    vexans 387, 389
aerobic endospore-forming bacteria
    (AEFB) 2-4, 7, 10, 12-13, 16-17
α-exotoxin 489
α-glucosidase 237, 245, 247, 249
Agree® 279, 358, 443
AgrEvo 450, 454, 464
agricultural productivity 441
agriculture 485-486
Agrobacterium 449
Agrotis ipsilon 454
α-helix 85-87, 89, 108, 207
alanine scanning 92
Alicyclobacillus 4
    acidocaldarius 5, 15
alkaline phosphatase 186, 190
alkaline proteinase 47, 50
allele 227-230
    frequency 401-402, 406
allergenicity assessment 339
allergy 44
alternative control method 441

Alternative Strategies and Regulatory
     Affairs Division (ASRAD) 342
American foulbrood 13
amino acid
     sequence repeat 470
     substitution 88, 91-92
aminopeptidase N (APN) 87, 90, 93, 182,
     185-188, 190, 207
Ampelomyces quisqualis 341
Amphibia 384, 389-390
Amphibians 384
Anagrapha falcifera 341
Anarsia lineata 362
Anas platyrhynchos 261
Aneurinibacillus aneurinolycticus 5
Annelids 384
anoetid mite 33
Anopheles 102-104, 109, 324, 420, 434,
     465, 476
     albimanus 46, 125, 325, 392
     anthropophagus 391
     dirus 475
     gambiae 125, 240-243, 249, 384, 423,
          475
     nigerrimus 392
     rangeli 392
     sinensis 390, 391
     stephensi 16, 104, 107, 123-125, 238,
          240, 439, 476-477
     sundaicus 392
Anthonomus grandis 454
antibiotic 47, 51, 55, 137, 489, 495, 498
     resistance 47, 51
          gene 445
Anticarsia gemmatalis 324
anti-evaporant 306
anti-foam 300
Apanteles fumiferanae 258
aphid 455
apoptosis 177
Aporrectodea caliginosa 54
Aquabac® 386
aquatic vertebrate 493
arbovirus 396
armyworm 357-359
Arthroplea bipunctata 256
Aspergillus fumigatus 113-114
asporogenous
     mutant 471
     strain 56, 445, 497

autoagglutinated strain 123
Autographa californica 341

B

Bacillus 2-4, 7-8, 10, 15
     anthracis 7-8, 42-43, 137, 144, 152,
          154, 156, 260, 486-488
     cereus 3, 7-8, 17, 24-26, 28-29, 35, 42-
          56, 83, 128, 137, 138-139, 144-145,
          150, 152-155, 167, 177-178, 260,
          265, 486-490, 495, 499
     circulans 3, 5, 7
     cycloheptanicus 4
     firmus 5
     licheniformis 5, 51
     megaterium 3
     mycoides 7-8, 24, 42, 44, 144, 152,
          486, 487
     popilliae 69, 95, 131, 257-258, 263-
          264
          melolonthae 146, 151
     sotto 27
     sphaericus 5, 7, 10-12, 16-17, 102-
          104, 109-113, 115, 125, 146, 150,
          153, 157, 178, 257, 262, 264, 266,
          281, 297-299, 301-302, 306, 311,
          313, 384-386, 388-391, 393-394,
          396, 419, 420-434, 439, 462-467,
          474-479
          high-toxicity strain 109
          low-toxicity strain 108, 111-112
          mode of action *See* Chapter 3.5
     stearothermoplilus 5, 137
     subtilis 5, 78, 128-129, 131-132, 136,
          176
     thuringiensis 3, 5, 7- 8, 16-17, 29-30,
          32, 42, 102, 144-147, 149-156, 165-
          166, 297, 485
          aizawai 25, 27-28, 51, 54-55, 109,
               123-124, 146, 149, 153, 156,
               166, 222-223, 277-278, 279,
               322, 335, 341, 358-359, 407,
               422, 469
          alesti 27, 45, 145
          canadensis 123
          darmstadiensis 108, 123, 151, 165
          dendrolimus 146, 149, 323
          entomocidus 9, 123, 149, 153, 156,
               222, 277, 322

finitimus 151, 165
fukuokaensis 108, 123-124, 153, 166
galleriae 24, 28, 34, 45, 109, 123, 125, 286, 323, 357
higo 108, 123-124
indiana 146
israelensis 9, 25, 27, 30, 32, 34, 46, 48-50, 54-55, 101-107, 113, 123-124, 145-146, 149, 151- 152, 155, 165-166, 169-173, 175-176, 178, 256-258, 260- 262, 266-278, 302-303, 305- 306, 310, 313, 322-325, 335, 337, 341, 384-390, 392, 393- 396, 419, 422, 429, 432-433, 462-467, 472, 474-479
japonensis 258
jegathesan 107, 123-124, 153-154, 166, 169, 288, 432, 434, 466, 476
kenyae 27-28, 123
konkukian 44, 259, 261, 495-496
kurstaki 24-25, 27-28, 34, 43, 45- 46, 48-51, 54-55, 82-84, 94, 109, 123-124, 145-146, 149, 153, 155-156, 165, 222-223, 255, 258-261, 263-265, 278-279, 282-284, 322-325, 335, 341, 355-359, 361, 370-371, 400, 407, 491, 495
kyushuensis 108, 123-124, 169
malaysiensis 123
medellin 107, 123-124, 152-153, 165-166, 169, 288, 432, 434
morrisoni 106, 123, 145, 154, 166, 169, 277, 278, 281, 289, 466, 469
  tenebrionis 27, 34, 55, 83, 154, 156, 169, 256, 335, 341, 357, 361, 469, 471
pakistani 25
shandongiensis 123
sotto 49, 171, 178
thompsoni 77, 123
thuringiensis 51, 145, 176, 322, 341
toumanoffi 145
wuhanensis 68, 149, 323

weihenstephanensis 42
backbone structure 90-91
Bactecide® 386
bacterial persistence 496
bacteriocin 51
Bactimos® 464
  G® 386
  PP® 386
  WP® 386
Bactospeine® 357
baculovirus 363
bagasse 323, 326
Bancroftian filariasis 420
bceT gene 47, 49
Beauveria bassiana 341
Becker Microbial Products 386, 464
beet armyworm 92, 454
Berliner *See Bacillus thuringiensis thuringiensis*
β-exotoxin 47, 51, 83, 137, 150, 254, 299, 355, 362, 489
β-galactosidase 133
bin gene 11, 111
Bin toxin 11, 103, 109-111, 150-151, 385, 423, 426-429, 432, 434, 464-479
  Bin1 111
  Bin2 111
  Bin3 111
  BinA 109-111, 125, 237, 238-242, 247-248, 423
  BinB 109-111, 125, 237, 238, 239-243, 246, 248, 423
  mode of action *See* Chapter 3.5
  protein-protein interaction 110
binding
  assay 182, 186, 189
  regionalisation 110
  site 223-224, 232, 407
    concentration (R_t) 223, 241
    dissociation constant (K_d) 223, 241
bioassays 276, 278-279, 282-283, 285- 289
Biobit® 279, 357
biocontrol agent 442
BioDalia 357
bioencapsulated bacteria 54
Biological and Pollution Prevention Division (BPPD) 334
biological control agent 485
biopesticide 255, 441, 443, 445

biopesticide registration *See* Chapter 5.4
Biotech International Ltd. 386
biotechnological
    approach 442
    product 455
biotic factor 33, 35
Bio-Ti® 357
Biotouch® 386
bird 493
Bitayon® 357
black cutworm 454
black larva 26, 31
blackfly 101, 256, 266, 384, 386, 389,
    395-396, 462-466, 494
black-throated blue warbler 264
β-lactam 498
β-lactamase 52, 137
blepharocerid 256
Blepharoceridae 256
blood pH 45
bluegill sunfish 261
blue-green alga 497
Boarnia (Ascotis) selenaria 362, 363
bollworm 360, 450
Bombyx mori 25, 27, 32, 81, 170-172,
    185-186, 295, 448
Brachedra amydraula 357
Braconidae 258
Brassicaceae 498
breeding site 420-421, 425-426, 430-431,
    433-434, 439
Brevibacillus 16
    borstelensis 5
    brevis 3, 5
    coshinensis 5
    laterosporus 5, 16, 17, 102, 104, 115,
        146
Brevibacterium 2
broad spectrum mosquito control 477
broadleaf forest 24
brook trout 262
brown trout 262
brush border membrane vesicles (BBMV)
    87, 91-94, 176, 181-191, 202-208, 222,
    223-226, 230-231, 240-243, 245-247
β-sheet 207
β-strand 85, 90-91, 108
Bt 8010 Rijin 357
Bt formulation 448
budworm 373, 376, 378, 450

Bufo 390
Bupalus piniaria 375
Burkholderia cepacia 341
burn wound 29, 44, 495

C

cabbage root fly 54
cadaver 32, 46, 54, 257, 258, 491
cadherin 185, 188, 190
cadherin B 207
cadherin-like receptor 87, 91
caged rock bass 262
Candida oleophila 341
carbon source 317, 320, 326
carrier 309
Cbm17 113, 114
    Cbm17.1 113-114, 125
    cbm17.2 113-114, 125
CDC-light trap 389
cell
    culture 175-176
    surface protein 137
    viability 202, 205
    wall hydrolase 139
CellCap® 312, 358, 456
CerB 138
cereolysin 49-50
cereulide 47
channel formation 202, 209
chemical insecticide 356, 360, 362-365,
    485
chestnut-backed chickadee 263
chironomid 256-257, 266, 384, 396
    midge 462
Chironomidae 102
Chi-square 279, 283
chitinase 47, 51, 137, 455
chloroplast
    genome 452
    transformation 452
cholesterol oxidase gene 455
Choristoneura
    fumiferana 45, 51, 175, 227, 258, 295,
        370-371, 374, 381
    fumiferana 288
    occidentalis 372-374
    pinus pinus 372
chromophore 361
chromosomal

location 150
map 488
sequence 156
chromosome 82, 113, 144, 149-151, 156, 487-488, 498
plasticity 144, 156
Chrysomela scripta 227, 469
Chrysoperla carnea 256
chymotrypsin 85
Ciba 319
cigarette beetle 45
Clavibacter xyli 497
var. *cynodontis* 444
clinical
case 486, 495
infection 29
syndrome 494
Clostridium 3, 104, 488
bifermentans 102, 104, 115, 136, 466
malaysia 67, 69, 104, 113, 124-125, 146, 151, 154, 165-166
paraiba 104
kluyveri 6
perfringens 112, 210
cluster 73-76, 150
Cnidaria 384, 389
co-aggregation 155
COAX 360
codon usage 449
Coefficient of Variation (CV) 279
cointegrate 154
Coleomegilla maculata 256
Coleoptera 26, 27, 42, 69, 83, 85, 169-170, 254, 319, 357-358, 384, 487, 491
colicin A 87
Colinus virginianus 261
collagen 47, 50, 52
collagenase 137
collagenolytic proteinase 47, 50
Colletotrichum gloeosporioides 341
colloid osmotic lysis 189
colonisation 492
colony forming unit (cfu) 259-262, 492-493, 496
Colorado potato beetle 208, 256, 281, 355, 357, 360-361, 363-364, 409-411, 413, 443, 450, 471
combination of toxin genes 445
commercial formulation 465
common cockchafer 16

Condor® 358, 442, 443
conjugation 82, 144, 153-156, 442, 455
transfer 155
conjugation-like
mechanism 145
process 149
conjugative plasmid 443
consensus
block *See* conserved block
sequence *See* conserved block
conservative transposition mechanism 152
conserved block 70-71, 75-76, 78, 86, 106-107, 109, 113
continuous
centrifugation 303-304
fermentation 301
corn meal 323, 325-326
corn rootworm 454
corneal ulcer 43, 495
Corynebacterium 2
Costar® 357
Costelytra zealandica 2
Cotesia
marginiventris 364
plutellae 364
cotton boll weevil 454
cotton bollworm 402, 408, 410
cottonwood leaf beetle 469
CP-51 155
CP-54 155
crane fly 26
Crop Genetics International (CGI) 444
cross-hybridisation 498
cross-resistance 427-429, 431-432
Crustacea 384
cry2A operon 77, 469, 470
cry11A operon 467-468
cry gene 9, 17, 86, 96, 149-150, 156, 178, 326, 356, 359, 432, 442-445, 447, 449, 450, 454-455, 490, 496-498
cry1 149, 152, 449
cry1A 24, 82, 84, 96, 129-131, 133, 149, 151, 153, 156, 448, 468, 491
cry1C 84, 114, 153, 156, 448
cry1D 153
cry1E 84
cry2 95
cry2A 136, 153, 472

cry3
   cry3A 96, 130-134, 135, 154, 471,
      447-449
   cry3B 132, 135
   cry3C 135
cry4
   cry4A 150, 152
   cry4B 106, 475
cry10
   cry10A 106
cry11
   cry11A 106, 131, 136, 152, 467,
      468, 472-473, 475-476
   cry11B 152-153
cry16
   cry16A 113-114, 154
cry17
   cry17A 113-114
cry18
   cry18A 131, 151
cry19
   cry19A 153-154
cry27
   cry27A 108
cry29
   cry29A 153
cry30
   cry30A 152
   expression 128, 443
   nomenclature 82
Cry protein *See* Cry toxin
Cry toxin 9, 170, 254, 255, 326, 454, 490,
  497
   activation 85
   activity spectrum 72, 78
   Cry1 178, 245, 255, 288
      Cry1A 46, 68, 70, 76, 83-96, 109,
         124, 181-191, 207-210, 220-
         231, 256, 295, 400-401, 407,
         447-448, 470
      Cry1B 68, 70, 76, 96, 220-221, 223-
         226, 295
      Cry1C 46, 68, 76-77, 83-84, 91-96,
         109, 124, 220-226, 229-232,
         295, 400, 407-448, 468
      Cry1D 68, 295, 407
      Cry1E 68, 76-77, 84, 295
      Cry1F 68, 76, 220, 221, 223-229,
         231, 295, 401, 407
      Cry1G 68, 75

      Cry1H 68
      Cry1I 68, 136
      Cry1J 68, 76, 220-221, 224-225
      Cry1K 68
   Cry2 74, 169, 254, 407
      Cry2A 14, 68, 77, 85, 91, 94-95,
         109, 124, 136, 207, 221-224,
         226, 230-231, 295, 468, 470,
         472-473
   Cry3 169, 256
      Cry3A 68, 73, 85, 88, 90, 95, 136,
         207-208, 209, 446-447, 469,
         471-472

         three-dimensional structure 136
      Cry3B 68, 208, 446
      Cry3C 68
   Cry4 256
      Cry4A 68, 73, 106, 124, 385, 463,
         466, 468
      Cry4B 68, 73, 106, 124, 385, 463,
         466, 475
   Cry5
      Cry5A 68, 295
      Cry5B 68, 73
   Cry6
      Cry6A 68
      Cry6B 68
   Cry7
      Cry7A 68
   Cry8
      Cry8A 69, 73, 75
      Cry8B 69, 73, 75
      Cry8C 69, 73
   Cry9
      Cry9A 69, 73, 75
      Cry9B 69
      Cry9C 69, 73
      Cry9D 69
      Cry9E 69
   Cry10
      Cry10A 69, 73, 106, 107-108, 124
   Cry11 74
      Cry11A 69, 106-107, 124, 136, 385,
         432, 463-464, 466, 468, 472-
         473, 475-476
      Cry11B 69, 107, 124, 432, 466, 476
   Cry12
      Cry12A 69, 74, 76

Cry13
    Cry13A 69, 73, 74, 76
Cry14
    Cry14A 69, 73, 74, 76
Cry15A
    Cry15A 69, 77
Cry16
    Cry16A 69, 113-114, 124, 136
Cry17
    Cry17A 69, 113-114, 124
Cry18
    Cry18A 69, 73-74, 76, 94-95
Cry19
    Cry19A 69, 107-108, 124
    Cry19B 69, 108, 124
Cry20
    Cry20A 69, 75, 108, 124
Cry21
    Cry21A 69, 74, 76
Cry22
    Cry22A 69
Cry23
    Cry23A 69
Cry24
    Cry24A 69, 75, 107, 124
Cry25
    Cry25A 69, 75, 107, 124
Cry27
    Cry27A 108, 124
Cry29
    Cry29A 107, 124
Cry30
    Cry30A 107, 124
domain 207 *See* Chapter 2.2
domain function *See* Chapter 2.2
evolutionary relationships 66
hybrid protein 84, 93-94
nomenclature 65
receptor *See* Chapter 3.2
three dimensional structure 70, 85
    domain I 70, 74
    domain II 70, 75
    domain III 70, 76
    loop 73
CryMax® 358, 443
cryptic plasmid 145, 153
crystal 25, 32-33, 54
    protein overproduction 131
crystallisation 72, 77

crystal-minus
    mutant 149
    strain 45, 248
CryStar® 443
Culex 102-104, 178, 257, 325, 390, 420, 465, 479, 498
    molestus 125, 388, 389
    pipiens 30, 104, 107-108, 123-125, 237-243, 246-249, 281, 284, 384, 386-387, 389, 394, 421, 423, 429, 439, 474, 476-477
        L-SEL resistant colony 423, 427, 429
        Montpellier strain 281, 294
        SPHAE resistant colony 423-425, 427, 429, 431
    quinquefasciatus 16, 108, 110-112, 125, 227, 239, 241, 281, 284, 384-385, 391, 393-394, 420-425, 433-434, 439, 464-476, 478
        Geo resistant colony 423-425, 429
        Kochi resistant colony 421, 423, 425, 427, 429, 439
    tritaeniorhynchus 391
Culicidae 27, 102, 104, 114
Culinex® 385-386, 389, 392
culture
    collection 298
    maintenance 299
    selection 298
curing experiment 149
cutaneous infection 448
cutaneous inflammatory lesion 495
Cutlass® 279, 358, 443
Cyanobacteria 432, 434
Cydia pomonella 341
Cyprionodon variegatus 261
cyt gene 149, 156
    cyt1
        cyt1A 106, 152, 467, 471-473, 476, 478
    cyt2
        cyt2B 106, 152
    expression 128
Cyt toxin 66, 169, 423, 432
    Cyt1
        Cyt1A 69, 106-108, 124, 136, 385, 432, 463-469, 475-476
        Cyt1B 69

Cyt2 169
    Cyt2A 69, 108, 124
        three-dimentional structure 108
    Cyt2B 69, 106-108, 124
cytolysosome 238
cytolytic activity 66, 102, 105-106, 463, 469
cytosol 114
cytotoxic effect 495
cytotoxin 112

D

D-α-alanine 45
DDT 391
DEAE ion-exchange column 84
degradative enzyme 137
dehydrochlorinase 431
Dekalb 450
Delia radicum 54
delivery system 486, 492, 496-498
deltamethrin 391
δ-endotoxin 27, 29, 32-33, 35, 42-43, 45-46, 54, 145, 151, 156, 168-169, 171-179, 254, 276, 287, 400, 441, 487-488, 490, 493, 496, 499
    mode of action *See* Chapter 3.1
    nomenclature 68
    origin 177
    phylogram 67
    specificity 68
Dendroica caerulescens 264
Dendrolimus 375
    sibiricus 31
    superans sibericus 375
dengue 101, 385, 392-393, 461
dermal exposure 494
Design® 358
Diabrotica 83, 282, 454
diamondback moth 182, 224-225, 399, 400, 402, 404-405, 407, 410
diarrhoea 46, 48, 52
diarrhoeal
    enterotoxin 490, 495
    symptom 494
diet 279, 280, 282, 284-285, 288-289
digestive tract 175
Dimilin® 373-375
Dipel® 149, 220-227, 230-231, 279, 286, 357, 499

Diptera 25-27, 42, 69, 85, 94, 101, 254, 319, 487
disease control 461-462
dispersing agent 282
dissolved oxygen (DO) 302
distance matrix 66
disulphide bridge 83, 85, 87, 88, 136
DNA
    homology 487
        group I 146
        group II 146
    reassociation 3-4, 8, 10, 14, 16
    transfer 447
domain
    exchange 93, 94
    swapping 77, 78
dominance 401
dominant autosomal gene 499
dose
    /response regression line 430
    acquisition 371, 376, 378
    management 400
Drosophila
    melanogaster 243
    virilis 243

E

early sporulation phase 488
earthworm 54
Ecogen Inc. 358, 442-443, 446
ecological
    niche 143
    risk 450, 452
ecosystem 255-256, 264, 266
ecotoxicity 441
ectoperitrophic space 206
effective dosage 388, 392
efficacious dose 376-377
electroporation 444, 448-449
ELISA 313
elm bark beetle 357
emergence trap 390
emetic toxin 44, 47-48, 53
encapsulation 360-361, 447-448, 456
endogenous proteinase 112
endophthalmitis 44
endophylic
    host 497
    bacterium 444

endoplasmatic reticulum 173
endotoxin
    combinations 477
    synthesis enhancing 467
engineered toxin gene 486
enhancing factor 467, 479
Entente Interdépartementale de
    Démoustication (EID) 421
Enterobacteriaceae 2
Enterococcus
    faecalis 176
    faecium 154
enterotoxin 28-29, 44, 47-50, 53, 55, 83,
    137-138, 150, 260
    gene (*entFM*) 47, 49
    production 487
environment 48, 51-53, 56, 485-487, 489-
    492, 494-495, 497-500
environmental
    benefit 499
    impact 486, 494
Environmental Protection Agency (EPA)
    280, 285, 334, 400, 454, 493
enzootic 26, 30-32, 54
enzyme 487, 489
Ephemeroptera 255-256, 384
Ephestia
    cautella 322
    elutella 45
    kühniella 27-28, 31, 82, 170, 178, 283,
    294
epiphylic host 497
epizootic 26, 30-33, 35, 54, 265, 491-492
Escherichia coli 49, 82, 110-114, 176,
    211, 444-445, 467-468
esterase 431
ε-toxin 112
Eubacterium tenue 6
European corn borer 402, 411, 442, 450
European foulbrood 2, 13
European sunflower moth 31
European Union (EU) 334
evolution of resistance 453
exosporium 109, 464
expression system 447, 472
extrachromosomal
    element 144
    prophage 146

F

F value 283
fall armyworm 91
fathead minnow 262
fatty acid content analysis 487
Federal & State Co-operative Suppression
    Program 373
Federal Food, Drug and Cosmetic Act
    (FFDCA) 337
Federal Insecticide, Fungicide, and
    Rodenticide Act (FIFRA) 337
feeding
    behaviour 360, 377
    inhibition 376, 378
fermenter 299-303, 321, 441
fertiliser 440
fibronectin 52
field resistance
    to *B. sphaericus See* Chapter 6.2
    to *B. thuringiensis See* Chapters 6.4
    and 3.4
filariasis 385, 391, 396, 461-462
filter feeding larva 110
fingerprinting 9, 10
fitness 401, 404
fitness cost 404, 412
flagella antigen 69
flagellar agglutination 123
flagellin 137
flocculant 303
Florbac® 221, 279, 357
flour moth 178
flow cytometry 191
Foil® 279, 442, 443
foliage 24-26, 29, 32-34
food 28-29, 34
Food and Agricultural Organisation
    (FAO) 334
Food and Drug Administration (FDA) 260
Food Quality Protection Act (FQPA) 337
Foray® 279, 357
forespore 128-129
forestry 485
formulation 304, 308- 311, 313
    additive 309
    aqueous suspension 304
    briquette 304, 310, 324, 385
    component 306
    concentrated liquid 324

corncob granule 304, 393
diluent 309
dispersant 306
donut 304, 311
dry flowable *See* water dispersable
    granules
dust 304
fluid concentrate 385, 393
granule 310, 324, 385, 388-389
liquid concentrate 385, 389, 395
oil emulsifiable suspension 304, 311
pellet 304, 310, 385
sand granule 304
suspending agent 306
tablet 385, 389, 392
thickener 306
water dispersible granule (WDG) 304,
    307-308
wettable powder (WP) 304, 307-308,
    385, 388-389, 392
Fourier transform infrared spectroscopy
    487
fungus 33, 318, 320, 326- 327, 363
Fusobacterium necrophorum 44
Futura® 357

G

Galleria mellonella 53
gastro-enteritis 43, 490, 495
gastro-intestinal disease 44
gene
    copy number 471
    flow 452, 453, 498
    hybridisation 149
    inheritance 402
    mapping 155
    multiplicity 143
    transfer 168, 177
genetic
    manipulation 255, 444-445, 447, 451
    mapping 150
    transfer 115
genetically
    engineered *Bt* strain 445
    modified *Bt* products 358
    modified organism (GMO) 497-498,
    500
        release 497
genome size 144, 487-488

genomic
    environment 143
    flexibility 144
Geratec 386
German Mosquito Control Association
    (KABS) 387-388, 397, 420
germination 33, 35, 44-46, 48, 54-55
γ-irradiation 385, 499
    mutagenesis 471
Gliocladium
    catenulatum 341
    virens 341
Global Positioning System 374
glutathione-S-transferase 431
glycocalix 206
glycolipid 183, 185, 188-191
glycoprotein 184-186, 191
glycosyl-phosphatidylinositol (GPI) 182,
    186-187, 243, 245
Golgi apparatus 173
Good Laboratory Practices (GLP) 288
grain
    dust 28, 491
    store 28, 491
grass foliage 24-26
grass grub 2
growth regulator 434
gut 46
Gynaikothrips ficorum 326
gypsy moth 255, 263, 265-267, 371, 373
gyrB gene 487

H

haemolysin 47, 50, 53, 137, 150, 211
haemolytic
    activity 66, 102-103, 106-108, 489
    enterotoxin (Hbl) 47-48, 138, 150
Halobacillus halophilus 5
hbl operon 139
heat-sensitive vector 445
helicopter 387-388
Helicoverpa 287-289, 410
    armigera 363, 402
    zea 341, 408
Heliothis 287, 288, 289
    armigera 31, 33, 295
    irritans 33

virescens 16, 89, 93, 170-171, 184-186, 221-224, 226, 228-229, 231-232, 295, 360, 364, 402, 424
    zea 33, 424
helper protein 466-467, 469, 471, 479
Hemiptera 27, 256
hemocoel 137, 175, 177
heritability (h^2) 228
Heterodera glycines 7
Heteroptera 384
Heterorhabditis bacteriophora 258
heterozygote 401, 403-405
high dose strategy 400, 403, 404-406
high dose/refuge strategy 499
high pressure liquid chromatography (HPLC) 83, 84, 285, 286, 313
His$_6$-tagged protein 113
histopathology 168, 172-173, 175, 177
holotype 69
Homoesoma nebulella 31
homologous recombination 432, 445
Homoptera 319
homozygote 403-405
honey bee 2, 5, 13, 493
hooded warbler 264
horizontal gene transfer 451
host plant 453
Huazhong Agricultural University 298
human
    bacterial infection 495
    population 317
Hyalophora cecropia 50
hydrogen bond 86
hydrolytic enzyme 47-49
Hyla arborea 390
Hymenoptera 27, 42, 69, 319

I

ice nucleating protein 470
immunoassay 284-285
immunosuppressed animal 259, 496
improved recombinant mosquitocidal bacteria 474, 479
improvement of dipteran larvicide 465
in vivo
    gene deletion 114
    recombination 447
InCide® 444
inclusion body 487

Indianmeal moth 399, 402, 406
industrial
    fermentation *See* Chapter 5.2
    production *See* Chapter 5.2
infection 492-493, 495-496
infectivity 493
inflammatory
    process 492
    reaction 496
inhibition ELISA factor 285
initiation of translation 133
inoculum 324-325
    preparation 299
insect
    damage 455
    pest control 486
    rearing 26
    resistance 46
    resistance management (IRM) 340
insecticidal
    activity 298, 301-302, 308-309, 313
    crystal protein (ICP) 219-220, 223-227, 231-232, 422
insecticide market 486
insertion sequence (IS) 144, 151-153
    IS*3* 153, 166
    IS*4* 151, 165
    IS*6* 152, 166
    IS*21* 152, 165
    IS*231* 151, 152-154, 156, 165-166
    IS*232* 151, 152, 165
    IS*233* 151, 153, 165
    IS*240* 151, 152, 166
    IS*240*-like element 152
    IS*982* 153, 165
    IS*Bt1* 151, 153, 166
    IS*Bt2* 151, 153, 166
Institut fur Biologische Schadlingsbekampfung 298
Institut Pasteur 356
integrase 154
integrated
    control programme 384, 388, 392, 394, 431, 434
    pest management (IPM) 318, 334, 356, 408
integrational vector 497
integrative plasmid 475
intercellular mobility 144
internal resolution sites (IRS) 446

International Entomopathogenic Bacillus Centre (IEBC) 103

International Service for the Acquisition of Agri-biotech Applications (ISAAA) 336

International Toxic Unit (ITU) 388

International Unit (IU) 371

intracellular mobility 143

intraperitoneal exposure 493

invertebrate 493

inverted repeated sequence (IR) 112, 151-152

ion
    channel 203, 204, 205, 210
        activity 87, 88
    flux 176

irreversible binding 91

Isaria 326

ISO 9002 288

iso-element 152, 156

IstA 152

IstB 152

Ixodes scapularis 54

J

jacalin 90

jack pine budworm 263

jackpine budworm 372

Japanese beetle 14, 95

Javelin® 221, 225, 230, 232, 357

Junco hyemalis 262

K

KABS (Kommunale Aktionsgeinschaft zur Bekämpfung der Schnakenplage) *See* German Mosquito Control Association)

kanosamine 47, 51

kingdom transfer 498

Kurthia 5

Kyushu University 298

L

L-α-alanine 45

Lactobacillus 137

lacZ reporter gene 131, 133

Lagenidium giganteum 341

laminin 52

large plasmid 145, 149, 150, 155

LarvXSG® 310

Lasioderma serricorne 45

LC$_{50}$ 421, 423, 429-430, 439

lectin 455

Lepidoptera 27, 31, 33, 42, 45, 69, 83, 85, 94-95, 228, 238, 254-255, 264-265, 267, 319, 357-371, 424, 487, 491

Lepomis macrochiurus 261

Leptinotarsa decemlineata 227, 230, 232, 256, 281, 289, 294, 356, 443, 450, 471

Leptinox® 358

lesion 492

lethal
    dose 376, 378
    effect 493, 496

life cycle of *B. thuringiensis* 32

ligand blotting 182-189

light-scattering 202, 204

liposome 191

Lymantria
    dispar 185-187, 227, 263, 288, 295, 371-372, 374-375

M

Mainfeng pesticide 357

malaria 101, 387, 390, 391, 396, 420, 461-462, 485

mallard duck 261

Mallophaga 319

maltase 243, 245

Mamestra 287-289

mammal 260, 262-264, 384

mammalian toxicity 492

Manduca sexta 45, 77, 87, 88, 91, 93, 109, 171, 176, 182-191, 205, 207, 295

Mamestra brassica 295

Mansonia 103

mating
    behaviour 453
    experiment 154

Mattch® 443

Mcap® 312

Mediterranean flour moth 82

Melissococcus pluton 2, 13

Melolontha melolontha 16

membrane
    binding 108
    insertion 201

membrane-spanning domain 87
meningitis 44
metabolite 486, 489
Metarhizium anisopliae 341
microbial
    insecticide 254, 257, 263, 266
    pest control agent (MPCA) 253, 258-259, 261, 264
micro-filtration 303-304
Microplitis croceipes 363
midgut
    anterior stomach 238-239
    cell microvilli 172-173, 206
    columnar cell 172, 174, 177
    epithelium 169, 171-173, 175-178
    gastric caecum 110
    goblet cell 172-173, 177
    microvilli 465
    pH 105, 170
    posterior stomach 110
    proteinase 85, 105
    tract 175
migration 424-425, 433
milky disease 14
mite 319
mitochondria 55, 173-174, 239-240
mob gene 155
mobile element 115, 145, 147, 151, 153
Mobile Insertion Cassette (MIC) 153
mobilisable plasmid 155
mobilisation 155-156
modified *cry* gene 497
molecular chaperone 467, 470
mollusc 5, 7, 13
Mollusca 384
monarch 493
Mono-Q ion-exchange column 84
Monsanto 450
mosaic 451
    pattern 387
mosquito 27, 30, 33, 35, 101, 103-104, 108, 110, 167, 169, 171, 173, 176, 178, 256-257, 264, 266, 384-386, 388-389, 393, 419-422, 424-427, 430-434, 439, 461-462,467, 475-476
mosquitocidal strain 102-103
mother cell compartment 128-129, 448
mRNA
    stabiliser 133, 135

stabilising sequence 467, 471-472, 478
stability 133-135
M-Trak® 443
mtx gene
    mtx1 11, 111-112
    mtx2 11
    mtx3 11
Mtx toxin 111
    Mtx1 11, 103, 112, 125, 151, 465
    Mtx2 11, 103, 112, 125, 465
    Mtx3 11, 103, 112, 125
mulberry silkworm 448
multigenic family 488
multilocus enzyme electrophoresis (MLEE) 9, 10, 146, 487
multiplication 487, 492
MVP® 443
Mycogen Corp. 443, 450, 456
Myiopharus doryphorae 363
myonecrosis 44
Myotis daubentoni 390

N

N-acetyl-D-glucosamine 240
N-acetylgalactosamine (GalNAc) 186, 188-189
National Registration Authority (NRA) 344, 400
natural *Bt* products 357
nausea 494
neem 423, 425-426, 432
Nematocera 101, 256, 257
Nematoda 42
Nematode 69, 319, 363
neonate 359, 364
Neuroptera 326
neutral metalloproteinase immune inhibitor 50
neutral proteinase 50
nomenclature committee for *Bt* toxin genes 66, 78
non-flagellated *Bacillus thuringiensis* 44
non-haemolytic enterotoxin (NHE) 47, 48, 138
non-living recombinant *Bt* insecticide 448
non-mobilisable plasmid 155
non-target
    insect 493

organism (NTO) 254-257, 265-267
non-transgenic plant 451
Nordic Committee on Food Analysis 29
North American Free trade Association
    (NAFTA) 336
Northern bobwhite quail 261
Novartis 319, 450, 454
Novodor® 279, 357, 471-472
nuclease 133, 137-138
nun moth 375
nutrient deprivation 128-139
Nutrile Products Inc. 325

O

Odonata 384
Onchocerca volvulus 394-395, 465
onchocerciasis 101, 394, 464-465, 485
Onchocerciasis Control Programme (OCP)
    266, 394-395, 420, 464
Oncorhynchus mykiss 261
Operophtera brumata 375
opportunistic human pathogen 486
ORF2 (from *Bt jegathesan*) 107
Organisation for Economic and Co-
    operative Development (OECD) 334
Orgyia
    leucostigma 372
    thyellina 374
oriT 155
Orthoptera 319
Ostrinia nubilalis 227, 361, 363, 402,
    442, 444, 450
outbreak 490-491, 498
overlapping promoter 129
oxidase 431
Oxoid 48
oxygen uptake rate (OUR) 302

P

Paecilomyces fumorsoroseus 341
Paenibacillus 4, 12, 14-15
    chibensis 5
    glucanolyticus 5
    larvae 13
        larvae 5, 13
        pulvifasciens 5
    lentimorbus 5, 14-15, 17
    polymyxa 5
    popilliae 5, 14-17

        melolonthae 16
        rhopaea 16
    pulvifasciens 13
        pulvifasciens 13
palindromic sequence 138
pAMβ1 153
Panolis flammea 375
paralysis 46, 171
parasite 412
parasitism 258
parasitoid 257
    ecosystem 453
parasporal
    body 463-466, 478
    inclusion (crystal) 66
particle bombardment 449
particle size 282-283, 289, 308-310
Parus rufescens 263
patch-clamp 202-203
pathogenicity 493, 496
pathogenicity island 150
pBU4 477
Pectinophora gossypiella 27, 31
Pentatomidae 364
Perillus bioculatus 256
peritrophic membrane 171, 206
persistence 492, 497
Pest Control Practices Act (PCPA) 342
Pest Management Regulatory Agency
    (PMRA) 342
pest mortality 453
pesticide 440-441, 445
    resistance 441
pGI2 145
pGI3 145
pH control 302
Phaedon cochleariae 289
phage 145-146, 155
    genome 145
    infection 145
phage-typing 487
phagostimulant 360-361
phagostimulant mixture 360
pHD2 145
pheromone 409
phosphatidylcholine 47
phosphatidylcholine phospholipase C
    (PC-PLC) 49, 137-138, 489
phosphatidylinositol 47

phosphatidylinositol phospholipase C (PI-PLC) 49, 53, 137, 150, 186, 243, 245, 489
phosphoinositol phospholipase C 183
phospholipase 150, 275
  A 51
  C 49, 137, 489
phosphorylation 128-129
Photorhabdus 2
  luminescens 312
pHT1000 145
pHT1030 145
pHT3101 467, 468, 472, 478
pHT73 145, 155
phyllogenetic analysis 74-76
phylloplane 24-26, 34, 56, 491, 494
phytophagous insect 455
pI 84, 85
Pieris brassicae 170-172, 174-178, 182, 295
Pimephales promelas 262
pine looper 375
Pisces 384
planar lipid bilayer (PLB) 190, 202-205, 207-210, 239
Planococcus 5
plant
  chloroplast 449
  foliage 491
  transformation 449
Plant Genetic Systems 449
plasmid 43, 51, 54, 82, 288-289
  acquisition 488
  curing 359
plcA gene 137
PlcB 138
PlcR 53, 489, 500
plcR gene 137-138
Plecoptera 255-256, 258
pleiotropic
  regulator 138, 489, 496
  transcriptional activator 53
Plodia interpunctella 220, 222-223, 228-229, 231-232
Plutella xylostella 46, 96, 182, 185-186, 220-221, 224-226, 228-232, 280, 295, 356, 358, 363-365, 400, 424
pMK3 475-476
pneumonia 44
pollen 452

polluted water 465, 476-477
Popillia japonica 14, 95
population dynamics 453
pore formation 108, 112, 182, 183, 190, 191, 201-207, 211, 239
positive
  regulator (PlcR) 137-139
  retroregulator 133
potency 278
  bioassay 276
  determination 276-277, 280, 284-285
  standardisation 287
  unit 281
pPL603E 475
predator 256-258, 266-267, 386, 388, 389, 392, 412
preservative 306
Prillus bioculatus 364
primary
  powders 282
  production 485
  structure 102
proactive plan 408
processionary caterpillar 375
programmed cell death 177-178
promoter 82, 112, 114, 128-132, 136, 138-139, 468, 471-473, 477-478
  activity 131
Prostoia completa 258
protein export 136
proteinase 73, 137, 139
  inhibitor 451, 455
  K 87
  negative strain 497
proteolysis 422
proteolytic activation 70, 108
protoplast transformation 475
protoxin 85, 105, 108
  activation 169, 170, 171
Protozoa 42
protozoan 497
Pseudomonas 312, 407
  aeruginosa 112, 176
  aureofasciens 341
  cepacia 443
  fluorescens 358, 443-444, 448, 456, 470, 497
  syringae 341
Psorophora 103, 257
psychodid 384

Psychodidae 384
pTX14-3 145, 155
Puccinia canaliculate 341
pulsed-field electrophoresis 487
pXO1 145, 154
pXO2 154
pXO11 145
pXO12 145, 154
pXO13 145
pXO14 145
pXO15 145
pXO16 145, 155
pyemotid mite 33

Q

quasi-species 488

R

Rana 390
random amplified polymorphic DNA
    (RAPD) 10, 16, 146
random mutation 89
RAPD (random amplified polymorphic
    DNA) fingerprinting 487
Raven® 358, 443, 446
reactive plan 408
receptor 385
    binding 86, 90-91, 93, 95, 171-172,
    207
    -toxin interaction 428
recombinant
    DNA technology 486
    plasmid 359
recycling 491
reference standard 276-278, 283, 287
refuge 403, 405-406, 409-410, 425-426,
    432, 434, 451, 453
registration procedure 493 *See* Chapter 5.4
regulatory agency 497, 498
repeated DNA sequence 144
replication origin 145
replicative mechanism 154
replicon 149
rep-PCR 146
resident plasmid 144, 149
residue decay 400
resistance 385, 388, 390-395, 462-466,
    469, 474, 476, 478-479
    cost 404

dominance 401, 403, 412
fitness cost 227, 231
gene 220, 227, 228, 230-231
inheritance 220, 227, 229-230, 246
management 232-233, 422, 430-431,
    499 *See* Chapter 6.4
monitoring 412
risk 453
to *B. sphaericus* Bin toxin *See*
    Chapter 3.5
to *B. thuringiensis* Cry toxin *See*
    Chapter 3.4
resistant
    genotype 401, 406, 413
    homozygote 451
resolvase 154
restriction fragment length polymorphism
    (RFLP) 146
restriction mapping 145
restriction-modification system 444
reverse phase column 84
reversible binding 91
Rhizobium leguminosarum 443
ribonuclease activity 133
ribotyping 9, 487
rice bran 326
Rimi 357
risk
    assessment 42, 492 *See* Chapter 7.3
    /benefit balance 497
river llindness *See* onchocercosis
rolling circle mechanism 145
Romanomermis 432
rotation 406, 408, 410-412, 431, 433
Rotatoria 384
rRNA gene sequence 146
ruthenium red 175

S

Saccharomyces cerevisiae 176
Salmonella 313
salt bridge 136
Sandoz 319
sanitary void 421, 426
satellite inclusion 155
scaffolding protein 469
scattered light assay 190-191
sciarid 384
secondary metabolite 455

secretion 136
self-transmissible plasmid 145, 155
septicaemia 137, 490
sequence
   alignment 66, 94
   similarity 66, 72
serine proteinase 170
serological procedure 487
Serratia entomophila 2, 301
sex-linkage 424
sheepshead minnow 261
Shine-Dalgarno sequence 133
short-circuit current 88, 202, 204
Shuangdu 357
shuttle expression vector 462, 467
shuttle vector 497
Siberian moth 375
sigma factor 82
   gene *sigK* 448
   σ^{28} 448
   σ^A 128-129, 132, 136, 139
   σ^E 128-129, 131-132
   σ^F 128-129
   σ^G 128-129
   σ^H 128-129, 131
   σ^K 128-129, 131-132
signal transduction system 128
silkmoth 50
silkworm 26, 27, 31, 81, 172, 344, 448
simuliid 258, 266
Simuliidae 102
Simulium 104, 286
Simulium Control Program (PCS) 420
Simulium damnosum 394, 465, 494
site-directed mutagenesis 88, 90, 93, 111
site-specific recombination 359, 445, 497
site-specific-recombinase 153
σ^K-strain 448
slate-colored junco 262
S-layer 52, 137, 138, 155
small acid-soluble protein 145
small intestine 48
small plasmid 145, 153, 155
soil sample 491
solid state fermentation (SSF) 298 *See*
   Chapter 5.3
solubilisation 83, 87, 422
solute flux 203
somatic infection 44, 48, 50
soybean meal 323-326

specific activity 96
specificity 254, 258, 260, 263
Spherico® 386
Spherimos® 386
sphingomyelinase 47, 49-50, 53, 137-138,
   489
Spicturin® 357
Spo0A 128-129, 447-448
spo0A mutant 130-131
Spodoptera 83, 278, 287-289, 294, 359
   exempta 295
   exigua 46, 51, 92-94, 222, 224, 226,
     295, 341, 454
   frugiperda 33, 45, 91, 182, 190, 295
   littoralis 170-171, 222, 226, 230, 232,
     280, 284, 295
spoIIID mutant 131
spontaneous insertion 154
sporangium 109
spore 54, 82, 257, 260-263, 265
   appendage 52
   coat 45
   hydrophobicity 52
   persistence 261, 487, 496
   recycling 257, 265
   -crystal mixture 45, 55
Sporolactobacillus 5
sporulation 32-33, 35, 43, 46, 54, 487-
   488, 499
   initiation 128-129, 447
   process 82, 128-129, 131
   -deficient 447-448
   -dependent promoter 468, 472
spray drier 307
spruce budworm 45, 175, 258, 370-373,
   376-378
stable zone strategy 433
STAB-SD 471, 472-473, 478
standard powder
   635- S-1987 294
   HD1-S-1971 294
   HD1-S-1980 294
   HD-968-S-1983 281, 294
   IPS78 294
   IPS80 294
   IPS82 281, 282, 284, 294, 386
   RB80 294
   SPH84 294
   SPH88 281, 294, 386
standardisation 276, 279, 286-288

Staphylococcus aureus 176, 313
stationary phase 42-43, 46, 50, 53, 82-83,
    128-129, 131, 136-137, 139, 471
steelhead trout 261, 262
Steinernema glaseri 258
stem-loop structure 133, 473
sterilised plant 498
Steward® 357
sticker 361
stored product 28, 30-32, 491
Streptomyces 454, 455, 488
    griseoviridi 341
Strongwellsea castrans 33
Sturnus vulgaris 262
SU-11 146
sublethal dose 376
sudden-death *Bacillus* 31
sugar utilisation 487
sugarcane 326
Summit Chemicals 311
sunscreen 361
surfactant 309
susceptible heterozygote 451
synergism 385, 465, 475
synergistic
    effect 286
    interaction 106, 463, 488
synergy 498

T

Taeniopteryx nivalis 255
target organism 66, 70
taxonomic relationships 486
taxonomy 486, 487
Technical Grade Active Ingredient (TGAI)
    307, 308, 310-313
Tecra Diagnostics 48
Teknar® 394, 464
    G® 386
    HP-D® 386, 388, 395
    TC® 386
teletransposition 153
temephos 392, 394-395
Tenebrio molitor 27
Tennessee warbler 264
terminal peptidase 85
terminator 82
Tetrahymena pyriformis 54
tetrazolium salt 55

Thaumetopoea pityocampa 375
thermostable exotoxin *See* β-exotoxin
Thermo-Trilogy, Inc. 464
thin layer chromatography (TLC) 183,
    185, 188, 189
thiol-dependent haemolysin 489
Thuricide® 260, 279, 357, 371
thuricin 47, 51
thuringiensin *See* β-exotoxin
thuringiolysin 489, 496
Ti plasmid 449
Tipula 26
tipulid 256, 384
Tipulidae 102
titration 277-278, 283, 286
tobacco beetle 26
tobacco hornworm 87
Tortricidae 370
Tortrix viridana 375
toxaemic effect 139
toxic metabolite 493
toxicity spectrum 442
toxicology assessment 339
toxin
    complementation 241
    insertion 86-88, 90-91, 171
        hairpin (penknife) model 87-88, 209
        umbrella model 87-88
    loop 89-93, 95
    mixture 400, 406
    mutation 88-89
    oligomerisation 201-202, 206-208,
        210, 241
    postbinding events 211
    processing 220, 222, 232
    solubilisation 102, 109
    stability 109
Toxorhynchites 103, 256, 432
TP21 146
transconjugant 497
transconjugation 359
transcription 112, 128-139
    start site 129
transcriptional fusion 130-131
transcript-ribosome complex 471
transducing phage 146, 155
transduction 359
transepithelial potential 204
transformation 442, 444, 456
transgenic

plant 96, 336, 338, 400, 403, 405-406,
441, 449-453, 456, 490, 500
corn 96
cotton 96
potato 96
transition state 128, 131
transmissible plasmid 70
transposable element 70, 144, 151, 153,
156
transposase 152-154
transposition 144, 152
transposon (Tn) 144, 475
Tn*3* 153, 154, 166
Tn*917* 475
Tn*4430* 151-154, 166
Tn*5401* 151, 154, 166
Tn*Bth1* 151, 154, 166
Trematoda 42
Trichoderma
harzianum 341
polysporum 341
Trichogramma
caoeciae 363
platneri 363
Trichoplusia ni 33, 45, 51, 186, 227, 279,
280, 282, 284, 287-289, 294-295, 326
Trichoptera 255-256, 384
Trident® 279, 357
tropical disease 485, 499
trypsin 84-85, 93
Turbellaria 384

U

ultra-low volume (ULV) 372, 375
application 362
United Nations Environment Programme
(UNEP) 334
United States Department of Agriculture
(USDA) 280, 298
UV (ultra violet)
inactivation 32, 361
irradiation 444
protectant 35
resistance 145

V

vacuole 173, 177, 238-239
Valent BioScience Corp. 386

vascular permeability reaction (VPR) 52,
53
Vault® 357
Vectobac® 464
12AS® 386, 388, 395
TP® 386
WDG® 386
Vectolex® 386, 465
vegetative
cell 43-44, 46-48, 50-52, 54-55
division 487
gene promoter 128
growth 32, 35, 465, 469, 471
insecticide protein *See* Vip toxin
Vermivora peregrina 264
Vero cell 44, 48-49, 53
vip gene 96, 136, 156, 455
expression 136
vip1 83
vip2 83
Vip toxin 77, 136, 139, 490
Vip1 83
Vip2 83
Vip3 83, 490
Vip3A 47, 50, 454
viral encephalitis 461
Virgibacillus pantothenicus 5
virulence 42, 44, 46, 47, 50-53, 137, 139,
488, 497, 499
factor 46, 51, 138-139, 423, 488, 490,
496, 499-500
expression 52
gene 143, 150
VVM (air volume / fermenter volume / min)
302, 325

W

war wound 44, 448
Western spruce budworm 372, 373
wetter 361
white crowned sparrow 262
white grub 258
whitemarked tussock moth 372
Wilsonia citrina 264
World Health Organization (WHO) 262,
264, 277, 280, 282, 284, 289, 306, 314,
393-394, 397, 430, 464

X

Xenorhabdus 2
Xentari 357
X-ray crystallography 85, 95

Y

yellow fever 101
yellow fever mosquito 91

Yersinia pestis 78
ynzF gene 78
yobEF gene 78
yokGF gene 78

Z

Zohar Dalia 386
Zonotrichia leucephrys 262
zwittermicin 47, 51